FUNDAMENTALS OF COLLEGE MATHEMATICS

Sandra Pryor Clarkson *Hunter College*
Barbara J. Barone *Hunter College*

HOUGHTON MIFFLIN COMPANY Boston Toronto
Geneva, Illinois Palo Alto Princeton, New Jersey

Dedication

This book is dedicated to my large, supportive, and wonderful family and especially to my daughter, Sansi. —*S.P.C.*

This book is dedicated to my friends who gave me constant support throughout this endeavor, and to my best friend, my husband, James Patrick, without whose support and encouragement this project would never have come to fruition. —*B.J.B*

Cover design by Linda Manly Wade
Cover image by G.V. Faint/Image Bank
Interior design by George McLean

Photo credits: Chapters 1, 2, 4, 5, 6, and 8: Stock/Boston; Chapters 3, 7, and 9: Image Bank/Boston

Sponsoring Editor: Maureen O'Connor
Development Editor: Robert Hupp
Project Editor: Maria A. Morelli
Editorial Assistant: Tracy Theriault
Production/Design Coordinator: Sarah Ambrose
Senior Manufacturing Coordinator: Priscilla Bailey
Marketing Manager: Michael Ginley

Printed in the U.S.A.

Library of Congress Number: 93-78640

ISBNs:
Text: 0-395-48403-0
Instructor's Annotated Edition: 0-395-69133-8

234567-B-02 01 00 99 98 97

Contents

Preface

Our purpose in writing this text was to provide our students and yours with a well-written accessible text that presents mathematics as do-able, interesting, and relevant. Between us, we have a total of more than 50 years experience teaching developmental mathematics to college students, in both classroom and laboratory settings. This experience includes choosing texts, constructing texts, and ancillary materials, analyzing student errors, selecting appropriate pedagogical approaches and individualizing instruction. Having worked with a broad range of students—including bilingual students, students with learning and perceptual difficulties, and adult students returning to school—we know the importance of clear, concise language; well-chosen examples and notation; and relevant, real-life problems. This text, as well as the *Introduction to Algebra* and *Intermediate Algebra* texts that follow it, are the results of our desire to provide these students with a serious text that incorporates those features.

Fundamentals of College Mathematics is primarily for students who need a solid foundation in arithmetic to move comfortably and confidently into algebra. It can also be used as a terminal course for a student who wants to gain the mathematics skills needed daily in our society. The text is designed for use in a classroom setting, as a self-study text, or in a laboratory.

We have written *Fundamentals of College Mathematics* in a friendly, conversational style that reflects the way we talk to our students. The kind of text that works best for these students has certain characteristics: These students need a text they can read—one that gives them many worked examples as well as many practice exercises for them to work. Students need continual exposure to the correct language of mathematics but in a form and a setting that is supportive not threatening. They also need continual exposure to the visualization and organization of mathematics. Students want to see a connection between the mathematics they are learning and the real world. The best text for them should have useful and interesting real-life applications and should emphasize the interconnections among geometry, statistics, and many other mathematics and non-mathematics topics. We believe that this is such a text.

For background information in this text, we combed newspapers, hundreds of source books, and collected every scrap of statistical information that we could find. We used that information to write a book that we hope is interesting—both to students who learn from it and also to the teachers who use it in their classroom. It was certainly an interesting text to write.

Approach

Visualizing Mathematics

We give students many opportunities to "see" mathematics in multiple formats—such as graphs, diagrams, tables, charts, and illustrative art—and to use these formats in solving problems.

Mastery

Section goals are identified and the tests monitor achievement of those goals. However, concepts and skills are integrated into a smoothly flowing text and not simply presented as isolated objectives.

Motivation

We present problem situations from many areas of interest and concern that reflect the impact of mathematics in all aspects of modern life. Therefore, answers are not always artificially "nice" numbers but are usually from real situations. This realism helps students develop genuine number sense.

Sensitivity to Learning Styles

We are sensitive to students with certain learning and perceptual difficulties by providing appropriate pedagogy and notation.

Assessment Features

We have incorporated a variety of assessment features into *Fundamentals of College Mathematics,* based on current research in developmental math.

Skills Check

A short quiz introduces each chapter. Its purpose is to help students identify any weaknesses in those skills prerequisite in the new chapter. Students are given answers and sections to review if necessary. Passing this test does not allow a student to "skip" the chapter; instead, it lets them know if they are ready to begin.

Margin Exercises

Two margin exercises accompany each example in the book. These exercises reinforce and extend the procedure used in the example. All answers appear at the back of the book. The first exercise usually mirrors the example; the second extends and clarifies the work.

End-of-Section Exercises

These exercises provide review and practice for all skills taught in the section. Answers to odd-numbered problems are at the back of the book.

Mixed Practice

In all sections except the first section in a chapter, there are practice problems that review and reinforce skills previously taught *in that chapter.* Answers to the odd-numbered problems are at the back of the book.

Chapter Review

A three-part review section appears at the end of each chapter, prior to the chapter test. Answers to the odd-numbered problems are at the back of the book.

Error Analysis This section can be assigned or used as a catalyst for classroom discussion. Incorrect solutions are given to problems. Students are asked to identify and correct the errors. We ultimately want students to be able to identify and correct the errors they make in their own work. Answers to the odd-numbered problems are at the back of the book.

Interpreting Mathematics These exercises emphasize mathematics as communication, and require students to write about mathematics. The student explains, defines, clarifies and interprets mathematical ideas, terms and procedures. We do not include any answers to this feature in the student text to encourage students to use their own words in writing about mathematics.

Review Exercises These exercises are divided into two sets. The first group of exercises provides a section-by-section review of the chapter's work. The second set is a mixed review that is not section referenced. Answers to the odd-numbered problems are at the back of the book.

Chapter Test

This test follows the chapter review material. It is mastery-based and contains three questions for each section goal in a chapter. Questions are organized by objective with answers in the Instructor's Resource Material. A student who answers two questions correctly out of each group of three has most likely mastered the objective. The chapter test takes approximately one hour. Additional chapter tests, and suggestions for Assessment Alternatives—in keeping with the standards recommended by the National Council of Teachers of Mathematics (NCTM)—appear in the Instructor's Resource Manual. All the answers are at the back of the book.

Cumulative Review

Beginning with Chapter 2, a Cumulative Review section appears at the end of each chapter that reinforces skills taught in all previous chapters including the current one. The exercises are mixed. Answers to the odd-numbered problems are at the back of the book.

Areas of Special Emphasis

In this text, we emphasize a number of areas recommended by the National Council of Teachers of Mathematics (NCTM) in the Curriculum and Evaluation Standards for School Mathematics.

Applications

Problems involving numbers gathered from contemporary business, sports, current events, and virtually all real-life situations appear early in the text and throughout all chapters. These problems involve biology, physics, astronomy; buildings and other man-made structures; and questions about popular culture and movies, just to name a few. The wide variety and realism of the applications will appeal to students. In addition, students are introduced to applications in sections called Geometry and Measurement and An Application to Statistics (optional) as soon as they have the numerical skills to understand them. However, the instructor can—without compromising the flow of the text material—introduce these topics toward the end of the course, if a more concentrated approach is desired.

Problem-Solving Preparation

There is an emphasis on problem solving throughout the text. A four-step procedure is introduced early and is used in solving all word problems. A useful set of problem-solving strategies is developed in the text, especially in these optional, unnumbered sections, so that students learn to approach problems in an organized, efficient manner. Students are taught to reason mathematically. Whether the strategy is to draw a diagram or read a table, the student is exposed to interesting applications that make use of the strategies.

Algebra Preparation

Students are introduced to the real numbers and are thoroughly prepared for algebra by the time they complete this text.

Extra

In this special feature titled "Extra" we have included extended applications in unusual, interesting areas; for example, using fractions to identify fingerprints (Chapter 5). Exercises are included in some of these special text sections.

Supporting Features

Section Goals

Section goals are identified in the margin on the first page of each section. Those goals are the "outcomes" desired from a student studying the section.

Calculators

Material (identified by a calculator logo) is included to introduce students to the use of calculators early in the text and throughout. Exercises indicated for use with the calculator appear throughout the text.

Estimation

Estimation is taught and used throughout the text to promote number sense.

Reasonable Answers

Students use estimation and checking, as well as reviewing the problem, to determine whether an answer is reasonable.

Graded Examples

Examples in each section begin with the most basic and, as successive examples appear, additional skills are incorporated and the level of complexity increases.

Key Words and Procedures

Key words are defined and rules (or procedures) are clearly delineated in boxes set off from the rest of the material. These features build vocabulary and aid students in communicating mathematics.

Study Hints

Study hints that aid a student in learning the material appear throughout the text.

Ancillary Materials

Instructor's Annotated Edition

Teaching suggestions to the instructor appear in red in the Instructor's Annotated Edition. Some examples are suggestions for teaching students with learning or perceptual difficulties, cautions about possible student misunderstandings, and information about multi-cultural influences on student learning.

Answers are printed in red below each problem.

Instructor's Resource Manual and Printed Test Bank

Information is given about organizing a laboratory course using this text. Hints are also given about the use of Cooperative Learning Groups, and for working with students with certain types of perceptual or learning disabilities.

The Printed Test Bank consists of items contained in the Computerized Test Generator, as well as two tests for each chapter drawn from test-bank items.

A cumulative exam is provided for use after Chapters 3, 6, and 9. Questions appear in random order, not by chapter, and test key objectives previously assessed in the Chapter Tests.

Computerized Test Generator

This user-friendly software permits the instructor to construct a customized test selecting from over 1500 items. Test items are organized by chapter and section goal. Available for IBM PC and compatible, and Macintosh.

Solutions Manual

Complete solutions to all margin, end-of-section, and end-of-chapter exercises, chapter tests and cumulative reviews.

Student Solutions Manual

Complete solutions to the odd-numbered end-of-section and end-of-chapter exercises, cumulative reviews, and all margin exercises and chapter tests.

"Mathabilities" Tutorial Software

This powerful new, user-friendly, algorithmically-driven tutor is linked to the text topic by topic. Basic concepts are presented, explained, and drilled using color and animation to graphically illustrate mathematical operations.

Videos

A complete set of text-linked videotaped lessons is available to instructors for use in classrooms, learning labs, and resource centers.

Acknowledgments

Many people have helped us with this project. In particular, we wish to thank in alphabetical order: the late Tom Clarke, Mike Ginley, Rob Hupp, Maria Morelli, Maureen O'Connor, and Greg Tobin. Special thanks must go to Ed Millman. His tireless work, and endless red ink, always kept us thinking. Without his extremely helpful criticism this book would never have been completed.

Additionally, many people influenced us in our professional life. We wish to list some of those people and thank them here. Mrs. Bohan, Mrs. Breiner, Mac Callaham, the late Mary P. Dolciani and her husband James Halloran, C.H. Edwards, Jr.; George Grossman, Eleanore Kantowski, Mrs. Leonardi, the late Len Pikkart, Henry Pollack, Donna Shalala, Andre Thibodeau, Mr. Towson, Zalman Usiskin, W. H. Williams, J. W. Wilson, and Gloria Wolinsky.

We would also like to thank the following reviewers for their helpful suggestions:

Leonard Andrusaitis, *University of Lowell, MA*
Donna J. Brann, *Moraine Valley Community College, IL*
Thomas A. Carnevale, *Shawnee State University, OH*
Ellen Casey, *Massachusetts Bay Community College*
James Coleman, *Community College of Baltimore, MD*
William Elliott, *Keene State College, NH*
Barbara Ton Ferullo, *Quinsigamond Community College, MA*
Jerry Frang, *Rock Valley College, IL*
Don Harris, *South Puget Sound Community College, WA*
Carol Hay, *Northern Essex Community College, MA*
Chris Kodaczewski, *University of Akron, OH*
Regina Massare, *Suffolk County Community College, NY*
Jill McKenney, *Lane Community College, OR*
Anthony D. Monteith, *College of Marin, CA*
Linda J. Murphy, *Northern Essex Community College, MA*
Nancy K. Nickerson, *Northern Essex Community College, MA*
Charles Reinauer, *San Jacinto College Central, TX*
John G. Rose, *Miami Dade Community College, FL*
Ned Schillow, *Lehigh County Community College, PA*
Marvin Schlichting, *Triton College, IL*
Patricia Stoltenberg, *Sam Houston State University, TX*
Diane Tischer, *Metropolitan Community College, NE*
John Tobey, *North Shore Community College, MA*
Susan F. Wagner, *Northern Virginia Community College*
Charles N. Walter, *Brigham Young University, UT*
Julie Wilson, *Aims Community College, CO*

S.P.C.

B.J.B.

1 Adding and Subtracting Whole Numbers

There is a constantly changing sign in Times Square in New York City that shows the National Debt. It has a number on it that is larger than one trillion. Have you thought about just how big one trillion is? If you write it, it is 1,000,000,000,000. If you started counting from one to one trillion and counted one number per second, do you know how many years it would take? Ten years? Fifty? One thousand? It would take more than 31,000 years.

- *In what other situations are numbers larger than one million used?*

Skills Check

Take this short quiz to see how well prepared you are for Chapter 1. The answers are given at the bottom of the page.

1. In the number 745, which digit has the place value hundreds? tens? ones?

2. Add: $9 + 8$

3. Add: $8 + 7$

4. Add: $\begin{array}{r} 7 \\ \underline{+6} \end{array}$

5. Subtract: $\begin{array}{r} 17 \\ \underline{-\ 9} \end{array}$

6. Subtract: $\begin{array}{r} 15 \\ \underline{-\ 6} \end{array}$

ANSWERS: 1. hundreds = 7; tens = 4; ones = 5 2. 17 3. 15 4. 13 5. 8 6. 9
These items are basic place value, addition, and subtraction facts that we will cover in Chapter 1.

1.1 Place Value: Naming Numbers

SECTION GOAL

- To write any number from one to one billion in words, in numerical form, and in expanded form

All of mathematics, and much of our daily life, is concerned with numbers. In this chapter, we discuss the meanings and uses of whole numbers and two of the basic operations on them. In this first section, we deal mainly with names and symbols for numbers.

When you were young, you learned to count things, saying "one, two, three, four, five, six," and so on. The numbers that you used to count with and the number zero are called **whole numbers.** All whole numbers can be written using the ten **digits**

0, 1, 2, 3, 4, 5, 6, 7, 8, 9

The digits themselves are the symbols, or numerals, for zero and the first nine counting numbers. We have to use two or more digits to symbolize the remaining numbers, as in 57 or 163. The numeral 57 has two digits; the numeral 163 has three digits.

In everyday language, the words *number* and *numeral* are often interchanged. Technically, however, a **number** specifies an amount; a **numeral** is a symbol, written with digits, that represents a number.

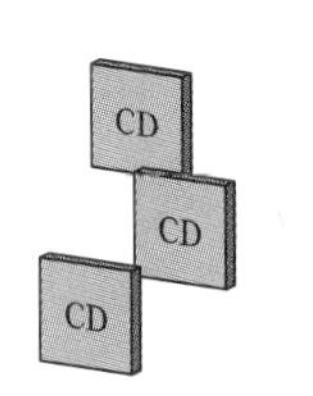

We will use the word *number* unless we are discussing the digits or properties of a numeral.

Each digit in a numeral has a **face value** (the name of the digit) and a **place value** (which depends on its place in the numeral). The meaning of the digit depends on both its face value and its place value. For example, the digit 3 has a different place value—and a different meaning—in each of the following numbers:

293	Here, 3 means three ones, or 3.
1,234	Here, 3 means three tens, or 30.
378	Here, 3 means three hundreds, or 300.

The following chart shows the place values of the digits in any number up to almost a quadrillion. The farthest digit to the right in any whole number is always taken to be in the *ones* or *units* place.

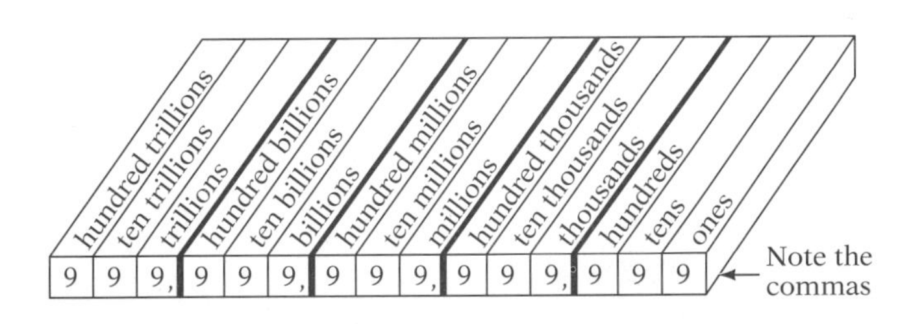

1a. What does the digit 6 mean in 76, in 163, and in 6,894?

1b. What is the hundreds digit in 3275, in 653227, and in 5,841,464?

2a. Write the number 3,089,406 in expanded notation.

2b. Write the number expressed by 200,000 + 6,000 + 300.

EXAMPLE 1

What does the digit 7 mean in each of the following whole numbers?

a. 27 **b.** 173 **c.** 7,894

SOLUTION

We use the chart on page 3 to find the place value of each 7.

a. Because 7 is in the ones place in 27, here it means seven ones, or 7.
b. Because 7 is in the tens place in 173, here it means seven tens, or 70.
c. Because 7 is in the thousands place in 7,894, here it means seven thousands, or 7,000.

■ *Now do margin exercises 1a and 1b.*

The correct answers are given at the back of the book. Be sure you understand each example and its margin exercises before you go on.

The number 2,697 is said to be in **standard notation.** This same number in expanded notation is

$$2{,}000 + 600 + 90 + 7$$

Expanded notation is the sum of the meanings of all the digits in the number.

EXAMPLE 2

Write the number 3,406,089 in expanded notation.

SOLUTION

We first need to find the meaning of each digit.

9 means nine ones, or 9

8 means eight tens, or 80

0 means no hundreds

6 means six thousands, or 6,000

0 means no ten thousands

4 means four hundred thousands, or 400,000

3 means three millions, or 3,000,000

This list enables us to write the non-zero meanings with plus signs to indicate addition.

$$3{,}000{,}000 + 400{,}000 + 6{,}000 + 80 + 9$$

■ *Now do margin exercises 2a and 2b.*

To make large numbers readable, we use commas to separate groups of three digits, starting from the right, as in 3,000,000 and 5,617,089,312. Each group of three digits is called a **period.** Each period gets its name from the lowest place value it includes. The following figure shows the names of the first (from right to left) five periods.

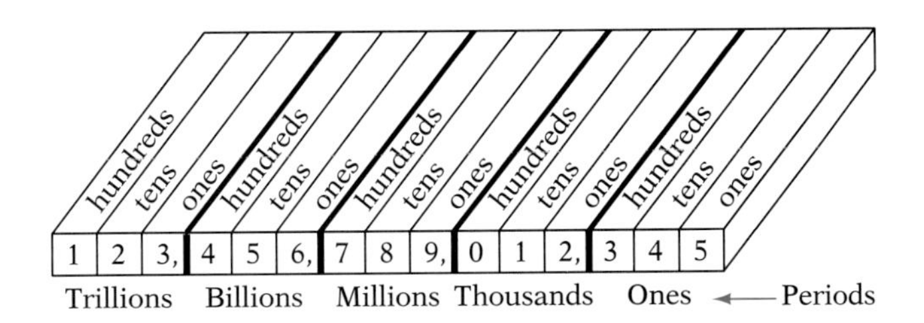

The number 1,234,567,890 has four periods—billions, millions, thousands, and ones. To mark off the periods, we start from the right and group the numbers in threes, using commas or spaces. Note that 1,234,567,890 and 1 234 567 890 represent the same number

To read a number, we read each period separately, beginning at the left. The number 1,234,567,890 is read

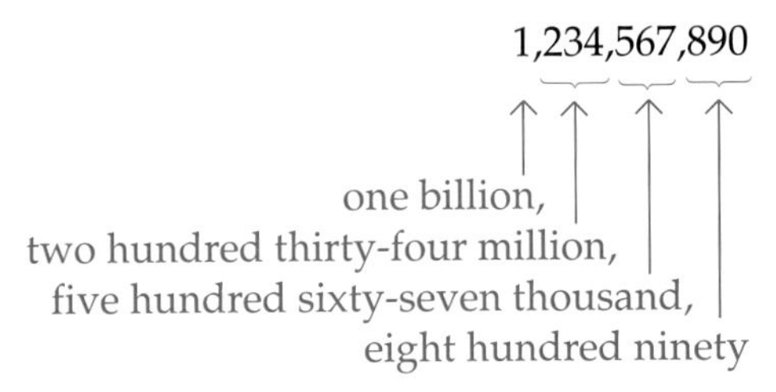

Once you can read and name a three-digit number, you can read and name any number. First group the digits in threes from right to left. Then, moving from left to right, read each three-digit group, ending with the name of the period.

EXAMPLE 3

Write 463047 in words. (That is, name the number.)

SOLUTION

First group the digits.

$$463047 = 463\ 047$$

Then read 463 as four hundred sixty-three; its period is thousands.

Next read 047 as forty-seven; its period is ones. (We don't say or write the "ones.")

Therefore, 463047 is written as four hundred sixty-three thousand, forty-seven.

■ *Now do margin exercises 3a and 3b.*

3a. Write 47604 in words.

3b. Write three hundred five thousand, six in numerals.

Note that we also use commas to separate the periods when we write out the name of a number in words. Note too that years, such as 1995, are an exception.

Work the exercises that follow.

The answers to selected exercises are given at the back of the book.

1.1 EXERCISES

Identify the place value of each underlined digit in the numbers in Exercises 1 through 8.

1. 2143578

2. 3726983765

3. 188275

4. 1092800007

5. 175283960

6. 513372039

7. 1029384756

8. 1928473625

In Exercises 9 through 24, rewrite the underlined number using words or digits. Then write it in expanded form.

9. The depth of the Marianas Trench in the Pacific Ocean is thirty-six thousand, one hundred ninety-eight feet below sea level.

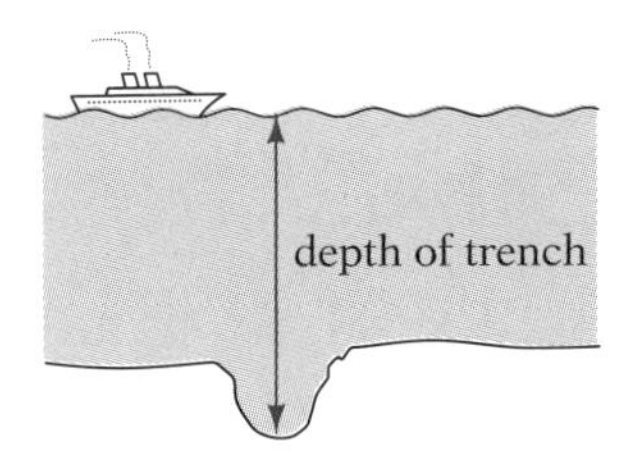

10. The average distance from the Earth to the Moon is three hundred eighty-four thousand, three hundred sixty-five kilometers.

11. The surface temperature of molten lava is 1730 degrees Celsius.

12. Halley's comet and the sun are about 32814000000 miles apart at their greatest distance.

13. The surface area of Jupiter is about 24717000000 square miles.

14. The surface area of the Earth is five hundred ten million, seventy thousand square kilometers.

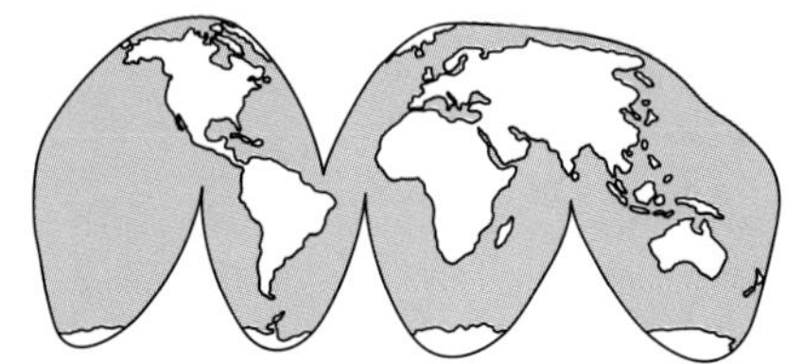

15. The area of the world's largest ocean, the Pacific, is sixty-three million square miles.

16. The Verrazano Narrows Bridge in New York is forty-two hundred sixty feet long.

17. The area of Canada is 9976139 square kilometers.

18. The surface area of the Sun is two trillion, three hundred fifty-three billion, three hundred million square miles.

19. The estimated volume of the Earth itself is one trillion, eighty-three billion, two hundred eight million, eight hundred forty thousand cubic kilometers.

20. The estimated volume of the Earth's oceans is one billion, two hundred eighty-five million, six hundred thousand cubic kilometers.

21. The capacity of the John F. Kennedy Space Center's Vehicle Assembly Building is 3666500 cubic meters.

22. The weight of a blue whale is 307000 pounds.

23. The area of an ice hockey rink is 2222 square yards.

24. The height of Mauna Kea in Hawaii is thirty-three thousand, four hundred seventy-six feet.

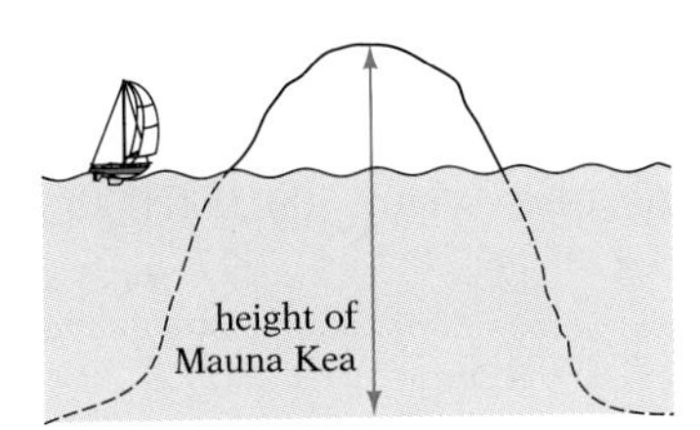

In Exercises 25 through 30, determine whether the number or sum in parentheses is the correct representation for the underlined number.

25. The orbital speed of an electron in a uranium atom is eight million, seven hundred thousand kilometers per hour. (8,007,000)

26. The heat of the Sun's interior is estimated to be twenty million degrees Kelvin. (20,000,000,000)

27. The elevation of Kayenta, Arizona, near Monument Valley is a little more than 1 mile above sea level, or about sixty-seven thousand, nine hundred twenty inches above sea level. (67,900)

28. In 1987 the population of Phoenix, Arizona, was eight hundred sixty-six thousand, six hundred eight and growing. (860,680)

29. Americans make 87600000 pounds of licorice candy in one year. (80,000,000 + 7,000,000 + 600,000)

30. In 1986 the attendance at all major league baseball games was 48452000 spectators. (40,000,000 + 8,000,000 + 400,000 + 50,000 + 2,000)

1.2 Place Value: Ordering Numbers

SECTION GOAL

■ To use < and > and write whole numbers in order

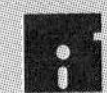

Sometimes it is important to be able to tell which number is the greater of two numbers or which number is the lesser. Mathematicians often think of numbers as points on a line called a **number line.** One number is said to be greater than another if it is to the right of that number along the number line.

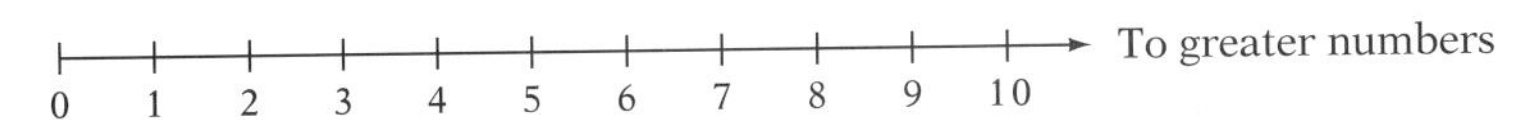

On the accompanying number line, we can see that 7 is to the right of 5; therefore 7 is greater than 5. The sign > is used to symbolize this, and we write 7 > 5.

A number is said to be less than another if it is to the left of that number along a number line. The sign < is used to indicate that one number is less than another. For example, 5 < 7 means "5 is less than 7." This is true because 5 is to the left of 7 on the number line.

We can also say that "5 is between zero and 7" which we can write as either 0 < 5 < 7 or 7 > 5 > 0.

On the number line we locate numbers in order from least to greatest, looking from left to right. For this reason, determining which of two numbers is greater or lesser is often called **ordering** the numbers.

We order whole numbers with many digits by writing them one above the other, aligned at the right. We then compare them period by period. If they do not have the same number of digits, the number with more digits is the greater. If they have the same number of digits, the number with the first greater digit *from the left* is the greater.

EXAMPLE 1

Which is greater, 177362998 or 177371110?

SOLUTION

We first group the digits, using commas, and align the numbers at the right.

177362998 is 177,362,998
177371110 is 177,371,110

Both numbers have nine digits; both are in the hundred millions. To find which is greater, we compare the digits from left to right until they differ:

```
177,362,998
↕↕↕ ↕ *
177,371,110
```

The numbers first differ in the ten-thousands place. Because 7 ten thousands is greater than 6 ten thousands, 177,371,110 is greater than 177,362,998. So we write 177,371,110 > 177,362,998.

■ *Now do margin exercises 1a and 1b.*

1a. Which is greater, 773612998 or 77371110?

1b. Which is less, 317762998 or 317771110?

2a. Order from greatest to least: 2839485, 204958, and 2039438.

2b. Which would be the middle number if these numbers were ordered: 2303948, 2928374, or 2928375?

EXAMPLE 2

Order from least to greatest: 11246375, 11337469 and 112378.

SOLUTION

First we group the digits, using commas, and align them at the right.

11,246,375
11,337,469
112,378

It is clear that the third number is less than the first two. Comparing only the first two numbers, from left to right, we find that they differ in the hundred-thousands place. The 2 hundred thousands of the first number is less than the 3 hundred thousands of the second number. Using the < sign, we write

$$112{,}378 < 11{,}246{,}375 < 11{,}337{,}469$$

■ *Now do margin exercises 2a and 2b.*

Study Hint

To compare numbers, try using graph paper and writing the digits one to a box.

Make sure you align ones digits.

1	1	2	4	6	3	7	5
1	1	3	3	7	4	6	9
		1	1	2	3	7	8

3a. Use < to write a statement about 2,371,146,985 and twenty-eight billion, three. Insert commas as necessary.

3b. Determine which is larger: two billion, nine hundred thousand, forty or 2,000,000,000 + 90,000 + 8000 + 200.

EXAMPLE 3

Use the symbols < and > and write two true statements comparing the underlined numbers.

The rotation period of the Earth is eighty-six thousand, one hundred sixty-four seconds, and the rotation period of Mars is 88643 seconds.

SOLUTION

We write the first number with numerals and group both numbers.

eighty-six thousand, one hundred sixty-four is 86,164
88643 is 88,643

They both have five digits, so we must compare them digit by digit.

8 6, 1 6 4
↕ *
8 8, 6 4 3

Because 8 thousands is greater than 6 thousands, the 88,643 is greater.

Therefore 88,643 > 86,164 and 86,164 < 88,643. If you get confused about which symbol to use, < or >, remember that the open side faces the greater number.

■ *Now do margin exercises 3a and 3b.* Work the exercises that follow.

1.2 EXERCISES

Table 1 gives standard word descriptions for wind speeds. Use it to classify the wind speeds given in Exercises 1 through 8.

TABLE 1 Beaufort Wind Scale

Description	Wind Speed, km/hr	Description	Wind Speed, km/hr
Calm	less than 1	Moderate gale	51 to 61
Light air	1 to 5	Fresh gale	62 to 74
Light breeze	6 to 12	Strong gale	75 to 87
Gentle breeze	13 to 20	Whole gale	88 to 102
Moderate breeze	21 to 29	Storm	103 to 120
Fresh breeze	30 to 39	Hurricane	greater than 120
Strong breeze	40 to 50		

1. 55 km/hr

2. 42 km/hr

3. 38 km/hr

4. 19 km/hr

5. 102 km/hr

6. 120 km/hr

7. 83 km/hr

8. 99 km/hr

In Exercises 9 through 14, a number is underlined in each of two statements. Use $<$ or $>$ to write a true statement comparing the first number to the second number. Rewrite numbers as needed.

9. a. The number of valentines handled yearly by the U.S. Post Office is about two hundred thousand.

b. The number of telephone calls made per year: 292,000,000,000.

10. a. The number of pairs of running shoes bought each year is about 18,980,000.

b. The number of new automobiles purchased each year is about nine million, one hundred twenty-five thousand.

11. a. The weight of the world's largest pizza is 298624 ounces.

b. The amount of food eaten yearly by an average teenage girl is approximately twenty-seven thousand, four hundred seventy-two ounces.

12. **a.** A person with good hearing can hear 375000 different sounds.

b. A person has about nine thousand taste buds on the tongue.

13. **a.** The number of record singles produced per year in the United States is about 124830000.

b. In one year, two hundred nine million, fifty-one thousand record albums are produced in the United States.

14. **a.** The distance from Blytheville, Arkansas, to Texarkana, Texas, is one million, seven hundred thirty-seven thousand, one hundred twenty feet.

b. The distance from Yukon to Fairbanks, Alaska, is eight million, twenty-five thousand, six hundred feet.

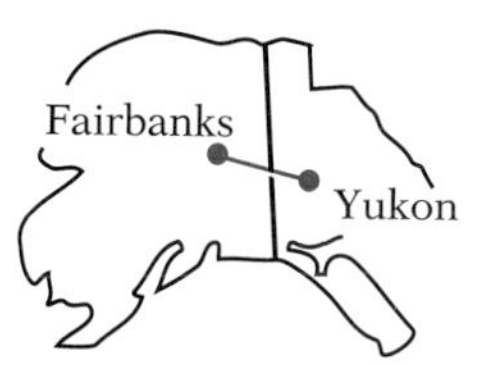

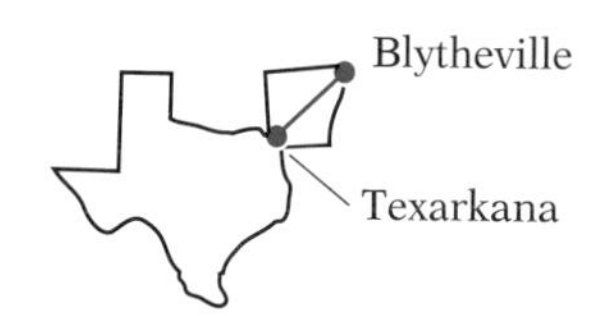

1.2 EXERCISES

Table 1 gives standard word descriptions for wind speeds. Use it to classify the wind speeds given in Exercises 1 through 8.

TABLE 1 Beaufort Wind Scale

Description	Wind Speed, km/hr	Description	Wind Speed, km/hr
Calm	less than 1	Moderate gale	51 to 61
Light air	1 to 5	Fresh gale	62 to 74
Light breeze	6 to 12	Strong gale	75 to 87
Gentle breeze	13 to 20	Whole gale	88 to 102
Moderate breeze	21 to 29	Storm	103 to 120
Fresh breeze	30 to 39	Hurricane	greater than 120
Strong breeze	40 to 50		

1. 55 km/hr

2. 42 km/hr

3. 38 km/hr

4. 19 km/hr

5. 102 km/hr

6. 120 km/hr

7. 83 km/hr

8. 99 km/hr

In Exercises 9 through 14, a number is underlined in each of two statements. Use < or > to write a true statement comparing the first number to the second number. Rewrite numbers as needed.

9. a. The number of valentines handled yearly by the U.S. Post Office is about two hundred thousand.

b. The number of telephone calls made per year: 292,000,000,000.

10. a. The number of pairs of running shoes bought each year is about 18,980,000.

b. The number of new automobiles purchased each year is about nine million, one hundred twenty-five thousand.

11. a. The weight of the world's largest pizza is 298624 ounces.

b. The amount of food eaten yearly by an average teenage girl is approximately twenty-seven thousand, four hundred seventy-two ounces.

12. **a.** A person with good hearing can hear 375000 different sounds.

b. A person has about nine thousand taste buds on the tongue.

13. **a.** The number of record singles produced per year in the United States is about 124830000.

b. In one year, two hundred nine million, fifty-one thousand record albums are produced in the United States.

14. **a.** The distance from Blytheville, Arkansas, to Texarkana, Texas, is one million, seven hundred thirty-seven thousand, one hundred twenty feet.

b. The distance from Yukon to Fairbanks, Alaska, is eight million, twenty-five thousand, six hundred feet.

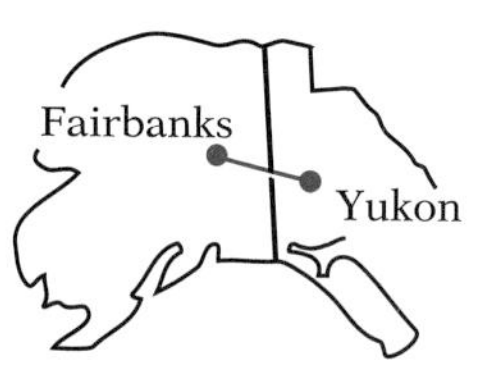

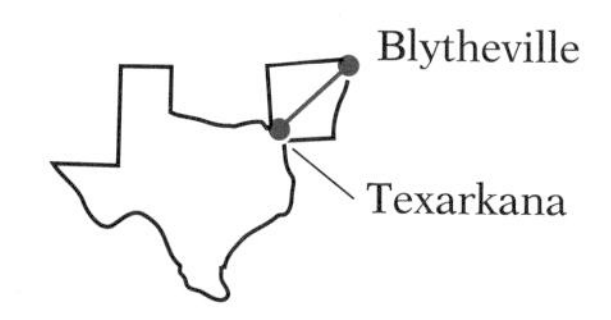

In Exercises 15 though 26, place the numbers in order, first from greatest to least and then from least to greatest, using < and >.

15. 1223, 1526, and 1173

16. 2031, 2295, and 2117

17. 12726, 19287, and 27365

18. 196456, 18763, and 11431

19. 152859384 and 19928704

20. 25388493 and 28104098

21. 19283746 and 4227368

22. 13243546 and 1152636

23. 19987632, 19987667, and 19908767

24. 75309828, 75238098, and 75382098

25. 6368429, 6364289, and 6362984

26. 902756460, 90276469, and 902746560

Use the information given in Table 2 to determine the color we would see for light of each of the wavelengths given in Exercises 27 through 34.

TABLE 2 Visible Light

Color	Wavelength of light, angstroms (Å)	Color	Wavelength of light angstroms (Å)
Violet	3900 to 4550	Yellow	5770 to 5970
Blue	4550 to 4920	Orange	5970 to 6220
Green	4920 to 5770	Red	6220 to 7700

27. 3928 Å

28. 5775 Å

29. 7629 Å

30. 3996 Å

31. 4639 Å

32. 6205 Å

33. 5927 Å

34. 6277 Å

1.2 MIXED PRACTICE

By doing these exercises, you will practice all topics up to this point in the chapter.

35. What does the digit 0 mean in the number 708,635?

36. Write 54,540,004 in words.

37. Order 1,098, 1,908, and 1809, using the symbol >.

38. Write 11,256 in expanded form.

39. Write the underlined words as a number in standard form: Mount Everest is <u>twenty-nine thousand, twenty-eight</u> feet tall.

40. Order 999,999,999 and 899,999,999 and 1,898,898 using the symbol <.

41. Write 11,000,000 + 6,000 + 50 in standard form.

42. Write the number expressed by 3,000,000,000 + 800,000 + 90,000 + 700 in words.

1.3 Place Value: Rounding Numbers

The re-entry speed of the Apollo command module was reported to be 25,000 miles per hour. This speed was obtained by *rounding* the actual speed, 24,791 miles per hour, to the nearest thousand. Numbers are usually rounded when the exact number isn't really needed or isn't really known. The rounded number is an estimate.

When we round a number, we are simply indicating which of two numbers it is closer to. Suppose we wanted to round 1503 to the nearest thousand. We could show it on a number line as follows:

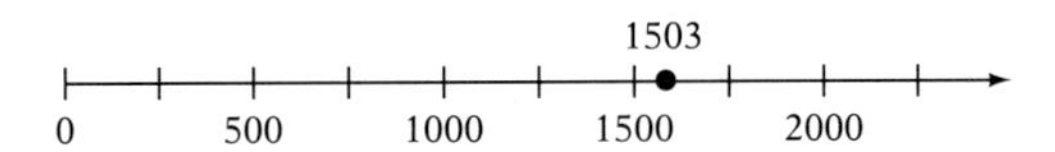

On this number line, 1503 appears to be halfway between 1000 and 2000. Should we round it to 1000 or to 2000?

It's tough to tell how to round some numbers. So mathematicians have agreed to always "round up" when the digit to the right of the place they are rounding to is 5 or more. Otherwise they "round down." Here we are rounding to the nearest *thousand*, and the digit to the right of the *thousands* place is a 5.

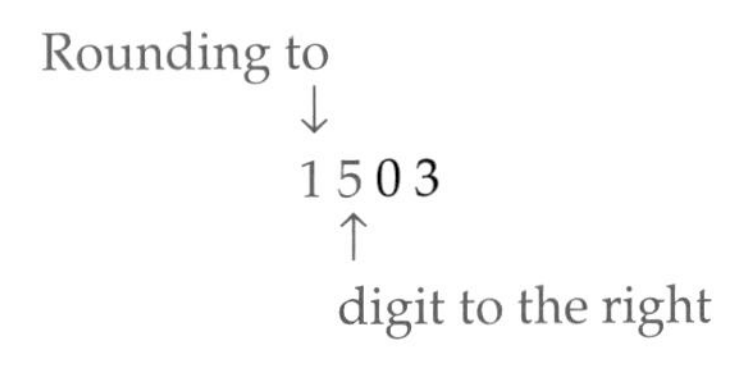

According to the agreement, we round up to 2000.

Here is a procedure you can use to round any number.

To round a whole number

1. Write the number to be rounded. Underline all the digits from the left through the place you are rounding to.
2. Mark the digit just to the right of the last underlined digit.
3. If the digit to the right of the last underlined digit is 5 or greater, increase the entire underlined number by 1 and replace all the digits to its right with zeros.
4. If the digit to the right of the last underlined digit is less than 5, replace it and all the digits to its right with zeros.

Section Goal

- To round whole numbers to the nearest ten, hundred, thousand, and so on

1a. Round 74,469 to the nearest thousand.

1b. Rounded to the nearest ten thousand, which of these numbers would be 10,000: 9,654 or 11,436?

2a. Round 3465125 to the nearest ten thousand.

2b. Which number, when rounded to the nearest million, will have an increase in the millions place: 3,645,934 or 8,321,799?

EXAMPLE 1

Round 24,796 to the nearest thousand.

SOLUTION

Underline the number through the thousands place and mark the next digit to the right.

<u>24</u>,796
↑

Because 7 is greater than 5, increase 24 by 1, making it 25. Then replace all digits to its right with zeros: 25,000.

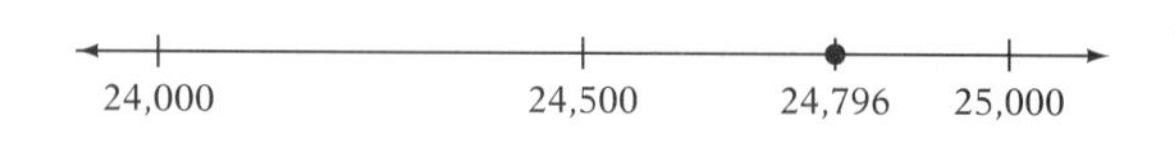

Thus 24,796 rounded to the nearest thousand is 25,000.

■ *Now do margin exercises 1a and 1b.*

Now we will round a larger number.

EXAMPLE 2

Round 1,253,465 to the nearest ten thousand.

SOLUTION

We write the number and underline the digits through the ten-thousands place. We mark the next digit.

<u>1,25</u>3,465
↑

Because 3 is less than 5, we replace it and all digits to the right with zeros:

1,250,000.

Thus 1,253,465 rounded to the nearest ten thousand is 1,250,000.

■ *Now do margin exercises 2a and 2b.*

EXAMPLE 3

Round 19546576 to the nearest million.

SOLUTION

Grouped, 19546576 is 19,546,576.

We underline through the millions place and mark the next digit.

19,546,576
↑

Because the marked digit is 5, we increase 19 by 1 and replace all digits to the right with zeros: 20,000,000.

Thus 19,546,576 rounded to the nearest million is 20,000,000.

■ *Now do margin exercises 3a and 3b.*

3a. Round 8,529,000 to the nearest million.

3b. Which number will give 1,000,000 when rounded to the nearest million: 956,607 or 1,500,000?

4a. Round 863 and 499 to the nearest thousand.

4b. Round 499,552 to the nearest million.

In this next and last example, you will see how to round when there is no digit in the place you are rounding to.

EXAMPLE 4

Round 683 to the nearest thousand.

SOLUTION

Underline the thousands place and mark the next digit.

$$\underline{}683$$
$$\uparrow$$

We can think of all the places to the left of the 6 as zeros.

Because 0683 and 683 are the same number, we have

$$\underline{0}683$$
$$\uparrow$$

Because 6 is greater than 5, we increase the thousands place from 0 to 1 and replace the digits to the right with zeros: 1000.

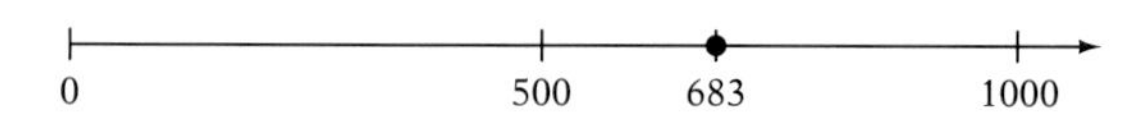

Thus, 683 rounded to the nearest thousand is 1000.

■ *Now do margin exercises 4a and 4b.*

Work the exercises that follow.

1.3 EXERCISES

In Exercises 1 through 8, round each number to the nearest hundred, thousand, and ten thousand.

1. 19786

2. 2647324

3. 39846

4. 11285636

5. 96745

6. 198663

7. 99999

8. 1957632

Table 1 shows the population of the United States from 1810 to 1900. Use this information to answer the questions in Exercises 9 through 12.

TABLE 1 U.S. Population from 1810 to 1900

Year	Population	Year	Population
1810	7,239,881	1860	31,443,321
1820	9,638,453	1870	39,818,449
1830	12,866,020	1880	50,155,783
1840	17,069,453	1890	62,947,714
1850	23,191,876	1900	75,994,575

9. In which year did the population, rounded to the nearest million, first reach forty million?

10. Which three years have the same population, rounded to the nearest ten million?

11. When did the population, rounded to the nearest hundred million, first reach one hundred million?

12. When did the population, rounded to the nearest ten million, first reach 10 million?

In Exercises 13 through 28, round each of the underlined numbers to the indicated place.

13. In 1920 there was $<u>5,698,214,612</u> in circulation. (millions)

14. The number of ways to get dealt three of a kind from a deck of 52 playing cards is <u>54912</u>. (thousands)

15. The diameter of the sun is 1392900 kilometers. (hundred thousands)

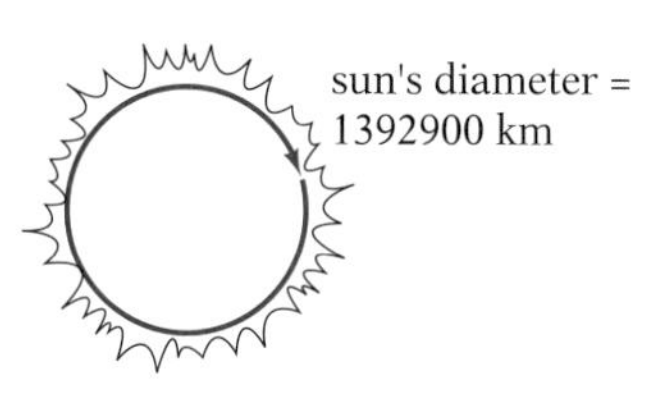

16. The number of Americans injured by needles or pins in one year is 40150. (hundreds)

17. Each year Americans eat about sixty-two billion, fifty million pounds of canned tuna. (hundred millions)

18. Each year Americans buy about one hundred million, three hundred seventy-five thousand pounds of yarn. (hundred thousands)

19. Each day Americans eat about eight hundred fifteen billion calories of food. (billions)

20. The population of the United States in 1986 that was under 5 years of age was one hundred eighty-one million. (hundred millions)

21. It is estimated that by the year 2000, there will be 34882000 people over the age of 65 in the United States. (hundred thousands)

22. The federal acreage of the national parks system in 1986 was 75862484 acres. (millions)

23. In 1985 the population of Arkansas was 2286435. (thousands)

24. Monument Valley is 29816 acres in size. (ten thousands)

25. Arizona's highest mountain, Humphreys Peak, is 12655 feet high. (thousands)

26. Alaska has an area of five hundred eighty-six thousand square miles. (millions)

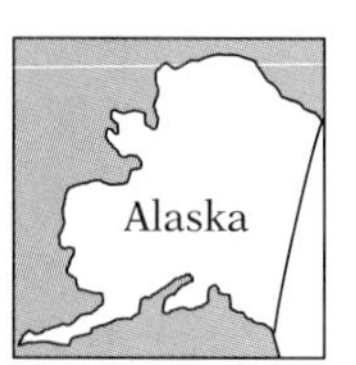

27. About five hundred ninety-eight million dollars was collected on the sale of gym shoes and sneakers in 1987 in the United States. (billions)

28. The length of Boca Chica bridge in the Florida Keys is two thousand, seven hundred thirty feet. (thousands)

In Exercises 29 through 31, use what you have learned about rounding to find the numbers.

29. In 1986 fifty-three million compact discs were sold in the United States. If this number was rounded to the nearest million before being reported, what is the greatest and the least number of compact discs that could have been sold?

30. Waimea Falls Park is a park in Oahu eighteen hundred acres in size. If this number was rounded to the nearest hundred before being reported, what is the least area and the most area it could have?

31. Mount Haleakala is a volcano with a crater large enough to hold Manhattan. It is 10,000 feet tall. If this number was rounded to the nearest ten thousand before being reported, what is the tallest and shortest height it could be?

1.3 MIXED PRACTICE

By doing these exercises, you will practice all topics up to this point in the chapter.

32. Write 9038 in expanded form.

33. Round 427 to the nearest ten.

34. Round 8,775 to the nearest hundred.

35. Write 632, 090 in words.

36. Round 9,635 to the nearest thousand.

37. Write the underlined words as a number in standard form: The number of rentable square feet in the Twin Towers in New York City is four million, three hundred seventy thousand square feet.

38. Arrange 2,098,908 and 2,990,998 so that you can use the symbol <.

39. Round 2,988,000 to the nearest million.

40. Write 12,973 in expanded form.

41. What does the digit 8 mean in the following number? 3,987,541

42. Write 100,005 in words.

43. Arrange 463,875; 4,063,875; and 4,006,387 in order from least to greatest.

1.4 Adding Whole Numbers

SECTION GOALS

- To find the sum of two or more whole numbers
- To use estimation to find approximate sums

In this section, we review the procedure for adding whole numbers.

When you first learned to add one-digit numbers, say 4 and 3, you may have joined two sets of objects together (a set of four objects and a set of three) and then counted the total, 7.

Counting			*Adding*
○ ○ ○ ○	and	○ ○ ○	4
1 2 3 4		5 6 7	+3
			7

This "joining together" is called **addition.** The numbers you added are called the **addends,** and their total is called the **sum.**

Thus, for example, the addends 4 and 3 give the sum 7.

On a number line, we would start at 4 and "count on" three more to the right, "five, six, seven," to find the sum 7.

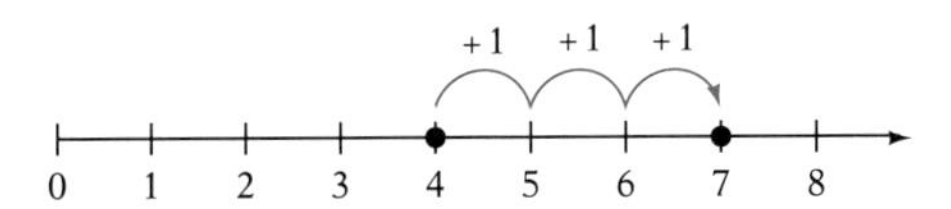

You probably learned that you could start at 4 and then add 3, or start at 3 and then add 4, and the total would be the same. This property of whole numbers is called the **commutative property of addition.** The following definition expresses it in symbols.

Commutative Property of Addition

If a and b are whole numbers, then

$$a + b = b + a$$

This property tells us that the order in which we add numbers is not important. We get the same sum either way. The study of number properties will be important in algebra.

Table 1 shows the basic addition facts, which you should memorize. Actually, you need to memorize only the highlighted facts. The others can be found by applying the commutative property.

TABLE 1 Addition Facts

+	0	1	2	3	4	5	6	7	8	9
0	0	1	2	3	4	5	6	7	8	9
1	1	2	3	4	5	6	7	8	9	10
2	2	3	4	5	6	7	8	9	10	11
3	3	4	5	6	7	8	9	10	11	12
4	4	5	6	7	8	9	10	11	12	13
5	5	6	7	8	9	10	11	12	13	14
6	6	7	8	9	10	11	12	13	14	15
7	7	8	9	10	11	12	13	14	15	16
8	8	9	10	11	12	13	14	15	16	17
9	9	10	11	12	13	14	15	16	17	18

To use the table, find one addend at the left side of the table and the other at the top. Where their row and column intersect, you will find their sum. For example, to add 7 and 5, read across the row labeled 7 and down the column labeled 5 to find the sum 12.

Note, in Table 1, that when zero is added to any number, the sum is that number. For example, $2 + 0 = 2$ and $0 + 2 = 2$. For this reason, we call zero the **identity element for addition.** In symbols,

Identity Element for Addition

If a is any whole number, then

$$a + 0 = a \quad \text{and} \quad 0 + a = a$$

Another property is the **associative property of addition.** In symbols,

Associative Property of Addition

If a, b, and c are whole numbers, then

$$(a + b) + c = a + (b + c)$$

In other words, when we add, we can group numbers together however we wish and not change the sum. For example, we may add $3 + 5 + 7$ in either of these two ways:

$$3 + 5 + 7 = (3 + 5) + 7 = 8 + 7 = 15$$

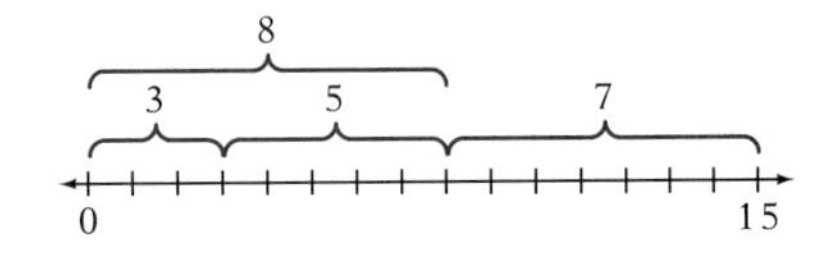

$$3 + 5 + 7 = 3 + (5 + 7) = 3 + 12 = 15$$

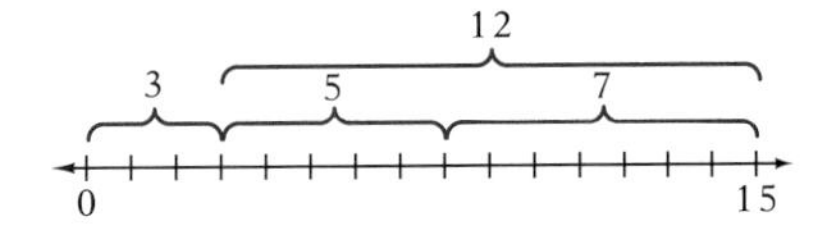

Combining the commutative and associative properties, we can even write

$$3 + 5 + 7 = (3 + 7) + 5 = 10 + 5 = 15$$

There are several other orders you can use in adding these three numbers. Can you find them?

The Addition Procedure

Now we need to discuss the addition of numbers with more than one digit. First we'll use expanded form.

EXAMPLE 1

Add: $\begin{array}{r} 26 \\ +\ 37 \\ \hline \end{array}$

SOLUTION

Writing each addend in expanded form and then adding the tens and the ones, we have

$$\begin{array}{rcrcr} 26 & = & 20\ (2\text{ tens}) & + & 6\ (6\text{ ones}) \\ \underline{+\ 37} & = & \underline{+30\ (3\text{ tens})} & + & \underline{+7\ (7\text{ ones})} \\ & & 50\ (5\text{ tens}) & + & 13\ (13\text{ ones}) \end{array}$$

We have 5 tens and 13 ones, and we cannot write that as 513. We must break the 13 ones into 1 ten + 3 ones. Then

$$\begin{aligned} &5\text{ tens} + 13\text{ ones} \\ &= 5\text{ tens} + 1\text{ ten} + 3\text{ ones} \\ &= 6\text{ tens} + 3\text{ ones} = 63 \end{aligned}$$

So the sum of $26 + 37$ is 63.

■ *Now do margin exercises 1a and 1b.*

Can we do this addition without using expanded form? If we begin by adding the ones column, we'll get 13, a two-digit number. There's no place to put it, so we must *regroup*, or *carry*.

Whenever a column is added and the result is a two-digit number, we regroup it into tens and ones, write the "ones" digit, and carry the "tens" digit

1a. Add in expanded form:

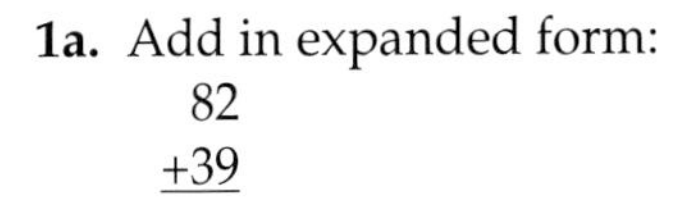

$\begin{array}{r} 82 \\ +39 \\ \hline \end{array}$

1b. Add in expanded form: 35 and 347

1c. Add:
$$\begin{array}{r} 37 \\ \underline{+49} \end{array}$$

1d. Find the total of 328 and 55.

2a. Add:
$$\begin{array}{r} 236 \\ \underline{+745} \end{array}$$

2b. Add: five hundred, eighty-four and ten thousand, nine.

to the next column to the left. This works regardless of which column we are adding. Let's use it in our example.

Add:
$$\begin{array}{r} 26 \\ \underline{+37} \end{array}$$

Adding the ones column gives 6 + 7 – 13. We place the 3 beneath the 6 and 7, and we "carry" the 1 to the next column to be added in.

$$\begin{array}{r} {\scriptstyle 1} \\ 26 \\ \underline{+37} \\ 3 \end{array}$$

Adding the tens column gives $1 + 2 + 3 = 6$.

$$\begin{array}{r} {\scriptstyle 1} \\ 26 \\ \underline{+37} \\ 63 \end{array}$$

Thus, the sum of 26 and 37 is 63.

■ *Now do margin exercises 1c and 1d.*

In the next example, we will add two 3-digit numbers.

EXAMPLE 2

Add:
$$\begin{array}{r} 368 \\ \underline{+284} \end{array}$$

SOLUTION

Starting from the right side, first add the ones to get $8 + 4 = 12$. Write the 2 and carry 1 to the next column.

$$\begin{array}{r} {\scriptstyle 1} \\ 368 \\ \underline{+284} \\ 2 \end{array}$$

Then add the tens ($1 + 6 + 8 = 15$); write the 5 and carry 1 to the next column.

$$\begin{array}{r} {\scriptstyle 1\,1} \\ 368 \\ \underline{+284} \\ 52 \end{array}$$

Then add the hundreds ($1 + 3 + 2 = 6$), and write the 6 in the sum.

$$\begin{array}{r} {\scriptstyle 1\,1} \\ 368 \\ \underline{+284} \\ 652 \end{array}$$

Therefore, $368 + 284 = 652$.

■ *Now do margin exercises 2a and 2b.*

We are given the next problem in horizontal form. We just rewrite the numbers vertically, making sure to align the ones columns.

3a. Add: 26 + 85 + 153

EXAMPLE 3

Add: 3602 + 394 + 8427 = ?

3b. Find the total of 8425 and 1856 and 75897.

SOLUTION

Rewrite the addition problem in vertical form:

$$\begin{array}{r} 3602 \\ 394 \\ +\,8427 \\ \hline \end{array}$$

Align the units digits.

Then add the ones column:

$$\begin{array}{r} {\scriptstyle 1} \\ 3602 \\ 394 \\ +\,8427 \\ \hline 3 \end{array}$$

2 + 4 + 7 = 13

Write 3 and carry 1.

Add the tens column, including the carried 1.

$$\begin{array}{r} {\scriptstyle 11} \\ 3602 \\ 394 \\ +\,8427 \\ \hline 23 \end{array}$$

1 + 0 + 9 + 2 = 12

Write 2 and carry 1.

Then add the hundreds column, including a carried 1.

$$\begin{array}{r} {\scriptstyle 111} \\ 3602 \\ 394 \\ +\,8427 \\ \hline 423 \end{array}$$

1 + 6 + 3 + 4 = 14

Write 4 and carry 1.

The thousands column contains no digit for the number 394. But this number can also be written 0394, so we include the zero to make the addition smoother.

$$\begin{array}{r} {\scriptstyle 111} \\ 3602 \\ 0394 \\ +\,8427 \\ \hline 12423 \end{array}$$

1 + 3 + 0 + 8 = 12

Write the 12.

Thus, 3602 + 394 + 8427 = 12,423.

■ *Now do margin exercises 3a and 3b.*

Estimating Sums

When you are solving an addition problem, it is important to obtain an estimate of the sum. Then, if you make an error—maybe by not carrying properly—you will know about it.

4a. Round to the nearest hundred thousand and add:

273,613
985,274
673,780
+ 11,427

4b. Which two numbers, *when added,* will give a rounded answer of 2,000,000?

563,213
523,320
964,596

From our work with rounding, we know that in the previous example 3602 is close to 4000, 394 is close to 400, and 8427 is close to 8000. Adding these rounded numbers gives an estimate of 12,400 for the exact sum. Because we rounded *up* in two cases, our estimate is probably a little large. Thus our answer, 12,423 is reasonable.

EXAMPLE 4

Estimate the following sum, and then find the exact total.

341,287
274,384
12,967
+986,586

SOLUTION

We round to the greatest place in each number and add the rounded numbers.

341,287 rounds to	300,000
274,384 rounds to	300,000
12,967 rounds to	10,000
986,586 rounds to	1,000,000
Estimate of sum:	1,610,000

The actual addition looks like this:

2 1 2 3 2
341,287
274,384
012,967
+986,586
1,615,224

The estimate of 1,610,000 shows that this answer is reasonable.

■ *Now do margin exercises 4a and 4b.*

Calculators

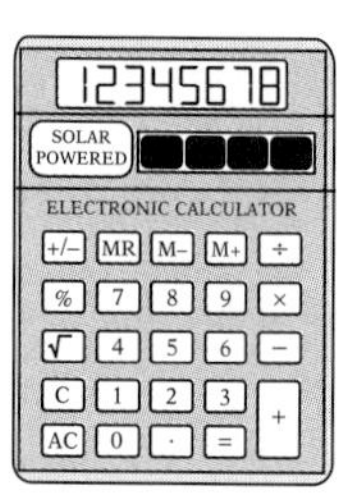

Most hand-held calculators have at least the basic keys shown on the keyboard in the accompanying figure: the digits from 0 through 9; the decimal point; the operations of addition, subtraction, multiplication and division; the square root key; and the percent key. These may, however, be located in different positions.

In our work with calculators, we shall show number keys as numbers and shall show other keys as signs in boxes. Thus

5 [+] 2 [−] 6 [=]

will mean to press the 5 key, then the + key, and then the 2, −, 6, and = keys in that order. You should use your calculator (if you have one) only for examples and exercises marked with the symbol 🖩.

Work the exercises that follow. Use estimation to see if your answers are reasonable.

1.4 EXERCISES

In Exercises 1 through 20, find the sums:

1. $\begin{array}{r} 36 \\ +52 \\ \hline \end{array}$

2. $\begin{array}{r} 385 \\ +\ 14 \\ \hline \end{array}$

3. $\begin{array}{r} 61 \\ +38 \\ \hline \end{array}$

4. $\begin{array}{r} 27 \\ +62 \\ \hline \end{array}$

5. $\begin{array}{r} 84 \\ +67 \\ \hline \end{array}$

6. $\begin{array}{r} 96 \\ +45 \\ \hline \end{array}$

7. $\begin{array}{r} 73 \\ +28 \\ \hline \end{array}$

8. $\begin{array}{r} 262 \\ +\ 39 \\ \hline \end{array}$

9. $\begin{array}{r} 784 \\ +213 \\ \hline \end{array}$

10. $\begin{array}{r} 2193 \\ +\ 706 \\ \hline \end{array}$

11. $\begin{array}{r} 628 \\ +537 \\ \hline \end{array}$

12. $\begin{array}{r} 3457 \\ +\ 521 \\ \hline \end{array}$

13. $\begin{array}{r} 59 \\ 730 \\ +451 \\ \hline \end{array}$

14. $\begin{array}{r} 827 \\ 1644 \\ +\ 390 \\ \hline \end{array}$

15. $\begin{array}{r} 938 \\ 2863 \\ +\ 672 \\ \hline \end{array}$

16. $\begin{array}{r} 1726 \\ 357 \\ +\ 564 \\ \hline \end{array}$

17. $\begin{array}{r} 9534 \\ 6820 \\ +\ 416 \\ \hline \end{array}$

18. $\begin{array}{r} 4926 \\ 18515 \\ +\ 3702 \\ \hline \end{array}$

19. $\begin{array}{r} 901 \\ 8435 \\ +2786 \\ \hline \end{array}$

20. $\begin{array}{r} 6935 \\ 71 \\ +8104 \\ \hline \end{array}$

In Exercises 21 through 28, check each answer to see if it is correct. If it is not, write the correct answer.

21. $\begin{array}{r} 154 \\ 720 \\ +698 \\ \hline 1562 \end{array}$

22. $\begin{array}{r} 235 \\ 806 \\ +794 \\ \hline 1735 \end{array}$

23. $\begin{array}{r} 814 \\ 325 \\ +670 \\ \hline 1819 \end{array}$

24. $\begin{array}{r} 204 \\ 165 \\ +782 \\ \hline 1051 \end{array}$

25. 120
769
354
+879
2022

26. 528
736
164
+905
2332

27. 103
782
516
+429
1820

28. 916
342
780
+453
2391

In Exercises 29 through 36, find the missing digits.

29. 1_4
320
+993
_417

30. 465
26_
+537
1_67

31. 287
_39
+495
17_1

32. 153
49_
+6_7
1_59

33. 67
59
2_
+43
198

34. 13
6_
98
+_2
251

35. 34
1_
85
+_6
207

36. 26
_3
5_
+45
221

In Exercises 37 through 44, find the sums.

37. Add: 27 + 36 + 1856

38. Add: 94 + 37 + 84

39. Add: 253 + 49 + 30

40. Add: 87 + 62 + 1907

41. Add: 137 + 99 + 884

42. Add: 1876 + 224 + 35

43. Add: 946 + 89 + 26

44. Add: 47 + 85 + 4653

In Exercises 45 through 48, estimate the sums.

45. 36 + 59

46. 24 + 18

47. 32 + 47 + 26

48. 35 + 42 + 11

Match one of these estimates to each of Exercises 49 through 52. Do not add.

a. two hundred thousand
b. one hundred sixty thousand
c. twenty-two thousand
d. ten thousand

49.
```
  63,194
  67,186
 +65,223
```

50.
```
  5,644
  8,859
 +7,235
```

51.
```
  47,927
  80,005
 +26,172
```

52.
```
  1,345
  6,149
  2,378
 +1,100
```

In Exercises 53 through 56, use paper and pencil to find the sum. Use estimation to see if your answers are reasonable. *Check your answers with a calculator.*

53.
```
  24,987
  19,956
  98,876
 +93,598
```

54.
```
  38,697
  12,324
  98,657
 +23,274
```

55. 543,876 + 12,387 + 4,342,123 + 93,876

56. 138,765 + 31,564 + 3,364

Of the following numbers, find the pair whose sum is closest to each estimate given in Exercises 57 through 60. Use a calculator.

1598762	59887	1217788	1968777
312161	883260	32597	18212

57. 2,500,000

58. 2,000,000

59. 1,500,000

60. 90,000

1.4 MIXED PRACTICE

By doing these exercises, you will practice all topics up to this point in the chapter.

61. Round 366 to the nearest ten.

62. Write four hundred eighty-seven million, four in standard form.

63. Find the sum: 11,603 + 427 + 864

64. Write the underlined number in words: The closest the Moon comes to us is 221,463 miles.

65. Round 236,851 to the nearest thousand.

66. Write 5,837 in expanded form.

67. Round 9,988 to the nearest hundred.

68. Write the underlined words in standard form: The number of people attending Expo '70 was sixty-four million, two hundred eighteen thousand, seven hundred seventy.

69. What does the digit 2 mean in the number 2,004,006?

70. Add: 3456 + 9088 + 90,007

71. Round 17,299 to the nearest thousand.

72. Estimate the sum: 324,543 + 981,234 + 8,298,791

1.5 Subtracting Whole Numbers

SECTION GOALS

- To find the difference between two whole numbers
- To use estimation to find approximate differences

Subtraction is often called the opposite of addition (or the inverse operation) because subtraction "undoes" addition. For example, we add 4 + 3 to get 7, and we subtract 3 from 7 to get back to 4.

$$4 + 3 = 7$$
$$7 - 3 = 4$$

The following figure shows this on a number line.

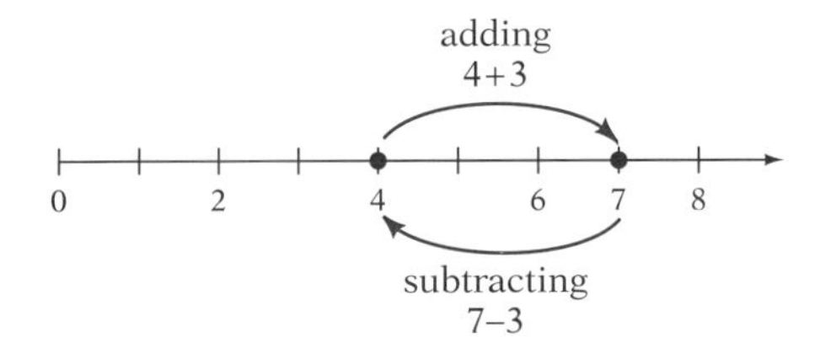

Subtraction

When a, b, and c are whole numbers and $a + b = c$, then

$$c - a = b \text{ and } c - b = a.$$

The number we are subtracting from is called the **minuend**; the number we are subtracting is called the **subtrahend.** The result is called the **difference.**

In the subtraction 7 – 3 = 4, 7 is the minuend, 3 is the subtrahend, and 4 is the difference.

Because addition and subtraction are opposite operations, we can use the *addition table* (on page 26) to review the basic subtraction facts. To read the table for a subtraction fact, say 12 – 7, we first find the row of the subtrahend (7). We follow this row into the table until we locate the minuend (12). The number that heads the minuend's column is 5, which is the difference: 12 – 7 = 5. Note that the commutative property does not hold for subtraction, because 12 – 7 and 7 – 12 cannot give us the same answer. You can see this in the following figure.

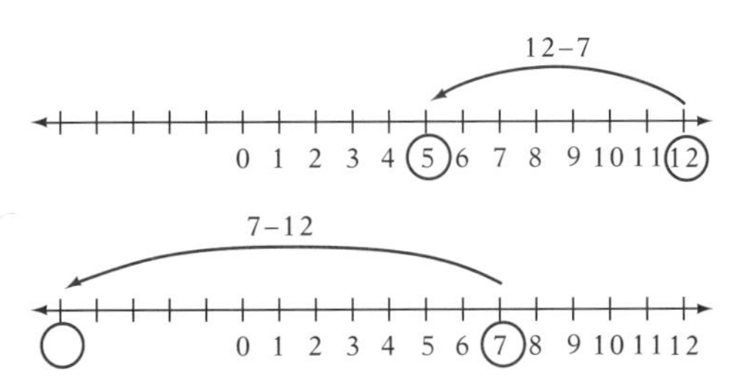

The Subtraction Procedure

We cannot use a table to subtract large numbers, but fortunately there is a procedure that works for all subtraction problems. To explain it, we'll use expanded notation, as we did for addition.

1a. Subtract:
$$\begin{array}{r} 56 \\ \underline{-23} \end{array}$$

1b. Subtract in expanded form:
49 – 18

2a. Subtract:
$$\begin{array}{r} 54 \\ \underline{-29} \end{array}$$

2b. What number subtracted from 97 will give 38?

EXAMPLE 1

Subtract: $\begin{array}{r} 37 \\ \underline{-26} \end{array}$

SOLUTION

To work this problem, we need only to know these facts: 7 – 6 = 1 and 3 – 2 = 1. We subtract the digits in the ones column and then those in the tens column.

$$\begin{array}{r} 37 \\ \underline{-26} \\ 11 \end{array} \quad \text{The difference is 11.}$$

■ *Now do margin exercises 1a and 1b.*

Sometimes, however, we cannot subtract place by place. The next example shows what actually happens and why.

EXAMPLE 2

Subtract: $\begin{array}{r} 84 \\ \underline{-36} \end{array}$

SOLUTION

In expanded form, we have

$$\begin{array}{rcrcr} 84 & = & 80 \text{ (8 tens)} & + & 4 \text{ (4 ones)} \\ \underline{-36} & = & \underline{-30 \text{ (3 tens)}} & + & \underline{-6 \text{ (6 ones)}} \end{array}$$

In this problem we cannot immediately subtract in the ones column. We cannot subtract 6 ones from 4 ones. However, we can rewrite the minuend's 8 tens as 7 tens and 10 ones; then we can combine that with the 4 ones to get 7 tens and 14 ones. The problem can then be solved.

$$\begin{array}{rcr} 70 \text{ (7 tens)} & + & 14 \text{ (14 ones)} \\ \not{80} \text{ (8 tens)} & + & \not{4} \text{ (4 ones)} \\ \underline{-30 \text{ (3 tens)}} & + & \underline{-6 \text{ (6 ones)}} \\ 40 \text{ (4 tens)} & + & 8 \text{ (8 ones)} \end{array}$$

This gives us 40 + 8, or 48.

Now see if you can follow along in this shorter version of the same problem.

Subtract: $\begin{array}{r} 84 \\ \underline{-36} \end{array}$

In the ones column, 6 is larger than 4, so we borrow from the tens to get 14 and then subtract the ones.

$$\begin{array}{r} \scriptstyle 7 \\ \not{8}{}^{1}4 \\ \underline{-3\,6} \\ 8 \end{array} \quad \begin{array}{l} 14 - 6 = 8 \\ \text{Write 8.} \end{array}$$

Then we subtract tens.

$$\begin{array}{r} \scriptstyle 7 \\ \not{8}{}^{1}4 \\ \underline{-3\,6} \\ 4\,8 \end{array} \quad \begin{array}{l} 7 - 3 = 4 \\ \text{Write 4.} \\ \text{The answer is 48, as before.} \end{array}$$

■ *Now do margin exercises 2a and 2b.*

The next example is three-digit subtraction. The method does not change; the subtraction just takes a little longer.

EXAMPLE 3

Subtract: 436
−189

SOLUTION

We cannot subtract the ones because we cannot subtract 9 from 6. So we borrow from the tens to get 16 ones and then subtract.

$$\begin{array}{r} 4\,\overset{2}{\not{3}}\,{}^{1}6 \\ -1\,8\,9 \\ \hline 7 \end{array}$$

16 − 9 = 7
Write 7.

Next, we have to borrow from the hundreds to subtract the tens.

$$\begin{array}{r} \overset{3}{\not{4}}\,\overset{12}{\not{3}}\,{}^{1}6 \\ -1\,8\,9 \\ \hline 4\,7 \end{array}$$

12 − 8 = 4
Write 4.

Finally, we subtract the hundreds.

$$\begin{array}{r} \overset{3}{\not{4}}\,\overset{12}{\not{3}}\,{}^{1}6 \\ -1\,8\,9 \\ \hline 2\,4\,7 \end{array}$$

3 − 1 = 2
Write 2.
The difference is 247.

■ *Now do margin exercises 3a and 3b.*

Does anything change if there is a zero in the problem? Let's look at an example.

EXAMPLE 4

Subtract: 19002
− 185

SOLUTION

As always, we start with the ones place. We need to borrow from the tens place, but there is nothing to borrow from because the digit is a zero. There is also a zero in the hundreds place. We continue to the next place, thousands, and borrow 1 thousand or 10 hundreds.

$$\begin{array}{r} 1\,\overset{8}{\not{9}}\,{}^{1}0\,0\,2 \\ -\quad 1\,8\,5 \\ \hline \end{array}$$

Borrow 1 from 9. Write 8 and make 0 a 10.

This does not immediately solve the problem, because we still cannot subtract in the ones place. We must borrow from the hundreds and from the tens and subtract.

$$\begin{array}{r} 1\,\overset{8}{\not{9}}\,\overset{9}{\not{1}\not{0}}\,\overset{9}{\not{1}\not{0}}\,{}^{1}2 \\ -\quad 1\,8\,5 \\ \hline 7 \end{array}$$

Borrow 1 from 10. Write 9 and make 0 a 10. Repeat and make 2 a 12. Subtract: 12 − 5 = 7. Write 7.

3a. Subtract:
247
−158

3b. Find the difference between 1613 and 367.

4a.
```
  18009
-   632
```

4b.
```
  400000
-   8182
```

5a. Subtract using your calculator: four million minus 6,521

5b. Subtract using your calculator:

```
  8,000,001
-   493,782
```

Next, we subtract the tens digits. We don't have to borrow further to subtract in the tens column.

```
  8 9 9
 1 9 0 0 2
-    1 8 5
       1 7      9 − 8 = 1 Write 1.
```

We subtract the hundreds, thousands, and ten thousands and we've finished.

```
  8 9 9
 1 9 0 0 2      9 − 1 = 8 Write 8.
-    1 8 5      8 − 0 = 8 Write 8.
 1 8 8 1 7      1 − 0 = 1 Write 1.
```

Check: To check a subtraction problem, add the result (the difference) to the number you subtracted (the subtrahend). You should get the number you started with (the minuend).

```
Check: 18817 difference
     +   185 subtrahend
       19002 minuend
```

The answer checks.

■ *Now do margin exercises 4a and 4b.*

EXAMPLE 5

Subtract 19782 from 236874 on your calculator.

SOLUTION

Key in, on your calculator,

236874 [−] 19782 [=]

The display should show the difference, 217092.

Because subtraction is not commutative, the order in which the numbers are keyed in makes a difference. The first number keyed in must be the minuend, the second the subtrahend. See what happens if you reverse the order.

■ *Now do margin exercises 5a and 5b.*

Work the exercises that follow.

Name Section Date

1.5 EXERCISES

In Exercises 1 through 28, subtract. Check your answers by adding.

1. 237 − 116

2. 462 − 251

3. 958 − 345

4. 763 − 142

5. 547 − 69

6. 9325 − 48

7. 172 − 87

8. 1463 − 76

9. 524 − 435

10. 633 − 254

11. 814 − 323

12. 3261 − 192

13. 307 − 63

14. 5604 − 858

15. 509 − 93

16. 103 − 84

17. 4060 − 187

18. 3050 − 264

19. 2010 − 182

20. 7040 − 155

21. 6000 − 27

22. 4000 − 63

23. 2000 − 896

24. 5000 − 432

25. 21345 − 9657

26. 22648 − 8351

27. 17632 − 9485

28. 32415 − 9762

In Exercises 29 through 32, fill in the digits that make the problems correct.

29. 8625_
− 9_35
_6616

30. 462_3
− _327
3988_

31. 5421_
− 18_2
_2351

32. 642_3
− _854
6035_

In Exercises 33 through 36, check each subtraction to see if the answer is correct. If not, give the correct answer.

33. 7,090,800
−5,735,162
1,355,638

34. 6,040,070
−3,305,007
2,735,963

35. 2,030,060
−1,245,321
784,739

36. 4,060,020
−1,998,635
2,072,385

1.5 MIXED PRACTICE

By doing these exercises, you will practice all topics up to this point in the chapter.

37. Write the number expressed by 50,000 + 600 + 40 + 3

38. Insert the proper symbol (< or >) for the question mark. 878786565 ? 878785664

39. Approximate the difference: 59,875 − 4,286

40. Find the sum: 162 + 8741 + 760

41. Subtract: 704,006 − 1,375

42. Find the sum: 234,567 + 9086 + 364

43. Round 35,664 to the nearest hundred.

44. Add: 234 + 923 + 123,762

1.6 Applying Addition and Subtraction of Whole Numbers: Word Problems

SECTION GOAL

- To solve problems involving the addition and the subtraction of whole numbers

Mathematics problems involving situations in the "real world" are usually presented as word problems. Before you can solve a word problem, you must understand the problem situation and what is being asked. Then you must decide which operation or operations to use on which numbers. After using an operation, you need to see if your answer makes sense in the problem situation.

Most word problems that require addition represent one of three types.

Addition applications involve

a. *Combining* separate groups of numbers (such as populations, or dollar amounts)
b. *Joining* one item to another item (such as lengths of items)
c. *Extending* a measure (such as distance traveled or age)

Most word problems requiring subtraction also represent one of three types.

Subtraction applications involve

a. *Taking away* (such as elements from a group)
b. *Cutting off* a portion (such as a length from an object)
c. *Comparing* (such as larger than, shorter than, or bigger than)

In this first example, we will show how addition is used to *combine* groups of numbers—in this case, numbers of students.

EXAMPLE 1

In 1987, a census showed that 155,917 students attended colleges in Puerto Rico. In the Virgin Islands, 2,572 attended, and in Guam, 4,601 attended. Find the total number of students attending colleges on these islands.

SOLUTION

We are asked to "find the total number" of students. "Total" indicates that we must add.

155,917	in Puerto Rico
2,572	in Virgin Islands
+ 4,601	in Guam
163,090	total

■ *Now do margin exercises 1a and 1b.*

1a. At a certain pool, for the seven days in one week, the attendance was 346, 22, 185, 226, 104, 307, and 280 people. How many people went to the pool that week?

1b. I have boards of length 42 feet, 37 feet, 53 feet, and 40 feet. How far would these boards stretch if they were joined end to end?

2a. The area of Brazil is 3,286,470 square miles. The area of Bulgaria is 428,234 square miles. How much bigger is Brazil than Bulgaria?

2b. *Rocky IV* grossed $75,782,000 in the United States and Canada. This was $1,682,000 more than *Saturday Night Fever.* How much did *Saturday Night Fever* gross? How much did the two movies gross together?

Many problems—like many "real world" problems—require more than one step to solve. Usually, each step calls for a different operation.

In the next example, we use subtraction to take away. But first we must use addition to combine.

EXAMPLE 2

A ream of paper (500 sheets) was loaded into a copy machine. Then 235 copies of an advertising brochure, 57 copies of a bill, and 90 copies of a menu were made. How many sheets remain in the copy machine?

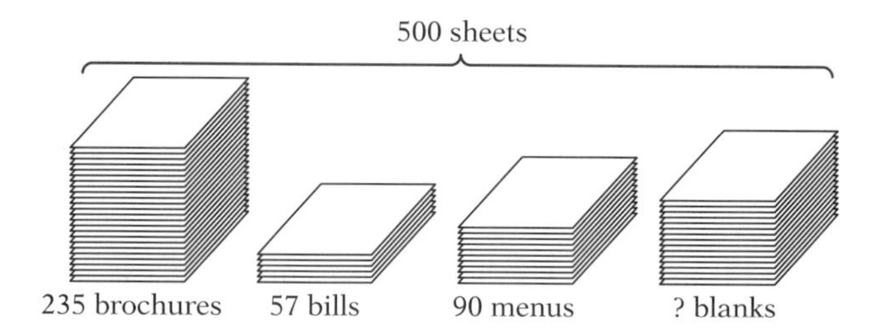

SOLUTION

To solve this problem, we first need to find the total number of copies made. Then we can subtract (take away) this total from the 500 sheets loaded into the machine.

Step 1: We add to find the total number of copies made.

$$\begin{array}{rl} 235 & \text{brochures} \\ 57 & \text{bills} \\ +\ 90 & \text{menus} \\ \hline 382 & \text{sheets used} \end{array}$$

Step 2: We subtract to find the number that remain.

$$\begin{array}{rl} 500 & \text{to start} \\ -382 & \text{used up} \\ \hline 118 & \text{remain} \end{array}$$

There are 118 sheets of paper remaining in the machine.

■ *Now do margin exercises 2a and 2b.*

In this next example, addition is used to *extend* years (dates).

EXAMPLE 3

Louisiana joined the Union in 1812. Twenty-four years later Arkansas joined the Union, and 53 years after that Montana joined the Union.

In what year did Montana join the Union?

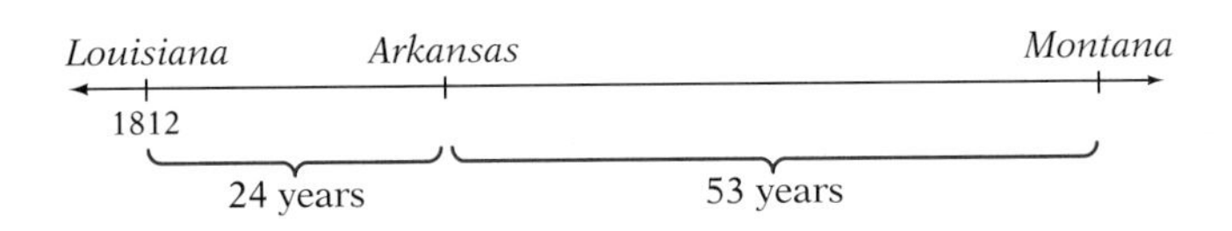

SOLUTION

We are given the date 1812 and events happening 24 years and 53 years later.

We are asked to find the latest date. This situation requires adding.

$$\begin{array}{rl} 1812 & \text{Louisiana joins} \\ +\ \ 24 & \text{to Arkansas} \\ +\ \ 53 & \text{to Montana} \\ \hline 1889 & \text{Montana joins} \end{array}$$

Montana joined the Union in 1889.

■ *Now do margin exercises 3a and 3b.*

Remember to use the four problem-solving steps—read and interpret the problem, decide which method to use, use the method and then look back and reason—as you work the exercises that follow.

3a. Jackie Robinson was born in 1919. He died at the age of 53. In what year did he die?

3b. Susan was 30 in the year 1992. How old will she be in the year 2004?

Finding Palindromes

The word DAD is a **palindrome,** a word that reads the same forward and backward. A number that reads the same forward and backward is also called a palindrome. Examples are 343, 1001, and 30303. Here's how to get one:

Start with any number	687
Is it a palindrome?	NO
Then reverse the digits:	786
Add it to the original number:	687 +786 1473
Is the sum a palindrome?	NO
Reverse and add again:	1473 +3741 5214
Is the sum a palindrome?	NO
Reverse and add again:	5214 +4125 9339

9339 is a palindrome.

It took three reversals and additions, but the answer is a palindrome. Try the procedure with these numbers:

263 87 123

This process may take several steps, but it will give you a palindrome every time.

1.6 EXERCISES

In Exercises 1 through 45, solve each word problem.

1. Delaware became a state in 1787, Michigan entered the Union 50 years later, and Arizona 75 years after that. In what year did Arizona enter the Union?

2. In the late 1800s the United States purchased Alaska for $7,200,000; Puerto Rico, Guam, and the Philippines for $20,000,000; and the Virgin Islands for $25,000,000. How many millions of dollars were spent on these purchases altogether?

3. In 1770 the French built a steam car, 44 years later the steam locomotive was invented, and 90 years later Henry Ford started to mass produce cars. In what year did Ford begin the mass production of cars?

4. The population of the Dominican Republic in 1987 was 6,500,000, and the population of Trinidad was 1,300,000. How much larger was the population of the Dominican Republic than that of Trinidad?

5. The world's deepest cave is Pierre St-Martin, located in the Pyrenees. It is 4370 feet deep. That is 1906 feet deeper than Ghar Parau in Iran. How deep is the Ghar Parau cave?

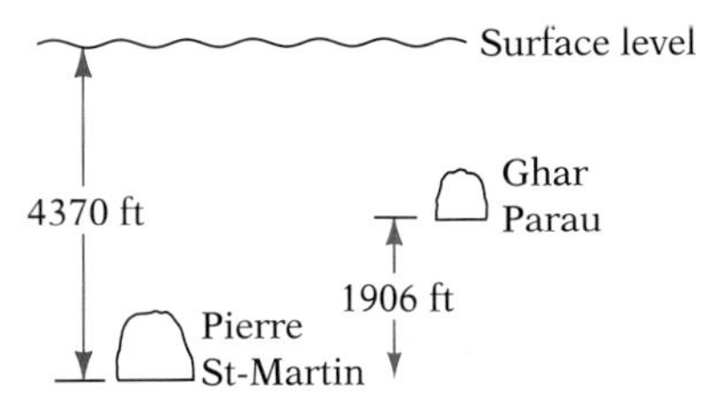

6. The play *Fiddler on the Roof* had 3388 Broadway performances, whereas *Harvey* lasted 1775 performances. How many more performances did *Fiddler on the Roof* have?

7. The flying distance from London to Rome is 1427 kilometers. The driving distance is an additional 470 kilometers. What is the total traveling distance between these two cities?

8. A washer was originally priced at $498, and a dryer at $398. How much will you save if you buy the combination for $700?

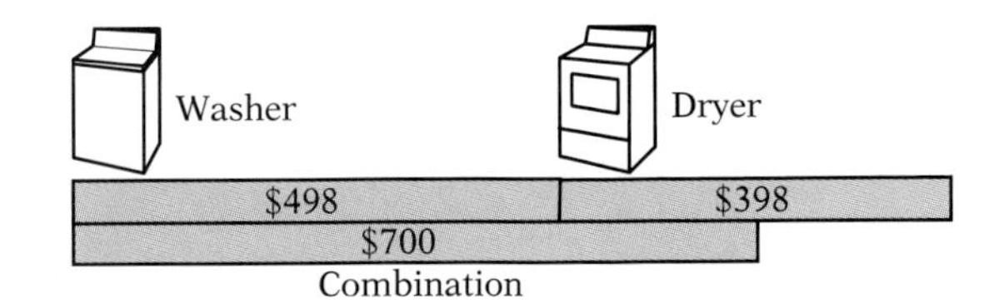

9. In 1986, the number of phonograph records purchased was 219,100,000. This was 294,700,000 fewer than were purchased in 1979. How many phonograph records were purchased in 1979?

10. The world's shortest continuous border is between Spain and Gibraltar and is only 1672 yards long. The longest continuous border is 7,015,448 yards longer than that and lies between the United States and Canada. How long is that border?

11. Traveling east, the distance from Tokyo to New York City is 10839 kilometers, and the distance from New York City to London is 5567 kilometers. What is the distance from Tokyo to London by way of New York?

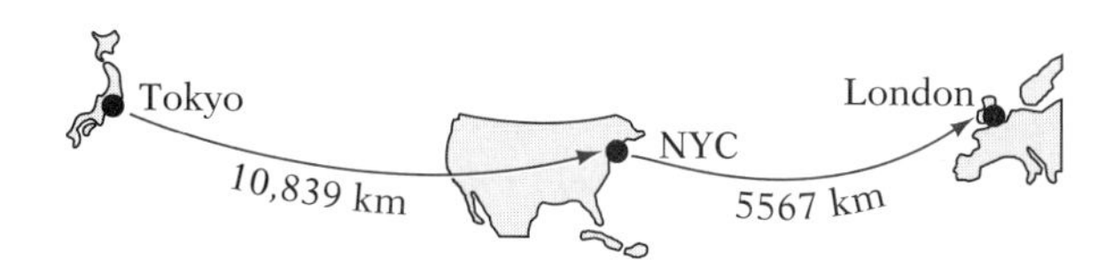

12. If Mt. Everest's base were at the bottom of the Marianas Trench, deep under the Pacific Ocean, its peak would be 7,169 feet below sea level. Mt. Everest is 29,028 feet tall. How deep is the Marianas Trench?

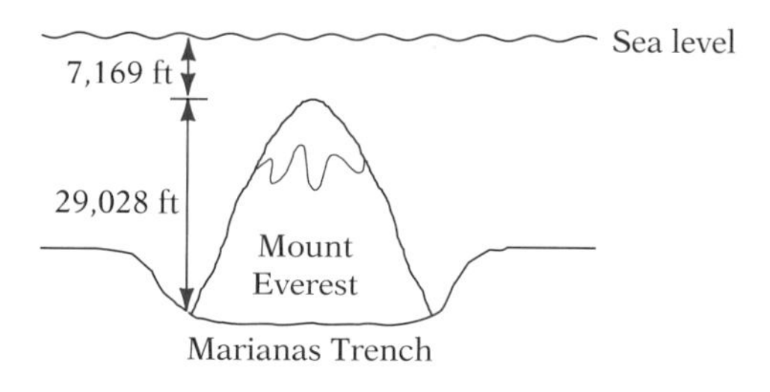

13. There are 139,500,000 square miles of salt-water oceans in the world. This is 139,170,000 more than the number of square miles of fresh-water lakes. How much area is covered by fresh-water lakes?

14. The Empire State building is 1472 feet tall, counting its mast. Mt. Kilimanjaro is 19,340 feet tall. How much taller is the mountain than the building?

15. In 1984 there were 8,401,000 members in the American Bowling Congress. This was 868,000 fewer than in 1983. What was the enrollment in 1983?

16. The win records for the New York Yankees from 1979 through 1982 are as follows:

Year	1979	1980	1981	1982
Wins	89	103	59	79

What was the total number of wins in the four years 1979 through 1982?

17. The highest recorded tsunami (tidal wave) was 220 feet high. The Statue of Liberty is 305 feet tall with its pedestal. How much taller is the statue than this wave?

18. A ribbon worm can grow to 180 feet in length; a giant squid can reach up to 57 feet in length. How much longer is the ribbon worm?

19. The area of the largest island, Greenland, is 839,999 square miles. The total area of the next largest island, New Guinea, is 316,615 square miles. Find the difference in size between the two islands.

20. In a certain year, the number of elementary school teachers in Indiana was 653,554, compared to 1,225,311 in Illinois. How many more elementary school teachers were there in Illinois?

21. The Sears Tower in Chicago is the tallest office building in the world. It stands 1454 feet without its mast and 1559 feet with it. How tall is the mast?

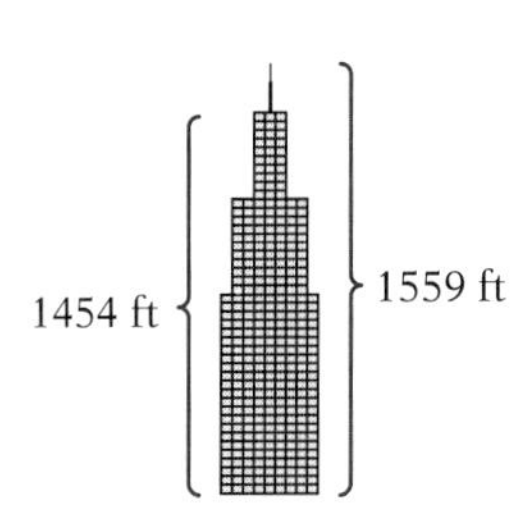

22. In May 1987, the median resale price for houses in the South was $81,800. That was $1300 less than the median resale price in May 1988. What was the median resale price in May 1988?

23. The world's deepest mine is Western Deep in South Africa. It reaches 12,600 feet deep. The world's deepest drilling site, the Kola peninsula in what was formerly the Soviet Union, is 31,911 feet deep. How much deeper is the Kola site?

24. The highest point in Minnesota is Eagle Mountain (2,301 feet), and the highest point in Nevada is Boundary Park (13,140 feet). What is their difference in elevation?

25. The elevations of three different landings of Carlsbad Caverns are 754 feet, 900 feet, and 1,320 feet below sea level. Find the difference between the highest point and the lowest point.

26. There are 7,748 windows in the Pentagon Building. The Sears Tower has 16,000 windows. How many more windows are there in the Sears Tower?

27. The total number of rental units in New York City that were converted to cooperatives and condominiums from 1981 to 1984 was 35,405. That is 8098 fewer than in the period 1984 to 1987. How many were converted in that period?

28. The first modern motorcycle was made in 1901, exactly 32 years after the first motorcycle was patented. In what year was the first motorcycle patented?

29. During 1981, the number of people who attended the National Hockey League regular-season games was 10,726,000. By 1986, attendance was up by 895,000 people. What was the 1986 attendance?

30. A California redwood tree has been measured at 366 feet tall. The longest recorded strand of seaweed measured 196 feet. How much longer is the redwood?

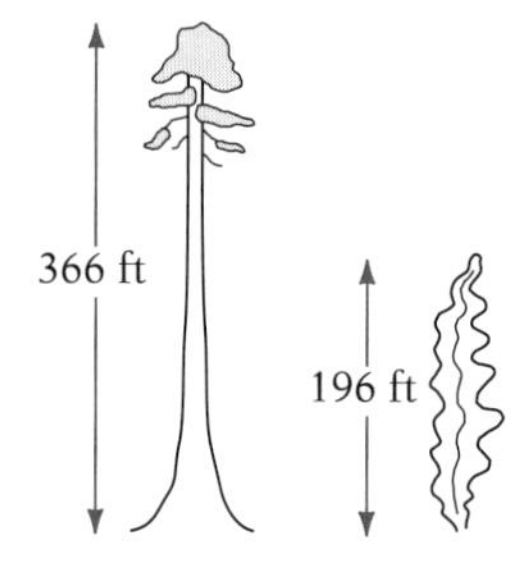

31. There were 1100 passengers booked on a certain cruise ship to Bermuda. A hurricane hit the ship. When some people decided to fly home on the return voyage, there were 923 passengers on the ship. How many people did not continue on the cruise?

32. The flying distance from Los Angeles to New York is 2451 miles. The driving distance is 464 miles more. What is the driving distance?

33. The first organized horse race was run in 1664. That was 124 years after the first horses were brought to North America. In what year were horses first brought to North America?

34. The year 1970 was the last year in which more men than women were issued U.S. passports. In that year 1,095,000 women were issued passports. This was 29,000 fewer than the number of passports issued to men. How many were issued to men?

35. A used car and truck dealer advertised a sale. Before the sale, he had 259 cars, 15 trucks, and 35 vans. He sold a total of 139 vehicles. How many vehicles were left on the lot at the end of the sale?

36. When it closed on Broadway, the *Wiz* had 1,672 performances compared to *Grease* which had 3,388 performances. How many more performances did *Grease* have?

37. The airship *Hindenburg* was about 809 feet long. A Boeing 747 jumbo jet is about 231 feet long. About how much longer was the airship than the jumbo jet?

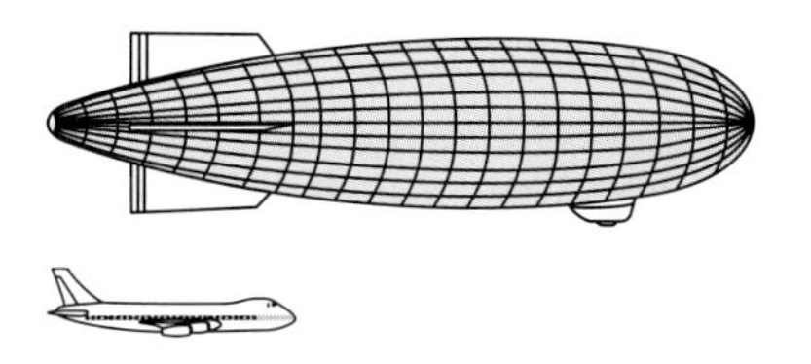

38. In 1869 Mendeleyev designed the periodic table of the elements. The modern idea of an element was first introduced in 1661. How many years later was the periodic table developed?

39. Iceland has an area of 39,709 square miles. Finland has an area of 130,119 square miles. How many more square miles are there in Finland?

Use estimation to solve Exercises 40 through 45.

40. In 1987 total motor vehicle registrations in British Columbia, Québec, and Ontario were 2,175,032, 2,974,099, and 5,179,918, respectively. Estimate how many million vehicles were registered.

41. The 1987 estimated revenue for the Republic of China was $16,349,000,000. Estimated expenses were $16,329,200,000. Estimate the difference to the nearest ten million.

42. There were 953,000 college graduates in the academic year 1981–1982 and 930,684 in the academic year 1971–1972. Estimate the change in the number of college graduates to the nearest ten thousand. Use your calculator to check your answer.

43. The greatest official altitude reached by an occupied balloon is 113,740 feet, whereas the greatest unofficial altitude reached by an occupied balloon is 123,800 feet. How much higher is the unofficial record, estimated to the nearest thousand? Use your calculator to check your answer.

44. Estimate the total population of these three cities to the nearest hundred thousand: Atlanta 426,090; Austin 397,001; Dallas 974,234. Use your calculator to check your answer.

45. The 18-and-over viewing audience showed in one year that it had the following preferences:

Drama: 12,810,000	Suspense: 10,350,000
Comedy: 14,220,000	Adventure: 8,150,000

Estimate the greatest and the least number of people who could have been surveyed, to the nearest million. (*Hint:* What if each person could choose only one category? What if each person could choose as many as 4 categories?)

1.6 MIXED PRACTICE

By doing these exercises, you will practice all topics up to this point in the chapter.

46. Subtract: 1000 – 981

47. Round 4683 to the nearest ten.

Use the following information to solve Exercises 48 and 49: The Yukon River is 1979 miles long; The Danube River is 1766 miles long; and The Congo River is 2716 miles long.

48. How many miles longer is the Congo River than the Danube River?

49. What is the total length of the three rivers?

50. Write 4,000,011,284 in words.

51. Three ounces of brook trout broiled contains 216 calories, whereas three ounces of lake trout broiled contains 74 more calories. How many calories are there in three ounces of broiled lake trout?

52. Add: 210,283 + 1,837,867 + 27,892,232

53. Write one hundred five thousand, four as a number expressed in standard form.

54. I am thinking of a number. The number is 345 more than 500. What is the number?

55. Round 496,473 to the nearest thousand.

56. Subtract: 1,000,001 – 999,998.

57. Order the numbers 5621, 5631, and 5611 using >.

58. Portsmouth, Virginia, was founded in 1752. One hundred fifty-four years later, Virginia Beach was founded. In what year was Virginia Beach founded?

59. Add 5,066 acres and 46,469 acres, and express the sum to the nearest hundred acres.

??? Problem-Solving Preparation: Reading A Table

You may sometimes have to refer to a complex table or chart to obtain information you need. Table 1 is such a table, and it can be confusing. The color notations have been added to help you interpret the information in the table, above and below it, and at the left. Note, especially, the *"total"* and *"average"* rows, which can get in the way of reading the table.

1 **TABLE 1**

2 **U.S. Travel to Foreign Countries—Travelers and Expenditures: 1975 to 1985**

3 Travelers in thousands; expenditures in millions of dollars, except as indicated. Covers residents of United States and Puerto Rico.

ITEM AND AREA	1975	1979	1980	1983	1984	1985 (4)
Total overseas travelers	**6,354**	**7,835**	**8,163**	**9,628**	**11,252**	**12,309**
Region of destination:						
Europe and Mediterranean	3,185	4,068	3,934	4,780	5,760	6,457
Caribbean and Central America	2,065	2,533	2,624	2,989	3,313	3,497
South America	447	434	594	535	557	553
Other	657	800	1,011	1,324	1,622	1,802
Total Expenditures abroad	**6,417**	**9,413**	**10,397**	**13,556**	**15,449**	**16,482**
Canada	1,306	1,599	1,817	2,160	2,416	2,694
Mexico	1,637	2,460	2,564	3,618	3,599	3,531
Total overseas areas	3,474	5,354	6,016	7,778	9,434	10,257
Europe and Mediterranean	1,918	3,185	3,412	4,201	5,171	5,857
Average per trip (dollars)	602	783	867	882	897	(NA)
Caribbean and Central America	787	1,019	1,134	1,428	1,786	1,830
South America	242	288	392	408	357	365
Japan	131	142	185	276	400	458
Other	396	720	893	1,465	1,720	1,747

(Brackets: 5 = rows Total overseas travelers to Other; 6 = rows Total Expenditures abroad to Other; 7 = right side of travelers rows; 8 = right side of expenditures rows.)

9 NA Not available
10 Adapted from *Statistical Abstracts of the United States 1988*

1 table number
2 title: describes the information in the table
3 units
4 years for which data was compiled
5 places visited
6 places visited
7 number of travelers (in thousands)
8 travel costs (in millions of dollars)
9 explanatory note
10 source of data

EXAMPLE 1

a. What was the total amount spent by people traveling from the United States to Canada during 1983?

b. How many people traveled to South America from the United States in 1979?

SOLUTION

a. Look across the *row* that starts "Canada" to where it *intersects* the *column* headed "1983." The number in that intersection is 2160. This is the num-

1a. How many US travelers went to South America from 1983 through 1984?

1b. An average vacation for two people in Europe cost about how much in 1983?

ber of millions of dollars spent, so the answer is 2160 million dollars, or $2,160,000,000.

b. In the section that refers to travelers, look across the row that starts "South America" until you are under the column headed "1979." The number there is 434, so the answer is 434 thousand (or 434,000) people.

■ *Now do margin exercises 1a and 1b.*

Work the exercises that follow.

EXERCISES

In Exercises 1 through 5, use Table 1 to answer the questions.

1. How much more was spent for U.S. travel to Japan in 1985 than was spent 10 years before that?

2. How many people traveled from the United States to the Caribbean and Central America in 1985?

3. What were the expenditures by travelers in Japan during 1984?

4. What was the average spent per trip in Europe and the Mediterranean during 1984?

5. An average vacation to Europe and the Mediterranean would have cost how much more in 1983 than in 1975?

1.7 Geometry and Measurement: Applying Addition and Subtraction of Whole Numbers

There are geometry sections like this throughout the book. In these sections, you will be applying your computational skills to geometry and geometric measures. We begin here with a review of geometric figures and three measures.

SECTION GOALS

- To identify polygons
- To distinguish among area, volume, and perimeter problems

Introduction to Geometric Figures

A **polygon** is a closed figure made up of three or more line **segments,** or parts of lines. (See the accompanying figure.) The segments are called the **sides** of the polygon. The point where two sides meet is a **vertex** and the sides form an **angle** at the vertex.

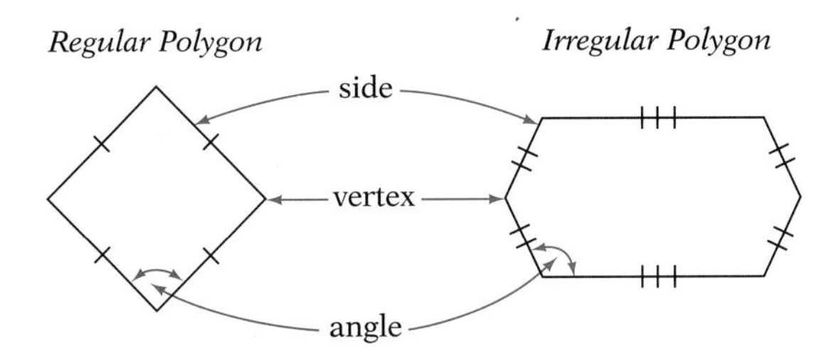

The measure of a side is called its **length.** If the sides of a polygon all have the same length, and its angles all have the same measure, the polygon is called a **regular polygon.** If the lengths or angles differ, the polygon is **irregular.** Some polygons that you are probably familiar with, and some new ones, are shown on pages 53 and 54.

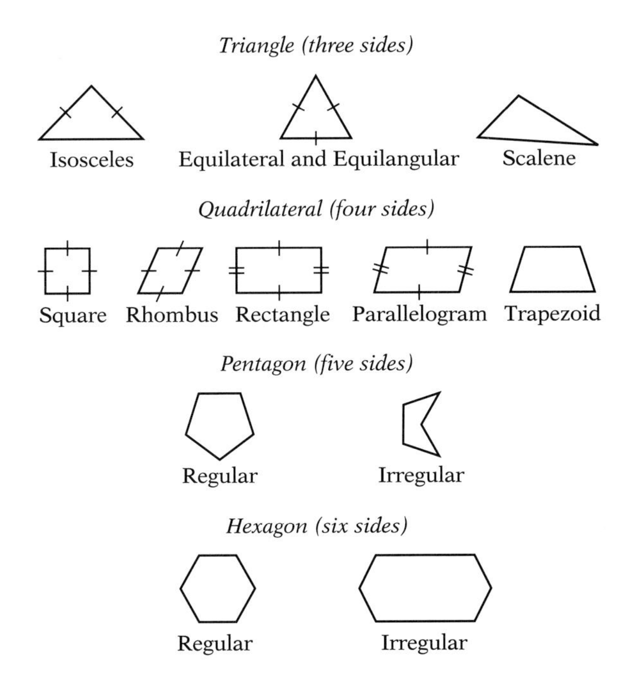

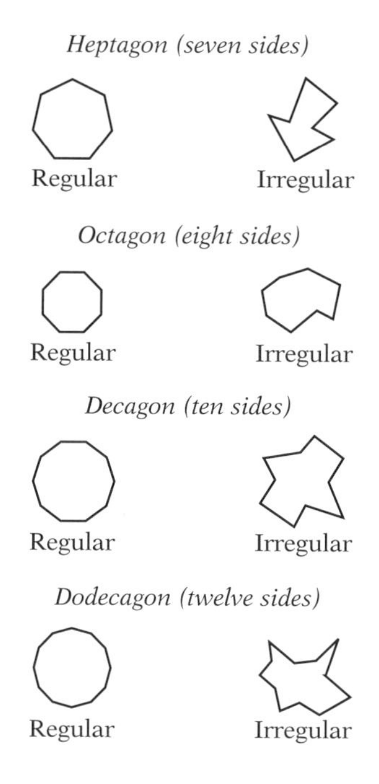

Triangles that have *two* sides of equal length are called **isosceles triangles.** Triangles that have *three* sides of equal length are called **equilateral triangles;** their angles are all of equal measure too. Those that have *no* sides equal in length are called **scalene triangles.**

Quadrilaterals are a little more difficult to classify. **Rectangles** are quadrilaterals that have two pairs of opposite sides of equal length and all four angles of equal measure; the angles are called **right angles,** or 90-degree angles. **Squares** are quadrilaterals that have all four sides equal in length; their angles too are all right angles. If the angles are not right angles, the figure is called a **rhombus. Parallelograms** are quadrilaterals that have opposite sides equal in length and parallel. **Parallel** means that the distance between the two lines is the same no matter where on the line the comparison is made. Parallel lines never intersect. See the following figure.

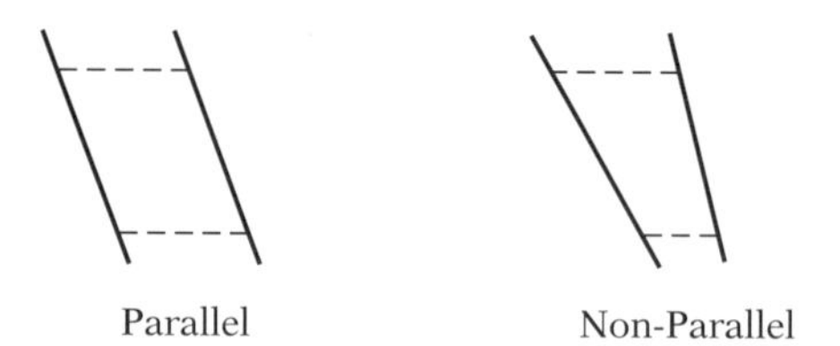

Trapezoids are quadrilaterals that have exactly one pair of opposite sides parallel.

All the common polygons shown on pages 53 and 54 are **plane figures;** they exist only on a flat surface. Note, in the figure, that we use tic marks to show equal measure. Lines with the same number of marks have equal lengths.

In example 1, you are asked to name geometric figures. Keep in mind that a figure's "orientation," or the way it is turned, does not change its geometric classification.

EXAMPLE 1

Name each of these figures.

a.  b. 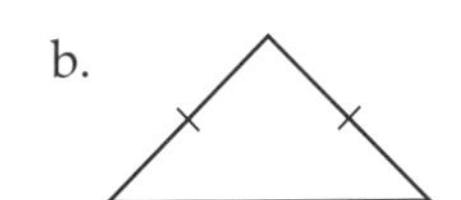 c. 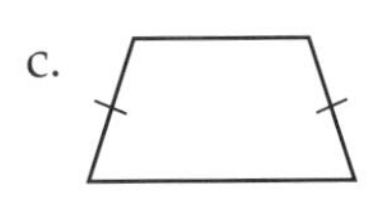

SOLUTION

Figure (a) has six sides, so it is a hexagon; because the sides are not equal, it is irregular. Figure (b) has three sides, of which two are equal; it is an isosceles triangle. Figure (c) has four sides. The opposite sides appear to be parallel; one pair of sides is equal. It is an isosceles trapezoid.

■ *Now do margin exercises 1a and 1b.*

1a. Name this figure.

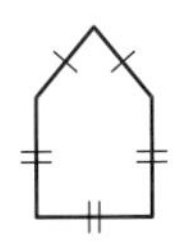

1b. Name this figure.

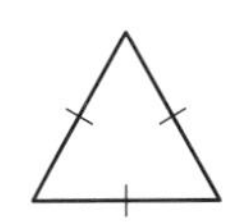

Introduction to Perimeter, Area, and Volume

Now let's look at two measures of plane figures.

Perimeter

The **perimeter** of a closed geometric figure is the distance around the figure.

There are formulas for finding the perimeters of *regular* geometric figures. However, the perimeter of *any* closed polygon (regular or not) can be found by just adding up the lengths of all its sides.

Area

The **area** of a closed plane figure is the amount of surface within the figure.

Area measures the flat surface inside a plane figure. In later chapters we shall use formulas to find the areas of polygons.

Our last measure applies only to three-dimensional figures such as balls and boxes, not to plane figures.

Volume

The **volume,** or **capacity,** of a three-dimensional object is the amount of space contained within the object.

2a. Is this a perimeter, area, or volume problem? A pool table surface measures 2.74 meters by 1.37 meters. If I want to replace the cloth on the table, how much cloth do I need?

2b. Is this a perimeter, area, or volume problem? A lawn measures 145.3 yards by 108.67 yards. How much fencing will I need to enclose it?

The volume of a water glass, for example, is the amount of space in the glass—the amount of water we could pour into it. In later chapters we shall also use formulas to find volumes.

EXAMPLE 2

Determine whether each of the following is an area, perimeter, or volume.

a. The amount of carpet needed for a room

b. The amount of fencing needed to surround a piece of property

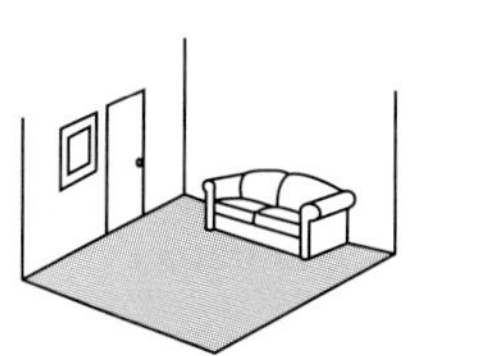

c. The size of a tablecloth required to cover a table

d. The amount of coffee needed to fill a can

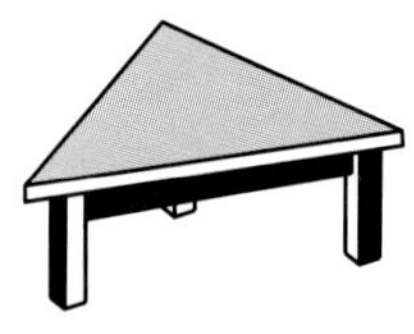

SOLUTION

In each of the foregoing examples, we need to distinguish among distance around, surface within, and space inside a three-dimensional figure. Carpets and tablecloths both cover a flat surface, so we are dealing with areas in (a) and (c). Fencing goes around a piece of property, so (b) involves perimeter. The amount of coffee needed to fill a can is the volume of the can, so (d) involves volume. The answers are then: a. area; b. perimeter; c. area; d. volume or capacity.

■ *Now do margin exercises 2a and 2b.*

Work the exercises that follow.

1.7 EXERCISES

In Exercises 1 through 8, identify the figures and determine whether they are regular or irregular figures.

1. 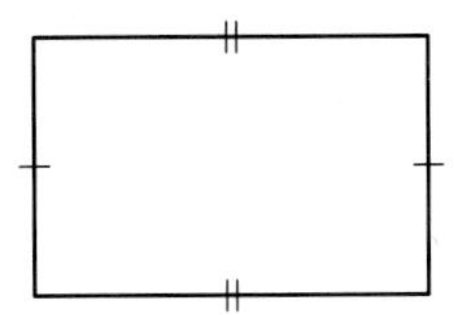

2. 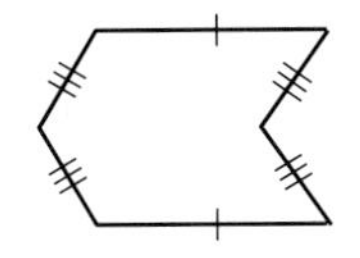

3. 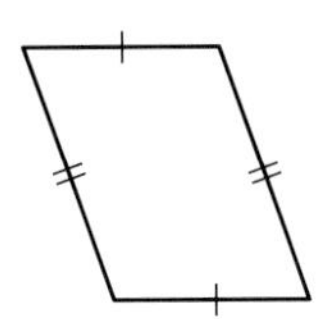

4. 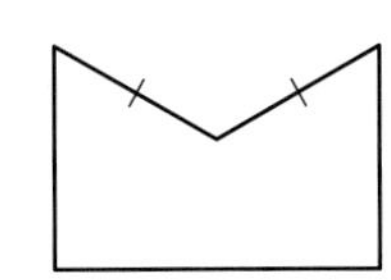

5. 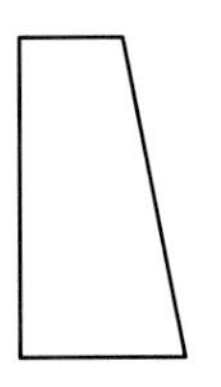

6. 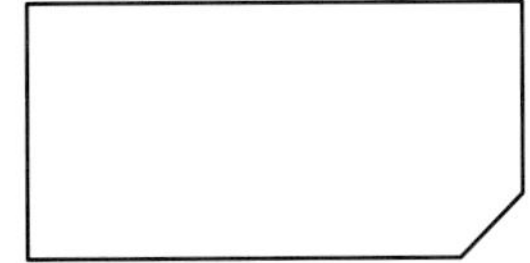

7. 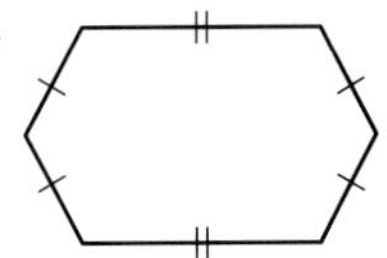

8. 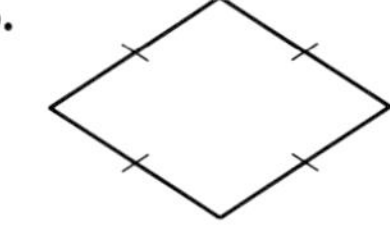

In Exercises 9 through 16, determine whether you would have to find an area, a perimeter, or a volume to solve the problem. DO NOT SOLVE.

9. In a game called netball, the dimensions of the court are 30.5 meters by 15.25 meters. How many 0.25-meter-wide boards are needed to enclose the court?

10. A baseball field is a square measuring 27.45 meters on one side. How much astroturf must be used to cover the field?

11. How much sawdust can be placed in a container that measures 4.3 inches wide, 5.6 inches long, and 3.8 inches high?

12. A rectangular pool table has dimensions of 2.74 meters by 1.37 meters. How much wood is needed to frame the table?

13. In professional football, the goal posts are 5.64 meters apart and the cross bar is 3.05 meters above the ground. They form a rectangular region of what size?

14. A croquet lawn measures 25.6 meters by 32 meters. How many square meters does this lawn cover?

15. A license plate measures 30 centimeters by 15.8 centimeters. Find the amount of rubber edging that is needed to enclose the license plate.

16. Find the amount of a liquid that can be put into a container that has a height of 5 inches, a width of 4 inches, and a length of 8 inches.

1.7 MIXED PRACTICE

By doing these exercises, you will practice all topics up to this point in the chapter.

17. Find the sum: 12,998 + 384,685 + 90,394

18. Find the difference: 98,286 – 8,007

19. Barbara is thirteen years older than her cousin. Her cousin is fifteen years older than Barbara's brother. How old is Barbara if her brother is 25?

20. Find the sum: 13,293 + 45,987 + 200

21. Write 31,234,496 in words.

22. Round 29,876 to the nearest ten.

23. Order 56576, 45656, and 45556 using the symbol <.

24. Find the difference: 89,000,000 – 34,588,765

25. Classify the following triangles.

a.

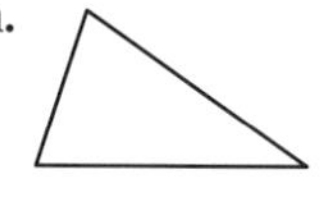

b. **c.**

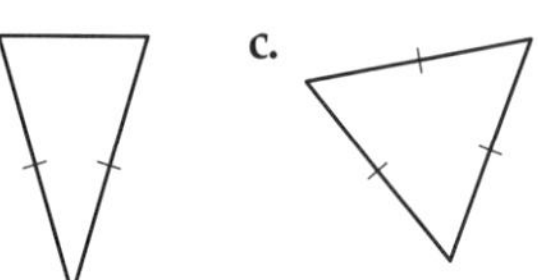

26. Round 456,212 to the nearest ten thousand.

27. Find the difference: 90,000 – 45,687

28. Is this an area question or a perimeter question? What will be the distance that a runner has run if he has run 5 times around a track that measures 34 yards by 20 yards? DO NOT SOLVE.

29. Write the number for <u>fourteen billion, eighty-four.</u>

30. Find the amount of money that I have in the bank if I originally had $235 and wrote out checks for $123 and $64.

Name Section Date

Chapter 1 Review

ERROR ANALYSIS

These problems have been worked incorrectly. Tell what the error is and then write the correct solution.

1. Subtract: 2,000,000 – 123

Incorrect Solution

```
  2,000,000
–       123
  1,999,987
```

Correct Solution

Error ______________________________

2. Write the number fifty-six thousand, eight.

Incorrect Solution

56,000,8

Correct Solution

Error ______________________________

3. Add: 47 + 123 + 276

Incorrect Solution

```
  47
 123
+276
 869
```

Correct Solution

Error ______________________________

4. Find the sum: 46 + 69

Incorrect Solution

```
 46
+69
105
```

Correct Solution

Error ______________________________

5. Subtract: 426 – 293

Incorrect Solution

```
 426
–293
 273
```

Correct Solution

Error ______________________________

6. Subtract: 3000 – 89

Incorrect Solution

$$\begin{array}{r} 3000 \\ -\ \ 89 \\ \hline 3089 \end{array}$$

Correct Solution

Error ______________________________

____

INTERPRETING MATHEMATICS

By working these exercises, you will test and strengthen your mathematics vocabulary.

1. Describe 49,861 using the words digits, numeral, number, place value, standard notation, and periods.

2. Explain the difference between place value and face value.

3. Discuss the process of addition by using the terms regroup, addends, sum, and identity element for addition. Refer to problem

$$\begin{array}{r} 276 \\ +\ 80 \\ \hline 356 \end{array}$$

4. In the subtraction 23 – 7 = 16, identify the difference, subtrahend, and minuend.

5. Here are the names of some geometric figures:

parallelogram rectangle square

triangle hexagon trapezoid

The following terms might be used to describe these figures. Next to the name of each figure, place the letter(s) of the term(s) that could be used to describe it always.

a. regular polygon **b.** irregular polygon
c. equilateral **d.** equiangular
e. parallelogram **f.** quadrilateral
g. polygon **h.** isosceles

6. What happens to a number when it is rounded?

7. Use a non-mathematical example to illustrate the commutative property of addition.

8. Discuss area, perimeter, and volume in relation to a child's building block.

REVIEW PROBLEMS

The exercises that follow will give you a good review of the material presented in this chapter. Work through them and check your answers at the back of the book.

Section 1.1

1. Write 206,122 in words.

2. Write 4560 in expanded form.

3. Write 1,000,765 in words.

4. Write two hundred million, four thousand, five as a numeral.

5. Write 8000 + 600 + 5 in standard form.

6. Write ten thousand, four in expanded form.

Section 1.2

7. Which underlined number is larger?

a. Mt. Everest is twenty-nine thousand, twenty-eight feet high.

b. The volcano Chimborazo is 20,560 feet tall.

In Exercises 8 and 9, use < to write the numbers in order.

8. 14827, 18826, 1342, 199872

9. 192730000, 274880000, 182760000, 1772660000

10. Which underlined number is smaller—twenty-four thousand, seven hundred ninety-one or 24,099?

11. Arrange 198742; 1938; 927; and 199273 in order using >.

12. Use < to arrange 1970220, 1773015, 198218, and 1990632 in order.

Section 1.3

Round each of the following to the nearest ten, hundred, thousand, ten thousand, and hundred thousand.

13. 1256

14. 59,994

15. 129,835

16. 3,249,986

17. 75,234

18. 9899

Section 1.4

19. Add: 126 + 54 + 97

20. Find the sum of 987 and 903 and 816.

21. What is the total of 118 + 202 + 48?

22. Add: 118,878 + 999 + 42,926

23.
$$\begin{array}{r} 10{,}005 \\ 932 \\ +\ 1{,}789 \\ \hline \end{array}$$

24. Find the sum of 123,456 and 3,687,234 and 908.

Section 1.5

25. What is the difference between 86,003 and 984?

26. Subtract: 9,000,008 – 8457

27. Subtract:
$$\begin{array}{r} 5376 \\ -\ 984 \\ \hline \end{array}$$

28. Find the difference between 10,007 and 239.

29. From 1,230,405, subtract 7,438.

30. Subtract 19,874 from 54,368.

Section 1.6

31. A look at your monthly paycheck shows that your gross pay was $1876. You had taxes deducted totaling $675, union dues of $56, and social security payments of $127. What was your net pay after the deductions?

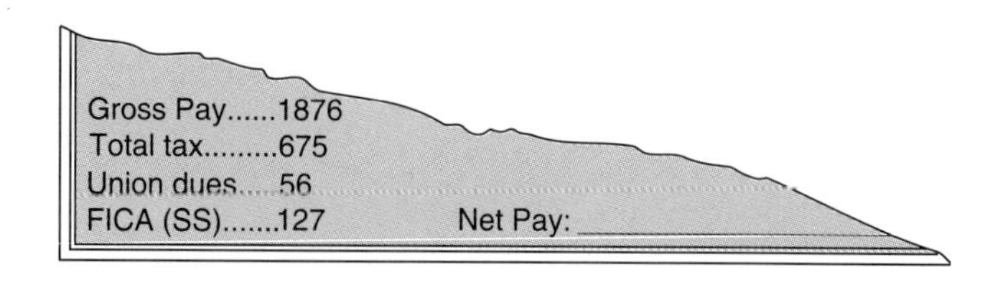

32. The CN Tower in Toronto is the tallest free-standing tower in the world, standing 1815 feet tall. How much taller is it than the 1353-foot-tall World Trade Center towers?

33. A standard competition ice rink for ice dancing has an area of 19,375 square feet. For U.S. ice hockey, the rink is 2460 square feet smaller. What is the size of a U.S. ice hockey rink?

34. Three water towers hold 3475 gallons, 8299 gallons, and 2839 gallons, respectively. How much more is held by the largest of these than by the two smallest put together?

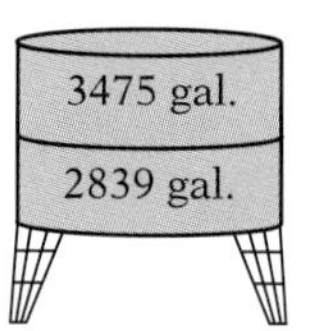

35. According to the 1986 census report, 5,786,000 families had incomes between $25,000 and $49,000 annually, whereas 4,035,000 made between $10,000 and $14,999 annually. Estimate, to the nearest ten thousand, how many more families earned the higher annual incomes.

36. The first successful electric typewriter was sold in 1902. This was 92 years after the first known typewriter was made. In what year was the first typewriter made?

Section 1.7

37. Match the description of each polygon with the correct name.

__hexagon a. eight-sided polygon
__octagon b. quadrilateral
__heptagon c. six-sided polygon
__rhombus d. seven-sided polygon

In Exercises 38 through 44, identify each situation as an area, perimeter, or volume problem.

38. The size of a football field

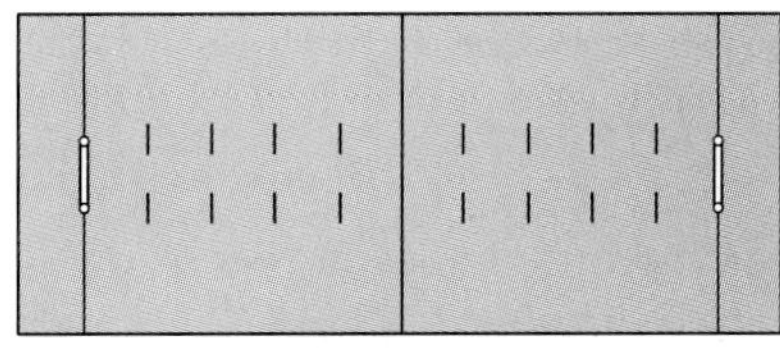

39. The distance around a hockey puck

40. The number of yards of material to cover a sofa

41. The distance around a mirror

42. The size of a lot on which you will build a house

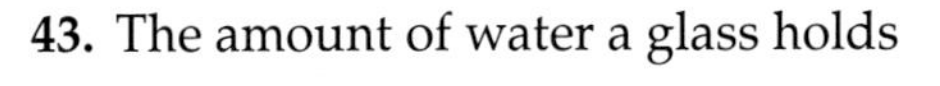

43. The amount of water a glass holds

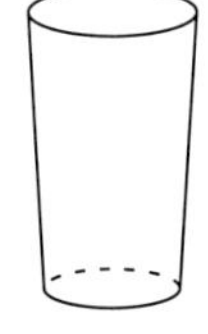

44. The amount of paint left in a bucket

Mixed Review

45. An architect is drawing up floor plans for a house. The area of the living room is 275 square feet; of the dining room, 150 square feet; and of the kitchen, 130 square feet. How much larger than the kitchen are the living room and dining room together?

46. In May 1988, the lowest median resale price of houses was $68,300 in the Midwest. That was $80,400 less than the highest median value reported in the Northeast. What was the highest Northeast median price?

47. Your gas meter for your house read 6344 ccf/kwh in March. It increased 141 ccf/kwh in April and increased 62 ccf/kwh more in May. What was the reading in May?

48. Two million tons of gravel, 122,000 tons of clay, and 2,000 tons of peat are dug up daily in the United States. How many tons of gravel, clay, and peat is that altogether? Give your answer to the nearest hundred thousand tons.

49. Energy production in the Middle East was approximately one billion, six hundred eighty-eight million metric tons in 1976. North America produced 2,309,000,000 metric tons. How many million more tons were produced by North America than by the Middle East?

50. In a certain year, the populations, rounded to the nearest thousand, of the states touching Lake Michigan were as follows:

Wisconsin	4,807,000	Illinois	11,582,000
Michigan	9,200,000	Indiana	5,531,000

What was the total population?

Chapter 1 Test

This exam tests your knowledge of the topics in Chapter 1.

1. Write the numeral for

a. Two million, five hundred thousand, six

b. Seven hundred thousand, eleven

c. Write in words: 62,030,007

2. Use the symbol < to write the following numbers in order.

a. 1422, 1234, and 1002

b. 1,000,030 and 1,000,031

c. Use the symbol > to write 1029399 and 277738 in order.

3. Solve:

a. Round 845 to the nearest ten.

b. Round 9843 to the nearest thousand.

c. Round 345,439,085 to the nearest ten thousand.

4. Solve:

a. Find the sum of 98 and 234 and 983.

b. What is the total of 23,944 and 182,123 and 34,127?

c. Add: 8906 + 74 + 19,230,576

5. Solve:

a. Subtract 823 from 1000.

b. Find the difference between 24,547 and 13,278.

c. How much more than 185 is 92,000,007?

6. Solve:

a. You have a balance of three thousand, eighty-five dollars in your bank account. If you deposit two checks for $415 each and write a check for $58, how much is left in your account?

b. Six hundred eighty-three is how much less than the sum of 187,227,963 and 35,472?

c. Jeffie is 53 years old. She had a child when she was 25, another 5 years later, and a third 4 years later. Her fourth child was born 9 years ago. Based on this information, give the difference in ages between the oldest and the youngest children.

7. Tell what each question is asking for—area, volume, or perimeter.

a. What is the distance around a triangle with sides 14 inches, 23 inches, and 28 inches?

b. A regular pentagon with a side of length 7 inches is being covered with fabric. How much is needed?

c. How much water can be poured into a can 16 centimeters tall and 10 centimeters in diameter?

2 Multiplying and Dividing Whole Numbers

In the National Hockey League, there are two conferences and four divisions. To determine which teams are in the playoffs, teams are given 2 points for each game they win, 1 point for each tie game and no points for a loss. So a team with a Win-Tie-Lose record of 39-10-3 has

$$\begin{aligned} 39(2) + 10(1) + 3(0) &= 78 + 10 + 0 \\ &= 88 \text{ points.} \end{aligned}$$

Then the teams are ranked according to those points. The regular season division champs (two in each conference) qualify. Then, the next six teams with the highest number of points qualify.

- *What other numerical methods are used to choose playoff positions for sports teams?*

Skills Check

Take this short quiz to see how well prepared you are for Chapter 2. The answers are given at the bottom of the page, along with the sections to review if you get any answers wrong.

1. Divide: $9\overline{)63}$

2. Multiply: 9×8

3. Divide: $56 \div 7$

4. Add:
$$\begin{array}{r} 1256 \\ \underline{+\,290} \end{array}$$

5. Subtract:
$$\begin{array}{r} 105 \\ \underline{-\;\;98} \end{array}$$

6. Fill in the missing numbers:
___ inches = 1 foot
___ minutes = 1 hour
___ ounces = 1 pound

ANSWERS: 1. 7 2. 72 3. 8 4. 1546 [Section 1.4] 5. 7 [Section 1.5] 6. 12, 60, 16
Items 1, 2, and 3 are basic multiplication and division facts and item 6 deals with measurement. We will cover this material in Chapter 2.

2.1 Multiplying by One-Digit Whole Numbers and Multiples of Ten

SECTION GOAL

- To multiply by one-digit whole numbers

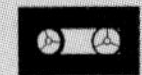 

Multiplication provides a quick way of calculating the result of adding the same number many times. For example, suppose you spend $3 a day on transportation. You can find how much you spend during the 5-day work week with either

Repeated Addition		*Multiplication*
$\$3 + \$3 + \$3 + \$3 + \$3 = \$15$	or	$5 \times \$3 = \$15$

     = 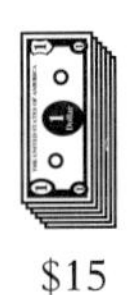

$3 $3 $3 $3 $3 = $15

Multiplication, then, is the same as repeated addition of the same number. But it allows us to find answers more quickly than we could with several additions.

The multiplication of two numbers a and b can be written in different ways:

$$a \times b = c \qquad \begin{array}{r} a \\ \underline{\times b} \\ c \end{array} \qquad (a)(b) = c \qquad a \cdot b = c$$

> In the multiplication $a \times b = c$, a is the **multiplicand,** b is the **multiplier,** and c is the **product;** a and b are also called **factors.**

Thus the multiplication of 2 and 3 can be written as

$$2 \times 3 = 6 \qquad \begin{array}{r} 2 \\ \underline{\times 3} \\ 6 \end{array} \qquad (2)(3) = 6 \qquad 2 \cdot 3 = 6$$

In each of these, 2 and 3 are the factors (3 is also the multiplier, and 2 the multiplicand). The result of the multiplication (here, 6) is the product.

A multiplication table shows the products of whole numbers with each other. If you memorize the products in Table 1 you will be able to multiply numbers faster.

TABLE 1 Multiplication Facts

×	*0*	*1*	*2*	*3*	*4*	*5*	*6*	*7*	*8*	*9*
0	0	0	0	0	0	0	0	0	0	0
1	0	1	2	3	4	5	6	7	8	9
2	0	2	4	6	8	10	12	14	16	18
3	0	3	6	9	12	15	18	21	24	27
4	0	4	8	12	16	20	24	28	32	36
5	0	5	10	15	20	25	30	35	40	45
6	0	6	12	18	24	30	36	42	48	54
7	0	7	14	21	28	35	42	49	56	63
8	0	8	16	24	32	40	48	56	64	72
9	0	9	18	27	36	45	54	63	72	81

Properties of Multiplication

Like addition, multiplication has several important and useful properties. One of them tells us that the result of the multiplication 5×3, which is

$$5 \times 3 = \underbrace{3 + 3 + 3 + 3 + 3}_{\text{5 times}} = 15$$

is exactly the same as the result of the multiplication 3×5:

$$3 \times 5 = \underbrace{5 + 5 + 5}_{\text{3 times}} = 15$$

In other words, changing the order of the factors does *not* change the product. This property is called the **commutative property of multiplication.** In symbols, we write

Commutative Property of Multiplication

If a and b are whole numbers, then

$$a \times b = b \times a$$

This property will save you time in learning the multiplication table, because you will have to learn only half the table, as you did for addition. The portion you should learn is highlighted in Table 1.

Whenever we multiply a number by 1, we always get the original number as the product. For this reason, the number 1 is called the **identity element for multiplication.** That is, $2 \times 1 = 2$ and $1 \times 23 = 23$ and $6{,}000{,}007 \times 1 = 6{,}000{,}007$ and so on. In symbols,

Identity Element for Multiplication

If a is any whole number, then

$$a \times 1 = 1 \times a = a$$

Recall that the number 0 (zero) acted as an identity for addition. In multiplication, the number 0 acts very differently. When we multiply by zero, we always get zero as an answer. That is, $4 \times 0 = 0$ and $0 \times 4 = 0$ and $2875 \times 0 = 0$ and so on. This constitutes the **multiplication property of zero.** In symbols,

Multiplication Property of Zero

If a is any whole number, then

$$a \times 0 = 0 \times a = 0$$

When we need to multiply more than two numbers together, we can multiply them together two by two in any order, and the final product will always be the same. We can illustrate this property with the multiplication $2 \times 4 \times 5$.

$$2 \times 4 \times 5 = (2 \times 4) \times 5 = 8 \times 5 = 40$$
$$2 \times 4 \times 5 = 2 \times (4 \times 5) = 2 \times 20 = 40$$

This property is called the **associative property of multiplication.** In symbols,

Associative Property of Multiplication

If a, b, and c are any whole numbers, then

$$(a \times b) \times c = a \times (b \times c)$$

Multiplying Numbers

Now let's use these properties (and the multiplication table) to multiply numbers that end in zeros—such as 10, 20, 30, . . ., 100, 200, and 1100.

EXAMPLE 1

Multiply: 800×7

1a. Multiply: 90×7

1b. Multiply: 6×5000

2a. Multiply: 700×60

2b. How many zeros will be in the product of 610 and 20,000?

SOLUTION

If we write 800 in expanded form, the problem becomes

$$\begin{array}{rcrcrcr} 800 & = & 8 \text{ hundreds} & + & 0 \text{ tens} & + & 0 \text{ ones} \\ \underline{\times\ 7} & & \underline{\times\ 7} & & \underline{\times\ 7} & & \underline{\times\ 7} \\ & & 56 \text{ hundreds} & + & 0 \text{ tens} & + & 0 \text{ ones} \end{array}$$

So $800 \times 7 = 5600$.

Because we know that zero times any number gives zero, we could have concentrated on the non-zero numbers only. We could have multiplied 7 times 8 and then "tagged on" two zeros to get 5600.

Study Hint

To multiply any number *ending* in zeros, simply multiply the non-zero digits and "tag on" the final zeros afterwards.

■ *Now do margin exercises 1a and 1b. Answers are at the back of the book.*

Use the short-cut method that's in the study hint.

EXAMPLE 2

Find the product: 900×60

SOLUTION

Here we have 9 hundreds times 6 tens. Our short cut allows us to multiply 9×6 and then "tag on" the three ending zeros.

$$\begin{array}{r} 900 \\ \underline{\times\ 60} \\ 54{,}000 \end{array}$$

So $900 \times 60 = 54{,}000$.

■ *Now do margin exercises 2a and 2b.*

Next we will multiply some numbers that do not end in zeros. We will use the study hint (about zeros) to estimate answers.

EXAMPLE 3

Multiply: $\begin{array}{r} 23 \\ \underline{\times\ 3} \end{array}$

SOLUTION

We can estimate the product as follows: Round 23 to 20 and estimate that the product should be about 20 times 3, or 60.

To do the actual multiplication, let's rewrite the multiplicand using expanded notation.

$$\begin{array}{rcrcr} 23 & = & 2 \text{ tens} & + & 3 \text{ ones} \\ \underline{\times 3} & & \underline{\times 3} & & \underline{\times 3} \end{array}$$

Multiplying each of the numbers (expanded), we get

$$\begin{array}{rcrcr} 23 & = & 2 \text{ tens} & + & 3 \text{ ones} \\ \underline{\times\ 3} & & \underline{\times\ 3} & & \underline{\times\ 3} \\ & & 6 \text{ tens} & + & 9 \text{ ones} \end{array}$$

Thus, $23 \times 3 = 69$.

Comparing this product with our estimate of 60, we see that 69 is a reasonable result.

We can work this same problem using a shorter method. First we multiply the ones.

$$\begin{array}{r} 23 \\ \updownarrow \\ \underline{\times\ 3} \\ 9 \end{array}$$

Multiply: $3 \times 3 =$
9

Next we multiply the tens.

$$\begin{array}{r} 23 \\ \underline{\times\ 3} \\ 69 \end{array}$$

Multiply: $3 \times 2 = 6$
Write 6.

The result is again 69.

■ *Now do margin exercises 3a and 3b.*

When the multiplications involve greater numbers, we often have to regroup (or carry), just as we did in addition.

EXAMPLE 4

Multiply: $\begin{array}{r} 87 \\ \underline{\times\ 9} \end{array}$

SOLUTION

We first estimate the product: Eighty-seven is about 90, and 9×90 is 810. The answer will be about 810—but it will be slightly smaller because we rounded up.

For the actual multiplication, we again rewrite in expanded form.

$$\begin{array}{rcrcr} 87 & = & 8 \text{ tens} & + & 7 \text{ ones} \\ \underline{\times\ 9} & & \underline{\times\ 9} & & \underline{\times\ 9} \\ & & 72 \text{ tens} & + & 63 \text{ ones} \end{array}$$

3a. Multiply: 24×2

3b. Fill in the missing digit:
$31 \times \_ = 93$

4a. Multiply:

$$\begin{array}{r} 69 \\ \times\ 7 \\ \hline \end{array}$$

4b. Multiply eighty-four thousand, six by nine.

Then $720 + 63 = 783$. So $87 \times 9 = 783$.

Note that the product, 783, is slightly smaller than 810, as we predicted.

Working this example in a shorter form requires carrying.

First we multiply the ones.

$$\begin{array}{r} {}^{6} \\ 87 \\ \times\ 9 \\ \hline 3 \end{array}$$

Multiply: $9 \times 7 = 63$
Write 3 and carry 6.

Next we multiply the tens.

$$\begin{array}{r} {}^{6} \\ 87 \\ \times\ 9 \\ \hline 783 \end{array}$$

Multiply: $9 \times 8 = 72$
Add the 6: $72 + 6 = 78$
Write 78.

So $9 \times 87 = 783$.

■ *Now do margin exercises 4a and 4b.*

Work the exercises that follow. The answers to selected exercises are given at the back of the book.

2.1 EXERCISES

In Exercises 1 through 20, multiply.

1. $\begin{array}{r} 19 \\ \times\ 5 \\ \hline \end{array}$

2. $\begin{array}{r} 82 \\ \times\ 4 \\ \hline \end{array}$

3. $\begin{array}{r} 16 \\ \times\ 7 \\ \hline \end{array}$

4. $\begin{array}{r} 17 \\ \times\ 8 \\ \hline \end{array}$

5. $3000 \times 90{,}000$

6. $6{,}000{,}000 \times 7000$

7. $500{,}000 \times 80{,}000$

8. $40{,}000 \times 900{,}000{,}000$

9. $\begin{array}{r} 71 \\ \times\ 6 \\ \hline \end{array}$

10. $\begin{array}{r} 18 \\ \times\ 6 \\ \hline \end{array}$

11. $\begin{array}{r} 63 \\ \times\ 3 \\ \hline \end{array}$

12. $\begin{array}{r} 54 \\ \times\ 2 \\ \hline \end{array}$

13. $\begin{array}{r} 293 \\ \times\ \ 3 \\ \hline \end{array}$

14. $\begin{array}{r} 345 \\ \times\ \ 8 \\ \hline \end{array}$

15. $\begin{array}{r} 242 \\ \times\ \ 4 \\ \hline \end{array}$

16. $\begin{array}{r} 674 \\ \times\ \ 2 \\ \hline \end{array}$

17. $10 \times 30 \times 60$

18. $20 \times 40 \times 30$

19. $10 \times 200 \times 4000$

20. $20 \times 3000 \times 50$

In Exercises 21 through 24, state the property that is illustrated.

21. $17 \times 3 = 3 \times 17$

22. $26 \times 0 = 0$

23. $15 \times 1 = 15$

24. $(3 \times 5) \times 2 = 3 \times (5 \times 2)$

In Exercises 25 through 28, find

25. The weight of 1705 5-pound bags of flour.

26. The total age of 79 8-year-old boys.

27. The number of buttons on 973 cards if there are 3 buttons on a card.

28. The price of 9 subway tokens when each costs 115 cents.

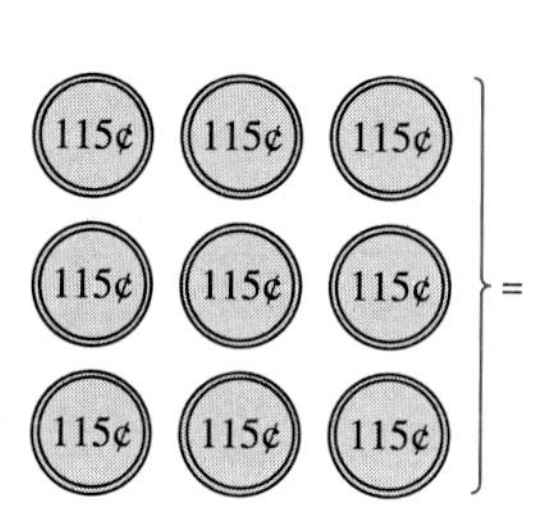

In Exercises 29 through 32, find the product of each pair of numbers.

29. Two million, twelve and 9.

30. 6 and five hundred thousand, sixteen.

31. Fifty thousand, eleven and 9.

32. Sixty-three thousand, five and 7.

In Exercises 33 through 40, multiply.

33. $10{,}660 \times 8$

34. 9×3569

35. $18{,}752 \times 7$

36. $29{,}735 \times 8$

37. 9898×6

38. 2059×5

39. $9 \times 19{,}191$

40. $7 \times 82{,}934$

In Exercises 41 through 48, find the product.

41. 2 and 17,392

42. 19,835 and 3

43. 45,002 and 7

44. 8 and 762,195

45. nine and 999,999

46. two and 893,798

47. seven and 18,796

48. 279,321 and eight

International Standard Book Numbers

During the year 1968, books began receiving an identification number consisting of 10 digits that are divided by dashes into 4 groups. Examples of these numbers include

0-8053-0865-2

0-395-29032-5

The first two groups identify the country and the publisher, the third is a sequence number, and the fourth is a check digit that can be used to ensure that the identification number has been recorded correctly. The numbers are known as International Standard Book Numbers, or ISBNs. To check that the digits in an ISBN are correct, use the following procedure.

Multiply the digits in order by 10, 9, 8, 7, and so on. Then add the products together and divide the sum by 11. If there is no remainder, the number is an actual ISBN.

If you want to see the ISBN number of this book, look at the outside back cover or the copyright page.

Let's check to see whether the number 0-395-29032-5 could be a correctly coded ISBN number.

First multiply the digits in order by 10, 9, 8, . . ., 1.

$0 \times 10 = 0$	0
$3 \times 9 = 27$	27
$9 \times 8 = 72$	72
$5 \times 7 = 35$	35
$2 \times 6 = 12$	12
$9 \times 5 = 45$	45
$0 \times 4 = 0$	0
$3 \times 3 = 9$	9
$2 \times 2 = 4$	4
$5 \times 1 = 5$	5
Add the products:	209

Divide the sum by 11: $209 \div 11 = 19$. There is no remainder, so this number could be an ISBN number.

Check the following numbers to see which could be ISBN numbers.

0-13-845918-2 no　　0-395-35368-8 yes　　85-982-3664-0 yes

2.2 Multiplying by Larger Whole Numbers

SECTION GOALS

- To find the product of two or more whole numbers
- To use estimation to approximate products

When we begin multiplying by numbers of two or more digits, we need to use another property of addition and multiplication, the **distributive property.** In symbols,

Distributive Property

If a, b, and c are any whole numbers, then

$$a(b+c)=a\times b+a\times c$$

This property allows us to do a multiplication like 71×63 by splitting the multiplication into two problems, 71×60 and 71×3, and adding the results. To show you how it works, we will first do the multiplication in two parts and then use the short form. When you actually do your multiplications, always use the short form.

EXAMPLE 1

Find the product: $\begin{array}{r} 71 \\ \times\ 63 \\ \hline \end{array}$

SOLUTION

Because 71 is close to 70 and 63 is close to 60, a reasonable estimate of the product is 70×60, or 4200. For the actual multiplication, we have

$$\begin{array}{r} 71 \\ \times\ 63 \\ \hline \end{array} = \begin{array}{r} 71 \\ \times\ 60 \\ \hline 4260 \end{array} + \begin{array}{r} 71 \\ \times\ 3 \\ \hline 213 \end{array} = 4473$$

This answer, 4473, is close to our estimate of 4200.

Follow along as we redo the same problem.

First we multiply by 3.

$$\begin{array}{r} 71 \\ \times\ 63 \\ \hline 213 \end{array}$$

Multiply: $3\times 71=213$
Write 213 with the last digit under 3.

The result, 213, is called a **partial product.** It is the result of an intermediate step in the multiplication process.

Next we multiply by 6 tens.

$$\begin{array}{r} 71 \\ \times\ 63 \\ \hline 213 \\ 426 \end{array}$$

Multiply: $6\times 71=426$
Write 426 with the last digit in the tens column.

1a. Multiply:
$$\begin{array}{r} 31 \\ \times\ 67 \\ \hline \end{array}$$

1b. Show the first partial product of 23 x 13.

To complete the problem, we add the partial products.

Note that we add as though there were a zero in the second partial product, under the 3. That's because 6 tens times 71 is 426 *tens*.

$$\begin{array}{r} 71 \\ \times\ 63 \\ \hline 213 \\ 4260 \\ \hline 4473 \end{array} \qquad \text{Add: } 213 + 4260 = 4473$$

■ *Now do margin exercises 1a and 1b.*

In the previous example, no carrying was needed. If you must carry, remember to add the carried number.

EXAMPLE 2

Find the product:
$$\begin{array}{r} 849 \\ \times\ 76 \\ \hline \end{array}$$

SOLUTION

A reasonable estimate here would be 800 times 80, or 64,000.

We begin by multiplying 849 by 6.

$$\begin{array}{r} {}^{5} \\ 849 \\ \times\ 76 \\ \hline 4 \end{array} \qquad \begin{array}{l} \text{Multiply: } 6 \times 9 = 54 \\ \text{Write 4 and carry 5.} \end{array}$$

$$\begin{array}{r} {}^{25} \\ 849 \\ \times\ 76 \\ \hline 94 \end{array} \qquad \begin{array}{l} \text{Multiply: } 6 \times 4 = 24 \\ \text{Add: } 24 + 5 = 29 \\ \text{Write 9 and carry 2.} \end{array}$$

$$\begin{array}{r} {}^{25} \\ 849 \\ \times\ 76 \\ \hline 5094 \end{array} \qquad \begin{array}{l} \text{Multiply: } 6 \times 8 = 48 \\ \text{Add: } 48 + 2 = 50 \\ \text{Write 50.} \end{array}$$

This number 5094 is the result of multiplying 6 × 849. It is a partial product.

Now we have to multiply by 7 tens.

$$\begin{array}{r} {}^{6} \\ 849 \\ \times\ 76 \\ \hline 5094 \\ 3 \end{array} \qquad \begin{array}{l} \text{Multiply: } 7 \times 9 = 63 \\ \\ \text{Write 3 and carry 6.} \end{array}$$

Because we were multiplying by 7 tens, we placed the 3 in the tens column. Continuing we have

```
 3 6
 849
  ↕
× 76     Multiply: 7 × 4 = 28
5094     Add: 28 + 6 = 34
 43      Write 4 and carry 3.

 3 6
 849
  ↘
× 76     Multiply: 7 × 8 = 56
5094     Add: 56 + 3 = 59
5943     Write 59.
```

The second partial product is 59430, even though we did not write the zero. The alignment takes care of it. (You can put in the zero, however, if you want to.)

Then we add the two partial products to get the product.

```
  849
×  76
 5094
59430
64524
```

This answer, 64,524, is very close to our estimate of 64,000, so it seems reasonable.

■ *Now do margin exercises 2a and 2b.*

When you work problems that contain zeros either in the multiplier or in the multiplicand, you must be extra careful to align your partial products with the correct columns. The next example illustrates some of the difficulties.

EXAMPLE 3

Multiply:

```
 4703
×2008
```

SOLUTION

Estimating, we get 5000 times 2000, which is 10,000,000. To multiply, we find the partial products and add.

```
   4703
 × 2008
  37624     4703 × 8
  0000      4703 × 0 tens
 0000       4703 × 0 hundreds
9406        4703 × 2 thousands
9443624     Adding
```

2a. Multiply:

```
 498
× 67
```

2b. Find the missing digit:

```
 126
× 3?
 882
378
4662
```

3a. Multiply:
$$\begin{array}{r} 4579 \\ \underline{\times 3008} \end{array}$$

3b. Multiply:
$$\begin{array}{r} 205{,}070 \\ \underline{\times \quad 4002} \end{array}$$

4a. Multiply 2356 by 1346 using your calculator.

4b. Multiply using your calculator: $459 \times 12 \times 46$

You can, if you like, eliminate almost all of the middle two rows of zeros.

$$\begin{array}{r} 4703 \\ \underline{\times 2008} \\ 37624 \\ \underline{940600} \\ 9443624 \end{array}$$

Because 9,443,624 is close to 10,000,000, our answer is reasonable.

Use the procedure you do best with, but always remember to align partial products in the correct place-value columns.

■ *Now do margin exercises 3a and 3b.*

EXAMPLE 4

Multiply: $416 \times 836 \times 27$ using your calculator.

SOLUTION

With pencil and paper, you would have to multiply two of these factors together and then multiply their product by the third factor. With a calculator, you can do it all at once.

Key in

416 [×] 836 [×] 27 [=]

Your display should read 9389952. The product is 9,389,952.

■ *Now do margin exercises 4a and 4b.*

Work the exercises that follow.

2.2 EXERCISES

In Exercises 1 through 28, multiply as indicated. Use your calculator to check your answers.

1. 305×23
2. 409×35
3. 607×58
4. 208×39
5. 789×27
6. 945×57
7. 687×98
8. 823×79
9. 323×53
10. 490×85
11. 670×68
12. 324×19
13. 9701×78
14. 5187×59
15. 3209×26
16. 7495×67
17. 518×94
18. 697×95
19. 736×87
20. 948×69
21. 5870×98
22. 2853×45
23. 3980×96
24. 3664×72
25. $76{,}513 \times 429$
26. $89{,}745 \times 863$
27. $118{,}354 \times 762$
28. $235{,}413 \times 726$

For Exercises 29 through 32, use your calculator. Rearrange the digits 6, 7, 8, and 9 into two 2-digit numbers whose product is the given number.

29. 5372 **30.** 6566 **31.** 6764 **32.** 6003

In Exercises 33 through 38, find the missing digits.

33.
```
   1_7
 × 52
   274
  685
  ----
```

34.
```
   207
 × __
  1449
  2__
  3__9
```

35.
```
   78_
 × 5_
  3__00
```

36.
```
   69_
 × 7_
  _8_0_
```

37.
```
   62_
 × _7
  43__
  18__
  23__5
```

38.
```
   5__
 × _9
  49__
  1_9_
  __805
```

In Exercises 39 and 40, use estimation to write a multiplication problem that will give you a product close to, but not equal to, each number.

39. seven million **40.** two hundred thousand

In Exercises 41 through 44, multiply. Be careful of the zeros.

41.
```
  1,001,001
 ×      603
```

42.
```
  500,600
 ×   6003
```

43.
```
  600,070
 ×   8050
```

44.
```
  2,000,060
 ×     7060
```

In Exercises 45 through 48, multiply. Estimate first, and use the associative property where you need to.

45. $22 \times 45 \times 9$ **46.** $18 \times 7 \times 19$ **47.** $62 \times 38 \times 90$ **48.** $72 \times 99 \times 56$

In Exercises 49 through 60, multiply.

49. (972) (55,289)

50. 74×97

51. (9774) (4264)

52. (4444) (50,000)

53. 23×45

54. (471) (33,729)

55. 85×87

56. 52×68

57. (19) (30) (20)

58. (30) (49) (50)

59. (2222) (5000)

60. (4132) (5644)

61. Do the first few of the following multiplications. Try to find a pattern and predict the answers to the last few. Use your calculator to find other patterns.

a. $7 \times 11 \times 13$

b. $7 \times 11 \times 13 \times 2$

c. $7 \times 11 \times 13 \times 3$

d. $7 \times 11 \times 13 \times 4$

e. $7 \times 11 \times 13 \times 7$

f. $7 \times 11 \times 13 \times 9$

62. Problems a through d exhibit a new pattern. What is it?

a. 25×11

b. 25×111

c. 25×1111

d. $25 \times 1{,}111{,}111{,}111$

In Exercises 63 through 66, test your multiplication skill.

63. $\begin{array}{r} 82{,}909 \\ \times\ \underline{\ 7{,}981} \end{array}$

64. $\begin{array}{r} 883{,}295 \\ \times\ \underline{\ 6{,}684} \end{array}$

65. $\begin{array}{r} 1{,}987{,}753 \\ \times\ \underline{\ 88{,}959} \end{array}$

66. $\begin{array}{r} 12{,}345{,}678 \\ \times\ \underline{\ 801{,}020} \end{array}$

2.2 MIXED PRACTICE

By doing these exercises, you will practice all topics up to this point in the chapter.

67. Multiply: 326×5

68. Multiply: $\begin{array}{r} 1685 \\ \underline{\times 2004} \end{array}$

69. Multiply: $2{,}376 \times 3000 \times 17$

70. Estimate the product: 105×27

71. Multiply: $\begin{array}{r} 361 \\ \underline{\times\ 402} \end{array}$

72. Multiply: $60{,}000 \times 800$

73. Estimate the product: 1104×54

74. Multiply: $4{,}608 \times 239$

75. Multiply: $157 \times 24 \times 31$

2.3 Dividing by One- and Two-Digit Whole Numbers and Multiples of Ten

SECTION GOAL

- To find the quotient when dividing by one- or two-digit numbers

Recall that addition and subtraction are "opposite" operations because one undoes, or reverses, the other. The next operation we discuss, **division,** undoes multiplication. For example, we multiply 3 by 4 to get 12, and we divide 12 by 4 to get back to 3.

$$3 \times 4 = 12$$
$$12 \div 4 = 3$$

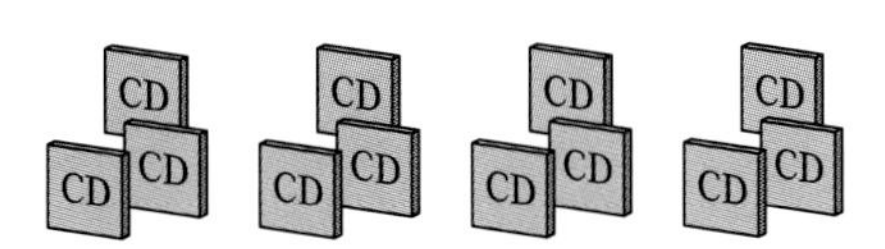

Division

If a, b, and c are non-zero whole numbers, and $a \times b = c$, then

$$c \div a = b \quad \text{and} \quad c \div b = a$$

Using the facts of multiplication, you can probably already divide many numbers mentally, such as $24 \div 8 = 3$ (because $3 \times 8 = 24$), $56 \div 7 = 8$, $81 \div 9 = 9$, and so on.

The commutative property does not hold for division. For example, $12 \div 2$ and $2 \div 12$ do not give the same answer.

We can write division in several ways:

$$a \div b \qquad a/b \qquad \frac{a}{b} \qquad b\overline{)a}$$

Each of these is read "a divided by b."

In the division $a \div b = c$, a is called the **dividend,** b is called the **divisor,** and c is the **quotient.**

Therefore, the division $32 \div 8 = 4$ may also be written

$$\begin{array}{r} 4 \\ 8\overline{)32} \end{array} \qquad 32/8 = 4 \qquad \frac{32}{8} = 4$$

In each of these, 8 is the divisor, 32 is the dividend, and 4 is the quotient.

Division by zero is undefined. That is, we cannot divide by zero.

Zero divided by any non-zero number is zero.

That is, $0 \div 6 = 0$, $0 \div 47 = 0$, and so on.

Let's look at an example of division by a one-digit whole number.

EXAMPLE 1

Divide: $369 \div 3$

SOLUTION

To solve the problem, we first rewrite it in the form

$$3\overline{)369}$$

which makes computing easier. We next estimate the size of the quotient. The quotient of this problem will be between 100 and 200. Why? Because 3 times 100 is 300 and 3 times 200 is 600 and the given dividend is between 300 and 600.

$$\begin{aligned} 3 \times 100 &= 300 \\ 3 \times \text{quotient} &= 369 \\ 3 \times 200 &= 600 \end{aligned}$$

Therefore, the first digit in the quotient will be 1. We place it over the hundreds digit in the dividend. Then we multiply it by the divisor and place the result under the 3 in the dividend.

$$\begin{array}{r} 1 \\ 3\overline{)369} \\ \underline{3} \end{array} \qquad \begin{array}{l} 3 \div 3 = 1 \\ \\ 1 \times 3 = 3 \end{array}$$

Now we subtract: 3 minus 3 is zero.

Usually, we do not write down the zero, but we will here. Then we bring down the next number in the dividend, here the 6. Be careful to bring down only one number at a time and to keep it in its own column.

$$\begin{array}{r} 1 \\ 3\overline{)369} \\ \underline{3}\downarrow \\ 06 \end{array} \qquad \text{Bring down the 6.}$$

Next we ask, what is 6 divided by 3?

The answer is 2 because $2 \times 3 = 6$. We place a 2 in the quotient over the 6 in the dividend; then we multiply it by the divisor and place the result, 6, under the 6 in the dividend.

```
   12      6 ÷ 3 = 2
3)369
  3
   6
   6       2 × 3 = 6
```

Now we subtract: 6 minus 6 is zero.

We bring down the 9.

```
   12
3)369
  3
   6
   6↓
   09      Bring down
           the 9.
```

Next we consider $9 \div 3$. This gives us 3, the last number in the quotient. Multiplying it by the divisor and subtracting leave us with a zero.

```
   123     9 ÷ 3 = 3
3)369
  3
   6
   6
    9
    9      3 × 3 = 9
    0      9 − 9 = 0
```

1a. Divide: $4\overline{)284}$

1b. Divide: $6\overline{)546}$

Because there are no other numbers to bring down, and because zero remains after subtraction, we have completed the problem.

Our estimate (between 100 and 200) indicates that the answer is reasonable. However, we can check it by multiplying the quotient and the divisor together to get the dividend.

$$3 \times 123 = 369$$

■ *Now do margin exercises 1a and 1b.*

The form we used in solving Example 1 is called **long division.** We could have divided using the **short division** form.

$$\begin{array}{r}1\\3\overline{)369}\end{array} \quad 3 \div 3 \quad \rightarrow \quad \begin{array}{r}12\\3\overline{)369}\end{array} \quad 6 \div 3 \quad \rightarrow \quad \begin{array}{r}123\\3\overline{)369}\end{array} \quad 9 \div 3$$

Note also that if we rewrote 369 as 300 + 60 + 9, divided each of those numbers by our original divisor 3, and added the **partial quotients,** 100 + 20 + 3, we would get the same answer. Try it with some other problems. The number property that makes this true is the *distributive property,* discussed in Section 2.2.

Let's look at an example with a two-digit divisor.

EXAMPLE 2

Divide: $3872 \div 36$

SOLUTION

Rewriting our problem, we have

$$36\overline{)3872}$$

We then estimate and find that the answer will be between 100 and 200, because 100×36 is 3600 and 200×36 is 7200 and our dividend, 3872, lies between 3600 and 7200.

$$\begin{aligned}36 \times 100 &= 3600\\36 \times \text{quotient} &= 3872\\36 \times 200 &= 7200\end{aligned}$$

We think, how many times does 36 go into 38? That is, how many 36's are there in 38? One. We place 1 over the 8 to indicate that this was the result of dividing into *38* and not into 3 or 387. Then we multiply 1 times 36, write the result under 38, and subtract.

```
    1        38 ÷ 36 = 1
36)3872
   36        Multiply: 1 × 36 = 36
    2        Subtract: 38 – 36 =2
```

Next we bring down the next number from the dividend, 7.

```
    1
36)3872
   36↓
    27
```

How many 36's are there in 27? Wait. 36 will not divide into 27 because 27 is smaller than 36. So we place a zero in the quotient over the 7, and bring down the next digit from the dividend, 2.

```
    10
36)3872
   36 ↓
    272
```

To complete the problem, we determine (by guessing and trying) that 36 will divide into 272 seven times. We write the 7, multiply 7 × 36, and subtract to obtain

```
    107
36)3872
   36
    272
    252      Multiply: 7 × 36 = 252
     20      Subtract: 272 – 252 = 20
```

2a. Divide: $12\overline{)6110}$

2b. Which will give the largest quotient?

$21\overline{)84021}$

$21\overline{)8421}$

$21\overline{)80421}$

Note that here we do not get zero as a result of the last subtraction. We get 20 as a **remainder.** Because this remainder is less than the divisor, and there is nothing left to bring down, we stop. We say that our answer is "107 with a remainder of 20." We write this as 107 r 20.

■ *Now do margin exercises 2a and 2b.*

Remember

You have completed the division when the remainder is smaller than the divisor and no more numbers remain to be brought down.

To check a division when there is a remainder, first multiply the quotient and the divisor. Then add the remainder to get the dividend. Try it for Example 2.

Work the exercises that follow. Be sure to check your answers by multiplying.

2.3 EXERCISES

In Exercises 1 through 32, divide. Be sure to include the remainder, if there is one, in your answer. Check your answers by multiplying.

1. $5\overline{)95}$ **2.** $8\overline{)96}$ **3.** $6\overline{)78}$ **4.** $7\overline{)91}$

5. $3\overline{)113}$ **6.** $2\overline{)413}$ **7.** $4\overline{)615}$ **8.** $5\overline{)477}$

9. $313 \div 8$ **10.** $315 \div 9$ **11.** $513 \div 9$ **12.** $659 \div 6$

13. 216/7 **14.** 511/8 **15.** 321/6 **16.** 822/7

17. $6\overline{)4734}$ **18.** $9\overline{)4563}$ **19.** $9\overline{)6183}$ **20.** $7\overline{)4886}$

21. $7\overline{)4256}$ **22.** $6\overline{)2430}$ **23.** $8\overline{)2448}$ **24.** $8\overline{)7736}$

25. $3\overline{)978,003}$ **26.** $5\overline{)935,025}$ **27.** $4\overline{)720,048}$ **28.** $2\overline{)290,006}$

29. $39,287 \div 28$ **30.** $69,834 \div 51$ **31.** $59,666 \div 72$ **32.** $76,821 \div 85$

In Exercises 33 through 38, divide to find the missing multiplier. (Use a calculator if you wish.)

33. $838500 = 12 \times$ ____

34. $1066658 =$ ____ $\times 79$

35. $12450332 =$ ____ $\times 83$

36. $283554 = 27 \times$ ____

37. ____ $\times 65 = 6283355$

38. $54 \times$ ____ $= 4457916$

In Exercises 39 through 44, find the quotient.

39. 999999 and 99

40. 253110 and 39

41. 2784782 and 91

42. 831577 and 83

43. 7112721 and 77

44. 2679803 and 29

2.3 MIXED PRACTICE

By doing these exercises, you will practice all topics up to this point in the chapter.

45. Multiply: 7650×90

46. Multiply: 274×19

47. Estimate: $36072 \div 18$

48. Divide: $5250 \div 42$

49. Multiply: $703 \times 800 \times 20$

50. Estimate: 12×7085

51. Divide four thousand, fifty by nine.

52. What is the remainder when one million is divided by ninety?

2.4 Dividing by Larger Whole Numbers

SECTION GOALS

- To find the quotient of two whole numbers
- To use estimation to approximate quotients

You will probably use a calculator when the divisor has three or more digits. Nonetheless, you should be able to do such division problems with pencil and paper.

EXAMPLE 1

Divide: $316\overline{)74490}$

SOLUTION

Dividing 75,000 by 300 gives us an estimated answer of about 250. Hence, we start the problem by placing 2 over the hundreds place (because there are 2 hundreds in the quotient). When we multiply by the divisor and subtract we get

```
      2
316)74490
    632       2 × 316 = 632
    112       744 − 632 = 112
```

We bring down the next number from the dividend, 9.

```
      2
316)74490
    632↓
    1129
```

Now, we know that 1129 divided by 316 is more than 2. We try 3.

```
      23
316)74490
    632
    1129
     948      3 × 316 = 948
     181      1129 − 948 = 181
```

We know the 3 is correct because 181 is less than 316. We bring down the 0. Continuing, we try 5 (the result of) for 1810 divided by 316.

```
      235
316)74490
    632
    1129
     948↓
     1810
     1580     5 × 316 = 1580
      230     1810 − 1580 = 230
```

1a. Divide: $524\overline{)23948}$

1b. Find the quotient when 96857 is divided by 249.

Because nothing remains to be brought down, and because 230 is less than the divisor, the division is completed.

So 74490 ÷ 316 is 235 r 230.

To check, we multiply the quotient (235) times the divisor (316) and add the remainder (230) to obtain the dividend (74,490).

■ *Now do margin exercises 1a and 1b.*

In Example 1 we always picked the right number for the quotient. But suppose we picked the wrong number—say 4 instead of 3—for the second digit. Look at the result:

$$\begin{array}{r} 24 \\ 316\overline{)74490} \\ \underline{632} \\ 1129 \\ \underline{1264} \end{array} \qquad \begin{array}{l} 4 \times 316 = 1264 \\ 1129 - 1264 = ? \end{array}$$

We cannot subtract because 4 times 316 is 1264, a number *greater* than the number we have to subtract from. This tells us that we have chosen too great a number for the quotient; we should try a smaller number.

Let's go back to the same step and choose a number that is too small—say 2. The result is

$$\begin{array}{r} 22 \\ 316\overline{)74490} \\ \underline{632} \\ 1129 \\ \underline{632} \\ 497 \end{array} \qquad \begin{array}{l} 2 \times 316 = 632 \\ 1129 - 632 = 497 \end{array}$$

This time the result after subtraction, 497, is greater than the divisor! This tells us that the partial quotient we chose is too low; we must try a greater one.

The next example has several zeros in the quotient. They require special care.

EXAMPLE 2

Find the quotient: $167\overline{)175,350}$

SOLUTION

Here the quotient is difficult to estimate. But the three-digit divisor is less than the first three digits of the dividend (that is, 167 < 175), so we can place 1 in the quotient, over the 5. Then we multiply 1×167 and subtract yielding 8.

$$\begin{array}{rl} 1 & \text{Write 1.} \\ 167\overline{)175350} & \\ \underline{167} & 1 \times 167 = 167 \\ 8 & 175 - 167 = 8 \end{array}$$

Bringing down the 3 gives us 83.

$$\begin{array}{r} 1 \\ 167\overline{)175350} \\ \underline{167}\downarrow \\ 83 \end{array}$$

83 cannot be divided by 167, so we place a zero in the quotient and bring down the next digit, 5.

$$\begin{array}{rl} 10 & \text{Write 0.} \\ 167\overline{)175350} & \\ \underline{167}\downarrow & \\ 835 & \text{Bring down 5.} \end{array}$$

835 divided by 167 is 5. Multiplying 167×5, we get 835, so subtracting gives zero.

$$\begin{array}{rl} 105 & \text{Write 5.} \\ 167\overline{)175350} & \\ \underline{167} & \\ 835 & \\ \underline{835} & 5 \times 167 = 835 \\ 0 & 835 - 835 = 0 \end{array}$$

We are not through yet; we still have to bring down the final zero in the dividend.

$$\begin{array}{rl} 105 & \\ 167\overline{)175350} & \\ \underline{167} & \\ 835 & \\ \underline{835}\downarrow & \\ 00 & \text{Bring down zero.} \end{array}$$

Because 167 divides into 0 exactly 0 times, we have

$$\begin{array}{rl} 1050 & \text{Write 0.} \\ 167\overline{)175350} & \\ \underline{167} & \\ 835 & \\ \underline{835} & \\ 00 & \end{array}$$

2a. Divide: $231\overline{)942{,}480}$

2b. Divide 240,000 by 3000.

3a. Divide with your calculator: $2536\overline{)4{,}757{,}549}$

3b. Use your calculator to find the missing number: $567{,}363 \div 342 = ?$ r 327

Our answer is 1050 with no remainder. You can check it by using your calculator.

■ *Now do margin exercises 2a and 2b.*

Sometimes your quotient may have a remainder. To find the remainder when using a calculator, there is a procedure that you can use. Follow along in the next example. We will first solve the problem using paper and pencil and then using the calculator.

EXAMPLE 3

Divide: $7792\overline{)7{,}040{,}072}$

SOLUTION

The quotient will be less than 1000 because 7792 > 7040. It will probably be greater than 900. Thus, we begin with a 9 in the quotient over the hundreds place and then continue the division through until the end.

$$\begin{array}{r l}
903 & \\
7792\overline{)7040072} & \\
\underline{70128} & 9 \times 7792 = 70128 \\
27272 & \\
\underline{23376} & 3 \times 7792 = 23376 \\
3896 &
\end{array}$$

The quotient is 903 r 3896.

To solve using the calculator, key in

7040072 [÷] 7792 [=]

Your display should read 903.5. The whole number quotient is 903.

To find the remainder, multiply 7792 by 903, the whole-number part of the quotient.

7792 [×] 903 [=]

Your display should read 7036176. Record your answer and clear the screen.

Then subtract

7040072 [−] 7036176 [=].

Your display should read 3896. This is your remainder; therefore, the answer is 903 r 3896.

■ *Now do margin exercises 3a and 3b.*

Work the exercises that follow. Be sure to check your answers by using multiplication.

2.4 EXERCISES

In Exercises 1 through 12, divide.

1. $170\overline{)17340}$
2. $180\overline{)54720}$
3. $506\overline{)105754}$
4. $308\overline{)218064}$
5. 624923 ÷ 689
6. 584672 ÷ 968
7. 386807 ÷ 769
8. 706056 ÷ 876
9. 240,000/60
10. 630,000/70,000
11. 27,000/9000
12. 210,000/700

In Exercises 13 through 24, divide. Show any remainder.

13. $117\overline{)6822}$
14. $287\overline{)6177}$
15. $395\overline{)8000}$
16. $821\overline{)9300}$
17. 38570 ÷ 203
18. 64800 ÷ 160
19. 247049 ÷ 407
20. 409354 ÷ 809
21. $\frac{200{,}000}{4000}$
22. $\frac{10{,}000{,}000}{50{,}000}$
23. $\frac{120{,}000{,}000}{60{,}000}$
24. $\frac{8{,}000{,}000}{20{,}000}$

In Exercises 25 through 28, use your calculator to divide.

25. 1,011,616 ÷ 2,504
26. 9,635,241÷ 11,111
27. 1,354,866 ÷ 12,121
28. 62,500,120 ÷ 190

In Exercises 29 through 32, each of the problems has missing digits. Find them.

29.
2_ _
_7)74_ _
_ _
_ _
00
0

30.
_ _ _
4_)92_ _
92
_ _
00
0

31.
_0_
7_)829_
_9
_ _5
3_ _
0

32.
_0_
8_)92_5
_5
_6_
7_ _
0

In Exercises 33 and 34, what is the quotient:

33. When one million is divided by 9278?

34. When 103,092 is the dividend and 9999 is divisor?

In Exercises 35 and 36, what is the remainder:

35. When 650,083 is divided by 10,902?

36. When 235,211 is divided by 61,133?

2.4 MIXED PRACTICE

By doing these exercises, you will practice all topics up to this point in the chapter.

37. Estimate: $954 \div 9$

38. Divide: 12)2760

39. Multiply: 1682×430

40. Divide: 28)32,400

41. Estimate: 789×103

42. Divide: $1792 \div 7$

43. Multiply: 2467×56

44. Divide: 129)93,968

45. Multiply: $98,045 \times 20,000$

2.5 Applying Multiplication and Division of Whole Numbers: Word Problems

Most word problems that require the multiplication of whole numbers represent one of four types.

Multiplication applications involve

a. *Repeated addition* (example: to find the total cost of 6 apples at 5 cents each)
b. *Scaled increase* (example: to find the height of a house that is 4 times taller than a second)
c. *Combinations* (example: to find the number of ways we can match 3 kinds of soda in an ice cream soda)
d. *Finding area*

We will look at the first three types in this section. The fourth type, area problems, is discussed in Section 2.7.

EXAMPLE 1

On September 16, 1988, the wind was blowing at a speed of 8 miles per hour in New York City. In Mexico, Hurricane Gilbert reached a wind speed of 20 times that. How fast were Gilbert's winds?

SOLUTION

This is a *scaled increase.* To find the wind speed of Hurricane Gilbert, we multiply the given speed (8 miles per hour) by the scale factor (20 times).

$$\begin{array}{r} 20 \\ \times\ 8 \\ \hline 160 \end{array}$$

The unit here is miles per hour. The wind was blowing at 160 miles per hour.

■ *Now do margin exercises 1a and 1b.*

Here is a *repeated addition* problem that would be solved in exactly the same way as Example 1: What is the total weight of 8 turkeys that weigh 20 pounds each?

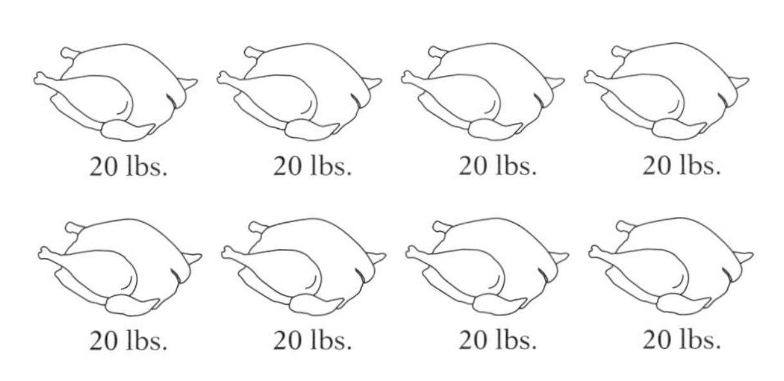

Here is a *combination* problem that would also be solved in exactly the same way: Rugged X pickup trucks are available in 20 different exterior colors and 8 different interior colors. How many combinations of one interior and one exterior color can I choose?

SECTION GOAL

■ To solve applications of multiplication or division of whole numbers.

1a. A tennis lesson costs about \$15 an hour in Orlando, Florida. In New York City, some people pay three times that for a piano lesson. What does a piano lesson cost in New York City?

1b. The original price of a 1955 Chevrolet Bel-Air hardtop was about \$2000. The restored value is six times that. What is the restored value?

1c. How much would you have to pay for a dozen (12) pens that cost $11 each?

1d. You have five pairs of jeans and 13 sweaters. How many different outfits can you make with a sweater and a pair of jeans?

2a. There are 144 pencils in a gross of pencils. How many packages of a dozen pencils each will that make?

2b. Apple boxes are made to hold 72 apples each. If you have 7272 apples, how many boxes can you pack to sell?

2c. A certain tank stores 17 times as much fuel oil as another. If the larger tank holds 9520 gallons, how much does the smaller tank hold?

2d. How many $11 pens can you buy with $132?

■ *Now do margin exercises 1c and 1d.*

Word problems that require division also describe four general types of situations.

Division applications involve

a. *Splitting up* a number of elements into parts (example: to divide 30 pens among 6 people)
b. *Repeated subtraction* (example: 45 balloons given away 9 at a time)
c. *Scaled decrease* (the opposite of scaled increase)
d. *Calculating rates* (such as number of miles driven on one gallon of gas)

We shall look at the first three types here and at the fourth type—rates—in Chapter 7.

EXAMPLE 2

At the Math Center library, fifteen shelves were set aside to hold 1800 books. If each shelf held the same number of books, how many books did each hold?

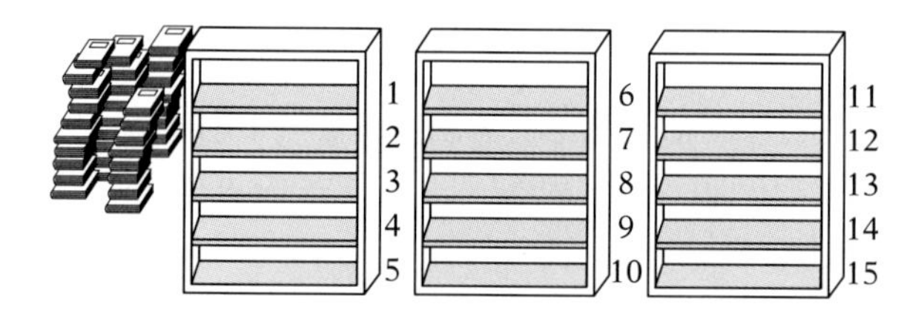

SOLUTION

This is a *splitting up* situation. To find the total number of books per shelf, we divide the total number of books (1800) by the number of shelves (15). Using long division, we obtain

$$\begin{array}{r} 120 \\ 15\overline{)1800} \\ \underline{15} \\ 30 \\ \underline{30} \\ 00 \end{array}$$

Therefore, each shelf will hold 120 books.

■ *Now do margin exercises 2a and 2b.*

Here is a *repeated subtraction* problem that would be solved in exactly the same way as Example 2: How many folding chairs, at $15 each, can you buy with $1800?

Here's a *scaled decrease* problem that is solved just like Example 2: Stewart's Peak is 1800 feet tall and is fifteen times taller than Red Hill. How tall is the hill?

■ *Now do margin exercises 2c and 2d.*

Work the exercises that follow. Remember to check that your answers make sense in each problem situation.

2.5 EXERCISES

Solve each of the following word problems. Be sure to read each problem carefully before starting your computations. Be careful; some addition and subtraction problems have been included.

1. A recent study of Cuba showed that there were approximately 10,000,000 people living in a country that had an area of roughly 40,000 square miles. About how many people is this per square mile?

2. The distance from Omaha to Buffalo is twice as far as the distance from Buffalo to Indianapolis. If the distance between Buffalo and Indianapolis is 510 miles, how far is it from Omaha to Buffalo?

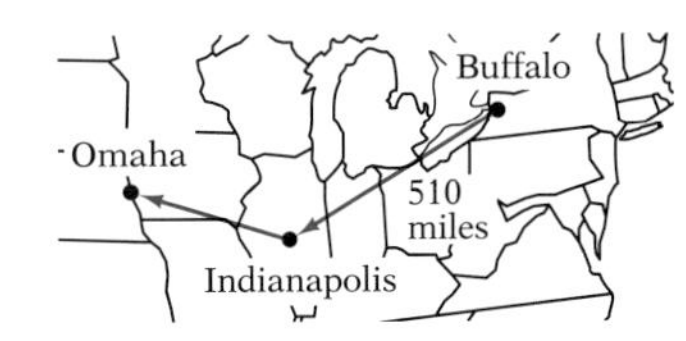

3. A certain game has chips with round, square, rectangular, and triangular shapes. If there are ten different colors for each shape, how many chips are there altogether?

4. Ellen and Shepherd's picnic set includes knives, forks, spoons, and iced tea spoons. If the set states on the package that there are enough utensils for 16 people, how many utensils are in the set?

5. King Kong was 50 feet tall. The Empire State Building is 1472 feet tall. How many apes of this size, standing feet to head, would it take to reach the height of the Empire State Building?

6. The Verrazano Narrows Bridge is how many times longer than the width of the Queen Elizabeth ocean liner if the bridge is 4260 feet long and the ocean liner is 119 feet wide? Round your answer to the nearest whole number.

7. If 1700 new corporations are started each day in the United States, how many are started in a year (365 days)?

8. A survey indicates that a husband and wife spend an average of 14 minutes a day in "meaningful" conversation. How many minutes of such conversation will they have in 50 years (about 18,263 days including leap years)?

9. The population in New York City was 7,071,639 in 1980 and 7,895,563 in 1970. How many more people were there in New York in 1970?

10. The heaviest weight ever lifted by a woman was lifted by Josephine Blatt in 1895. She lifted the equivalent of 26 women of average weight. If a woman of average weight weighs 135 pounds, how much weight did Blatt lift?

11. In 1985 a man playing golf with only one club made a score of 73 on the 6,037-yard course. If each of his 73 strokes caused the ball to go the same distance, about how many yards per stroke did the ball go? Give your answer to the nearest whole number.

12. In a certain city in Tennessee, $221,400 is paid out in council members' salaries. If each council member's salary is $5400, how many people are there on the council?

13. The most points ever scored by one team over another in a football game was scored by Georgia Tech in 1916. The total number of points scored was 222. If a touchdown gives a team 6 points, what is the greatest number of touchdowns that could have been scored by Tech?

14. The design of a coin may not be changed more often than every 25 years. What is the maximum number of design changes that there could have been for a penny if it was first minted 196 years ago?

Use the following information to answer Exercises 15 through 17.

The 1980 estimated population of the four largest cities in California was as follows: Los Angeles 2,966,850; San Francisco 679,974; San Jose 629,442; and Oakland 339,337.

15. How many fewer people were there in the smallest of these cities than in the largest?

16. Is the total population of San Francisco, Oakland, and San Jose larger than or smaller than the population of Los Angeles? The difference is ____________ people.

17. The population of San Jose and Oakland exceeds the population of San Francisco by ____________ people.

18. Your heart beats about 65 times a minute (93,600 times a day). If you live to be 70 (about 25,570 days including leap years), the number of times your heart will beat will be closest to which of the following: one billion, one million, one hundred thousand, or ten thousand?

19. The Canada "Gold Trail" men's curling tour offers \$1,115,100 to the winning team of four men. How much would each man on the winning team make?

20. It takes about 17 weeks to prepare a *new* denomination of currency. If the Bureau of Engraving and Printing prepared bills for every whole-dollar amount from 1 to 10 that does not now exist, how many weeks would it take altogether?

21. In the 1980 U.S. Presidential election, 43,899,248 people voted for Reagan, 35,481,435 people voted for Carter, and 5,719,437 voted for Anderson. How many people voted for one of these candidates in the election?

FOR PRESIDENT
Reagan 43,899,248
Carter 35,481,435
Anderson 5,719,437

Use the following chart to answer questions 22 through 24.

Serving Size	Calories
Beef, rib, lean roasted (4 oz)	273
Chicken, mixed meat, roasted (4 oz)	206
Carrots, cooked (4 oz)	20
Cabbage, cooked (4 oz)	16
Rice, boiled (4 oz)	400
Potatoes, baked (4 oz)	115
Apple pie (1 slice)	350
Chocolate chip cookie (1)	50

22. Plan a meal that consists of beef or chicken, a vegetable, potato or rice, and a dessert and has fewer than 390 calories.

23. Plan a meal that consists of beef or chicken, a vegetable, potato or rice, and a dessert and has more than 1000 calories.

24. Plan a meal that consists of beef or chicken, a vegetable, potato or rice, and a dessert and has fewer than 1000 calories and more than 950 calories.

25. The original price of a 1948 Tucker automobile was \$2,485. Fifty-one cars were produced altogether. If they had all sold at full price, how much would have been collected from these sales?

26. A Beechcraft H18 can hold 11 people altogether. How many of these would it take to transport the same number of people as a Boeing 747 that holds 500 passengers?

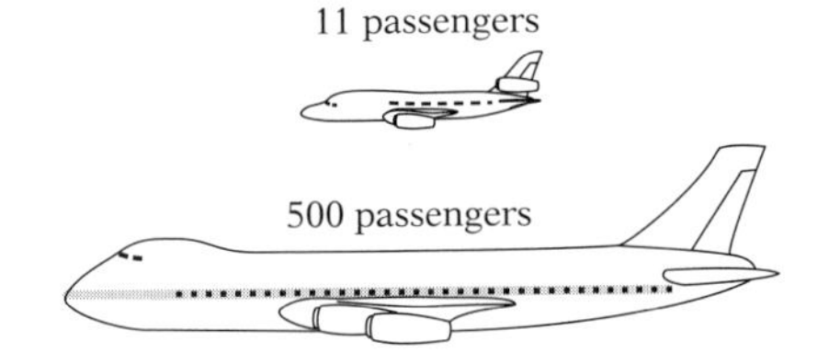

27. The orbital speed of an electron in a uranium atom is 50 times faster than the mean orbital velocity of Mercury around the Sun (172,404 kilometers per hour). What is the orbital speed of the electron?

Use the fact that the equator is 24,900 miles long to answer Exercises 28 and 29.

28. The longest distance a person has ever walked on his hands is 871 miles. How many times longer is the equator?

29. The longest distance a person has ever run is 5110 miles. How many times longer is the equator?

30. The area of what was formerly the Soviet Union is 8,649,496 square miles, and the area of the United States is three million, six hundred fifteen thousand, one hundred five square miles. How much larger is that area than the United States?

31. The area of the largest ocean, the Pacific Ocean, is more than 3 times that of the land area of Asia, the largest continent. If Asia's area is 17,085,000 square miles, what is the approximate area of the Pacific Ocean?

32. The Louisiana State Fair is held in Shreveport each year and attracts about 700,000 people. If the proceeds for the event are about 70 million dollars, about how much is spent per person?

33. You are constructing a graph showing the population for several cities. If the symbol * in your graph represents 200,000 people, how many whole symbols will you have to use to represent each of the following 1970 populations?

a. New York City 7,482,000

b. Moscow 6,942,000

c. Seoul 6,879,000

d. Mexico City 8,628,000

34. In 1986 about twice as many people bought *National Geographic* magazine as bought *Good Housekeeping*. If 5,221,575 bought *Good Housekeeping,* how many bought *National Geographic?*

35. The Des Allemands Catfish Festival takes place each July in Louisiana. During the festival weekend, some 300 people are needed to run the festival, cook, and sell 15,900 pounds of catfish. If each person sells the same amount, how many pounds of catfish are sold per person?

36. Anita has run 8 marathons, each slightly more than 26 miles in length. The actual distance of each is 46,112 yards. How many yards has she run in these marathons altogether?

37. The distance from Rio de Janeiro to Bombay, India, is about 39 times the distance from Paris to London. If the Rio/Bombay distance is 13,377 kilometers, what is the approximate Paris/London distance?

38. Only six 1930 Bugatti Royale open cars were produced. If the total sales from these cars was $270,000 and they all sold for the same amount, about how much did each car sell for? Round your answer to the nearest whole number.

39. A movie theater has its seats arranged so that there are 45 seats in a row. The theater has 42 rows. If 2000 people are waiting in line, how many will not be able to get in?

40. Rodan, a movie monster, had a wingspan of 500 feet. A MiG-25 fighter jet has a wingspan of almost 46 feet. How many MiG jets would have to be placed wingtip-to-wingtip to equal the monster's wingspan?

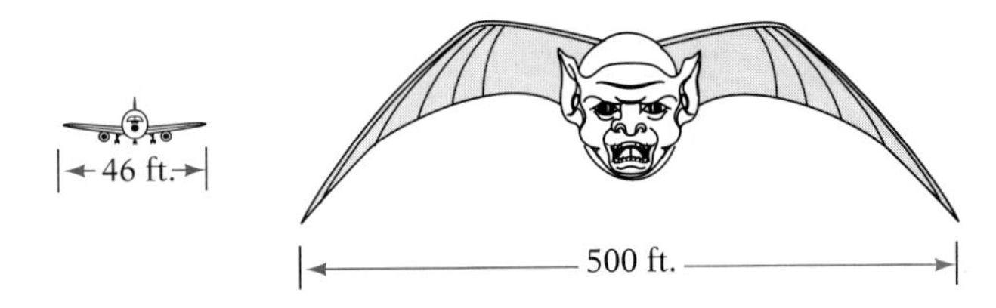

41. Howard commutes back and forth between Clinton Gardens and New York City once each weekend. There are 52 weekends in a year. How many hours does he spend commuting if each trip (one way) takes 57 minutes? (60 minutes = 1 hour)

42. In Georgia in 1989, a survey showed that there were approximately 750,000 elementary students in public schools being taught by approximately 30,000 teachers. How many students were there for each teacher?

2.5 MIXED PRACTICE

By doing these exercises, you will practice all topics up to this point in the chapter.

43. Divide: $6\overline{)6309}$

44. Exams are sent to the copy center to be duplicated. There are 79 students and each exam is 12 pages long. (The pages are printed on one side.) How many sheets of paper altogether will be used in order for each student to have an exam?

45. Multiply: 2374×190

46. Divide: $135\overline{)98{,}685}$

47. You have $1200 to pay your five charge account bills. If you want to send the same amount to each, how much should you send to each?

48. Divide: 24,600 by 24

49. Divide: $21\overline{)2{,}247}$

50. How many freezer bags are needed to freeze 14 frankfurters in each bag if a grocery store has 1325 frankfurters?

51. Multiply: $97{,}498 \times 2908$

52. A case of soda contains 24 cans. How many cans of soda did a group order for their party if they had 113 cases delivered?

53. The English hare can run 250 times faster than the giant tortoise. How many feet can the tortoise run in the time it takes the hare to run 1000 feet?

2.6 Geometry and Measurement: Using Denominate Numbers

SECTION GOALS

- To simplify denominate numbers
- To use denominate numbers in computation

Every day, we use or are involved with many measurements. Almost all our "real world" problems, for example, include measurements.

Denominate number

An amount, when used with its unit of measurement, is called a **denominate number.**

Thus 3 miles, 6 hours, and 12 ounces are all denominate numbers. In this section, we will perform the four basic operations on denominate numbers.

Listed in Table 1 are some measurement equivalents that you should already know. Review them, if necessary, because you will not be able to proceed without them. These are all used in the U.S. Customary system of measurement. Later, we will discuss the metric system.

TABLE 1 ***Equivalent Measurements*****–U.S. Customary System**

Length		
12 inches (in.)	=	1 foot (ft)
3 feet (ft)	=	1 yard (yd)
5280 feet (ft)	=	1 mile (mi)
Capacity		
8 ounces (oz)	=	1 cup (c)
2 cups (c)	=	1 pint (pt)
2 pints (pt)	=	1 quart (qt)
4 quarts (qt)	=	1 gallon (gal)
Time		
60 seconds (s)	=	1 minute (min)
60 minutes (min)	=	1 hour (hr)
24 hours (hr)	=	1 day (da)
7 days (da)	=	1 week (wk)
365 days (da)	=	1 year (yr)
Weight		
16 ounces (oz)	=	1 pound (lb)
2000 pounds (lb)	=	1 ton (t)

1a. Subtract:
10 pounds 9 ounces –
2 pounds 14 ounces

1b. A baby that weighed 8 pounds 9 ounces at birth has gained 11 ounces. How much does she weigh now?

To simplify a measurement means to restate it in the largest units possible, using whole numbers of units. For example, we would simplify 87 ounces of weight by dividing by 16 (the number of ounces per pound).

$$\begin{array}{r l} 5 & \text{pounds} \\ 16\overline{)87} & \\ \underline{80} & + \\ 7 & \text{ounces} \end{array}$$

We can perform all four operations on denominate numbers. Here is a subtraction—with units that are not easy to subtract.

EXAMPLE 1

Subtract:

8 years 3 months minus 5 years 4 months

SOLUTION

Line up the units and subtract each unit separately.

$$\begin{array}{r} 8 \text{ years } 3 \text{ months} \\ \underline{-5 \text{ years } 4 \text{ months}} \\ ? \end{array}$$

4 months cannot be taken away from 3 months, but 8 years is the same as 7 years 12 months. So rewrite 8 years 3 months as 7 years 15 months.

Our subtraction problem becomes

$$\begin{array}{r r} \overset{7}{\not{8}} \text{ years} & \overset{15}{\not{3}} \text{ months} \\ \underline{-5 \text{ years}} & \underline{4 \text{ months}} \\ 2 \text{ years} & 11 \text{ months} \end{array}$$

■ *Now do margin exercises 1a and 1b.*

In the next example, we multiply a denominate number by a pure (nondenominate) number.

EXAMPLE 2

Multiply 3 quarts 1 pint by 5.

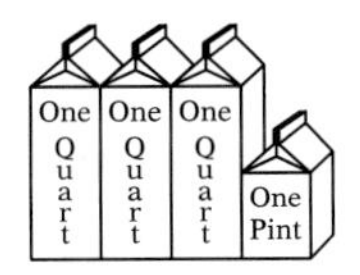

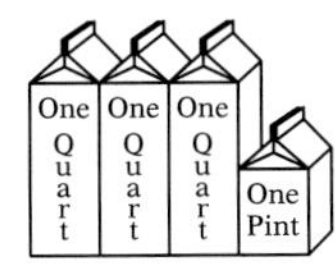

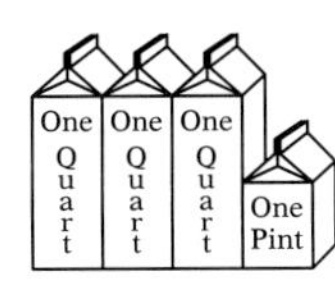

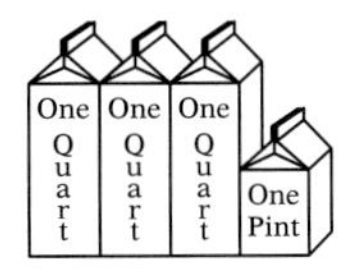

SOLUTION

We rewrite vertically and multiply each measure by the number 5.

$$\begin{array}{r} \text{3 quarts 1 pint} \\ \times 5 \\ \hline \text{15 quarts 5 pints} \end{array}$$

Next we simplify. 2 pints = 1 quart, so we rewrite 5 pints as 2 quarts 1 pint. Then we have

$$\text{15 quarts} + \text{2 quarts 1 pint}$$

That is 17 quarts 1 pint.

We can simplify again. Because 4 quarts = 1 gallon, our 17 quarts can be rewritten as 4 gallons 1 quart. When we combine the results, we get 4 gallons 1 quart 1 pint.

Therefore, our product is 4 gallons 1 quart 1 pint.

■ *Now do margin exercises 2a and 2b.*

The example on page 114 involves division.

2a. Multiply 3 hours 26 minutes by 8.

2b. A recipe calls for 11 ounces of a certain liquid. If I am making 8 batches of the recipe, how much of the liquid will I need?

3a. Divide 8 weeks 4 days by 3.

3b. If a portion of meat is considered to be 3 ounces, how many portions can I get from 4 pounds 5 ounces of meat?

EXAMPLE 3

Divide 26 feet 4 inches by 4.

SOLUTION

We set up the problem using the division symbol $\overline{)\quad}$ and divide each unit alone, from left to right.

$$\begin{array}{r} \text{6 feet} \\ 4\overline{)\text{26 feet 4 inches}} \\ \underline{\text{24 feet}} \\ \text{2 feet} \end{array}$$

Stop at the remainder.

Because the remainder (2 feet) is equal to 24 inches, we can add it to the 4 inches and continue.

$$\begin{array}{r} \text{6 feet 7 inches} \\ 4\overline{)\qquad \text{28 inches}} \\ \underline{\text{28 inches}} \\ 0 \end{array}$$

Combining the results of our divisions, we get 6 feet 7 inches as the answer.

■ *Now do margin exercises 3a and 3b.*

Work the exercises that follow.

2.6 EXERCISES

In Exercises 1 through 8, perform the indicated addition.

1. 3 yards 9 inches
+6 yards 2 inches

2. 5 miles 238 feet
+7 miles 146 feet

3. 5 yards 1 foot
+3 yards 1 foot

4. 10 feet 5 inches
+ 6 feet 4 inches

5. 7 gallons 3 quarts + 5 gallons 2 quarts

6. 6 pounds 8 ounces + 8 pounds 11 ounces

7. 4 pints 1 cup
+6 pints 1 cup

8. 14 tons 1518 pounds
+ 25 tons 163 pounds

9. You work 3 weeks and 5 days on a certain project. The next project takes you 4 weeks and 6 days. How many weeks and days have you spent on the two projects together?

10. Sunrise in Boston in March occurs at about 5:38 A.M. Lenora arrives at work at 1 minute after 11 A.M. How much later than sunrise does she arrive?

11. The first day of this week you worked 10 hours 25 minutes; the second day you worked 8 hours 15 minutes; and the third and fourth days you worked 5 hours 10 minutes and 7 hours 18 minutes, respectively. For how many hours and minutes should you be paid this week?

12. How many minutes and seconds are there in 365 seconds?

In Exercises 13 through 24, perform the indicated subtraction.

13. 26 days 23 hours
−14 days 11 hours

14. 19 hours 33 minutes
−12 hours 26 minutes

15. 37 minutes 45 seconds
−29 minutes 18 seconds

16. 9 years 11 months
−6 years 5 months

17. 8 feet 8 inches minus 6 feet 9 inches

18. 13 yards 1 foot minus 7 yards 2 feet

19. 11 miles 14 feet
− 4 miles 518 feet

20. 9 yards 12 inches
−2 yards 25 inches

21. 5 pints
− 1 cup

22. 9 quarts
− 1 pint

23. 6 gallons − 3 quarts

24. 3 pounds − 9 ounces

Solve the following word problems.

25. How many weeks, with how many days left over, are there in 1 million days?

26. Americans eat 2,800,000 pounds of fresh cucumbers each day. How many tons is this?

27. A quart container full of milk springs a leak. If it contains 11 ounces when the leak is found, how many ounces did it lose?

28. Two mothers are comparing the birth weights of their babies. Elizabeth's son Alan weighed 9 pounds 8 ounces, and Anne Marie's daughter Catherine weighed 6 pounds 15 ounces. What was the difference in their weights?

In Exercises 29 through 40, perform the indicated multiplication.

29. 4 feet 5 inches
$\underline{\qquad\qquad\times 2}$

30. 6 yards 1 foot
$\underline{\qquad\qquad\times 2}$

31. 18 miles 236 feet times 10

32. 25 yards 9 inches times 3

33. 15 pints 1 cup times 6

34. 14 quarts 1 pint times 9

35. 6 pounds 15 ounces times 7

36. 11 tons 185 pounds × 5

37. 9 weeks 3 days × 11

38. 18 days 22 hours times 4

39. 16 hours 26 minutes × 8

40. 11 minutes 7 seconds × 10

Solve the following word problems.

41. How many ounces are there in 100,000 gallons?

42. How many months are there in 1000 years?

43. You are making iced tea for a picnic. You have made 6 batches of tea; each batch is 4 gallons and 3 quarts. How much iced tea have you made?

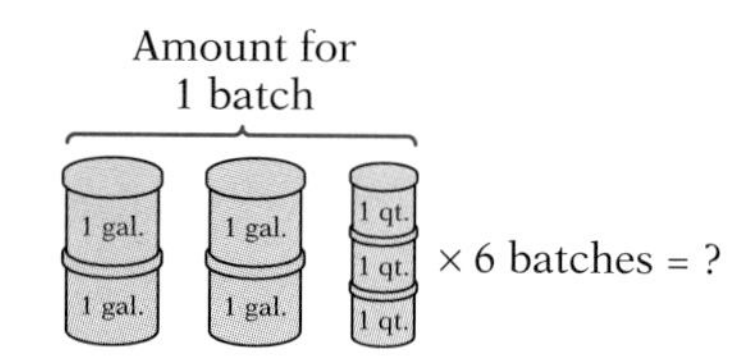

44. You are using vanilla ice cream in a recipe. The recipe calls for 3 quarts and 1 pint of vanilla for each batch. How much ice cream do you need if you are making three batches of the recipe?

In Exercises 45 through 56, perform the indicated division.

45. $9\overline{)27 \text{ feet } 9 \text{ inches}}$

46. $2\overline{)18 \text{ yards } 2 \text{ feet}}$

47. 75 miles 200 feet divided by 5

48. 8 yards 28 inches divided by 4

49. $3\overline{)7 \text{ quarts } 1 \text{ pint}}$

50. $3\overline{)19 \text{ quarts } 1 \text{ pint}}$

51. $6\overline{)13 \text{ gallons } 2 \text{ quarts}}$

52. 7 pints 1 cup divided by 3

53. $8\overline{)29 \text{ minutes } 20 \text{ seconds}}$

54. $5\overline{)17 \text{ hours } 30 \text{ minutes}}$

55. 22 weeks 8 days divided by 9

56. $4\overline{)19 \text{ years } 4 \text{ months}}$

Solve the following word problems.

57. How many tons are there in 20,000 pounds?

58. In Dallas on a day in March, the sun rises at 6:23 A.M. and sets at 6:47 P.M. How much daylight is there in Dallas on this day?

59. After returning from the supermarket, you have a package of hamburger that weighs 4 pounds 8 ounces. You want to freeze it in four equal packages. How much should you put in each package?

60. You have purchased 7 yards of ribbon for decoration on dresses that you are sewing. If this is enough ribbon for three dresses, how much ribbon will you use on each dress?

2.6 MIXED PRACTICE

By doing these exercises, you will practice all topics up to this point in the chapter.

61. Divide: $4\overline{)5 \text{ years } 4 \text{ months}}$

62. Multiply: 2450×307

63. Divide: $7\overline{)15 \text{ days } 4 \text{ hours}}$

64. Divide: $8928 \div 36$

65. The Earth makes one complete rotation every 23 hours 56 minutes and 4 seconds. How much time does it take to make 3 rotations?

66. The graduation class from a certain college had 1992 graduates. This was 3 times as many as another college. How many graduates were there at the other college?

67. Multiply: 4 weeks 3 days
$\times$ 5

68. Subtract: 23 days 5 hours
$-$ 12 hours

69. Divide: $11\overline{)11033}$

70. Multiply: 3,256
$\times$ 150

71. Four hundred fifty hot dogs have been bought for a company picnic. If there are 150 employees, how many hot dogs can each person have?

72. Multiply: $5{,}362 \times 1{,}008$

73. Add: 3 quarts 1 pint
+ 5 pints

74. Divide: $2392 \div 23$

2.7 Geometry and Measurement: Applying Multiplication and Division of Whole Numbers

Finding the Perimeter

We begin with a general procedure for finding the perimeter of any polygon.

To find the perimeter of a polygon

Add up the lengths of all of its sides.

EXAMPLE 1

A tomato patch is rectangular in shape. If the patch is 20 feet by 15 feet, how much wire fence will be needed to enclose it?

SOLUTION

To find the length of the fence, we must find the perimeter. By drawing a picture, we see that we need to add the lengths of the sides.

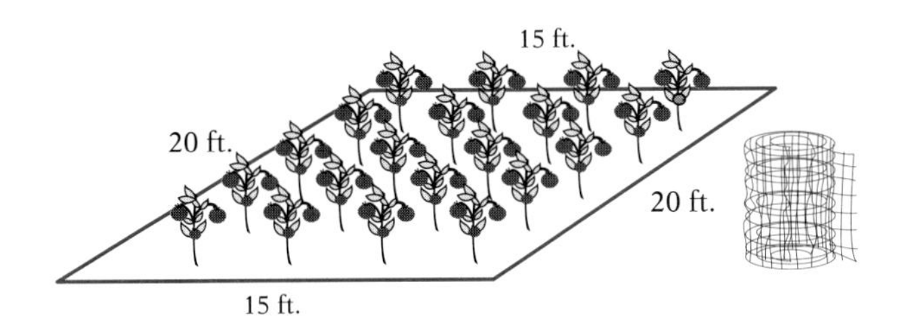

$$20 + 15 + 20 + 15 = 70$$

The perimeter is 70 feet.

■ *Now do margin exercises 1a and 1b.*

We can use formulas to find the perimeters of regular figures.

SECTION GOALS

- To find the perimeter of a polygon
- To find the area of a rectangle and of a parallelogram
- To use a formula to find the perimeter of a square

1a. What is the perimeter of a rectangular chicken coop with length 18 feet and width 12 feet?

1b. A dog pen is being constructed. If the length of the pen is 25 yards and the width is 12 yards, what is its perimeter?

To find the perimeter of a square

If s is the length of a side, then

$$\text{Perimeter} = 4 \times s$$

The statement $P = 4 \times s$ is called a **formula.** It shows what operations we must perform on what quantities to determine a particular value. This one tells us to multiply the length of a side by 4. It works because all four sides have the same length.

EXAMPLE 2

Find the perimeter of a square with a side of length 8 feet.

SOLUTION

We substitute 8 feet for s in the formula and then multiply.

$$P = 4 \times s = 4 \times 8 = 32$$

Thus the perimeter of the square is 32 feet.

■ *Now do margin exercises 2a and 2b.*

2a. Find the perimeter of a square with side 11 inches.

2b. Find the perimeter of a rectangle with sides of 13 feet and 18 inches.

Finding the Area

Measurements of area are always expressed in square units.

Square Unit

A **square unit** (square inch, square foot, square mile, and so on) is a square each of whose sides has a length of 1 such unit (inch, foot, mile, and so on).

The following figure shows that a rectangle with a length of 3 units and a width of 2 units has an area of 6 square units.

ℓ = 3 units

1	2	3
4	5	6

w = 2 units

1 unit

1 unit

Here is a formula for finding the area of any rectangle.

To find the area of a rectangle

If ℓ is the length and w is the width, then

$$\text{Area} = \ell \times w$$

EXAMPLE 3

Find the area of a rectangle that has width 16 inches and length 9 inches.

SOLUTION

We substitute the values for ℓ and w in the formula and multiply.

$$A = \ell \times w = 9 \times 16 = 144$$

So the area is 144 square inches.

■ *Now do margin exercises 3a and 3b.*

Please note that the length can be shorter, longer, or equal to the width.

There is also a formula for the area of a parallelogram.

To find the area of a parallelogram

If b is the length of the base and h is the height, then

$$\text{Area} = b \times h$$

3a. Find the area of a rectangle that has length 520 yards and width 124 yards.

3b. Find the area of a square that has a side of 23 inches.

4a. Find the area of a parallelogram with height 6 feet and base 12 feet.

4b. Find the height of a parallelogram with base 16 feet and area 368 square feet.

The base and height are shown in the following figure.

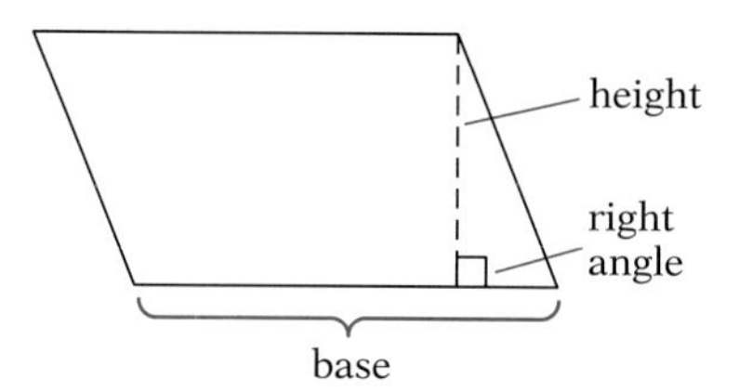

EXAMPLE 4

Find the area of a parallelogram with base 19 feet and height 12 feet.

SOLUTION

Substituting in the formula, we get

$$A = b \times h = 19 \times 12 = 228$$

So the area of the parallelogram is 228 square feet.

■ *Now do margin exercises 4a and 4b.*

Then do the exercises that follow.

2.7 EXERCISES

Work the following word problems.

1. Find the perimeter of an equilateral triangle that has a side of length 4 inches.

2. You bought a can of pre-made frosting for the top of a cake. The container says that it will cover a square cake that is 7 inches on one side. You have a rectangular cake pan that is 8 inches by 6 inches. Will you be able to use this can of frosting to frost the top of the cake?

3. Find the area of a rectangular pool that has a side of 23 yards and a side of 18 yards.

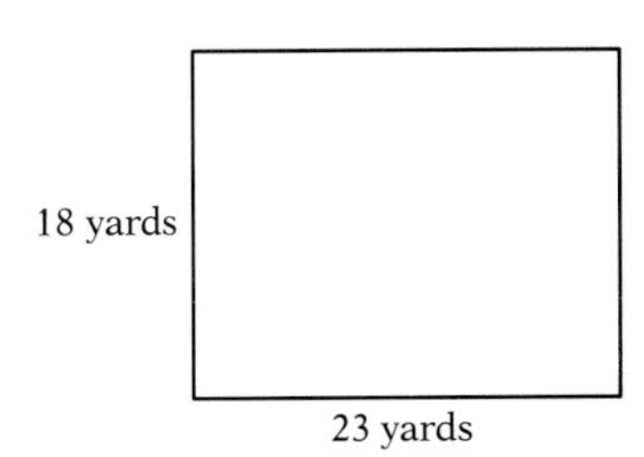

4. A square plot of grass is 15 feet on each side. How much fencing would you need to enclose it?

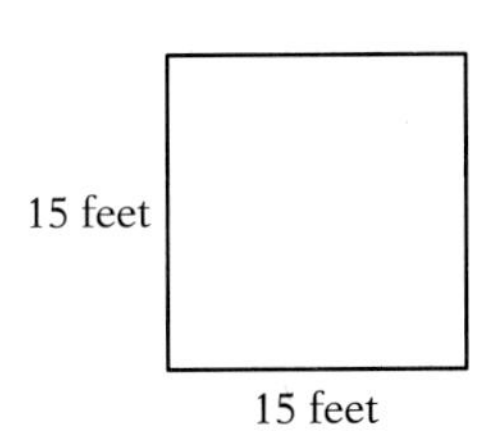

5. Find the perimeter of a regular pentagon that has a side of 12 feet.

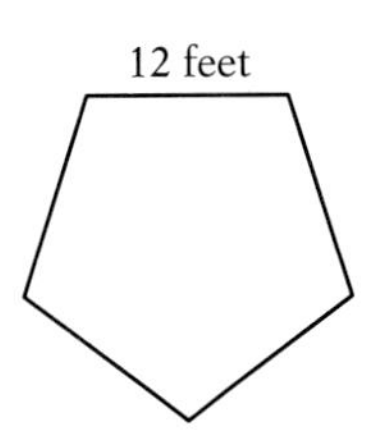

6. Find the perimeter of a scalene triangle that has sides 15 inches, 12 inches, and 14 inches.

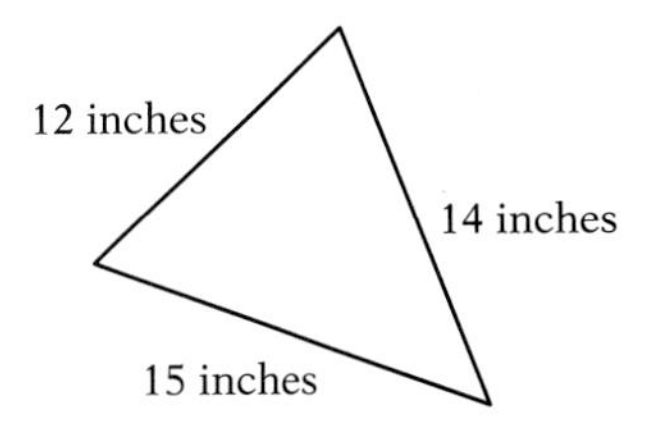

7. A neon sign outside of a store will direct people to a sale. The sign will be in the shape of a parallelogram with a base of 5 feet and a height of 4 feet. How much space will there be on the sign to write the message?

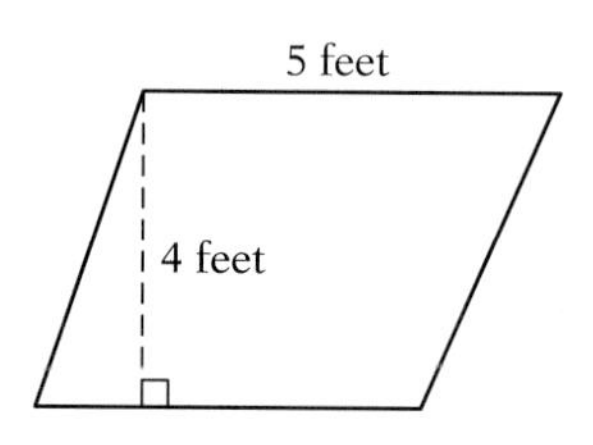

8. The dimensions of an auditorium are 200 feet by 325 feet. What is the area of the auditorium?

9. Find the perimeter of an isosceles triangle that has equal sides of 6 inches and a third side of 11 inches.

10. Find the perimeter of a square judo court 52 feet 6 inches on a side.

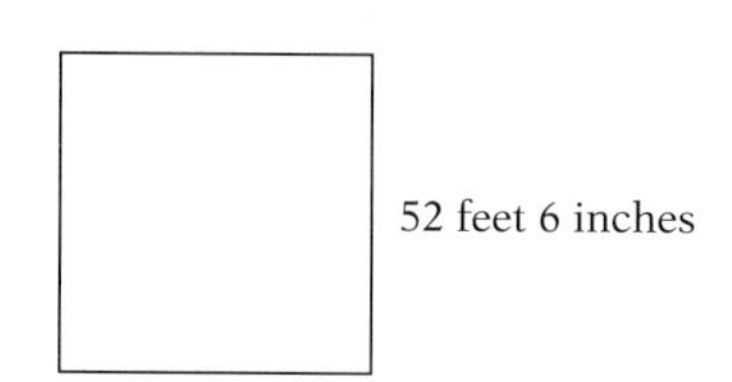

11. Find the area of the parallelogram shown in the accompanying figure.

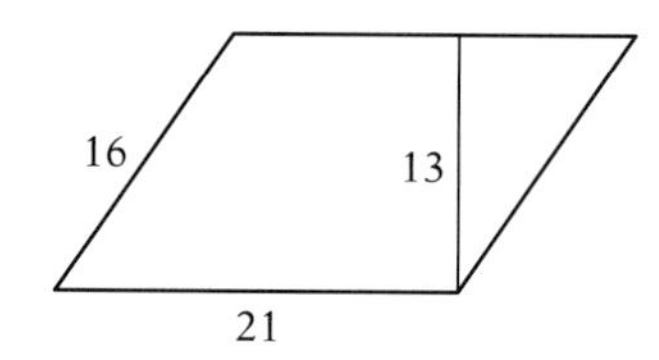

12. A regular undecagon has 11 equal sides. If the perimeter is 1628 feet, what is the measure of each side?

13. A regular nonagon has 9 equal sides. Each side is 140 yards. What is the perimeter of this figure?

14. Find the perimeter of the irregular house lot shown in the accompanying figure.

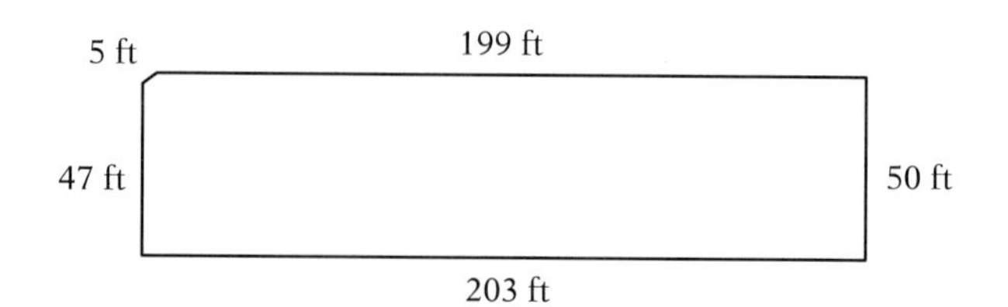

15. Jason runs forty-three times around the outside of the town pool every morning. If the pool is 20 yards long and 10 yards wide, how far does he run each day?

16. A contractor is trying to figure out how much material she will need to box a window for a house. The window is 12 feet by 5 feet. What is the perimeter of the window?

2.7 MIXED PRACTICE

By doing these exercises, you will practice all topics up to this point in the chapter.

17. Multiply: 3623 x 480

18. Divide: $23\overline{)283{,}485}$

19. Multiply: 3 pounds 4 ounces x 9

20. Find the perimeter of a triangle with sides of length 12 feet, 15 feet, and 10 feet.

21. Find the area of a parallelogram with base equal to 18 feet and height equal to 28 feet.

22. An employee earns $25,080 per year. The company she works for pays 346 times this amount in salaries. How much does the company pay in salaries?

23. Find the perimeter of a regular decagon with a side of length 18 units.

24. Add: 4 days 3 hours + 7 days 23 hours

25. Multiply: $83{,}485 \times 709$

26. Divide: $152\overline{)31{,}312}$

27. Subtract 35 weeks from 4 years 13 weeks.

28. Divide 14 gallons 1 quart by 3.

An Application to Statistics: Reading Bar Graphs and Double Bar Graphs

Reading a Bar Graph

Information is often presented in the form of a **graph,** a diagram that shows numerical data in visual form. Graphs enable us to "see" relationships that are difficult to describe with numbers alone.

The following graph is called a **bar graph.** This one shows the amount of electricity used daily, on the average, by a certain customer in each of 13 successive months.

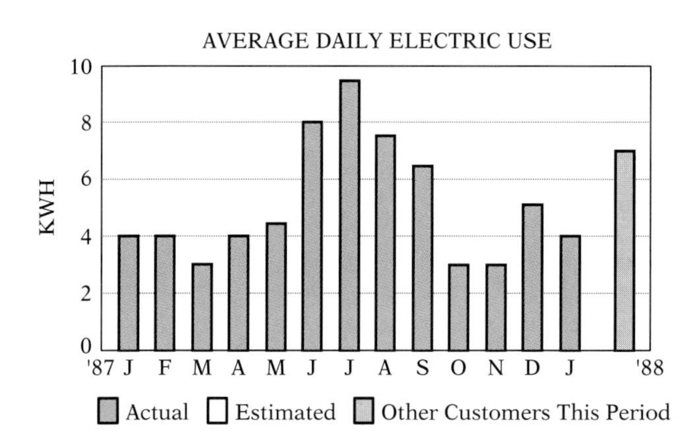

Information that helps you read the graph is given along the bottom and left-hand side of the graph. Along the bottom are the months covered by the graph, abbreviated, from January 1987 to January 1988. There is one bar for each month. The height of the bar gives the average amount of electricity used per day in kilowatt hours (kwh), for its month. The height is read on the scale at the left of the graph.

EXAMPLE 1

a. How many kilowatt hours did this customer use daily during March, to the nearest whole number?
b. During four of the months shown, the customer used the same number of kilowatt hours per day. Which months were these?

SOLUTION

a. The third bar from the left, the bar for March, ends halfway between 2 and 4. Thus, during March, this customer used 3 kilowatt hours per day.
b. We must find four bars that are the same height. These occur in January of both years, February, and April. They have a height of 4 kwh.

■ *Now do margin exercises 1a and 1b.*

Reading A Double Bar Graph

Comparative information is often shown on a **double bar graph** such as the following figure. This type of graph usually presents two measurements for each date or time. The reader compares the measurements by comparing bars.

1a. How many kilowatt hours were used in October?

1b. How many kilowatt hours were used altogether in the months April and June?

2a. In what year was the median sales price the highest for new private one-family houses?

2b. In what year was the median sales price of existing houses closest to that of new houses?

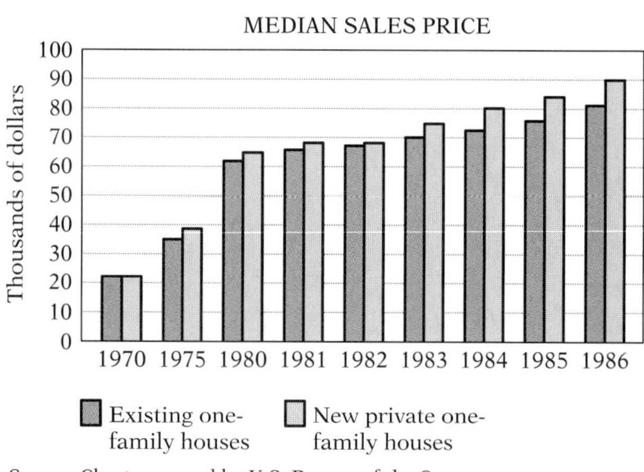

Source: Chart prepared by U.S. Bureau of the Census

The scale on the left side of this graph shows the median sales prices of one-family houses—existing or new—in thousands of dollars. The scale along the bottom shows the years from 1970 through 1986.

EXAMPLE 2

In what year was the difference in median sale prices about $10,000?

SOLUTION

The price scale on the right shows that $10,000 is the difference between any two adjacent horizontal bars. The only year for which the tops of the two bars are that far apart is 1986.

■ *Now do margin exercises 2a and 2b.*

Work the exercises that follow. Remember to see if your answers are reasonable.

EXERCISES

Use the following figure to answer Exercises 1 and 2.

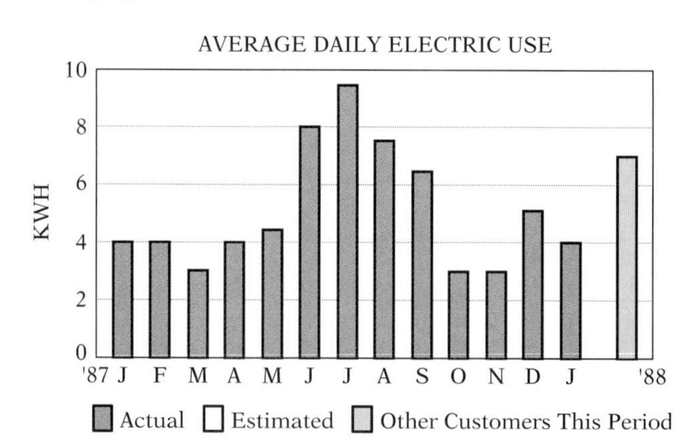

1. How many more kilowatt hours were used per day in June than in November?

2. How many kilowatt hours were used altogether in the months of October, November, and December 1987 and January 1988?

Use the following figure to answer Exercises 3 and 4.

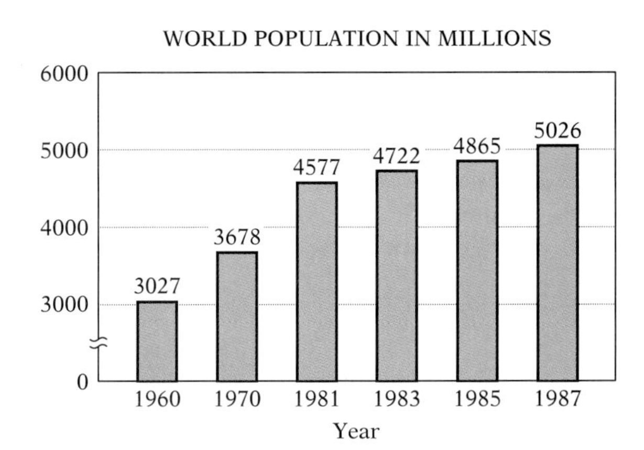

3. Was the increase from 1960 to 1970 more or less than the increase from 1970 to 1981?

4. Which two-year time period had the greatest increase?

Answer Exercises 5 and 6 by using the median sales price double bar graph from Example 2 on page 130.

5. What was the difference between the approximate median prices of new and existing one-family houses in 1983?

6. Was the jump in median prices of houses greater from 1970 to 1975 or from 1975 to 1980? How can you tell?

Use the following figure to answer Exercises 7 and 8.

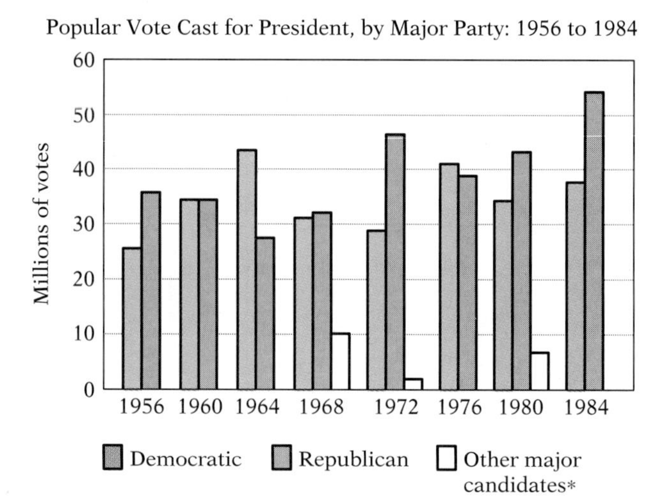

*1968 and 1972-American Independent; 1980-John Anderson

Source: Chart prepared by U.S. Bureau of the Census.

7. In what year was the difference in votes cast for the Democratic and Republican parties the greatest?

8. What year had the largest total vote?

Use the following figure to answer Exercises 9 and 10.

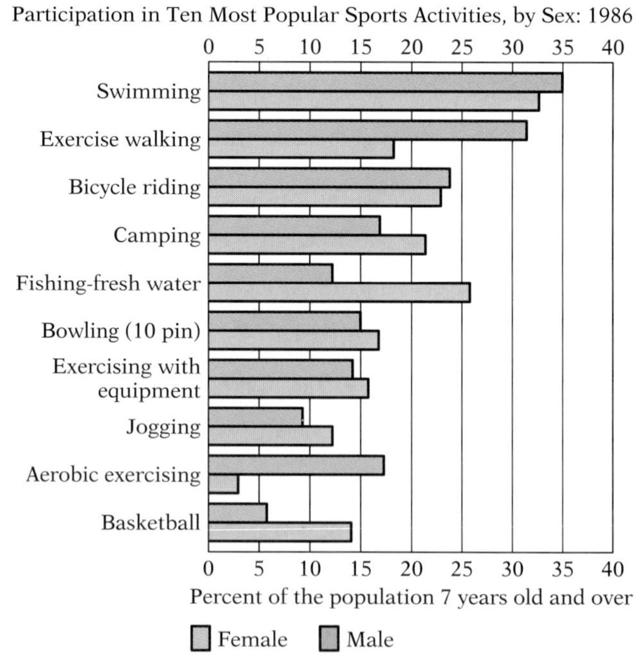

Source: Chart prepared by U.S. Bureau of the Census.

9. What activity shows the biggest difference in participation between men and women?

10. Which activity or activities are the least popular overall?

Chapter 2 Review

ERROR ANALYSIS

These problems have been worked incorrectly. Tell what the error is and then write the correct solution.

1. Multiply: 236×84

Incorrect Solution *Correct Solution*

$$\begin{array}{r} 236 \\ \times\ 84 \\ \hline 944 \\ 1888 \\ \hline 2832 \end{array}$$

Error ______________________________

2. Subtract: 3 days 4 hours − 1 day 12 hours

Incorrect Solution *Correct Solution*

3 days 4 hours
−1 day 12 hours
1 day 2 hours

Error ______________________________

3. Divide: $1648 \div 16$

Incorrect Solution *Correct Solution*

$$\begin{array}{r} 130 \\ 16\overline{)164} \\ \underline{16} \\ 48 \\ \underline{48} \end{array}$$

Error ______________________________

4. Multiply: 427×290

Incorrect Solution *Correct Solution*

$$\begin{array}{r} 427 \\ \times 290 \\ \hline 3843 \\ 854 \\ \hline 4697 \end{array}$$

Error ______________________________

5. Divide: $3\overline{)306}$

Incorrect Solution

$$\begin{array}{r} 12 \\ 3\overline{)306} \\ \underline{3} \\ 6 \\ \underline{6} \end{array}$$

Correct Solution

Error ______________________________

INTERPRETING MATHEMATICS

By working these exercises, you will test and strengthen your mathematics vocabulary.

1. In this multiplication problem,

$$\begin{array}{r} 798 \\ \underline{\times 635} \\ 3990 \\ 23940 \\ \underline{478800} \\ 506730 \end{array}$$

write the digits that form

a. the multiplier

b. the product

c. the factors

d. the multiplicand

e. the partial products

2. In this division problem,

$$\begin{array}{r} 635 \\ 798\overline{)506731} \\ \underline{4788} \\ 2793 \\ \underline{2394} \\ 3991 \\ \underline{3990} \\ 1 \end{array}$$

write the digits that form

a. the divisor

b. the dividend

c. the quotient

d. the partial quotients

e. the remainder

In your own words, explain the significance of:

3. The commutative property of multiplication.

4. The identity element for multiplication.

5. The associative property of multiplication.

6. The distributive property of multiplication over addition.

7. Separate the following units into groups according to whether they are units of length, mass, capacity, or time. Then list the units in each group from largest to smallest.

cup	day	foot	gallon	hour
mile	minute	ounce	pint	pound
quart	ton	week	yard	year

8. What do units and square units have in common? How do they differ?

REVIEW PROBLEMS

The exercises that follow will give you a good review of the material presented in this chapter. Work through them and check your answers at the back of the book.

Section 2.1

1. What is the product of 7 and 31,289?

2. Multiply 362,873 by 3.

3. Find 99,999 × 9.

4. Multiply: 100 × 40 × 200

5. Find the product: 19,876 × 7

6. What is the product of 3,000 and 3,000?

Section 2.2

7. Multiply: 106 × 24

8. Find the product: 220 × 35

9. Multiply 93,912 by 2007.

10. Multiply: 18,000 × 206

11. Multiply: 984 × 189

12. What is the product of 19 and 91 and 1991?

Section 2.3

13. Divide:

14. Divide 80,904 by 6.

15. What is the quotient of 198/9?

16. What is the result when 200,000 is divided by 40?

17. What is the remainder when 41,726354 is divided by 9?

18. Fill in the blank:

Section 2.4

19. Divide: 210,738 ÷ 103

20. Divide: $12\overline{)1272}$.

21. Divide 26,130 by 25.

22. Find the quotient of 3960/18.

23. Divide 10,040 by 26. What is the remainder?

24. Divide 39,700 by 15. What is the remainder?

Section 2.5

25. A company pays $12,000,000 a year to 500 employees, and each person earns the same salary. What is the monthly salary for each person?

26. Find the cost of renting a bicycle for 6 days if the cost for each day is $25.

27. Mt. Everest, which is about 29,000 feet tall, is about 20 times higher than the Sears Tower building in Chicago. What is the height of the building?

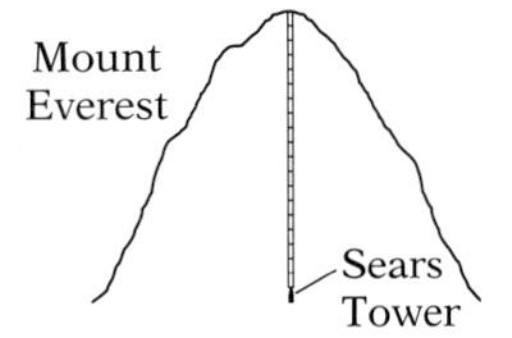

28. Blake, Bobby, Shareefa, Barrock, and Catherine each have 6 video games and 4 movies. How many video games and movies do they have altogether?

29. In 1990, the poverty level for a family of four was $12,675 and for a family of three it was $10,560. Find how much money is allowed per person in this system. Which level provides more per person? How much more (to the nearest dollar)?

30. In 1991, the minimum wage was raised from $3.85 (385 cents) an hour to $4.25 (425 cents). A person working 25 hours a week at the $3.85 level would make how much more per year at the new level?

Section 2.6

31. Multiply 12 miles 684 feet by 10.

32. Divide 13 days 20 hours by 4.

33. When your sister was born, she weighed 4 pounds 13 ounces. Your brother, who has just been born, weighs twice as much as your sister weighed. How much does your brother weigh?

34. Subtract 20 pounds 5 ounces from 25 pounds.

35. You have bought 2 pounds 3 ounces of hamburger meat for a barbecue for 7 people. Each person can have one hamburger. How many ounces will be in each hamburger if they are all the same weight?

36. Subtract 42 minutes 13 seconds from 56 minutes 5 seconds.

Section 2.7

37. Find the area of a parallelogram that has a height of 6 inches and a base of 24 inches.

38. Find the area of a rectangular playground that has dimensions 45 yards by 135 yards.

39. Find the perimeter of a regular heptagon (a 7-sided figure) with one side 1 foot 2 inches.

40. Rail the cat had a bed that measured 24 inches on one side and 2 feet 6 inches on the other. What was the area in square inches?

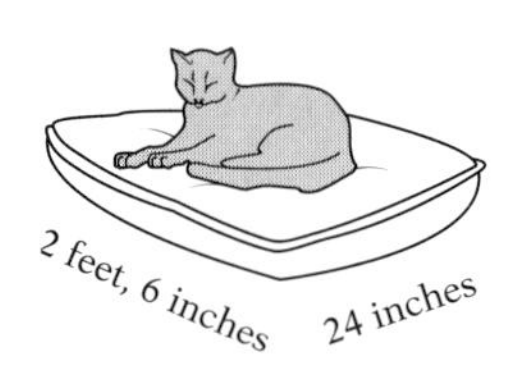

41. Patty Jeanne planted petunias, daisies and zinnias in a garden that had a perimeter of 12 yards. If the length was 12 feet, what was the width of this rectangular garden in yards?

42. Mattie Louise has a square quilting frame with a perimeter of 124 inches. What is the length of one side?

Chapter 2 Test

This exam tests your knowledge of the topics in Chapter 2.

1. a. Multiply: 2980×300

   b. What is the product of 98,506 and 1000?

   c. Multiply 9896 by six million.

2. a. Find the product of 108 and 685.

   b. Multiply 12,373 by 857.

   c. Multiply: $298{,}000 \times 3306$

3. a. Divide 1,258,500 by 500.

   b. What is the quotient of $3280 \div 40$?

   c. What is 90,000 divided by 3000?

4. a. Find the quotient of $10{,}675 \div 35$.

   b. Divide 248,040 by 120.

   c. What is the remainder when 24,992 is divided by 166?

5. Solve:

a. The number of pieces in an adult jigsaw puzzle is approximately 50 times the number of pieces in a child's jigsaw puzzle. If a child's puzzle has 25 pieces, how many pieces can you expect to have in the adult jigsaw puzzle?

b. An exam is to be timed. There are 56 questions. If you are allowed 4 minutes per question, how long should the exam take?

c. A set of 37 tools cost $185. How much does each tool cost, if each costs the same amount?

6. Perform the indicated operation.

a. 3 gallons 1 quart – 2 gallons 3 quarts

b. 34 pounds 12 ounces x 4

c. 6 yards 2 feet ÷ 4

7. Solve the following application problems.

a. Find the area of a rectangle with length 18 inches and width 23 inches.

b. Find the perimeter of a square with a side that measures 26 inches.

c. Find the area of a parallelogram with a base that measures 45 inches and a height that measures 16 inches.

Name Section Date

CUMULATIVE REVIEW

CHAPTERS 1–2

The following questions will help you maintain the skills you have developed.

1. Write the underlined expression as a number: The area of the Caribbean Sea is one million, forty-nine thousand, five hundred square miles.

2. Multiply: 203×196

3. Divide 30060 by 30.

4. Subtract: $7000 - 289$

5. $5\overline{)6 \text{ days } 6 \text{ hours}}$

6. Add: 5 pounds 13 ounces + 3 pounds 11 ounces

7. In a certain year, U.S. production of corn, wheat, and rice in thousands of tons was as follows: corn 212,238; rice 6,967; wheat 76,538. How much grain was produced during that year?

8. The city of Anchorage, Alaska, has approximately 10 times the population of Juneau, Alaska. If Juneau has a population of 25,964, what is the approximate population of Anchorage?

9. Find the area of a rectangle with a length of 12 inches and a width of 18 inches.

10. Add: $1004 + 234 + 18$

11. Subtract 689 from 90005.

12. Find the product of 106 and 98.

13. What is 5379 divided by 3?

14. Write the underlined expression in words: During Disneyland's first year, it attracted 10,700,000 visitors.

15. Subtract: 14 hours 32 minutes – 6 hours 49 minutes

16. Find the perimeter of a square with side 15 inches.

17. You have 402 balloons for a festival. If each child is given 3 balloons, to how many children will you be able to give balloons?

18. Estimate the total length of the following five rivers to the nearest thousand.

Mississippi	2,348 miles	Colorado	1,450 miles
Yukon	1,979 miles	Rio Grande	1,760 miles
Missouri	2,315 miles		

3

Other Operations with Whole Numbers

Many careers require the application of operations on whole numbers. For example, a real estate appraiser often makes use of geometry and measurement. For houses or lots that consist of simple rectangles, an appraiser determines the total area, or square footage, by measuring the length and width of the rectangular areas, calculating each area, and then adding them for the total. Determining the square footage of a lot that isn't square can be more difficult. Odd shapes or lack of information may require the appraiser to estimate perimeters, calculate the areas of rectangles and triangles, and then add the results for an estimated total.

- *How would you calculate the total square footage of your living quarters?*

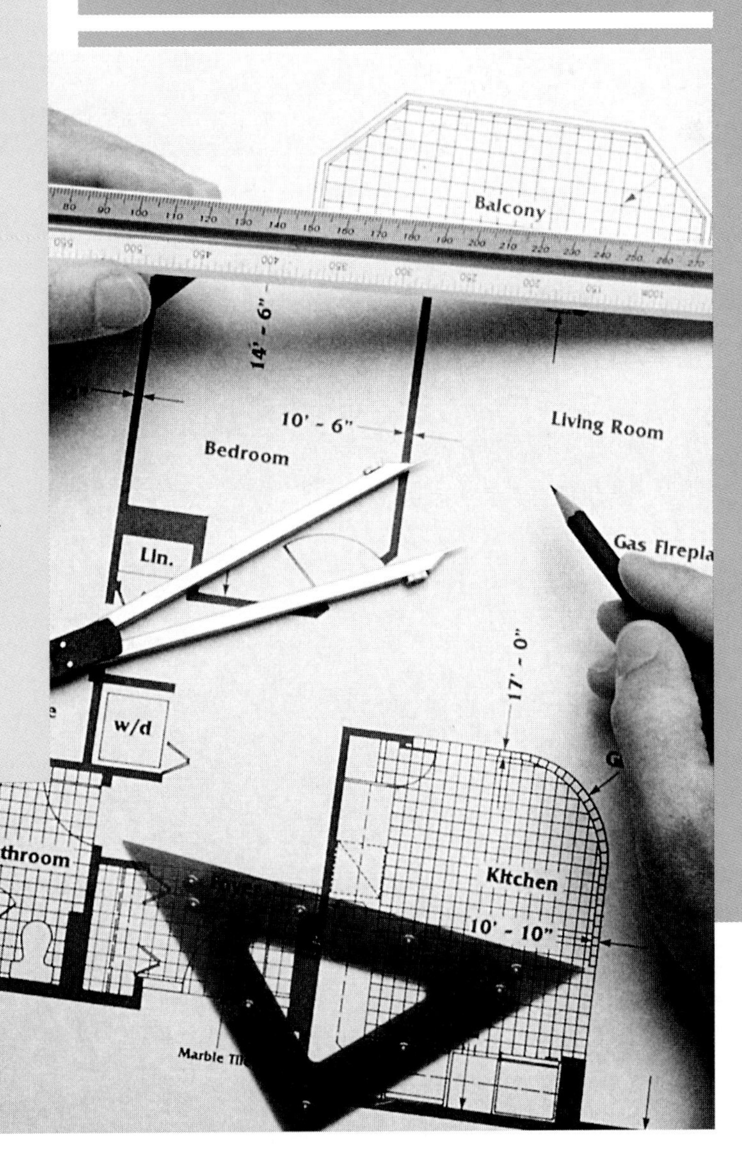

Skills Check

Take this short quiz to see how well prepared you are for Chapter 3. The answers are given at the bottom of the page. So are the sections to review if you get any answers wrong.

1. Divide: $5\overline{)5045}$

2. Multiply: 29×29

3. Multiply: $2 \times 2 \times 3 \times 5 \times 7$

4. Divide: $2079 \div 9$

5. Multiply: $10 \times 10 \times 10$

6. Multiply: $2 \times 2 \times 2$

7. Divide: $560 \div 10$

8. Divide: $6000 \div 60$

ANSWERS: 1. 1009 [Section 2.3] 2. 841 [Section 2.2] 3. 420 [Section 2.2] 4. 231 [Section 2.3] 5. 1000 [Section 2.1] 6. 8 [Section 2.1] 7. 56 [Section 2.3] 8. 100 [Section 2.3]

3.1 Using Divisibility Rules

SECTION GOAL

- To use divisibility rules to find factors of a number

In counting by fives, we would get 5, 10, 15, 20, ... Each of these numbers—5, 10, 15 and 20—is called a **multiple** of 5 because each number can be written as a product of the number 5 and a whole number. By definition,

Multiples

If a, b, and c are whole numbers and $a \times b = c$, then we say that c is a **multiple** of a and of b. Also, a and b are **factors** of c.

Thus if $5 \times 3 = 15$, then 15 is a multiple of both 5 and 3, and 5 and 3 are factors of 15.

$$\begin{array}{rl} 5 & \text{Factor} \\ \underline{\times 3} & \text{Factor} \\ 15 & \text{Multiple of 5 and 3} \end{array}$$

The word *factor* often means the same thing as *divisor.* But we use *divisor* mainly when we are referring to division. However, we do say a number is **divisible** by its factors.

Hence 15 is divisible by both 5 and 3. Note that this means there is no remainder.

Every whole number has 1 and itself as factors, because $1 \times n = n$ for all numbers n. There are rules that can help us find other factors. For example, when we count by twos, the numbers we count all end in 2, 4, 6, 8, or 0. Such numbers are often referred to as **even numbers.** This gives us our first rule.

Rule for Divisibility by 2

Every even number has 2 as a factor, or divisor.

Whole numbers that do not have 2 as a factor are called **odd numbers.**

When we count by fives, the last digit is always a zero or a five; counting by tens results only in numbers ending in 0. This gives us two additional rules for finding factors.

1a. Which of the following numbers have 5 as a factor but not 10?

225	5534
649	39,005
2791	4600

1b. Which of the following numbers have 5 and 10 as factors?

220	5034
468	3900
2812	4040

Rule for Divisibility by 5

Numbers that end in 5 or 0 have 5 as a factor.

Rule for Divisibility by 10

Numbers that end in 0 have 10 as a factor.

EXAMPLE 1

Which of the following numbers have 5 as a factor? Which have 10 as a factor?

105 1782 1559 10,101 1456 1,205,060 900 337,700

SOLUTION

The numbers that end in 0 or 5 have 5 as a factor. So 105; 1,205,060; 900 and 337,700 have 5 as a factor. The numbers that end in zero have 10 as a factor. So 1,205,060; 900; and 337,700 have 10 as a factor.

■ *Now do margin exercises 1a and 1b. Answers are at the back of the book.*

The correct answers are given at the back of the book. Be sure you understand each example and its margin exercises before you go on.

Numbers can have several factors. One of the things you might have observed in Example 1 is that a number that has 10 as a factor also has 2 and 5 as factors. The numbers 900; 337,700; and 1,205,060 were found to have 10 as a factor; they also have 5 and 2 as factors.

To determine whether a number has 4 or 8 as a factor, we use a different kind of test. Instead of merely looking at the last digit of the number, we have to do a little dividing. Here are the rules.

Rule for Divisibility by 4

If 4 is a factor of the last two digits of a number, then it is a factor of the number.

Rule for Divisibility by 8

If 8 is a factor of the last three digits of a number, then it is a factor of the number.

Note also that if a number ends in two zeros, it has 4 as a factor, because 100 is divisible by 4. If a number ends in three zeros, it has 8 as a factor, because 1000 is divisible by 8. And, if a number has 8 as a factor, it also has 2 and 4 as factors.

We have another rule for finding out whether a number has 3 or 9 as a factor.

Rule for Divisibility by 3 or 9

To check a number for divisibility by 3 or 9, add up its digits. If the sum of the digits is divisible by 3 or 9, then the number is also.

The number 3333 clearly has 3 as a factor, because $3 \times 1111 = 3333$. Adding the digits gives us 12. Because 12 is divisible by 3, the original number is also. Because 12 is not divisible by 9, we know that 9 is not a factor of 3333.

EXAMPLE 2

Which of the following numbers are divisible by 3, and which of those are also divisible by 9?

1,253,662 101,010 6363 1,111,111 99

SOLUTION

Adding the digits in these numbers gives the following:

$$
\begin{aligned}
1{,}253{,}662:&\quad 1+2+5+3+6+6+2=25\\
101{,}010:&\quad 1+0+1+0+1+0 \quad = 3\\
6363:&\quad 6+3+6+3 \quad =18\\
1{,}111{,}111:&\quad 1+1+1+1+1+1+1= 7\\
99:&\quad 9+9 \quad =18
\end{aligned}
$$

The numbers that are divisible by 3 are 101,010; 6363; and 99. Of these, the numbers that are also divisible by 9 are 6363 and 99.

■ *Now do margin exercises 2a and 2b.*

An interesting property has been illustrated again and again in our examples.

If a number a is a factor of another number b, then all the factors of a are also factors of b.

But is the reverse true? If a number is divisible by a and b, is it necessarily divisible by $a \times b$?

2a. Which of the following numbers are divisible by 3 but not by 9?

12,345,678 111
222,222 90,909
11,000,111,033,601

2b. Which of the following numbers are divisible by both 4 and 8?

2,310,000,000
3752 173
84 992,200

3a. Find out if 22,530 is divisible by

a. 12
b. 2 and 6
c. 3 and 9
d. 5 and 10

3b. Find out if 31,062 is divisible by

a. 12
b. 4
c. 3 and 9
d. 2 and 6

Suppose a number is divisible by 2. Is it divisible by multiples of 2? Not necessarily. We have seen numbers that are divisible by 2 but not by 4, such as 6 and 18. We have also seen numbers that are divisible by 3 and not by 9, such as 24 and 15. However, for some special pairs of numbers, we find that it is true. The following rules give two examples.

Rule for Divisibility by 6

If a number has both 2 and 3 as factors, then it has 6 as a factor.

Rule for Divisibility by 12

If a number has both 3 and 4 as factors, then it has 12 as a factor.

Let's use this information to work the next example.

EXAMPLE 3

Determine whether 25,332 has (a) 3, (b) 4, (c) 12, and (d) 6 as a factor.

SOLUTION

a. Adding the digits $2 + 5 + 3 + 3 + 2$ gives 15. Because 15 is divisible by 3, the number 25,332 is divisible by 3.
b. The last two digits of the given number are 32. Because 32 is divisible by 4, the number 25,332 is also.
c. Because the given number has already been shown to be divisible by 3 and by 4, we know it is also divisible by 12.
d. Because the number 25,332 is divisible by 12, it is also divisible by any factor of 12. It is therefore divisible by 6.

■ *Now do margin exercises 3a and 3b.*

Work the exercises that follow. The answers to selected exercises are given at the back of the book.

3.1 EXERCISES

In Exercises 1 through 4, answer true or false.

1. a. 4 is a factor of 12.

b. 25 is a factor of 5.

2. a. A factor can be greater than the number being factored.

b. A factor cannot equal the number being factored.

3. a. 1 is a factor of every number.

b. Zero is a factor of every number.

4. Every number has at least two factors, itself and 1.

In Exercises 5 through 16, decide whether the second number is a multiple of the first number.

5. 2 and 120

6. 9 and 837

7. 3 and 235

8. 10 and 2047

9. 4 and 457

10. 12 and 4172

11. 5 and 976

12. 3 and 972

13. 6 and 293

14. 9 and 8163

15. 10 and 650

16. 5 and 9825

In Exercises 17 through 32, decide whether the first two numbers are *both* factors of the third number.

17. 2, 3; 630

18. 3, 4; 480

19. 4, 5; 250

20. 2, 4; 675

21. 3, 5; 384

22. 2, 6; 1432

23. 2, 12; 600

24. 5, 2; 1200

25. 10, 5; 850

26. 8, 2; 682

27. 6, 3; 932

28. 12, 3; 600

29. 4, 8; 876

30. 2, 10; 3450

31. 3, 8; 284

32. 5, 6; 360

In Exercises 33 through 36, decide which of 2, 3, 4, 5, 6, 8, 9, 10, and 12 are factors of the given number.

33. 9009

34. 11,011

35. 5040

36. 3024

In Exercises 37 through 44, determine whether each statement is true or false, on the basis of what you have learned in this section. Justify your answers.

37. A number divisible by 4 is also divisible by 2.

38. A number divisible by 5 is also divisible by 10.

39. A number divisible by 6 is also divisible by 12.

40. A number divisible by 3 is also divisible by 9.

41. A number divisible by 10 is also divisible by 5.

42. A number divisible by 2 is also divisible by 4.

43. A number divisible by 12 is also divisible by 6.

44. A number divisible by 9 is also divisible by 3.

In Exercises 45 through 54 predict whether each of the following is true or false, on the basis of what you have observed in this section. Explain your reasoning in your own words. You may also use numerical examples to help in your explanation.

45. A number divisible by both 2 and 9 is also divisible by 18.

46. A number divisible by both 2 and 10 is always divisible by 20.

47. A number divisible by both 4 and 8 is guaranteed to be also divisible by 16.

48. A number divisible by both 3 and 5 is also divisible by 15.

49. A number divisible by both 3 and 6 is not necessarily also divisible by 18.

50. A number divisible by 3 or 8 is also divisible by 11.

51. A number divisible by both 2 and 7 is also divisible by 14.

52. A number divisible by both 2 and 8 is also divisible by 16.

53. Which of the following numbers are divisible by 4 but not by 8? 2644; 9000; 6,177,246; 2,776,351; 700

54. Which of the following numbers are divisible by both 3 and 9? 1221; 7,231,642; 111,111,111; 4,938,894

3.2 Factoring Whole Numbers

Finding Whole-Number Factors

For some operations on numbers, it is necessary to find all the whole-number factors of a number. To do so, we simply apply the divisibility rules.

EXAMPLE 1

List all the whole-number factors of 24.

SOLUTION

We start with the obvious factors, 1 and 24. For now, it is best to list them as a "multiplication."

$$1 \times 24$$

To continue, we apply the divisibility rules for 2, 3, . . . in order, as needed. We obtain

$$2 \times 12$$
$$3 \times 8$$
$$4 \times 6$$

Because 24 does not end in 5 or 0, 5 is not a factor. We stop here because the next possible factor, 6, is already listed.

So the factors of 24 are 1, 2, 3, 4, 6, 8, 12, and 24.

■ *Now do margin exercises 1a and 1b.*

In Example 1, we wrote 24 as the product of two of its factors in several different ways:

1×24 2×12 3×8 4×6

Each of these is called a **factorization** of 24. A factorization can have more than two factors, as in $24 = 2 \times 2 \times 6$, but their product must be the original number. The process of finding the factors of a number is called **factoring.**

Prime and Composite Numbers

We classify numbers as prime or composite depending on their factors.

Prime Number

If a whole number greater than 1 has only the factors 1 and itself, that number is called a **prime number** or is said to be **prime.**

Composite Number

If a whole number greater than 1 has factors other than itself and 1, that number is called a **composite number** or is said to be **composite.**

SECTION GOALS

- To find all the whole-number factors of a whole number
- To determine whether a number is prime or composite
- To write the prime factorization of a composite number using a tree diagram

1a. Find all the whole-number factors of 36.

1b. What are the whole-number factors of 148?

2a. Determine whether the number 189 is prime or composite.

2b. Determine whether the number 817 is composite or prime.

3a. Determine whether the number 1,001,001 is composite or prime.

3b. List the prime numbers between 110 and 120.

The one-digit numbers 2, 3, 5, and 7 are prime; the one-digit numbers 4, 6, 8, and 9 are composite. By convention,

> The whole numbers 0 and 1 are neither prime nor composite.

Example 2

Is 701 prime or composite?

SOLUTION

The divisibility rules eliminate 2, 3, 4, and 5 as factors. This means that all of their multiples are also eliminated. Next we check for divisibility by the primes 7, 11, 13, 17, 19, 23, and 29, in order, and we find that none of them divides 701. Because 29×29 is larger than 701, we need not check primes larger than 29.

Thus 701 is prime because it has only 1 and 701 as factors.

■ *Now do margin exercises 2a and 2b.*

Sometimes we can identify even very large numbers as composite by using the divisibility rules. Remember, in order to say that the number is composite, we only have to find *one* factor that differs from 1 and the given number.

Example 3

Is 100,110,111 prime or composite?

SOLUTION

This number is not divisible by 2 because it is an odd number. However, it is divisible by 3, because when we add the digits, we get

$$1+0+0+1+1+0+1+1+1=6$$

And because 6 is divisible by 3, the number 100,110,111 is also.

Thus 100,110,111 is composite.

■ *Now do margin exercises 3a and 3b.*

Prime Factorization of Whole Numbers

Factorizations that consist only of prime numbers are useful in much of the work that follows this chapter.

> **Prime Factorization**
>
> The **prime factorization** of a number is the expression of that number as the product of prime numbers (which are then called **prime factors**).

The easiest and most efficient way to find a prime factorization is to use a diagram called a **tree diagram.**

We use a tree diagram to work Example 4.

EXAMPLE 4

Factor 144 into prime factors, using a tree diagram.

SOLUTION

We begin by choosing any two numbers whose product is the number to be factored. (The divisibility rules are helpful.) We'll start here with 12 and 12, because we know that $12 \times 12 = 144$. We write these numbers under the 144, connected to it by "branches."

144

Then we break each of the 12s into any two of its factors and write them in the same way, forming a "tree."

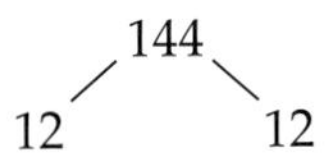

```
     144
    /   \
  12     12
```

We continue in the same way, factoring each of the latest factors. When a number cannot be factored further, we circle it and leave it. The process ends when every number that is on the end of a "branch" is circled.

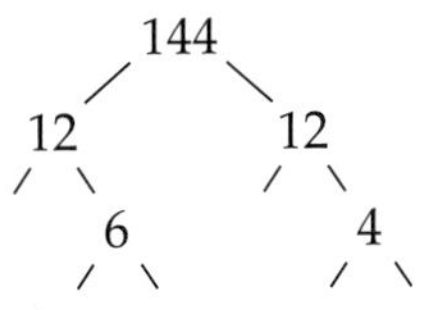

```
      144
     /   \
   12     12
   / \    / \
      6      4
     / \    / \
```

The circled numbers are the prime factors of the original number. We usually write them in numerical order, so we would write the prime factorization of 144 as

$$2 \times 2 \times 2 \times 2 \times 3 \times 3$$

You can check your work by multiplying the prime factors together to get the original number. Try it.

■ *Now do margin exercises 4a and 4b.*

Example 4 illustrates what is known as the Fundamental Theorem of Arithmetic.

4a. Factor 100 into a product of primes.

4b. Factor 2505 into a product of primes.

5a. Factor 192 into a product of primes.

5b. How many different prime factors does 8226 have?

The Fundamental Theorem of Arithmetic

Every whole number greater than 1 can be written as the product of one set, and only one set, of primes.

The order of the prime factors may be changed, but the factors themselves never change.

Next we will factor a number that is not even.

EXAMPLE 5

Factor 339 into a product of primes.

SOLUTION

The divisibility rules tell us that 339 is divisible by 3, so we start with the factors 3 and 113.

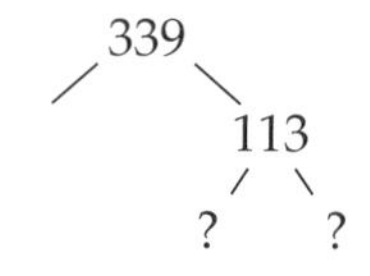

113 is not divisible by 2, 4, 6, 8, or 10, because it is not even.

113 is not divisible by 3 or 9, because the sum of its digits is 5 and 5 is not divisible by 3 or 9.

113 is not divisible by 5, because it doesn't end in 5 or 0.

113 is not divisible by 7, because $113 \div 7 = 16$ with a remainder of 1.

So 113 is *not* divisible by the numbers 2 through 10. We do not have to check 11, because 11×11 is larger than 113.

Thus 113 is prime. Our final tree diagram looks like this:

$339 = 3 \times 113.$

■ *Now do margin exercises 5a and 5b.*

Work the exercises that follow.

Name Section Date

3.2 EXERCISES

In Exercises 1 through 10, list all the whole-number factors for each number.

1. 36 **2.** 60 **3.** 132 **4.** 84 **5.** 308

6. 140 **7.** 220 **8.** 196 **9.** 72 **10.** 120

In Exercises 11 through 18 determine whether each is prime or composite.

11. 169 **12.** 571 **13.** 749 **14.** 207

15. 281 **16.** 313 **17.** 263 **18.** 441

In exercises 19 through 60, factor each into prime factors, using a tree diagram. List the factors in numerical order. If a number is prime, indicate this.

19. 180 **20.** 252 **21.** 108

22. 360 **23.** 396 **24.** 216

25. 504

26. 648

27. 324

28. 105

29. 135

30. 189

31. 405

32. 225

33. 175

34. 525

35. 675

36. 567

37. 275

38. 325

39. 1000

40. 2000

41. 1500

42. 1750

43. 3575

44. 1496

45. 1288

46. 2646

47. 1764

48. 4900

49. 10,000

50. 20,000

51. 15,000

52. 25,000

53. 1001

54. 3003

55. 5050

56. 2020

57. 100,000

58. 200,000

59. 900,000

60. 800,000

3.2 MIXED PRACTICE

By doing these exercises, you will practice all topics up to this point in the chapter.

61. Is 4 a factor of 109,876?

62. Is 703 prime or composite?

63. Write the prime factorization of 2460.

64. Is 643 prime or composite?

65. Is 982,543 divisible by 3?

66. Is 642 prime or composite?

67. Is 263,475 a multiple of 9?

68. Is 876,543,986 divisible by 5?

69. Write the prime factorization of 375.

3.3 Using Exponents

If we were to factor 32 into primes, we would get

$$32 = 2 \times 2 \times 2 \times 2 \times 2$$

There is a simpler way to express this factorization—namely as 2^5, meaning that 2 is a factor 5 times. In 2^5, the number 2 is called the base and the 5 is the exponent.

In the **exponential form** a^n, a is the **base,** and the superscript n is the **exponent.** An exponent indicates how many times the base is used as a factor.

As an example,

$$\underset{\substack{\uparrow \\ \text{base}}}{\overset{\substack{\text{exponent} \\ \downarrow}}{7^4}} = \underbrace{7 \times 7 \times 7 \times 7}_{\text{4 times}} = 2401$$

The expression 7^4 is read "seven to the fourth power," and a^n is read "a to the nth power."

Special names are often used for the exponents 2 and 3:

a^2 can be read "a squared" as well as "a to the second power."

a^3 can be read "a cubed" as well as "a to the third power."

In exponential notation, the exponent indicates how many times the base is used as a factor. Hence a^1 means one factor of a, or simply a.

For all whole numbers a,

$$a^1 = a$$

Here is another concept that we need to know:

For all whole numbers a not equal to zero,

$$a^0 = 1$$

Now let's try expressing a factorization in exponential form.

Section Goals

- To write the prime factorization of a composite number using exponential notation
- To rewrite numbers from exponential notation to standard notation

1a. Factor 216 and express your answer in exponential form.

1b. Stacie says that 729 expressed in exponential form is 3^6. Is she right?

2a. Rewrite as a whole number: $5^2 \times 3^4$

2b. Simplify: $7^1 \times 2^4 \times 3^2$

3a. Calculate 4^6 using your calculator.

3b. Find 3^{12} using your calculator.

EXAMPLE 1

Factor 405 and write your answer in exponential form.

SOLUTION

First we use a tree diagram to factor 405 into prime factors.

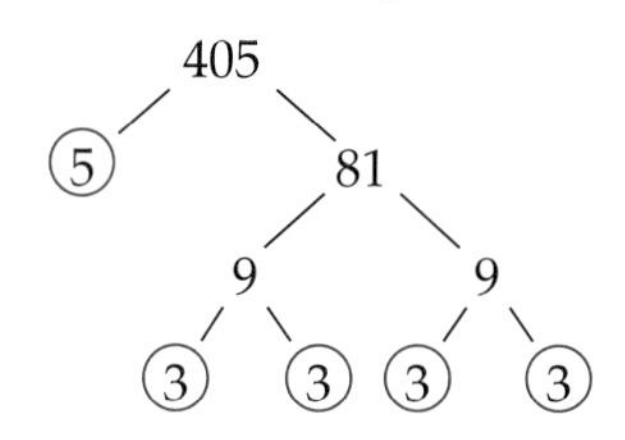

We find that $405 = 3 \times 3 \times 3 \times 3 \times 5$.

Because 3 is used as a factor four times and 5 is used as a factor one time, we get $405 = 3^4 \times 5$.

■ *Now do margin exercises 1a and 1b.*

To "translate" from exponential form to standard notation, simply multiply out as the next example shows.

EXAMPLE 2

Simplify: $2^4 \times 3^2$

SOLUTION

We first simplify 2^4 and then 3^2.

$$2^4 = 2 \times 2 \times 2 \times 2 = 16$$
$$3^2 = 3 \times 3 = 9$$

Then we combine.

$$16 \times 9 = 144$$

Caution: We cannot multiply without first raising the numbers to the given powers.

So

$$2^4 \times 3^2 = 144$$

■ *Now do margin exercises 2a and 2b.*

EXAMPLE 3

Find 3^5 using your calculator.

SOLUTION

First enter 3 [×] 3 [=]
This will give you three squared.

Then press [=]. This will give three cubed. Pressing [=] again will give three to the fourth power, and pressing it once more will give three to the fifth power, or 243.

■ *Now do margin exercises 3a and 3b.*

3.3 EXERCISES

In Exercises 1 through 12, write these prime factorizations in exponential form.

1. $480 = 2 \cdot 2 \cdot 2 \cdot 2 \cdot 3 \cdot 2 \cdot 5$

2. $336 = 3 \cdot 2 \cdot 2 \cdot 2 \cdot 2 \cdot 7$

3. $196 = 2 \cdot 2 \cdot 7 \cdot 7$

4. $630 = 3 \cdot 3 \cdot 7 \cdot 2 \cdot 5$

5. $400 = 2 \cdot 2 \cdot 2 \cdot 2 \cdot 5 \cdot 5$

6. $420 = 2 \cdot 3 \cdot 2 \cdot 5 \cdot 7$

7. $250 = 2 \cdot 5 \cdot 5 \cdot 5$

8. $900 = 3 \cdot 3 \cdot 2 \cdot 2 \cdot 5 \cdot 5$

9. $200 = 2 \cdot 2 \cdot 2 \cdot 5 \cdot 5$

10. $360 = 3 \cdot 3 \cdot 2 \cdot 2 \cdot 2 \cdot 5$

11. $405 = 5 \cdot 3 \cdot 3 \cdot 3 \cdot 3$

12. $700 = 7 \cdot 2 \cdot 5 \cdot 2 \cdot 5$

In Exercises 13 through 28, write the prime factorization for each number by using the exponential form.

13. 50 **14.** 80 **15.** 52 **16.** 40

17. 125 **18.** 81 **19.** 189 **20.** 28

21. 92 **22.** 98 **23.** 27 **24.** 68

25. 75 **26.** 100 **27.** 90 **28.** 60

In Exercises 29 through 42, write each expression in standard form.

29. 41^1 **30.** 29^4 **31.** 16^2 **32.** 11^0

33. 8^4 **34.** 2^5 **35.** 7^3 **36.** 5^4

37. 15^3 **38.** 5^3 **39.** 3^6 **40.** 4^4

41. 1^{12} **42.** 0^2

In Exercises 43 through 48, use your calculator to find the whole number represented by each of the following.

43. 27^4 **44.** 18^5 **45.** 32^3

46. 12^6 **47.** 2^{15} **48.** 15^5

In Exercises 49 through 56, find the whole number represented by each.

49. $18^2 \times 2$ **50.** $20^2 \times 3^2$ **51.** $10^2 \times 4^2$ **52.** $3^2 \times 5$

53. $8^2 \times 4^3$ **54.** $2^5 \times 6^2$ **55.** $12^3 \times 13^0 \times 11^3$ **56.** $3^4 \times 5^3 \times 6^5$

In Exercises 57 through 60, use your calculator to find the whole number represented.

57. $42^2 \times 28^3$ **58.** $21^3 \times 17^4$ **59.** $15^4 \times 9^3$ **60.** $11^3 \times 23^3$

3.3 MIXED PRACTICE

By doing these exercises, you will practice all topics up to this point in the chapter.

61. Write 2^4 in standard form.

62. Is 157 prime or composite?

63. Express the prime factorization of 220 in exponential form.

64. Write the prime factorization of 236.

65. Write $3^2 \times 2^4$ as a whole number.

66. Express the prime factorization of 1500 in exponential form.

67. Write 4^3 as a whole number.

68. Write the prime factorization of 480.

69. Determine whether 387 is prime or composite.

70. Write 5^3 without exponents.

71. Express the prime factorization of 720 in exponential form.

3.4 Using the Standard Order of Operations

SECTION GOAL

- To evaluate an expression using the standard order of operations

In previous chapters, we worked with one operation at a time. We never had to deal with several operations within the same expression. But suppose we wanted to simplify the expression

$$2+4\times 3+7$$

How would we begin? What operation should we perform first? We could start with multiplication, as in solution 1 below. Or we could work from left to right, as in solution 2. Or we could do all the additions first, as in solution 3.

Solution 1	*Solution 2*	*Solution 3*
$2+4\times 3+7$	$2+4\times 3+7$	$2+4\times 3+7$
$2+12+7$	$6\times 3+7$	$2+4\times 10$
$14+7$	$18+7$	6×10
21	25	60

The problem is that these three approaches give three different answers for the same problem! Obviously, we need a standard order for performing operations. Then everyone will understand the same thing when they write or read a series of operations such as $2+4\times 3+7$.

Mathematicians have already agreed on such an order; it is called the **standard order of operations.** We shall use it from now on.

Standard Order of Operations

1. Do all operations inside the grouping symbols—parentheses ( ), brackets [ ], and braces { }—working from the innermost grouping outward.
2. Simplify any expressions that contain exponents.
3. Do all multiplications and divisions, working from left to right.
4. Do all additions and subtractions, working from left to right.

EXAMPLE 1

Simplify: $2(16-4)-5+3$

SOLUTION

Using the standard order of operations, we first work inside the parentheses.

$$2(16-4)-5+3$$
$$2(12)-5+3$$

1a. Simplify:
$3(7-5)+2-3$

1b. Write the following as a numerical expression: Find the sum of 5 and 3, multiply it by 11, add 2, and subtract 3.

2a. Simplify:
$11-6(3-2)^2$

2b. Calculate:
$45-(18-12)^2 \div 3+4$

There are no exponents, so next we multiply.

$$2(12)-5+3$$
$$24-5+3$$

Then we do additions and subtractions from left to right.

$$24-5+3$$
$$19+3$$
$$22$$

Thus we find that $2(16-4)-5+3=22$.

■ *Now do margin exercises 1a and 1b.*

The next example includes an exponent as well as parentheses.

EXAMPLE 2

Simplify: $3^2-2+(16 \div 8)$

SOLUTION

We work inside the parentheses first.

$$3^2-2+(16 \div 8)$$
$$3^2-2+2$$

Then we simplify the exponential notation.

$$3^2-2+2$$
$$9-2+2$$

Finally, we do all additions and subtractions from left to right.

$$9-2+2$$
$$7+2$$
$$9$$

Thus we find that $3^2-2+(16 \div 8)=9$.

■ *Now do margin exercises 2a and 2b.*

Our last example involves the grouping symbols [ ] and { }, called *brackets* and *braces*. We use them like parentheses and always work with the innermost grouping symbols first.

3a. Simplify:
$(5-2)^3 + 2\,(7+4) \div 2$

3b. Simplify:
$[7\,(6-1) + 3\,(9-7) + 1]^2 \div 3$

EXAMPLE 3

Simplify: $\{[12-(4-1)^2]^2 + 6\,(8+5)\} \div 3$

SOLUTION

$$\{[12-(4-1)^2]^2 + 6\,(8+5)\} \div 3$$

$$\{[12-(3)^2]^2 + 6\,(13)\} \div 3$$ Work inside parentheses.

$$\{[12-9]^2 + 6\,(13)\} \div 3$$ Simplify exponent.

$$\{[3]^2 + 6(13)\} \div 3$$ Subtract inside brackets.

$$\{9 + 6(13)\} \div 3$$ Simplify exponent.

$$\{9 + 78\} \div 3$$ Multiply.

$$87 \div 3$$ Add.

$$29$$ Divide.

So $\{[12-(4-1)^2]^2 + 6\,(8+5)\} \div 3 = 29$.

■ *Now work margin exercises 3a and 3b.*

Note: Some calculators process information as it is entered; others use the standard order of operations. Check which you have before you start a series of calculations. Enter

8 [+] 3 [×] 5 [=]

If your display shows 23 as an answer, your calculator uses the standard order of operations.

Work the exercises that follow.

3.4 EXERCISES

In Exercises 1 through 63, simplify each expression.

1. $16 - 5 - 2$

2. $41 - 19 - 8$

3. $56 - 28 - 17$

4. $54 \div 3^2 - 3$

5. $52 - 30 + 12$

6. $38 - 16 + 15$

7. $26 - 14 + 9$

8. $3^3 + 24 \div 3$

9. $8 + 4 \times 3$

10. $26 + 8 \times 6$

11. $35 + 7 \times 5$

12. $17 + 8^2 \times 2$

13. $(11 + 5) \times 4 - 2$

14. $(15 + 7) \times 3 - 11$

15. $33 + (3 \times 2) - 17$

16. $64 - (39 + 5) + 3^4$

17. $(3 \times 5) + 4 \div 2$

18. $(12 \times 4) + 16 \div 4$

19. $(51 \times 2) + 27 \div 3$

20. $110 - (15 - 9)^2 + 65$

21. $13 + 9 \times 3 + 7$

22. $22 + 3 \times 2 + 5$

23. $17 + 4 \times 9 + 4$

24. $9^2 - 21 \div (14 - 11)$

25. $48 \div (4 + 2) - 8$

26. $36 \div (12 + 6) - 1$

27. $104 \div (7 + 1) - 5$

28. $7(9 - 2) + 9^2$

29. $2^2 + 8 \div 4$

30. $3^3 + 15 \div 3$

31. $4^2 + 84 \div 7$

32. $47 - 10 + 27$

33. $27 - 10 + 5^2$

34. $32 - 8 + 2^3$

35. $49 - 4 + 6^2$

36. $71 - 46 + 2^2$

37. $15 + 5^2 \times 9$

38. $11 + 4^3 \times 7$

39. $6 + 7^2 \times 5$

40. $9 + 6 \times 2$

41. $32 \div 2^3 - 4$

42. $80 \div 4^2 - 2$

43. $[(63 \div 9^1)^2 - 4]^2$

44. $42 - [13 - 9]^2$

45. $[6(8 - 4) + 3^2]^2$

46. $[5(12 - 5)^2 + 2^3]^2$

47. $[4(11 - 7)]^2 + 11^2$

48. $\{[72 \div (8 + 4)]^2 - 4\}^2$

49. $10^3 - [50 \div (3 + 2)]^2$

50. $[20^2 - 45 \div (21 - 12)]^2$

51. $5^3 - [30 \div (18 - 13)]^2$

52. $45 + 7 \times (8 - 6)^4$

53. $[96 - (24 + 5)]^2 + 8^2$

54. $[80 - (44 + 28)]^2 + 6^2$

55. $[34 - (14 + 8)]^2 + 4^3$

56. $\{[(29 + 4) \times 2]^2 - 5\}^2$

57. $95 - [(4 - 2)^2 + 9]$

58. $[300 - (8 - 3)^3 + 39]^2$

59. $[86 - (8 - 6)^2]^2 + 20$

60. $(11 \times 6) + (15 \div 5)^3$

61. $(\{[(2)^2 - 1]^2 + 6\}^2 - 5)^2$

62. $[(15 - 28 \div 2^2) \div 2^3]^4$

63. $\{[18 + 2 \times 5] - [16 \div (2 \times 4)]^2 + 7\}^2$

In Exercises 64 through 69, place +, −, ×, and ÷ signs between the numbers in each row to give the indicated answers. Use any grouping symbols you may need.

64. 0 1 2 3 4 5 6 7 8 9 = 10

65. 0 1 2 3 4 5 6 7 8 9 = 10

66. 0 1 2 3 4 5 6 7 8 9 = 10

67. 0 1 2 3 4 5 6 7 8 9 = 10

68. 10 9 8 7 6 5 4 3 2 1 = 0

69. 10 9 8 7 6 5 4 3 2 1 = 0

3.4 MIXED PRACTICE

By doing these exercises, you will practice all topics up to this point in the chapter.

70. Determine whether 153 is prime or composite.

71. Simplify: $(4 - 3)^2 + 3(12 \div 6)$

72. Write the prime factorization of 520 in exponential form.

73. Write the whole number represented by $2^5 \times 4^3$.

74. Simplify: $2^5 \div \{4 + 6^2 - [(3)(2)]^2\}^0$

75. Find all the divisors of 52.

76. Write the prime factorization of 2600 in exponential form.

77. Factor 36 in five different ways.

78. Simplify: $(18 \div 2) + [(12 - 10)^2 + 4]^2$

??? Problem-Solving Preparation: *Using Formulas*

Formulas are used in almost every area of study. Usually, we are given values for all the variables (letter symbols) in the formula except one, and we have to find the value of that variable.

To solve a problem with a formula

1. Substitute all the given numerical values into the formula.
2. Solve for the unknown value as indicated.

In a formula, when letters are written next to each other, multiplication is indicated. That is, ab means "a times b."

EXAMPLE 1

Let $C = IT$, where
C is the total number of ice cubes
I is the number of ice cubes in one tray
T is the number of trays

Find the total number of ice cubes in 288 trays of 14 cubes each.

SOLUTION

$$\begin{aligned} C &= IT \\ &= 14(288) \\ &= 4032 \end{aligned}$$

■ *Now do margin exercises 1a and 1b.*

Work the exercises that follow. Ensure that your answers are reasonable.

1a. Let $P = 20s$ where P is the total number of pecks a woodpecker makes in s seconds. Find P when s is 60.

1b. Let $S = C \div 2 + 4$, where S is the number of cups of sugar needed in a recipe, and C is the number of cups of flour needed. What is the value of S when the number of cups of flour is 38?

EXERCISES

1. Let $d = m \div v$, where

   d is the density of an object in grams per cubic centimeter
   m is the mass of the object in grams
   v is the volume of the object in cubic centimeters

   Find the value of d when the mass of a small plastic block is 44 grams and the volume is 11 cubic centimeters.

2. Let $V = at$, where

   V is the velocity of an object
   a is the rate of acceleration
   t is the elapsed time in seconds

   Find the value of V for a sparrow when a is 16 feet and the elapsed time is 28 seconds.

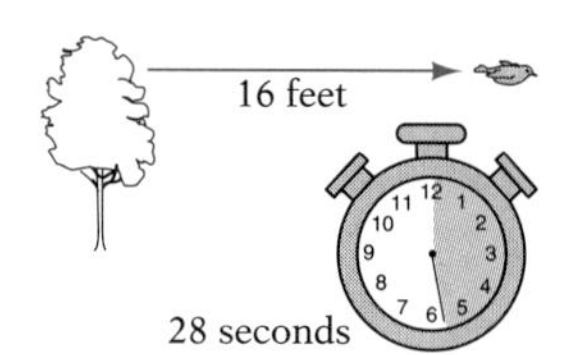

3. Let $S = O - D$, where

   S is the selling price
   O is the original price
   D is the discount

   Find the value of S when the original price is \$2420 and the discount is \$50.

4. Let $Q = L \div M$, where

   Q is the cost per unit
   L is the price of item
   M is the number of units in the item

   Find the value of Q when the price of the item is \$36 and there are 18 units in the item.

5. Let $F = ma$, where

F is the force on an object
m is the mass of the object
a is the rate of acceleration of the object

Find F when a mass of 904 grams has an acceleration of 35 units.

6. Let $C = A \div S$, where

A is the total number of calories
S is the number of servings
C is the number of calories per serving

Find the value of C when there are 157 servings and a total of 18,055 calories.

7. Let $F = D \div U$, where

F is the fuel in gallons used
D is the distance traveled
U is the rate of usage in miles per gallon

Find the amount of fuel used when you get 25 miles per gallon and you travel 1025 miles.

8. Let $P = I^2R$, where

P is the power loss in watts
I is the current in amperes
R is the resistance in ohms

Find the power loss when a current of 6 amperes flows through a resistance of 62 ohms.

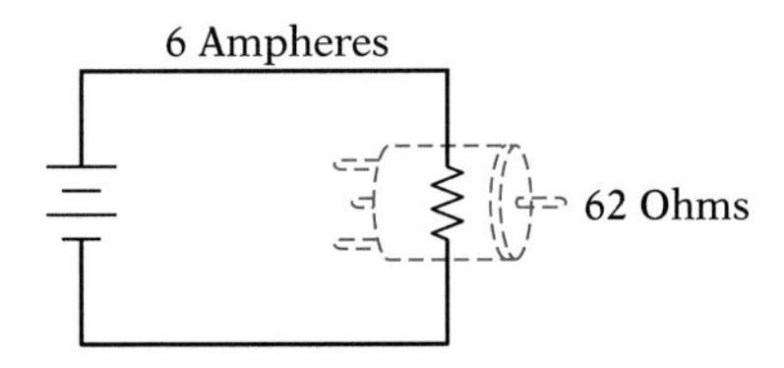

9. Let $rt = d$, where

d is the distance in miles
r is the rate or speed in miles per hour
t is the time in hours

How far will you travel if you move at 45 miles per hour for 17 hours?

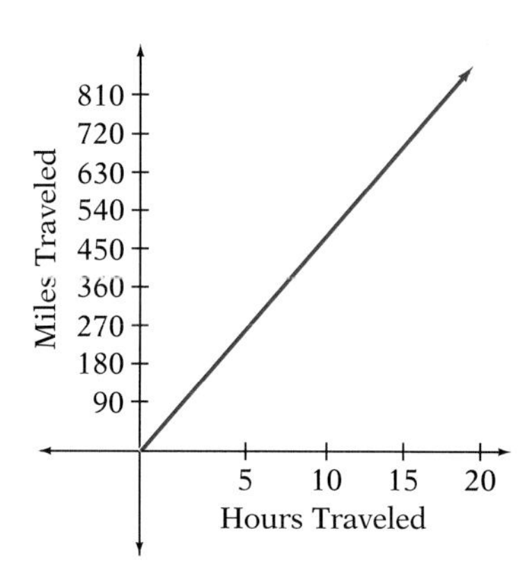

10. Let $A \div (B \times C) = H$, where

A is the annual salary
B is the number of weeks worked
C is the average number of hours worked per week
H is the hourly rate

What is the hourly rate when a person works 52 weeks at an average of 40 hours per week and earns a salary of $52,000?

11. Let $S = NV + X$, where

S is the score on an exam
N is the number of questions correct
V is the value for each question
X is the number of extra credit points

Find the score on an exam when 27 questions are answered correctly at 3 points each, and an additional 5 points are given for extra credit.

12. Let $c + 40 = F$, where

F is the temperature Fahrenheit
c is the number of cricket chirps in 15 seconds

What is the temperature when the crickets are chirping 55 times each 15 seconds?

3.5 Finding Square Roots of Perfect Squares

SECTION GOAL

- To find the square root of a perfect square

You will be working with squares and square roots throughout your study of mathematics.

In general,

Square Root

If $a^2 = a \times a = b$, then a is called the **square root** of b. The symbol for square root is $\sqrt{}$, so $a = \sqrt{b}$ if $a^2 = b$.

Thus $4 \times 4 = 16$ means that 4 is the square root of 16.

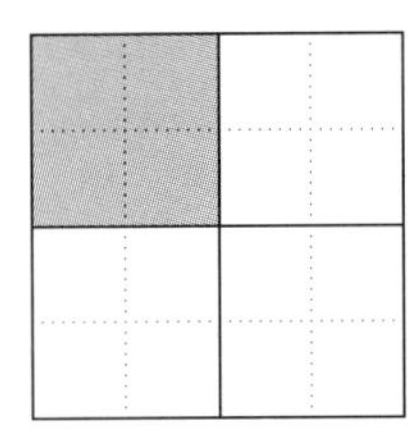

There is a complex algorithm (or rule) for finding square roots. But for some numbers, called perfect squares, we can guess and test to find their square roots.

Perfect Square

A number is a **perfect square** if its square root is a whole number.

EXAMPLE 1

Find $\sqrt{256}$.

SOLUTION

The square root of 256 is larger than 10 because $10^2 = 10 \times 10 = 100$. It is smaller than 20 because $20^2 = 400$. Thus we know that

$$10 < \sqrt{256} < 20$$

Because $\sqrt{256}$ lies between 10 and 20, it has a first digit of 1. Now we need the second digit.

The last digit of 256 is 6, so the last digit of its square root must be 4 or 6. This is true because $4 \times 4 = 16$ and $6 \times 6 = 36$ and no other single-digit number has a square that ends in 6. Try both 14 and 16.

$$14 \times 14 = 196? \quad \text{no}$$
$$16 \times 16 = 256? \quad \text{yes!}$$

Because $16 \times 16 = 256$, the square root of 256 is 16.

- *Now do margin exercises 1a and 1b.*

1a. Find $\sqrt{576}$.

1b. Find $\sqrt{961}$.

2a. Find $\sqrt{7396}$.

2b. Find $\sqrt{5929}$.

3a. Determine whether 46,574 is a perfect square using your calculator.

3b. Determine whether 183,784 is a perfect square using your calculator.

To find the square root of a perfect square

1. Find which two "tens" the square root lies between.
2. Find the possible last digits of the square root.
3. Test the possibilities.

EXAMPLE 2

Find $\sqrt{7569}$.

SOLUTION

The square root of 7569 is smaller than 90 because $90^2 = 8100$. It is larger than 80 because $80^2 = 6400$. Thus we know that $80 < \sqrt{7569} < 90$. Therefore, the first digit of the square root is 8.

The last digit of 7569 is 9, so the last digit of its square root must be 3 or 7. Why? Because $3 \times 3 = 9$ and $7 \times 7 = 49$, and no other digit has a square ending in 9. Try both.

$$83 \times 83 = 6889? \quad \text{no}$$
$$87 \times 87 = 7569? \quad \text{yes!}$$

Because $87 \times 87 = 7569$, the square root of 7569 is 87.

■ *Now do margin exercises 2a and 2b.*

EXAMPLE 3

Find the square roots of 18,496 and 578 using a calculator.

SOLUTION

First key in 18496 [√]. Your display should read 136. Note that you did not need to press the equals key to get the answer.

Now clear the screen and key in 578 [√]. Because the resulting display is not a whole number, 578 is not a perfect square.

■ *Now try margin exercises 3a and 3b.*

Work the exercises that follow.

3.5 EXERCISES

In Exercises 1 through 52, all the numbers are perfect squares. Find the square root of each and check by multiplying.

1. 36 **2.** 144 **3.** 169 **4.** 289

5. 484 **6.** 625 **7.** 784 **8.** 676

9. 324 **10.** 256 **11.** 529 **12.** 729

13. 841 **14.** 1936 **15.** 1024 **16.** 1225

17. 2116 **18.** 1089 **19.** 1681 **20.** 1849

21. 4489 **22.** 6724 **23.** 7744 **24.** 5329

25. 2601 **26.** 2209 **27.** 2916 **28.** 3481

29. 4225 **30.** 4761 **31.** 5625 **32.** 5184

33. 4356 **34.** 3969 **35.** 5041 **36.** 4096

37. 2704 **38.** 3249 **39.** 6241 **40.** 5476

41. 7056 **42.** 7569 **43.** 7921 **44.** 6561

45. 8836 **46.** 9409 **47.** 8464 **48.** 9801

49. 1156 **50.** 1369 **51.** 9604 **52.** 8649

In Exercises 53 through 60, use your calculator to determine whether each of these numbers is a perfect square.

53. 7228 **54.** 6089 **55.** 5776 **56.** 8286

57. 22,504 **58.** 20,739 **59.** 26,244 **60.** 14,404

3.5 MIXED PRACTICE

By doing these exercises, you will practice all topics up to this point in the chapter.

61. Find the square root of 484.

62. Determine whether 237 is prime or composite.

63. Write the prime factorization of 126.

64. Simplify: $[(7-2)^2 - 11]^2 \div 2$

65. Find the square root of 1521.

66. Write the whole number expressed by 3^3.

67. Simplify: $[6^2 \div (4-2)]^2 + 9(3)^2$

68. Find the square root of 1444.

69. Simplify: $[2(9) - 2^2]^4$

70. Find the square root of 576.

71. Find the prime factorization of 400 in exponential form.

72. Simplify: $\{[(40+5) \div 3]^2 - 6\}(2)^2$

73. Write $2^5 \times 3^4 \times 4^5$ in standard notation.

Generating Prime Numbers

Mathematicians have tried for a long time to find how to generate the prime numbers using rules for computation. Let's investigate one of the rules that was discovered using computers.

Start with any number from 0 through 15.	3
Square it.	$3^2 = 3 \times 3 = 9$
Add the number and its square and 17.	$\underline{17}$
The sum is a prime number.	29

In symbols, for any number n from zero through fifteen,

$$n^2 + n + 17 \text{ is a prime number.}$$

Try this with the numbers 1, 5, and 9. Use the rules for divisibility to prove that the number you get is prime.

This rule does not work for a number greater than 15, say 17. Follow along to see why.

$$17 + 17^2 + 17 = 17 + 289 + 17 = 323$$

323 is not prime. It has 17 as a factor as well as 19.

3.6 Geometry and Measurement: Applying Operations on Whole Numbers

In this section, we use formulas to find the perimeter of a rectangle and the area of a square. We also discuss and use the Pythagorean Theorem for right triangles.

SECTION GOALS

- To use a formula to find the perimeter of a rectangle and to find the area of a square
- To apply the Pythagorean Theorem

Perimeter of a Rectangle

Recall that the perimeter of any polygon can be found by adding up the lengths of the sides. For a rectangle, this can be stated in a formula:

To find the perimeter of a rectangle

If ℓ is the length and w is the width, then

$$P = 2\ell + 2w \text{ or } P = 2(\ell + w)$$

The length ℓ and width w are shown in the accompanying figure. (Actually, the labels ℓ and w could be reversed and the formula would still hold.)

ℓ

w $P = 2(\ell + w)$ w

ℓ

Note that in order to use either formula, we need to be familiar with order of operations. Both formulas give the same result when applied properly.

EXAMPLE 1

Find the perimeter of a rectangle with length 6 feet and width 7 feet.

SOLUTION

We use the second formula, substituting the actual length and width.

$$P = 2(\ell + w) = 2(6 + 7)$$

Then, using the standard order of operations, we have

$$P = 2(13) = 26 \text{ feet}$$

As a check, we use the first formula, obtaining

$$\begin{aligned} P &= 2\ell + 2w \\ &= 2(6) + 2(7) \\ &= 12 + 14 = 26 \end{aligned}$$

Therefore, the perimeter is 26 feet.

- *Now do margin exercises 1a and 1b.*

1a. Find the perimeter of a rectangle with length 13 inches and width 29 inches.

1b. Steven is making a picture frame for his wedding picture. The picture is 18 inches by 20 inches. How much wood will he need to buy to have enough to go around the picture?

2a. Find the area of a karate exhibition mat that is square and is 26 feet on each side.

2b. Find the area of a square that is $(5 + 4 \div 2)$ inches on each side.

Area of a Square

Because a square is also a rectangle, we can find its area with the rectangular area formula: Area $= \ell \times w$. But ℓ and w are the same for a square, and they both are equal to the length of a side s. Accordingly,

To find the area of a square

If s is the length of a side of a square, then

$$\text{Area} = s^2$$

This formula is what gives the name "squared" to the exponent 2.

EXAMPLE 2

Find the area of a square with a side measuring 5 inches.

SOLUTION

Substituting in the formula for area of a square, we have:

$$\begin{aligned} A &= s^2 \\ &= 5^2 \\ &= 25 \end{aligned}$$

s
s | Area $= s^2$ | s
s

The area is 25 square inches.

■ *Now do margin exercises 2a and 2b.*

Pythagorean Theorem

A **right triangle** is a triangle that contains a 90-degree angle. This is an angle of the same size as an angle in a square and in a rectangle. The side opposite the 90-degree angle is called the **hypotenuse** of the triangle, and the other two sides are called its **legs** as shown in the next figure. Note the small square that is used to denote the 90-degree angle.

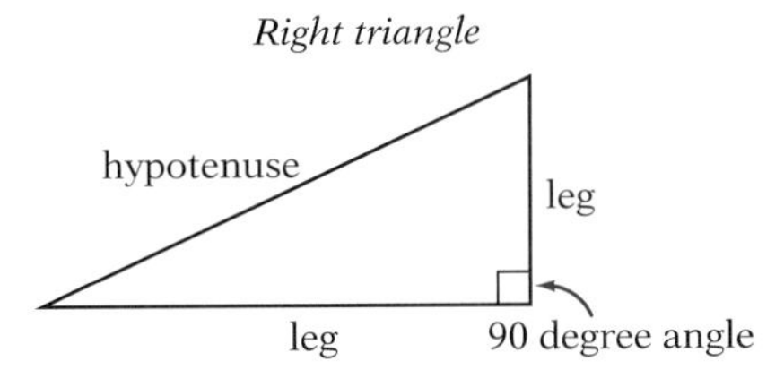

A certain relationship, known as the Pythagorean theorem, exists among the sides of every right triangle.

Pythagorean Theorem

For a right triangle with legs of lengths a and b and hypotenuse of length c,

$$a^2 + b^2 = c^2$$

In other words, the sum of the squares of the lengths of the two legs of a right triangle is equal to the square of the length of the hypotenuse. We can find the third side of any right triangle if we know the other two sides by using this theorem or its variations $c^2 - a^2 = b^2$ and $c^2 - b^2 = a^2$.

3a. Find the length of the longest side of a right triangle if the two shorter sides have lengths 12 in. and 9 in.

3b. Find the length of one of the shorter sides of a right triangle if the other leg has length 12 in. and the hypotenuse is 37 in. long.

4a. Fill in the missing length in the diagram.

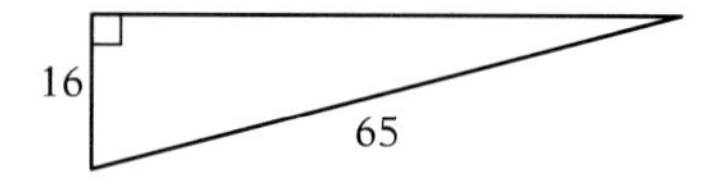

4b. Fill in the missing length in the diagram.

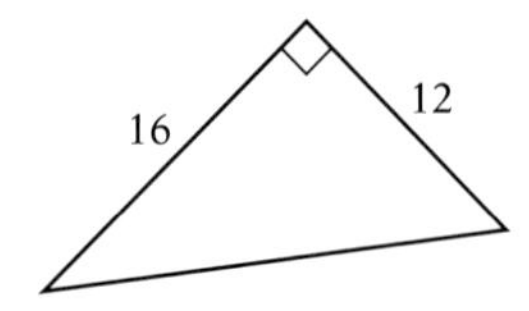

EXAMPLE 3

Find the length of a leg of a right triangle if the other leg is 6 inches and the hypotenuse is 10 inches.

SOLUTION

We substitute the known lengths of the hypotenuse and one leg into the formula

$$c^2 - b^2 = a^2$$

to obtain

$$10^2 - 6^2 = a^2$$

We then simplify according to the order of operations.

$$100 - 36 = a^2$$
$$64 = a^2$$

Remember that a^2 is $a \times a$. Because the square root of 64 is 8, we have $a = 8$.

So the length of the other leg is 8 inches.

■ *Now do margin exercises 3a and 3b.*

Let's look at another example.

EXAMPLE 4

Find the missing dimension.

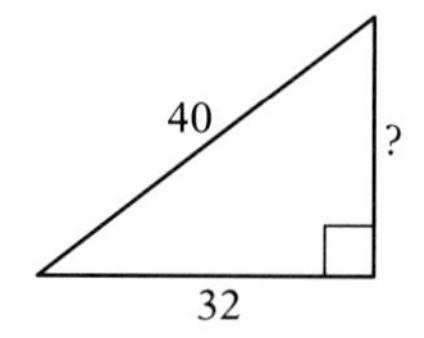

SOLUTION

$c^2 - a^2 = b^2$	The formula
$40^2 - 32^2 = b^2$	Substituting known values
$576 = b^2$	Simplifying
$\sqrt{576} = b$	
$\sqrt{576} = 24 = b$	Finding the square root

So the length of the missing leg is 24 units.

■ *Now do margin exercises 4a and 4b.*

Work the exercises that follow.

3.6 EXERCISES

1. Find the perimeter of a rectangular basketball court that has a length of 28 yards and a width of 15 yards 9 inches.

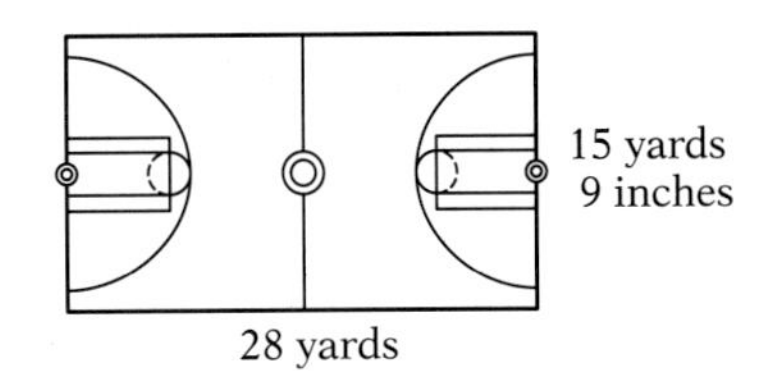

The playing areas described in Exercises 2 through 5 are rectangular. Find their perimeters.

2. U.S. football: 120 yards long, 53 yards 1 foot wide

3. Fencing: length 46 feet, width 6 feet 6 inches

4. Rugby: 160 yards by 75 yards

5. Soccer: 110 yards by 80 yards

6. In a square field the length of one side is 345 yards. Find the area.

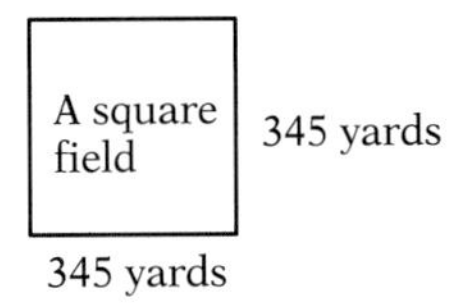

7. Find the area of a square rug if one side is 49 inches long.

8. Find the perimeter of a rectangular ice hockey rink that has length 66 yards 2 feet and width 33 yards 1 foot.

9. In a right triangle find the length of side a given that side $b = 40$ and the hypotenuse is 41.

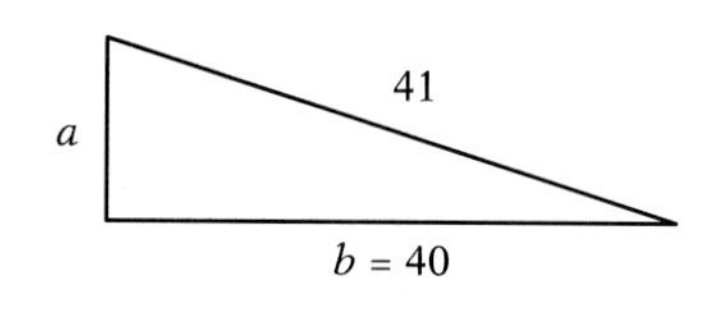

10. Find the perimeter of a square surface if its area is

a. 400 square feet (boxing ring)

b. 8100 square feet (baseball diamond)

11. Find the perimeter of a rectangular netball court that has a length of 33 yards 1 foot and a width of 16 yards 2 feet.

12. A rectangle has length 7 feet and perimeter 48 feet. Find its width if $w = \dfrac{P - 2\ell}{2}$.

13. A rectangle has perimeter 32 feet and width 2 feet. Find its length if $\ell = \dfrac{P - 2w}{2}$.

In Exercises 14 through 17, use the following figure and the given information to find the missing length.

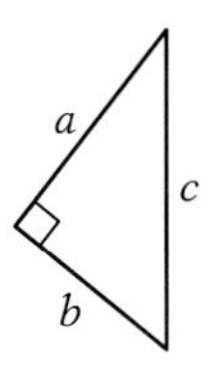

14. $a = 3, b = 4, c =$ ____

15. $a = 12, b =$ ____, $c = 13$

16. $a =$ ____, $b = 12, c = 15$

17. $a = 7, b = 24, c =$ ____

3.6 MIXED PRACTICE

By doing these exercises, you will practice all topics up to this point in the chapter.

18. Find the prime factorization of 108.

19. Simplify: $[(9 - 7)^3]^2 - 16 \div 4^2$

20. Find the square root of 529.

21. Find the square root of 1225.

22. Determine whether 137 is prime or composite.

23. Find the length of the hypotenuse of a right triangle when one leg is 18 units and the second leg is 24 units.

24. Simplify: $\{81 - [8^2 - 7(2)]\}^2$

25. Find the amount of linoleum needed for a square room that is 11 feet on each side.

26. Write the whole number represented by $8^3 \times 4^2$.

27. Find the length of the longest side of a right triangle when the two shorter sides measure 10 units and 24 units.

28. Write the prime factorization of 4500 in exponential form.

29. Simplify: $3(16 - 3) - [2^3 + 2(4)]$

30. Find the amount of fencing that is needed to enclose a rectangular field that is 200 yards by 120 yards.

3.7 Finding the Greatest Common Factor and the Least Common Multiple

Greatest Common Factor

The factors (or divisors) of a number divide evenly into that number without a remainder. Thus the factors of 6 are 1, 2, 3, and 6; and the factors of 8 are 1, 2, 4, and 8. Because 1 and 2 are factors of 6 *and also* are factors of 8, we call them the common factors of these two numbers. The greatest common factor of 6 and 8 is 2. It is the largest number that is a factor of both 6 and 8.

Common Factor

A **common factor** of two numbers is a number that is a factor of both.

Greatest Common Factor

The **greatest common factor** of two numbers is the largest number that is a factor of both.

The greatest common factor is abbreviated GCF. (In some textbooks, you will see it referred to as the greatest common divisor, or GCD.)

EXAMPLE 1

Find the greatest common factor of 24 and 36.

SOLUTION

First we use tree diagrams to find the prime factorizations of the two numbers.

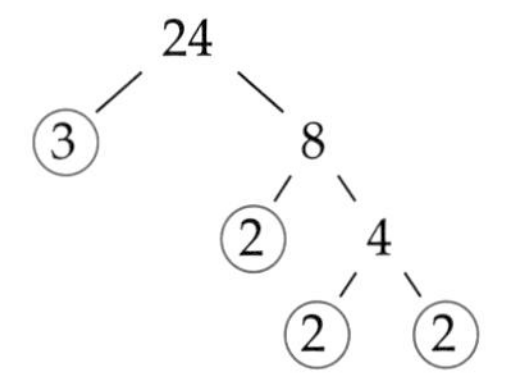

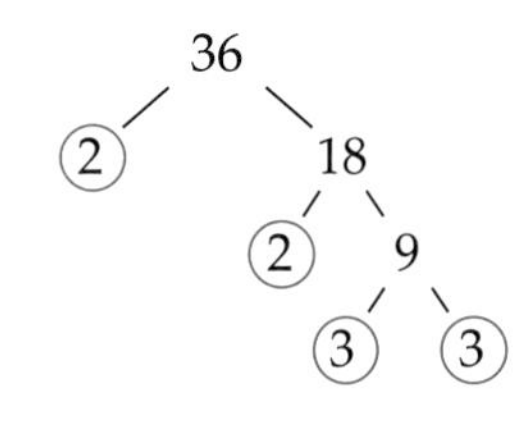

Next we compare the prime factorizations and circle the factors that they have in common.

$$24 = 2 \times 2 \times 2 \times 3$$
$$36 = 2 \times 2 \times 3 \times 3$$

The GCF is the product of the common prime factors: $2 \times 2 \times 3 = 12$.

The GCF of 24 and 36 is 12.

■ *Now do margin exercises 1a and 1b.*

SECTION GOALS

- To find the greatest common factor (GCF) of two or more whole numbers
- To find the least common multiple (LCM) of two or more whole numbers

1a. Find the GCF of 12 and 32.

1b. Find the GCF of 18 and 30.

2a. Find the GCF of 96 and 80.

2b. Find the GCF of 50 and 65.

EXAMPLE 2

Find the GCF of 168 and 126.

SOLUTION

Factoring both numbers completely yields

$$168 = 2 \times 2 \times 2 \times 3 \times 7$$
$$126 = 2 \times 3 \times 3 \times 7$$

The product of the common factors (circled) is $2 \times 3 \times 7 = 42$.

The GCF of 168 and 126 is 42.

■ *Now do margin exercises 2a and 2b.*

Sometimes two numbers have only one factor in common. For example,

$$25 = 5 \times 5$$
$$\text{and} \quad 36 = 2 \times 2 \times 3 \times 3$$

We cannot circle the common prime factors because there are none. However, all numbers have 1 as a factor, and here 1 is the only common factor.

Relatively Prime

Two numbers that have only the factor 1 in common are said to be **relatively prime**.

Hence 25 and 36 are relatively prime.

In this next example, we use exponents to simplify the search for a GCF.

EXAMPLE 3

Find the GCF of 100 and 75.

SOLUTION

If we factor 100 and 75 and write the results in exponential form, we get

$$100 = 2 \times 2 \times 5 \times 5 = 2^2 \times 5^2$$
$$75 = 3 \times 5 \times 5 = 3 \times 5^2$$

Circling the common factors (5×5) is the same as finding the common bases (only 5 here) and the least exponents of those bases (here 5^2).

Therefore, the GCF is 5^2, or $5 \times 5 = 25$.

■ *Now do margin exercises 3a and 3b.*

In general, then,

To find the greatest common factor

1. Factor the numbers and write them in exponential form.
2. List the common bases.
3. Give each base its least exponent from either factorization, and multiply.

We use this procedure to work one more example.

EXAMPLE 4

Find the GCF of 126 and 48.

SOLUTION

$$126 = 2^1 \times 3^2 \times 7$$
$$48 = 2^4 \times 3^1$$

The common bases are 2 and 3. The smallest exponent of 2 is 1, and the smallest exponent of 3 is also 1.

So the GCF is $2^1 \times 3^1 = 6$.

■ *Now do margin exercises 4a and 4b.*

3a. Find the GCF of 120 and 80.

3b. Find the GCF of 144 and 72.

4a. Find the GCF of 120 and 64.

4b. Find the GCF of 200 and 160.

Least Common Multiple

Recall that a whole number a is a *multiple* of another whole number b if b is a factor of a—that is, if b divides a without remainder.

Common Multiple

A number is a **common multiple** of two numbers if it is a multiple of both numbers.

For example, 36 is a common multiple of 4 and 6 because it is a multiple of 4 (it is 4×9) *and* a multiple of 6 (it is 6×6).

If we list the first several multiples of 4 and 6, we can see that they have common multiples other than 36.

Multiples of 4: 4, 8, (12), 16, 20, (24), 28, 32, (36)
Multiples of 6: 6, (12), 18, (24), 30, (36)

We are interested mainly in their least common multiple, 12.

Least Common Multiple

The *least common multiple* (LCM) of two numbers is the least whole number that is a multiple of both.

Therefore, 12 is the LCM of 4 and 6.

EXAMPLE 5

Find the least common multiple (LCM) of 3 and 5.

SOLUTION

First list the multiples of 3 and of 5.

Multiples of 3: 3, 6, 9, 12, 15, 18
Multiples of 5: 5, 10, 15

We stop at 15 because, as you can see, 15 is a multiple of both 3 and 5 and is the smallest whole number that is such a multiple.

So the LCM of 3 and 5 is 15.

■ *Now do margin exercises 5a and 5b.*

Example 5 illustrates an important property.

When two or more numbers are **relatively prime,** their least common multiple is equal to their product.

We need a different procedure when we are trying to find the least common multiples of larger numbers. We will use what we learned about the greatest common factor and exponential notation to help us.

EXAMPLE 6

Find the LCM of 24 and 36.

SOLUTION

We factor each number and write the results using exponential notation.

$$24 = 2 \times 2 \times 2 \times 3 = 2^3 \times 3$$

$$36 = 2 \times 2 \times 3 \times 3 = 2^2 \times 3^2$$

5a. Find the LCM of 7 and 9.

5b. Find the LCM of 5 and 7.

6a. Find the LCM of 15 and 35.

6b. Find the LCM of 63 and 60.

7a. Find the LCM of 36, 18, and 27.

7b. Find the LCM of 44, 11, and 66.

The LCM of these two numbers is the smallest number that has both $2^3 \times 3$ and $2^2 \times 3^2$ in its factorization. That is the number that has both bases, 2 and 3, to the *highest* exponents: It is

$$2^3 \times 3^2$$

Therefore, their LCM is $2^3 \times 3^2 = 8 \times 9 = 72$.

■ *Now do margin exercises 6a and 6b.*

To find the least common multiple (LCM)

1. Factor the numbers and write them in exponential form.
2. List all bases found in any of the numbers.
3. Give each of those bases its greatest exponent in either factorization, and multiply.

Let's do another example using this approach.

EXAMPLE 7

Find the LCM of 48, 64, and 10.

SOLUTION

$$48 = 2 \times 2 \times 2 \times 2 \times 3 = 2^4 \times 3$$
$$64 = 2 \times 2 \times 2 \times 2 \times 2 \times 2 = 2^6$$
$$10 = 2 \times 5$$

The bases are 2, 3, and 5. The greatest exponent of 2 is 6 (in 64), the greatest exponent of 3 is 1, and the greatest exponent of 5 is also 1.

Therefore, the LCM is

$$2^6 \times 3^1 \times 5^1 = 64 \times 3 \times 5 = 960.$$

■ *Now do margin exercises 7a and 7b.*

Work the exercises that follow.

3.7 EXERCISES

In Exercises 1 through 24, find the GCF of each group of numbers.

1. 22; 11

2. 81; 99

3. 94; 12

4. 56; 14

5. 68; 34

6. 72; 90

7. 95; 38

8. 70; 42

9. 85; 51

10. 16; 98

11. 64; 82

12. 84; 54

13. 92; 52

14. 78; 66

15. 80; 25

16. 55; 88

17. 54; 48; 6

18. 21; 84; 10

19. 46; 69; 9

20. 74; 32; 16

21. 100; 60; 25

22. 26; 69; 3

23. 51; 68; 42

24. 36; 81; 12

In Exercises 25 through 32, determine which pairs of numbers are relatively prime.

25. 18; 4

26. 27; 9

27. 28; 8

28. 39; 8

29. 60; 19

30. 15; 56

31. 57; 19

32. 66; 17

In Exercises 33 through 60, find the LCM of each group of numbers.

33. 76; 24

34. 77; 49

35. 50; 20

36. 88; 16

37. 14; 21

38. 42; 22

39. 24; 28

40. 98; 49

41. 32; 40

42. 25; 35

43. 52; 13

44. 30; 45

45. 82; 41

46. 96; 63

47. 34; 17

48. 45; 75

49. 65; 26

50. 90; 72

51. 44; 99

52. 75; 100

53. 38; 57; 19

54. 40; 64; 36

55. 48; 30; 20

56. 86; 43; 4

57. 35; 77; 28

58. 58; 29; 4

59. 87; 15; 10

60. 62; 31; 6

3.7 MIXED PRACTICE

By doing these exercises, you will practice all topics up to this point in the chapter.

61. Find the GCF of 84 and 128.

62. Simplify: $9^2 - 1(2 + 4)^2$

63. Find the GCF of 60 and 96.

64. Write the prime factorization of 1448 using exponents.

65. Find the LCM of 42 and 36.

66. Find the GCF of 248 and 31.

67. Simplify: $(13^3 - 8^2)^2 + 9(6)^2$

68. Find the square root of 676.

69. Find the LCM of 14 and 84.

70. Find the GCF of 90 and 400.

71. Is 1,029,389,478 divisible by 6?

72. Find the LCM of 25, 30, and 15.

73. Simplify: $[2(9 - 2^2)]^3 + [5(4)](3)^2$

74. Simplify: $(8^2 + 6^3)^2 - 5(8^2 - 4^3)^2$

75. Find the LCM of 40 and 56.

An Application to Statistics: *Reading Broken-Line Graphs*

In some graphs, points are plotted and connected by lines to present information. The following figure is such a **broken-line graph.** The height of each dot represents the average price of tea, in cents per pound, for the year shown directly beneath it. The dots are connected by lines so that readers can more clearly see the changes, up and down, in the data.

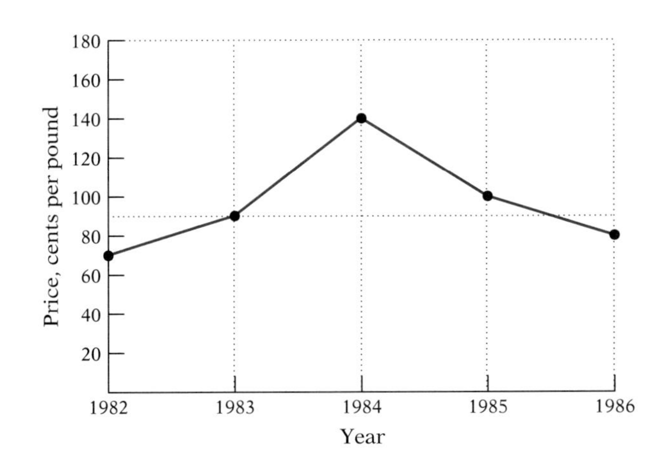

EXAMPLE 1

In what year did tea first average more than 90 cents per pound?

SOLUTION

We locate 90 cents on the price scale at the left. A horizontal line at 90 cents crosses the graph just about at the 1983 mark.

So tea first cost more than 90 cents a pound in 1983.

■ *Now do margin exercises 1a and 1b.*

We can sometimes use a broken-line graph to represent two or more sets of data. The next example shows such a use.

1a. Use the figure above Example 1 to find in what year tea was the most expensive.

1b. Use the figure above Example 1 to find the average cost of tea in 1985.

2a. Use the graph in Example 2 to determine what year had the least difference in prices for coffee and tea.

2b. Use the graph in Example 2 to determine which product had the highest price increase for one year.

EXAMPLE 2

This next graph shows the average prices in cents per pound of imported coffee, cocoa, and tea from 1982 to 1986.

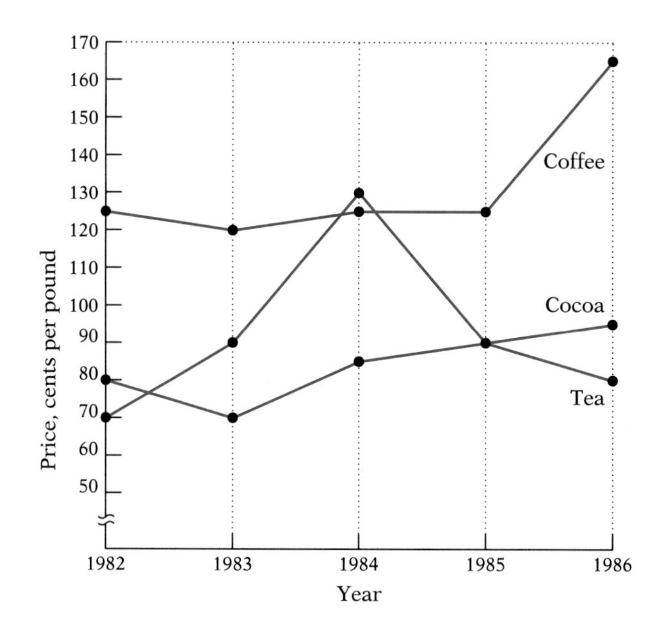

In what year did the price of tea exceed that of coffee and cocoa?

SOLUTION

As we look at this figure, we can see that the graph for tea is higher than that for coffee and cocoa in just one point. If we look directly beneath that point, we see the year 1984. Thus, the cost of tea exceeded that of coffee and that of cocoa in 1984.

■ *Now do margin exercises 2a and 2b.*

Work the exercises that follow. Remember to make sure that your answers are reasonable.

EXERCISES

Use the following graph to answer Exercises 1 through 5.

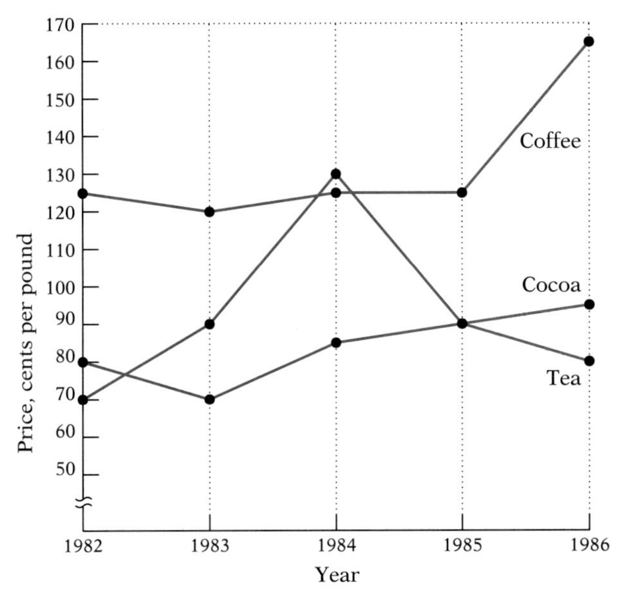

1. What is the difference in cost between the highest and lowest prices for 1986?

2. In what year were the costs of cocoa and tea the same?

3. In 1984, what was the difference in the prices of coffee and cocoa?

4. In 1985, what would you have paid altogether for a pound of cocoa and a pound of tea?

5. If you had bought a pound each of coffee, cocoa, and tea in 1986, about what would you have paid altogether (to the nearest 10 cents)?

The following graph shows the winning times for the Olympic 800-meter run for the years 1896 through 1960. Use this graph to answer Exercises 6 through 8.

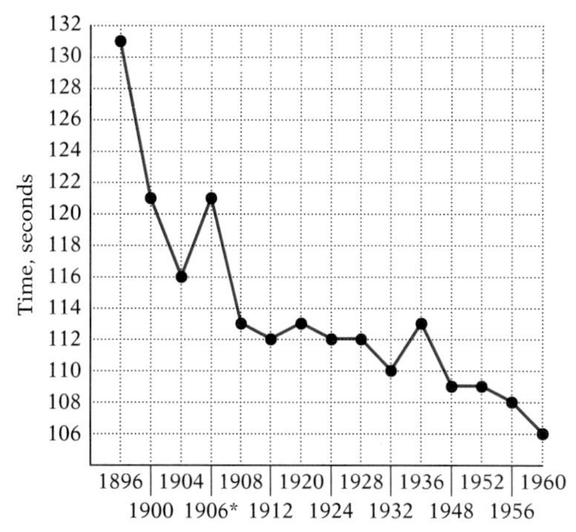

*Interim games in Athens but *not* official Olympics games.

6. What is the difference between the longest and the shortest record times shown?

7. During which two pairs of succeeding Olympics did the winning time remain the same?

8. Make up three additional questions about the graph and answer them.

CHAPTER 3 REVIEW

ERROR ANALYSIS

These problems have been worked incorrectly. Tell what the error is and then write the correct solution.

1. Write 3^5 as a whole number.

Incorrect Solution

$3^5 = 3(5) = 15$

Error ______________________________

____

Correct Solution

2. Find $\sqrt{144}$

Incorrect Solution

$\sqrt{144} = 72$

Error ______________________________

____

Correct Solution

3. Write 2^3 as a whole number.

Incorrect Solution

$2^3 = 2(2)(2) = 6$

Error ______________________________

____

Correct Solution

4. Simplify: $2 + 4 \times 5 + 6$

Incorrect Solution

$2 + 4 \times 5 + 6$

$6 \times 5 + 6$

$30 + 6$

36

Error ______________________________

____

Correct Solution

5. Simplify: $85 - (4 - 2)^3 + 8$

Incorrect Solution

$85 - (4 - 2)^3 + 8$

$85 - 64 - 8 + 8$

$21 - 8 + 8$

$13 + 8$

21

Error ______________________________

____

Correct Solution

6. Find the LCM of 2 and 4.

Incorrect Solution

Because $2 = 2^1$ and $4 = 2^2$, the LCM is 2^1, or 2.

Correct Solution

Error ________________________________

____

INTERPRETING MATHEMATICS

By working these exercises, you will test and strengthen your mathematics vocabulary.

Use the numbers 1 through 12 to write sentences containing the words underlined in Exercises 1 through 8.

1. multiple and divisor

2. divisible

3. even and odd

4. prime number and composite number

5. common multiple and common factor

6. least common multiple

7. greatest common factor

8. relatively prime

9. Make a tree diagram and use the results to write a sentence about each of the following terms: factor, prime factors, exponential notation, and the fundamental theorem of arithmetic.

10. Write the number 12 in exponential form and identify all bases and exponents.

Use the numbers 2, 3, 6, 8, 9, and 36 to help explain the words listed in Exercises 11 through 14.

11. squared

12. cubed

13. perfect square

14. square root

15. Use a diagram of a right triangle to identify the legs and the hypotenuse.

16. Restate the Pythagorean theorem in your own words.

REVIEW PROBLEMS

The exercises that follow will give you a good review of the material presented in this chapter. Work through them and check your answers at the back of the book.

Section 3.1

Use divisibility rules to determine which of the following numbers is divisible by 2, 3, 4, 5, 6, 8, 9, 10, or 12.

1. 1,121,314

2. 13,243,546

3. 111,111

4. 12,000

5. one million

6. 1,001,001

Section 3.2

7. Determine whether 223 is prime or composite.

8. Write all the whole-number factors of 230.

9. Factor 428 into a product of primes.

10. Write the prime factorization for 85.

11. Write the prime factorization for 12,000.

12. Is 1001 prime or composite? If it is composite, write its prime factorization.

Section 3.3

Write each of the following as a whole number.

13. $1^3 \times 2^3$

14. $1^{10} \times 6^2$

15. $2^2 \times 3^4 \times 5^0$

Write the prime factorization for each of the following in exponential form.

16. 6000

17. 445

18. 1060

Section 3.4

Simplify using the standard order of operations.

19. $5(42 \div 7) - 8 + 6$

20. $(8 - 4)^2 + 15 - 7(4)$

21. $16 \div 4 + 3 - 6$

22. $2^5 - [9(4) + 4^2] \div 4$

23. $12 - 8 + 16 \div 4 + 9$

24. $73 - 3^2 + 20 \div 4$

Section 3.5

Find the square root of each of the following perfect squares.

25. 3364

26. 1089

27. 1681

28. 729

29. Find the whole-number square root of 486 if it exists.

30. Find the whole-number square root of 3610 if it exists.

Section 3.6

31. Find the area of a square with a side equal to 14 feet.

32. Find the length of the longest side of a right triangle when the two shorter sides are 21 feet and 28 feet.

33. A rectangle with sides of 14 inches and 17 inches has what perimeter?

34. Is a triangle with sides of 60, 80, and 100 inches a right triangle?

35. A square has a perimeter of 64 feet. What is its area?

36. What are the area and the perimeter of a square with one side equal to 19 meters?

Section 3.7

In Exercises 37 and 38, find the least common multiple.

37. 48 and 56

38. 240 and 600

In Exercises 39 and 40, find the greatest common factor.

39. 125 and 375

40. 64 and 400

41. Find the least common multiple and the greatest common factor of 100, 75, and 15.

42. Are 35 and 50 relatively prime or not?

Mixed Review

43. Find the prime factorization of 540.

44. List all the whole-number factors of 650.

45. Factor 1000 into the product of primes.

46. Simplify: $5 + 6 \times 3^2 - 21 \div 7$.

47. Find the square root of 324.

48. A rental agency advertises that it has a square floor that is 120 feet on a side in an office for rent. How many square feet is this?

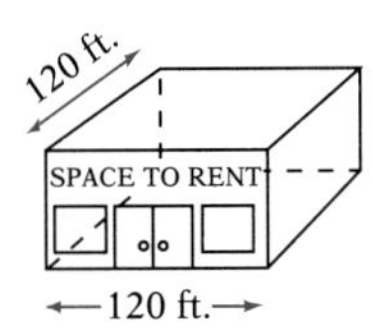

49. A right triangle has sides of 11 and 60 feet. What is the length of the hypotenuse?

50. Find the square root of 484.

51. Find the number of square feet of sod needed to cover a square plot 151 feet on a side.

52. Find the GCF of 48 and 56.

53. Find the least common multiple and the greatest common factor of 20, 50, and 75.

54. Factor 160 into the product of primes.

55. A triangle has sides of 24, 25, and 7 feet. Is it a right triangle? Justify your answer.

56. Find the square root of 361.

CHAPTER 3 TEST

This exam tests your knowledge of the topics in Chapter 3.

1. Use the divisibility rules to answer the following questions.

   a. Which of the following are divisible by 2, 4, and 8?

   20,000 1,888,872 244

   b. Is 1,234,567,890 divisible by 12?

   c. True or False? All numbers divisible by 10 are also divisible by 2 and 5.

2. a. Find all the whole-number factors of 56.

   b. Is 329 prime or composite?

   c. Write the prime factorization of 225 using a tree diagram.

3. a. Write the prime factorization of 245 using exponential notation.

   b. Express $5^1 \times 6^3$ as a whole number.

   c. Express $5^0 \times 9^1 \times 2^4$ as a whole number.

4. Simplify the following expressions.

   a. $[2(3-2)]^3 + (8-5)^2$

   b. $2(6 + 12 \div 4)^2 + 8^2$

   c. $14(8 \div 4)^2 - 7 + 5 \times 2$

5. Find the square root of each of the following numbers.
   a. 40,000

   b. 144

   c. 729

6. Solve the following application problems.
   a. Find the perimeter of a rectangular playing field with a side of 93 feet and a width of 39 feet.

   b. What is the area of a square mirror that has a side of 37 inches?

   c. Find the length of the hypotenuse of a right triangle given that the lengths of the two shorter sides are 8 inches and 15 inches.

7. Find the GCF of
   a. 18 and 42

   b. 24, 100, and 36

   c. 225 and 625

8. Find the LCM of
   a. 15 and 20

   b. 17 and 51

   c. 36, 48, and 12

CUMULATIVE REVIEW

CHAPTERS 1–3

The questions that follow will help you maintain the skills you have developed.

1. Add: $2365 + 984 + 1726 + 99$

2. Subtract: $10{,}003 - 875$

3. Multiply: 206×340

4. Divide: $4200 \div 15$

5. Subtract 23 hours from 4 days 10 hours.

6. Determine the place value for the digit 6 in the numeral 364,875.

7. Write the numbers 27,365 and 1,000,050 in words.

8. Write the prime factorization of 520 in exponential form.

9. Find the length of the hypotenuse of a right triangle with legs 20 inches and 21 inches.

10. Evaluate this expression: $2^6 - [9(3) + 18] - 4 \times 3$

11. Find all the factors of 200.

12. Find the square root of 1764.

13. Add: $263 + 815 + 974$

14. Multiply: $\begin{array}{r} 124 \\ \underline{\times\ 38} \end{array}$

15. According to *Variety* magazine, the top-earning video movie rentals in 1986 were *Top Gun,* which brought in $82,000,000 in rentals, and *The Karate Kid,* which brought in $56,936,752. How much more money was made from rentals of *Top Gun?*

16. Find the area of a parallelogram with a base 35 inches long and a height of 18 inches.

17. Rewrite 11^2 as a whole number.

18. Write this number as a numeral: five million, ninety thousand, three

19. Divide: $7254 \div 18$

20. Is 931 prime or composite? If it is composite, write its prime factorization.

4

Adding and Subtracting Fractions and Mixed Numbers

Listed Tuesday through Saturday in the daily paper are the results of the stock-market trading for the previous day. Typically, a stock might be worth $\frac{1}{8}$ of a point less, or $1\frac{1}{4}$ of a point more at the end of the day than at the beginning—the amount and direction of the change depends on the particular stock. The behavior of a certain stock over several months tells investors whether the stock is stable, gaining, or losing value.

- *Select a stock from those listed in the paper. Keep a record of the change in value from day to day over the next week.*

Skills Check

*T*ake this short quiz to see how well prepared you are for Chapter 4. The answers are given at the bottom of the page. So are the sections to review if you get any answers wrong.

1. Multiply: 603×85

2. Divide: $2280 \div 12$

3. List the whole-number factors of 36.

4. Write the prime factorization for 144.

5. Determine whether 10,986,532 is divisible by 2, 3, 4, 5, 6, 8, 9, and/or 10. Do not actually divide.

6. Find the least common multiple of 8 and 12.

7. Find the greatest common factor of 24 and 18.

8. Subtract: $103 - 78$

9. Add: $9 + 780 + 24$

10. Which is greater: 90732 or 90743?

ANSWERS: 1. 51,255 [Section 2.2] 2. 190 [Section 2.4] 3. 1, 2, 3, 4, 6, 9, 12, 18, 36 [Section 3.2] 4. $2 \times 2 \times 2 \times 2 \times 3 \times 3$ [Section 3.2] 5. divisible by 2 and 4 only [Section 3.1] 6. 24 [Section 3.7] 7. 6 [Section 3.7] 8. 25 [Section 1.5] 9. 813 [Section 1.4] 10. 90743 [Section 1.2]

4.1 Naming Fractions and Mixed Numbers

SECTION GOALS

- To use fractional notation for parts of a whole and part of a collection
- To write an improper fraction as a mixed number, and vice versa

As you have seen in the applications, whole numbers are useful for many situations. However, we often need to use numbers that represent an amount less than 1 or that convey the idea of a part of a whole. For example, few people can eat a whole pizza or are a whole number of feet tall.

Fractions

The numbers that we use to show parts of a whole are most often written in the form $\frac{a}{b}$, where a and b are whole numbers and b is not zero. The number $\frac{a}{b}$ is called a **fraction.** It is sometimes also written a/b. In either case, a is called the **numerator,** and b is called the **denominator.**

Fraction

fraction bar → $\frac{a}{b}$ ← numerator ← denominator

Some examples of fractions are $\frac{1}{2}, \frac{4}{3}, \frac{3}{4}, \frac{5}{19}, \frac{21}{197}$, and $\frac{601}{10{,}000}$. You may recognize the form $\frac{a}{b}$ as one we saw in division. In fact, the fraction $\frac{5}{19}$ can be interpreted to mean $5 \div 19$, and any fraction can be considered to indicate a division. Later, we will show that the positive fractions are part of a larger set of numbers called the rational numbers.

The accompanying figure shows several situations that are described by the fraction *one-half* ($\frac{1}{2}$). In these pictures, one out of every two parts is shaded in color. Also, one out of every two parts is not shaded. The fraction $\frac{1}{2}$ thus indicates that one of every two parts is of interest.

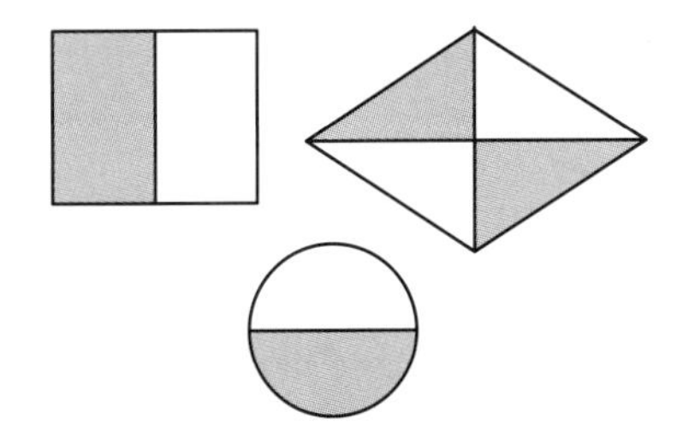

In the next figure, three of every four parts of each picture are shaded. We say that *three-fourths*, written $\frac{3}{4}$, are shaded in each case. Also, one of every four parts is not shaded, so *one-fourth* ($\frac{1}{4}$) is not shaded in each case.

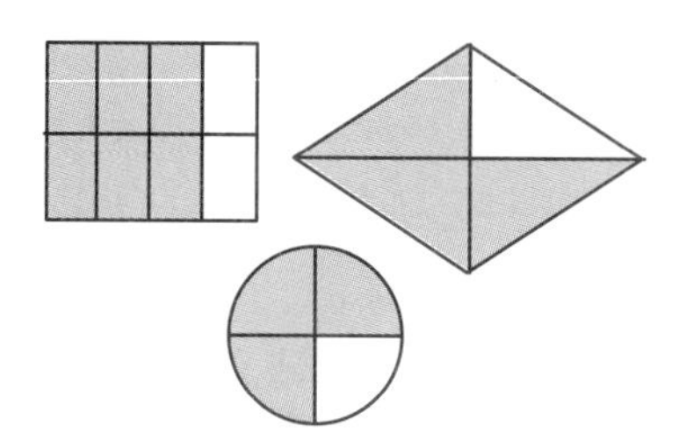

In a fraction, the **denominator** represents the total number of parts or elements available in the "whole"; the **numerator** represents the number of those parts that are of interest or that have a certain characteristic.

We can use fractions to represent amounts less than 1, equal to 1, and greater than 1.

In the following figure, all the parts of each *whole* are shaded. That is, in each case the entire figure is shaded. This is shown by the fractions beneath the pictures ($\frac{2}{2}$, $\frac{4}{4}$, and $\frac{6}{6}$), in which the numerator and denominator are the same. Such fractions represent the number 1. They are called **unit fractions.**

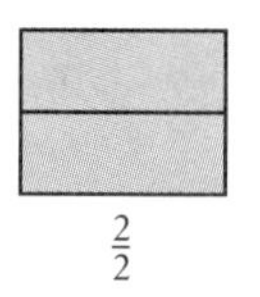

$\frac{2}{2}$

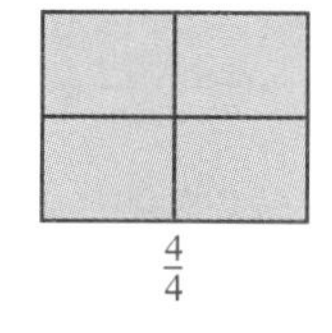

$\frac{4}{4}$

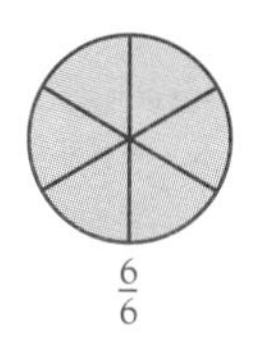

$\frac{6}{6}$

The fraction $\frac{a}{a}$, where a is any whole number except zero, represents the number 1. That is,

$$\frac{a}{a} = 1 \text{ for } a \text{ not equal to } 0$$

Look again at the previous figure. To represent the parts of the shape that are *not* shaded, we can use the fractions $\frac{0}{2}$, $\frac{0}{4}$, and $\frac{0}{6}$, respectively. Such fractions represent *no* parts of the figure, or the whole number zero.

The fraction $\frac{0}{a}$, where a is any non-zero whole number, represents the number 0. That is

$$\frac{0}{a} = 0 \text{ for } a \text{ not equal to } 0$$

We can use fractions to describe certain situations. Let's look at some examples.

EXAMPLE 1

There were 20 people running the Fifth Avenue Mile. Of these, 14 were expected to run the mile in less than 4 minutes. What fraction of the total field was expected to run the mile in less than 4 minutes?

SOLUTION

The total number of elements (people) is 20. The number of those expected to take less than 4 minutes is 14. The fraction, then, has 20 as its denominator and 14 as its numerator.

$$\frac{\text{Expected to take less than 4 min}}{\text{Total number of runners}} = \frac{14}{20}$$

■ *Now do margin exercises 1a and 1b. The answers for all margin exercises are at the back of the book.*

Let's try another one. This time, we will be looking for a number not given in the original problem.

EXAMPLE 2

A subway car has 80 people riding in it. Of these, 60 are males. What fraction of the car riders are female?

SOLUTION

80 is the total number of people in the subway car. 60 are males. Therefore, 80 – 60, or 20, people are female (not males).

$$\frac{\text{Females}}{\text{Subway riders}} = \frac{20}{80}$$

Thus the fraction that represents females is $\frac{20}{80}$.

■ *Now do margin exercises 2a and 2b.*

EXAMPLE 3

You flip two coins together 4 times and get two heads once. What fraction represents the times you did *not* get two heads?

SOLUTION

If you got two heads 1 time in 4 flips, then you did not get two heads 3 times.

The fraction that represents not getting two heads is $\frac{3}{4}$.

■ *Now do margin exercises 3a and 3b.*

1a. There will be about 22,000 runners in the New York City marathon. The number of runners expected to complete the 26.2 miles is 21,500. What fraction is expected to complete the run?

1b. Howard pays all his "consumer goods" bills each month. He now owes $300, and he will pay $200 of this. What fraction of his consumer goods bills will he pay?

2a. A bag contains 293 red and black markers. 200 of these are black. What fraction of the markers is red?

2b. Sansi buys mystery and science fiction books each month. Last month she bought 19 books. 17 of these are mysteries. What fraction of the books were science fiction?

3a. Students in the Math Learning Center study 60 objectives and must master 51. What fraction of the objectives do they *not* have to master?

3b. The chances of winning a raffle for a piece of luggage are 4 in 500. What fraction represents the chances of *not* winning the raffle?

Mixed Numbers

In most of the fractions we have looked at so far, the numerator has been less than the denominator. Such fractions are called proper fractions.

Proper fractions

Fractions in which the numerator is less than the denominator are called **proper fractions.** Proper fractions represent numbers that are less than 1.

The fractions in these figures are *not* proper fractions.

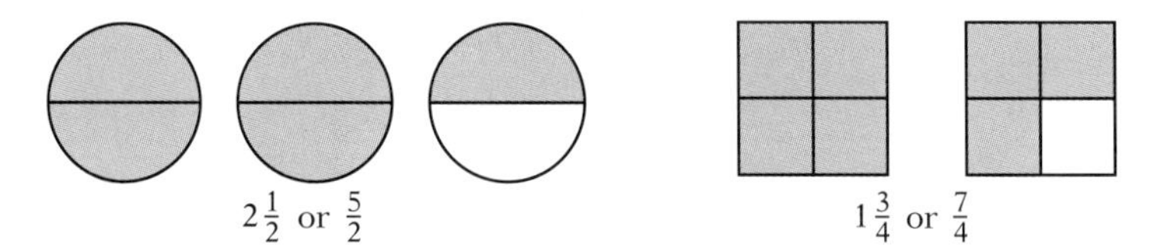

$2\frac{1}{2}$ or $\frac{5}{2}$ $1\frac{3}{4}$ or $\frac{7}{4}$

In the preceding figures, each grouping represents a number greater than 1. Two ways of showing this are indicated in the figure. One way is to combine a whole number and a fraction in a "mixed number." The other way is to write an "improper fraction," one in which the numerator is greater than the denominator. For example, in the grouping to the left, we can write the mixed number $2\frac{1}{2}$ to mean two wholes and one-half of a whole, or we can write the improper fraction $\frac{5}{2}$ to mean that each whole is broken into two halves and we have five of the halves. These two representations are equivalent.

Improper Fractions

Fractions in which the numerator is greater than or equal to the denominator are called **improper fractions.** Improper fractions represent numbers that are greater than or equal to 1.

Unit Fractions

Fractions that represent one are also called **unit fractions.**

Mixed Number

A proper fraction together with a whole number is called a **mixed number.**

Note that a mixed number represents the sum of the whole number and the fraction. Hence, $2\frac{1}{2}$ means $2 + \frac{1}{2}$.

EXAMPLE 4

Express the shaded portion of this figure as both an improper fraction and a mixed number.

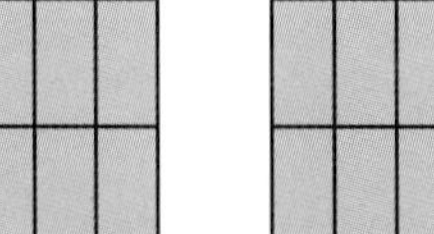 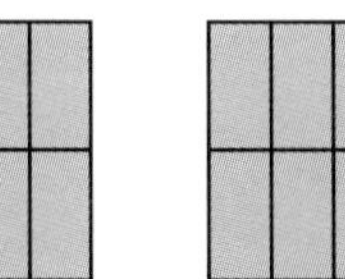 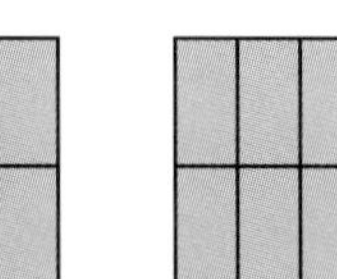 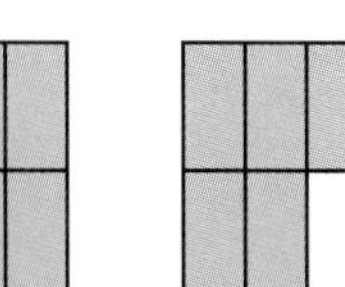

SOLUTION

Each rectangle is divided into six equal parts. Four rectangles are totally shaded, and five of the six parts of another are shaded.

For the mixed number, this gives us 4 wholes + $\frac{5}{6}$ shaded, or $4\frac{5}{6}$.

For the improper fraction, each whole is broken into six parts, so the denominator is 6. We have 29 parts shaded, which gives us $\frac{29}{6}$.

The shaded portion can thus be represented as the mixed number $4\frac{5}{6}$ or as the improper fraction $\frac{29}{6}$.

■ *Now do margin exercises 4a and 4b.*

As we have just shown, mixed numbers and improper fractions can provide two equivalent representations for the same number. We can use either representation in a calculation, depending on which is easier. We can also convert from one to the other fairly easily.

To write an improper fraction as a mixed number

1. Divide the denominator into the numerator.
2. The quotient of your division is the whole-number part of the mixed number. The remainder written over the original denominator is the fraction part of the mixed number.

To express $\frac{13}{2}$ as a mixed number, we would first divide 13 by 2.

$$\begin{array}{r} 6 \\ 2\overline{)13} \\ \underline{12} \\ 1 \end{array}$$

denominator → 2; 6 ← quotient; 13 ← numerator; 1 ← remainder

Then we would write this result as

$$6\frac{1}{2}$$

quotient → 6; 1 ← remainder; 2 ← denominator

Let's look at an example.

EXAMPLE 5

Change $\frac{29}{6}$ to a mixed number.

SOLUTION

Using the procedure, we first divide 29 by 6.

$$\begin{array}{r} 4 \\ 6\overline{)29} \\ \underline{24} \\ 5 \end{array}$$

4a. Express the shaded portion as both a mixed number and an improper fraction.

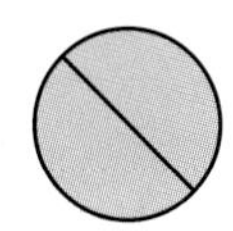

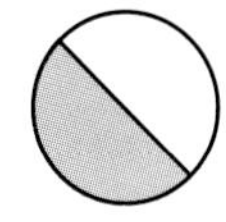

4b. Express the shaded portion as both a mixed number and an improper fraction.

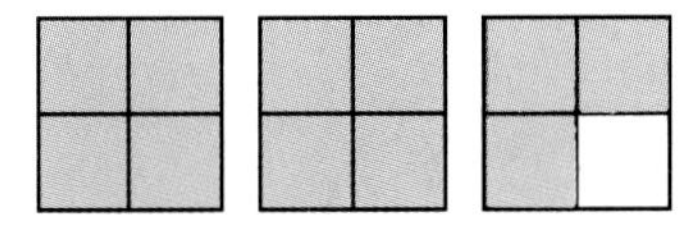

5a. Change $\frac{93}{4}$ to a mixed number.

5b. Change $\frac{683}{7}$ to a mixed number.

6a. Change $13\frac{4}{5}$ to an improper fraction.

6b. Change $17\frac{3}{4}$ to an improper fraction.

Then we write the result as $4\frac{5}{6}$.

So $\frac{29}{6}$ is the same as $4\frac{5}{6}$.

■ *Now do margin exercises 5a and 5b.*

To write $6\frac{1}{2}$ as an improper fraction, we could reverse the process. That is, we would first multiply the whole-number part (6) by the denominator (2) and add the numerator (1). Then we would write the result, 13, over the original denominator.

$$6\frac{1}{2} = \frac{6 \times 2 + 1}{2} = \frac{13}{2}$$

This gives us the following method:

To change a mixed number to an improper fraction

1. Multiply the whole-number part by the denominator and add the numerator to the result.
2. Write this sum over the original denominator.

Let's look at an example.

EXAMPLE 6

Write $110\frac{9}{11}$ as an improper fraction.

SOLUTION

$$110\frac{9}{11} = \frac{110 \times 11 + 9}{11} = \frac{1219}{11}$$

You can always check your work by simply changing the improper fraction back to a mixed number by division.

Here, $1219 \div 11 = 110\frac{9}{11}$.

■ *Now do margin exercises 6a and 6b.*

Any whole number n can be written in fractional form as $\frac{n}{1}$. Thus 6 is $\frac{6}{1}$, 73 is $\frac{73}{1}$, and so on.

Work the exercises that follow.

4.1 EXERCISES

In Exercises 1 through 3, write two proper fractions for each figure, one representing the portion that is shaded, and one representing the portion that is not shaded.

1.

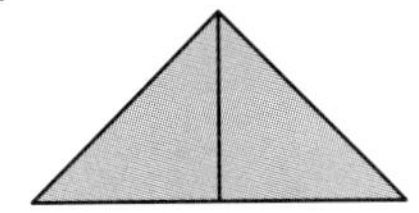

2.

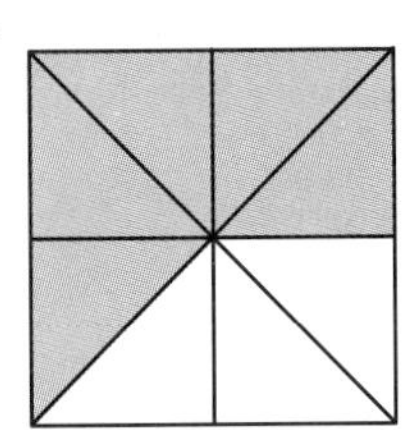

3.

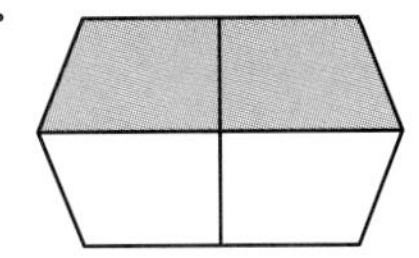

Use the information in Table 1 to answer Exercises 4 through 9.

Professional sports leagues have rules giving not only the number of players per team allowed on the field at any one time, but also the number of "dressed" players who can be used as starters or substitutes during the game.

TABLE 1 Team sizes, professional sports

Sport	Players Allowed on the Field	Dressed Players
American football	11	43
Soccer	11	13
Men's lacrosse	10	19
Ice hockey	6	17

4. What fraction of dressed football players are allowed on the field?

5. What fraction of dressed lacrosse players are not allowed on the field?

6. What fraction of dressed hockey players are allowed on the ice?

7. What fraction of dressed football players are not allowed on the field?

8. What fraction of dressed soccer players are not allowed on the field?

9. What fraction of dressed hockey players are not allowed on the ice?

10. Nine out of ten women will work outside the home at some time during their lives.

a. What fraction of women will work outside the home?

b. What fraction will not work outside the home?

11. Three out of four doctors recommend aspirin.

a. What fraction recommends aspirin?

b. What fraction does not recommend aspirin?

12. Eight out of eleven campers left their toothbrushes at home.

a. What fraction brought their toothbrushes?

b. What fraction left their toothbrushes at home?

13. Two out of seven children hate okra.

a. What fraction hates okra?

b. What fraction does not hate okra?

14. There are 206 bones in the body.

a. There are 32 bones in the arm (including the arm, wrist, and hand). What fraction of the total number of bones in your body do the bones in your two arms represent?

b. There are 31 bones in the leg (including the leg, ankle, and foot). What fraction of the total number of bones in your body do the bones in your two legs represent?

c. There are 29 bones in the skull. What fraction of the bones in your body are not in your skull?

d. There are 26 bones in your spine. What fraction of the total number of bones in your body are the bones in your spine?

e. There are 25 bones in your chest. What fraction of the total number of bones in your body are not in your chest?

15. There are 40 nickels in a roll of nickels.

a. If a certain roll contains 5 buffalo nickels, what fraction of the roll is buffalo nickels?

b. What fraction of the roll is not buffalo nickels?

16. What fraction of a leap year is the number of days in one week?

17. What fraction of a non-leap year are the number of days in September?

In Exercises 18 and 19, write both a mixed number and an improper fraction that describe the shaded portion.

18.

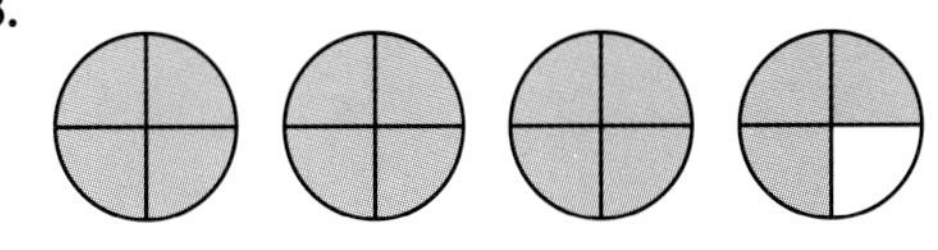

19.

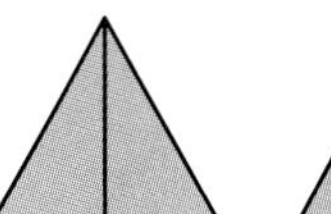

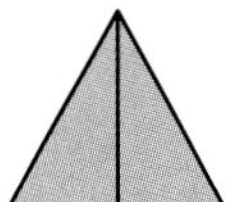

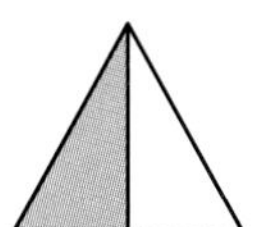

In Exercises 20 through 23, write the mixed number represented by the improper fraction.

20. $\frac{87}{9}$ **21.** $\frac{34}{7}$ **22.** $\frac{108}{5}$ **23.** $\frac{123}{8}$

In Exercises 24 through 35, write the improper fraction represented by the mixed number.

24. $11\frac{8}{9}$ **25.** $23\frac{1}{4}$ **26.** $12\frac{2}{3}$ **27.** $18\frac{3}{5}$

28. $6\frac{5}{8}$ **29.** $4\frac{4}{5}$ **30.** $7\frac{1}{9}$ **31.** $10\frac{7}{10}$

32. $14\frac{1}{8}$ **33.** $9\frac{2}{3}$ **34.** $17\frac{1}{4}$ **35.** $23\frac{1}{5}$

In Exercises 36 through 39, draw a picture representing each of the following mixed numbers.

36. $2\frac{3}{4}$ **37.** $1\frac{4}{5}$ **38.** $5\frac{1}{6}$ **39.** $3\frac{2}{3}$

In Exercises 40 through 43, write the mixed number represented by each improper fraction.

40. $\frac{245}{15}$ **41.** $\frac{93}{13}$ **42.** $\frac{622}{50}$ **43.** $\frac{129}{17}$

4.2 Finding Equivalent Fractions

Section Goals

- To find equivalent fractions
- To write fractions with their least common denominator

We saw in the previous section that the number 1 is represented by any fraction whose non-zero numerator equals its denominator. That is,

$$1 = \frac{1}{1} = \frac{2}{2} = \frac{17}{17} = \frac{522}{522} \cdots$$

Because all these fractions are equal to 1 (and therefore to each other), they are called **equivalent fractions**.

We also saw that the number zero is represented by any fraction whose numerator is zero and whose denominator is non-zero. Hence,

$$0 = \frac{0}{3} = \frac{0}{4} = \frac{0}{5} = \frac{0}{6} = \frac{0}{7} \cdots$$

All these fractions are equal to zero and are equivalent fractions.

Equivalent Fractions

Equivalent fractions are fractions that express the same quantity.

The following figure shows that $\frac{1}{2}$, $\frac{2}{4}$, and $\frac{3}{6}$ all represent the same portion of the rectangle. So these three fractions are also equivalent fractions.

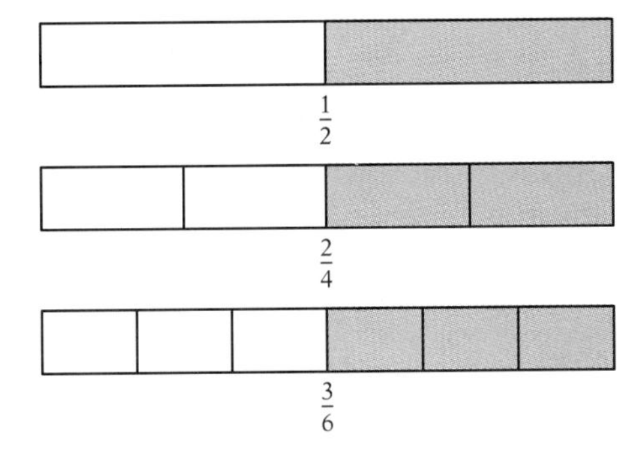

Another set of equivalent fractions is shown in the next figure, where $\frac{1}{3}$, $\frac{2}{6}$, and $\frac{3}{9}$ all represent the same shaded area.

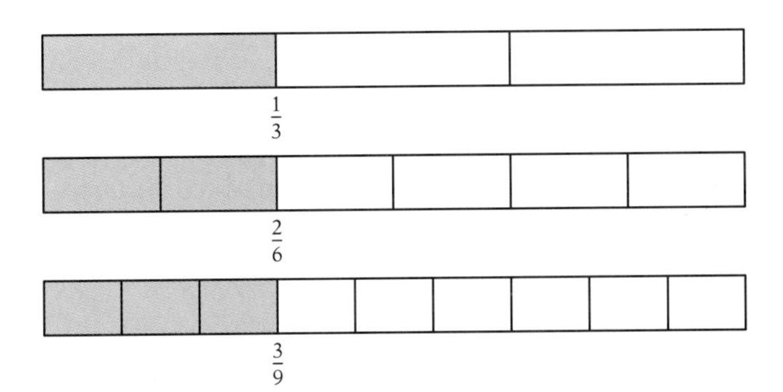

Finding Equivalent Fractions

Have you figured out the relationship between the equivalent fractions in each of the previous figures? An equivalent fraction can be formed from any fraction by multiplying the numerator and denominator by the same non-zero whole number.

Equivalent fractions

If $\frac{a}{b}$ is a fraction and n is any non-zero whole number, then

$$\frac{a \times n}{b \times n} \text{ is equivalent to } \frac{a}{b}$$

That is,

$$\frac{a}{b} = \frac{a \times n}{b \times n}$$

For example, we can begin with $\frac{1}{2}$ and write

$$\frac{1 \times 2}{2 \times 2} = \frac{2}{4} \text{ and } \frac{1 \times 3}{2 \times 3} = \frac{3}{6}$$

which gives us the equivalent fractions $\frac{1}{2}$, $\frac{2}{4}$, and $\frac{3}{6}$.

Any non-zero whole number can be used as the multiplier, so

$$\frac{1 \times 611}{2 \times 611} = \frac{611}{1222}$$

for example, is also equivalent to $\frac{1}{2}$.

We can check to see that two fractions are equivalent by **cross multiplying**—that is, multiplying the numerator of each fraction by the denominator of the other.

$$\frac{1}{3} = \frac{2}{6} \quad \begin{array}{l} \rightarrow 2 \times 3 = 6 \\ \rightarrow 1 \times 6 = 6 \end{array}$$

If the **cross products** are equal, as they are here, then the fractions are equivalent.

Example 1

Find three different fractions that are equivalent to $\frac{2}{3}$.

SOLUTION

If we multiply the numerator and denominator by any three non-zero whole numbers—say 7, 55, and 213—we will obtain three different equivalent fractions.

$$\frac{2 \times 7}{3 \times 7} = \frac{14}{21}$$

$$\frac{2 \times 55}{3 \times 55} = \frac{110}{165}$$

$$\frac{2 \times 213}{3 \times 213} = \frac{426}{639}$$

The fractions $\frac{14}{21}$, $\frac{110}{165}$, and $\frac{426}{639}$ are all equivalent to $\frac{2}{3}$ and are, therefore, equivalent to each other.

To check that, for example, $\frac{14}{21} = \frac{2}{3}$, we can cross multiply.

$$\frac{14}{21} = \frac{2}{3} \quad \begin{array}{l} \rightarrow = 2 \times 21 = 42 \\ \rightarrow = 14 \times 3 = 42 \end{array}$$

The cross products are equal, so the fractions are equivalent.

■ *Now do margin exercises 1a and 1b.*

1a. Find four different fractions that are equivalent to $\frac{3}{8}$.

1b. Find four different fractions that are equivalent to $\frac{5}{6}$.

In working with fractions, we often have to find an equivalent fraction with a specific denominator.

EXAMPLE 2

Find the fraction that is equivalent to $\frac{4}{9}$ and has the denominator 18.

SOLUTION

We can multiply the numerator and denominator of $\frac{4}{9}$ by some non-zero whole number to get an equivalent fraction. But we need to find the proper multiplier. We need

$$\frac{4 \times ?}{9 \times ?} = \frac{\square}{18}$$

Now we can see that if we multiply the denominator of the original fraction by 2, we get 18. That is, $9 \times 2 = 18$.

We have to multiply the numerator by the same number to obtain the equivalent fraction. Hence we write

$$\frac{4 \times 2}{9 \times 2} = \frac{8}{18}$$

You should check that $\frac{4}{9} = \frac{8}{18}$ by cross multiplying.

■ *Now do margin exercises 2a and 2b.*

2a. Find the fraction that is equivalent to $\frac{1}{7}$ and has the denominator 35.

2b. Find the fraction that is equivalent to $2\frac{1}{11}$ and has the denominator 22.

Least Common Denominator

Before we can add or subtract two fractions, we have to write them as equivalent fractions with the same denominator. We use the technique illustrated in the last example to do this. The denominator we seek is called the **least common denominator.**

> **Least Common Denominator**
>
> The **least common denominator** of two fractions is the least common multiple of the two denominators.

EXAMPLE 3

Rewrite the fractions $\frac{2}{3}$ and $\frac{4}{5}$ as equivalent fractions with their least common denominator.

3a. Rewrite the fractions $\frac{5}{6}$ and $\frac{2}{3}$ as equivalent fractions with the least common denominator.

3b. Rewrite the fractions $\frac{4}{7}$ and $\frac{1}{3}$ as equivalent fractions with a denominator of 42.

4a. Rewrite the fractions $\frac{7}{9}$ and $\frac{5}{6}$ as equivalent fractions with a denominator of 540.

4b. Rewrite the fractions $\frac{3}{7}$ and $\frac{9}{10}$ as equivalent fractions with their least common denominator.

SOLUTION

First we must find the least common multiple (LCM) of the denominators, 3 and 5. Because 3 and 5 are relatively prime, their LCM is their product. So the LCM is 3×5, or 15.

Next we find an equivalent fraction with denominator 15 for each original fraction.

$$\frac{2\times ?}{3\times ?} = \frac{\square}{15} \text{ and } \frac{4\times ?}{5\times ?} = \frac{\square}{15}$$

Multiplying by 5 and 3, respectively, gives us

$$\frac{2\times 5}{3\times 5} = \frac{10}{15} \text{ and } \frac{4\times 3}{5\times 3} = \frac{12}{15}$$

So $\frac{2}{3}$ and $\frac{4}{5}$ are equivalent to $\frac{10}{15}$ and $\frac{12}{15}$.

■ *Now do margin exercises 3a and 3b.*

Let's try an example where finding the least common multiple requires a little more work.

EXAMPLE 4

Rewrite the fractions $\frac{5}{6}$ and $\frac{2}{9}$ as equivalent fractions with their least common denominator.

SOLUTION

First we find the LCM of the denominators 6 and 9. Because they are not prime, we will need to find the prime factorizations of each.

The LCM is 2×3^2, or 18.

Next we find an equivalent fraction with denominator 18 for each original fraction.

$$\frac{5\times ?}{6\times ?} = \frac{\square}{18} \text{ and } \frac{2\times ?}{9\times ?} = \frac{\square}{18}$$

Multiplying by 3 and 2, respectively, we get

$$\frac{5\times 3}{6\times 3} = \frac{15}{18} \text{ and } \frac{2\times 2}{9\times 2} = \frac{4}{18}$$

So $\frac{5}{6}$ and $\frac{2}{9}$ are equivalent to $\frac{15}{18}$ and $\frac{4}{18}$.

■ *Now do margin exercises 4a and 4b.*

Work the exercises that follow.

4.2 EXERCISES

In Exercises 1 through 4, replace the ? with the proper value.

1. $0 = \frac{?}{5}$ **2.** $0 = \frac{?}{18}$ **3.** $1 = \frac{?}{27}$ **4.** $1 = \frac{?}{15}$

In Exercises 5 through 8, rewrite each fraction with the denominator indicated.

5. $\frac{2}{3}$, denominator 90 **6.** $\frac{4}{7}$, denominator 84 **7.** $\frac{5}{9}$, denominator 81 **8.** $\frac{3}{4}$, denominator 64

In Exercises 9 through 20, determine the correct numerator.

9. $\frac{11}{20} = \frac{?}{160}$ **10.** $\frac{2}{5} = \frac{?}{70}$ **11.** $\frac{7}{8} = \frac{?}{88}$ **12.** $\frac{5}{11} = \frac{?}{165}$

13. $\frac{9}{16} = \frac{?}{80}$ **14.** $\frac{15}{22} = \frac{?}{154}$ **15.** $\frac{8}{13} = \frac{?}{78}$ **16.** $\frac{7}{18} = \frac{?}{360}$

17. $\frac{3}{4} = \frac{?}{8}$ **18.** $\frac{5}{8} = \frac{?}{16}$ **19.** $2\frac{3}{4} = 2\frac{?}{16}$ **20.** $1\frac{8}{16} = 1\frac{?}{48}$

In Exercises 21 through 40, rewrite the given fractions as equivalent fractions with the least common denominator.

21. $\frac{1}{4}$ $\frac{3}{5}$

22. $\frac{11}{12}$ $\frac{5}{7}$

23. $\frac{2}{5}$ $\frac{4}{9}$

24. $\frac{5}{11}$ $\frac{6}{13}$

25. $\frac{7}{12}$ $\frac{5}{6}$

26. $\frac{6}{7}$ $\frac{17}{21}$

27. $\frac{5}{8}$ $\frac{3}{4}$

28. $\frac{7}{9}$ $\frac{35}{72}$

29. $\frac{4}{15}$ $\frac{7}{10}$

30. $\frac{8}{9}$ $\frac{11}{15}$

31. $\frac{5}{8}$ $\frac{9}{10}$

32. $\frac{7}{12}$ $\frac{13}{20}$

33. $\frac{3}{75}$ $\frac{7}{10}$ $\frac{17}{20}$

34. $\frac{5}{80}$ $\frac{9}{16}$ $\frac{4}{5}$

35. $\frac{11}{15}$ $\frac{2}{3}$ $\frac{9}{100}$

36. $\frac{23}{27}$ $\frac{7}{9}$ $\frac{3}{45}$

37. $\frac{1}{24}$ $\frac{2}{30}$ $\frac{3}{4}$

38. $\frac{4}{54}$ $\frac{1}{27}$ $\frac{2}{30}$

39. $\frac{7}{90}$ $\frac{1}{20}$ $\frac{3}{50}$

40. $\frac{4}{56}$ $\frac{6}{70}$ $\frac{1}{20}$

In Exercises 41 through 56, supply the missing value in each pair of equivalent fractions.

41. $\frac{9}{18}$ and $\frac{3}{?}$

42. $\frac{18}{36}$ and $\frac{9}{?}$

43. $\frac{2}{7}$ and $\frac{?}{42}$

44. $\frac{36}{50}$ and $\frac{?}{25}$

45. $\frac{5}{16}$ and $\frac{10}{?}$

46. $\frac{12}{60}$ and $\frac{?}{10}$

47. $\frac{6}{24}$ and $\frac{12}{?}$

48. $\frac{50}{125}$ and $\frac{?}{25}$

49. $\frac{6}{12}$ and $\frac{?}{240}$

50. $\frac{16}{24}$ and $\frac{?}{3}$

51. $\frac{5}{15}$ and $\frac{105}{?}$

52. $\frac{8}{64}$ and $\frac{160}{?}$

53. $\frac{24}{36}$ and $\frac{6}{?}$

54. $\frac{25}{100}$ and $\frac{?}{4}$

55. $\frac{9}{54}$ and $\frac{3}{?}$

56. $\frac{96}{120}$ and $\frac{?}{10}$

4.2 MIXED PRACTICE

By doing these exercises, you will practice all the topics up to this point in the chapter.

57. Change $\frac{11}{12}$ and $\frac{1}{4}$ into equivalent fractions with the least common denominator.

58. Write $\frac{201}{10}$ as a mixed number.

59. Write $20\frac{4}{9}$ as an improper fraction.

60. Write $\frac{9}{13}$ as a fraction with a denominator of 1300.

61. Rewrite 31 in fractional form with denominator 6.

62. Given the fractions $\frac{15}{14}$ and $\frac{2}{21}$, change them into equivalent fractions with the least common denominator.

63. Suppose 100 people are asked whether they support our economic policies. Forty-nine say yes, thirty-seven say no, and the rest aren't sure. What fraction of these people support the policies; what fraction doesn't support them; and what fraction isn't sure?

64. Write $\frac{27}{17}$ as an improper fraction with a denominator of 170.

4.3 Reducing Fractions to Lowest Terms

We can find an equivalent fraction by multiplying the numerator and denominator of a fraction by the same whole number. We can also reverse this procedure. We can *divide* the numerator and denominator by a common multiple to get an equivalent fraction. In Section 4.2 we saw that

$$\frac{5}{6} = \frac{5 \times 3}{6 \times 3} = \frac{15}{18}$$

By reversing the process, we can write

$$\frac{15}{18} = \frac{15 \div 3}{18 \div 3} = \frac{5}{6}$$

This is called *reducing* or *simplifying the fraction*. We cannot reduce a fraction when the numerator and denominator do not have a common factor greater than 1. Then the fraction is said to be **reduced to simplest (or lowest) terms.**

To reduce a fraction to lowest terms

Divide the numerator and denominator by their greatest common factor (GCF). If their GCF is 1, the fraction is already in lowest terms.

EXAMPLE 1

Reduce $\frac{10}{16}$ to lowest terms.

SOLUTION

We first find the GCF of the numerator and denominator. The GCF of 10 and 16 is 2. Then, dividing the numerator and the denominator by 2 gives

$$\frac{10 \div 2}{16 \div 2} = \frac{5}{8}$$

Because 5 and 8 have no common factor but 1, $\frac{10}{16}$ reduced to lowest terms is $\frac{5}{8}$.

The answer can be checked by cross multiplying:

$$10 \times 8 = 80 \quad \text{and} \quad 5 \times 16 = 80$$

■ *Now do margin exercises 1a and 1b.*

EXAMPLE 2

Reduce $\frac{288}{384}$ to lowest terms.

SOLUTION

First we find the greatest common factor.

$$288 = 2 \times 2 \times 2 \times 2 \times 2 \times 3 \times 3$$
$$384 = 2 \times 2 \times 2 \times 2 \times 2 \times 2 \times 2 \times 3$$

SECTION GOAL

■ To write fractions in lowest terms

1a. Reduce $\frac{12}{36}$ to lowest terms.

1b. A student thinks that $\frac{27}{420}$ is reduced to lowest terms. Tell him what divisibility rule convinces you that he is wrong, and reduce the fraction to lowest terms.

2a. Reduce $\frac{200}{5000}$ to lowest terms.

2b. Is $\frac{111}{252}$ written in lowest terms? How do you know?

We write the results of the prime factorization in exponential form to make it easier to find the GCF.

$$288 = 2^5 3^2 \quad \text{and} \quad 384 = 2^7 3$$

Therefore,

$$\text{GCF} = 2^5 3 = 2 \times 2 \times 2 \times 2 \times 2 \times 3 = 96$$

Then dividing by 96 gives us

$$\frac{288 \div 96}{384 \div 96} = \frac{3}{4}$$

So $\frac{288}{384}$ reduced to lowest terms is $\frac{3}{4}$.

■ *Now do margin exercises 2a and 2b.*

Sometimes, dividing by the greatest common factor means dividing by a large number, as you can see in Example 2. Instead, to reduce a fraction, you can divide the numerator and denominator by any common factor, not necessarily the greatest. But then you have to divide repeatedly until the numerator and denominator have no common factor other than 1. Let's take a look at Example 2 again and see how this alternative procedure works.

EXAMPLE 2 (alternative procedure)

Reduce $\frac{288}{384}$ to lowest terms.

SOLUTION

Because both numerator and denominator are even, they can be divided by 2.

$$\frac{288 \div 2}{384 \div 2} = \frac{144}{192}$$

Then, because the new numerator and denominator are even, they too can be divided by 2.

$$\frac{144 \div 2}{192 \div 2} = \frac{72}{96}$$

The numerator and denominator that result are still even, so we can divide by 2 again.

$$\frac{72 \div 2}{96 \div 2} = \frac{36}{48}$$

At this point, or sooner, we can see that the numerator and denominator are both multiples of 12. So we divide by 12.

$$\frac{36 \div 12}{48 \div 12} = \frac{3}{4}$$

The fraction is finally in lowest terms.

■ *Now do margin exercises 2c and 2d.*

Work the exercises that follow.

2c. Reduce $\frac{192}{512}$ to lowest terms.

2d. Is $\frac{150}{270}$ reduced to lowest terms? Prove it!

4.3 EXERCISES

In Exercises 1 through 20, reduce each fraction to lowest terms.

1. $\frac{36}{48}$ **2.** $\frac{78}{100}$ **3.** $\frac{48}{72}$ **4.** $\frac{84}{126}$

5. $\frac{63}{123}$ **6.** $\frac{39}{141}$ **7.** $\frac{84}{162}$ **8.** $\frac{75}{102}$

9. $\frac{17}{51}$ **10.** $\frac{13}{117}$ **11.** $\frac{19}{133}$ **12.** $\frac{9}{153}$

13. $\frac{105}{250}$ **14.** $\frac{85}{225}$ **15.** $\frac{65}{135}$ **16.** $\frac{40}{205}$

17. $\frac{226}{348}$ **18.** $\frac{108}{236}$ **19.** $\frac{200}{332}$ **20.** $\frac{148}{284}$

In Exercises 21 through 28, each fraction is equivalent to $\frac{1}{5}$, $\frac{1}{4}$, $\frac{1}{3}$, $\frac{3}{5}$, $\frac{3}{4}$, or $\frac{2}{3}$. Decide which, and write your answer under each fraction.

21. $\frac{13}{65}$ $\frac{13}{39}$ $\frac{13}{52}$ $\frac{39}{65}$ $\frac{26}{39}$ $\frac{39}{52}$

22. $\frac{20}{60}$ $\frac{60}{80}$ $\frac{20}{100}$ $\frac{20}{80}$ $\frac{60}{100}$ $\frac{40}{60}$

23. $\frac{54}{90}$ $\frac{36}{54}$ $\frac{18}{90}$ $\frac{18}{54}$ $\frac{18}{72}$ $\frac{54}{72}$

24. $\frac{17}{85}$ $\frac{51}{85}$ $\frac{34}{51}$ $\frac{17}{51}$ $\frac{17}{68}$ $\frac{51}{68}$

25. $\frac{12}{60}$ $\frac{24}{36}$ $\frac{12}{36}$ $\frac{12}{48}$ $\frac{36}{60}$ $\frac{36}{48}$

26. $\frac{51}{255}$ $\frac{153}{255}$ $\frac{102}{153}$ $\frac{153}{204}$ $\frac{51}{153}$ $\frac{51}{204}$

27. $\frac{25}{125}$ $\frac{75}{125}$ $\frac{50}{75}$ $\frac{75}{100}$ $\frac{25}{75}$ $\frac{25}{100}$

28. $\frac{33}{132}$ $\frac{99}{165}$ $\frac{33}{99}$ $\frac{66}{99}$ $\frac{33}{165}$ $\frac{99}{132}$

4.3 MIXED PRACTICE

By doing these exercises, you will practice all the topics up to this point in the chapter.

29. Write $\frac{235}{3}$ as a mixed number.

30. Given $\frac{13}{20}$ and $\frac{5}{8}$, change these into equivalent fractions with the least common denominator.

31. Write $14\frac{12}{23}$ as an improper fraction.

32. Reduce $\frac{48}{64}$ to lowest terms.

33. Rewrite $21\frac{5}{7}$ in fractional form.

34. Rewrite the fractions $\frac{16}{25}$ and $\frac{1}{3}$ as equivalent fractions with the least common denominator.

35. Reduce $\frac{224}{336}$ to lowest terms.

36. Four out of five cars on a street are less than 5 years old. What fraction of the cars on a street is not less than 5 years old? What fraction is less than 5 years old?

37. In the United States, 409 out of every 100,000 people die of heart disease each year. Write this as a fraction.

38. Change $46\frac{2}{9}$ to an improper fraction.

4.4 Comparing and Ordering Fractions and Mixed Numbers

Ordering Fractions and Mixed Numbers

If two different fractions have the same denominator, then the one with the greater numerator is the greater fraction. For example, $\frac{3}{4}$ is greater than $\frac{1}{4}$ because $3 > 1$. We can also show that $\frac{3}{4}$ is greater than $\frac{1}{4}$ by drawing the corresponding parts of the whole.

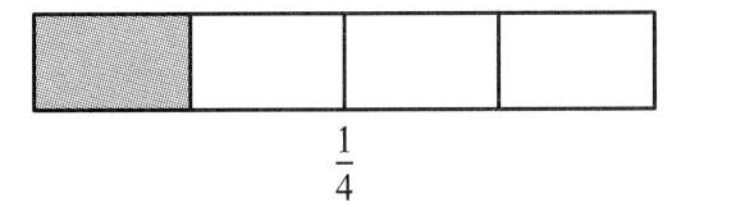
$\frac{1}{4}$

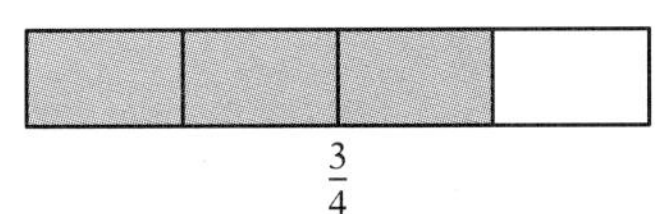
$\frac{3}{4}$

However, given the fractions $\frac{7}{12}$ and $\frac{3}{4}$, we cannot assume that $\frac{7}{12}$ is the greater just because 7 is greater than 3. In fact, the opposite is true: $\frac{3}{4} > \frac{7}{12}$! We cannot compare fractions with different denominators by inspection. Therefore, the key to comparing fractions is to make sure that the fractions have the same denominator. Then the fraction with the larger numerator is the larger fraction. Let's look at an example.

EXAMPLE 1

Which is greater, $\frac{1}{2}$ or $\frac{1}{4}$?

SOLUTION

We first have to find the least common denominator of the two fractions, which is 4. Because $\frac{1}{4}$ is already expressed in fourths, we have to make only $\frac{1}{2}$ into an equivalent fraction. To do so, we multiply numerator and denominator by 2.

$$\frac{1 \times 2}{2 \times 2} = \frac{2}{4} \quad \text{and} \quad \frac{1}{4} = \frac{1}{4}$$

We can now compare $\frac{2}{4}$ and $\frac{1}{4}$. Because 2 is greater than 1, we know that $\frac{2}{4}$ is greater than $\frac{1}{4}$.

So $\frac{1}{2} > \frac{1}{4}$.

We can also see this relationship on the following number line.

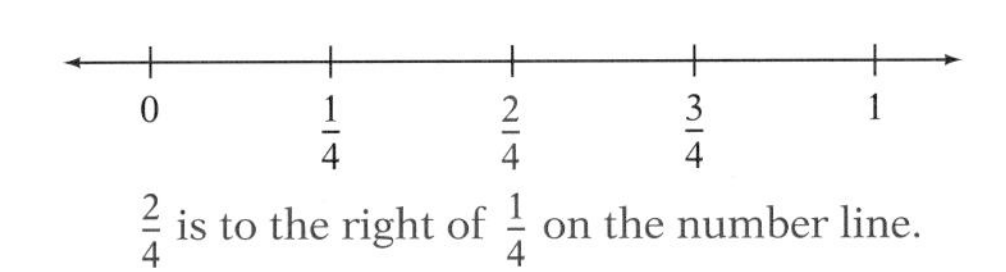

$\frac{2}{4}$ is to the right of $\frac{1}{4}$ on the number line.

■ *Now do margin exercises 1a and 1b.*

In this next example, we will see how to compare mixed numbers. Remember that a mixed number contains both a whole number and a fraction, as in $1\frac{3}{5}$.

SECTION GOALS

- To arrange two or more fractions or mixed numbers in order
- To round fractions to the nearest whole number

1a. Which is greater $\frac{1}{8}$ or $\frac{1}{10}$?

1b. Arrange in order using >: $\frac{1}{3}, \frac{1}{4}$ and $\frac{1}{6}$

2a. Replace the question mark with the appropriate symbol, < or >:
$15\frac{9}{10}$? $15\frac{7}{8}$

2b. Which is greater, $2\frac{5}{140}$ or $2\frac{3}{105}$? Write the two numbers correctly with < between them.

3a. Round to the nearest whole number: $2\frac{7}{12}$

3b. Round to the nearest whole number: $\frac{10}{21}$

EXAMPLE 2

Insert the appropriate symbol < or >: $13\frac{2}{9}$? $13\frac{7}{15}$

SOLUTION

First we compare the whole-number parts. If one is greater than the other, we can stop there. The larger whole number produces the greater mixed number.

In this case, both mixed numbers have the same whole number. Therefore, we must compare the fractions. Here, the greater fraction means the greater mixed number.

The least common denominator of $\frac{2}{9}$ and $\frac{7}{15}$ is 45. So we have the following equivalent fractions:

$$\frac{2}{9} = \frac{2 \times 5}{9 \times 5} = \frac{10}{45}$$
$$\frac{7}{15} = \frac{7 \times 3}{15 \times 3} = \frac{21}{45}$$

We compare the numerators. Because 10 is less than 21, $\frac{2}{9} < \frac{7}{15}$. So $13\frac{2}{9} < 13\frac{7}{15}$.

■ *Now do margin exercises 2a and 2b.*

Rounding Fractions

It is often possible to estimate the results of computations with fractions. But to do so, we need to be able to round fractions and mixed numbers to the nearest whole number.

To round a fraction to the nearest whole number

If a proper fraction is equal to or greater than $\frac{1}{2}$, round it up to 1; otherwise, round it to zero.

EXAMPLE 3

Round $23\frac{15}{29}$ to the nearest whole number.

SOLUTION

We need to compare $\frac{15}{29}$ to $\frac{1}{2}$. The LCD of these two fractions is 58. So we have

$$\frac{15 \times 2}{29 \times 2} = \frac{30}{58} \quad \text{and} \quad \frac{1 \times 29}{2 \times 29} = \frac{29}{58}$$

Because 30 > 29, we know that $\frac{15}{29} > \frac{1}{2}$. So we round it up to 1.

Then $23\frac{15}{29}$ rounded to the nearest whole number is 23 + 1 = 24.

■ *Now do margin exercises 3a and 3b.*

Work the exercises that follow.

4.4 EXERCISES

In Exercises 1 through 12, insert the appropriate symbol (< or >) that makes the resulting statement true.

1. $\frac{7}{9}$ __ $\frac{5}{9}$ **2.** $\frac{4}{13}$ __ $\frac{6}{13}$ **3.** $\frac{4}{15}$ __ $\frac{8}{15}$ **4.** $\frac{1}{2}$ __ $\frac{1}{3}$

5. $\frac{1}{24}$ __ $\frac{1}{12}$ **6.** $\frac{1}{11}$ __ $\frac{1}{5}$ **7.** $\frac{1}{8}$ __ $\frac{1}{10}$ **8.** $\frac{15}{26}$ __ $\frac{2}{3}$

9. $\frac{9}{64}$ __ $\frac{3}{4}$ **10.** $\frac{5}{6}$ __ $\frac{7}{48}$ **11.** $\frac{13}{36}$ __ $\frac{3}{5}$ **12.** $\frac{18}{30}$ __ $\frac{11}{33}$

In Exercises 13 through 16, decide which number is greater.

13. $\frac{3}{4}$ or $\frac{3}{8}$ **14.** $\frac{5}{12}$ or $\frac{5}{24}$ **15.** $12\frac{10}{11}$ or $12\frac{11}{12}$ **16.** $64\frac{4}{7}$ or $64\frac{11}{15}$

In Exercises 17 through 20, decide which fraction is the lesser.

17. $\frac{4}{7}$ or $\frac{7}{15}$ **18.** $\frac{2}{11}$ or $\frac{4}{9}$ **19.** $\frac{3}{14}$ or $\frac{5}{20}$ **20.** $\frac{9}{10}$ or $\frac{6}{7}$

In Exercises 21 through 24, decide which fraction is the least.

21. $\frac{11}{36}, \frac{5}{6}$ or $\frac{2}{3}$ **22.** $\frac{7}{8}, \frac{11}{24}$ or $\frac{23}{48}$ **23.** $\frac{15}{16}, \frac{35}{64}$ or $\frac{5}{8}$ **24.** $\frac{19}{48}, \frac{7}{12}$ or $\frac{13}{24}$

In Exercises 25 through 28, arrange the two mixed numbers so that you can use the symbol <.

25. $42\frac{3}{11}$ $42\frac{5}{44}$ **26.** $15\frac{2}{7}$ $15\frac{9}{14}$ **27.** $23\frac{9}{13}$ $23\frac{17}{26}$ **28.** $72\frac{19}{24}$ $72\frac{25}{48}$

In Exercises 29 through 32, arrange the three mixed numbers so that you can use the symbol >.

29. $33\frac{1}{3}, 33\frac{5}{8}, 33\frac{7}{24}$

30. $6\frac{7}{8}, 6\frac{11}{14}, 6\frac{56}{112}$

31. $11\frac{1}{4}, 11\frac{2}{5}, 11\frac{3}{20}$

32. $80\frac{5}{9}, 80\frac{13}{18}, 80\frac{31}{36}$

In Exercises 33 through 36, rearrange the fractions in order from least to greatest.

33. $\frac{3}{5}, \frac{2}{9}, \frac{1}{2}$

34. $\frac{2}{13}, \frac{3}{7}, \frac{15}{91}$

35. $\frac{7}{10}, \frac{5}{14}, \frac{8}{20}$

36. $\frac{2}{5}, \frac{11}{19}, \frac{3}{95}$

37. The mass of the Moon is approximately $\frac{1}{80}$ of the mass of Earth. The mass of Mercury is approximately $\frac{1}{16}$ of the mass of Earth. The mass of Venus is approximately $\frac{4}{5}$ of the mass of Earth. Arrange these four bodies in order from the smallest mass to the largest.

38. Edging for a tablecloth comes in long strips. If you bought strips totaling $115\frac{7}{8}$ inches and your tablecloth has a perimeter of $115\frac{3}{4}$ inches, did you buy enough edging?

39. The frame for a door measures $29\frac{3}{4}$ inches wide; the door itself measures $29\frac{11}{32}$ inches. Decide whether the door will fit in the frame. (If the door is smaller, it will fit the frame.)

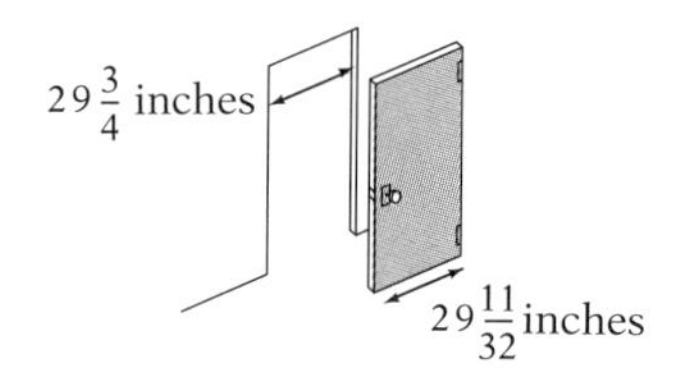

40. Three wrench sizes used by a certain mechanic are $\frac{3}{8}$, $\frac{7}{16}$, and $\frac{11}{32}$ inches. Arrange these three wrench sizes in order of decreasing size.

41. The gravity of Mars is $\frac{1}{2}$ that of Earth. The gravity of Mercury is $\frac{1}{3}$ that of Earth. Decide which planet—Mars or Mercury—has the stronger gravitational pull.

42. The wrapper on a curtain rod indicates that the rod will extend to a length of $36\frac{5}{16}$ inches. The window frame on which you are hanging the rod measures $36\frac{17}{32}$ inches. Will the curtain rod extend to the full width of the window?

43. An air conditioner is being installed in a sleeve (frame) that is $45\frac{5}{8}$ inches wide. The air conditioner itself is $45\frac{3}{32}$ inches wide. Will the air conditioner fit the frame? (Is the air conditioner narrower than the sleeve?)

44. Venus has a volume that is $\frac{9}{10}$ that of Earth. Mercury has a volume that is $\frac{1}{16}$ that of Earth. Mars has a volume equivalent to $\frac{1}{7}$ that of Earth. Which planet other than Earth has the largest volume? Which planet has the smallest volume?

45. A carpenter uses drill bits that measure $\frac{5}{8}$ inches, $\frac{9}{16}$ inches, and $\frac{15}{32}$ inches. Which of these bits is the largest?

46. Two students are working on large assignments for school. Both assignments are the same length. One student reports to the teacher that she has finished $\frac{2}{3}$ of the assignment. The other reports that he has completed $\frac{5}{8}$ of the assignment. Who has finished more work?

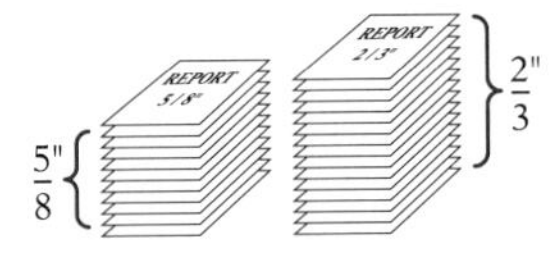

In Exercises 47 through 50, determine whether the numbers are greater than, equal to, or less than $\frac{1}{2}$.

47. $\frac{5}{8}$ **48.** $\frac{126}{218}$ **49.** $\frac{6}{13}$ **50.** $\frac{132}{316}$

In Exercises 51 through 55, round each mixed number to the nearest whole number.

51. $15\frac{3}{8}$ **52.** $29\frac{4}{9}$ **53.** $31\frac{17}{21}$ **54.** $26\frac{5}{21}$ **55.** $42\frac{19}{30}$

4.4 MIXED PRACTICE

By doing these exercises, you will practice all the topics up to this point in the chapter.

56. Insert the symbol < or > to make the following statement true. $\frac{11}{16} \ ? \ \frac{5}{8}$

57. Change $\frac{129}{5}$ to a mixed number.

58. Arrange $3\frac{5}{16}$, $3\frac{8}{9}$, and $2\frac{7}{16}$ in order from smallest to largest.

59. Write $\frac{13}{14}$ and $\frac{5}{8}$ as equivalent fractions with the least common denominator. Which fraction is larger?

60. There are 27 children in a class. Of these, 13 are boys. What fraction of the entire class is boys? What fraction is girls?

61. Reduce $\frac{930}{360}$ to lowest terms as an improper fraction.

62. There are 7 men and 20 women employees in a company. What fraction of the employees is male?

63. Reduce $\frac{160}{200}$ to lowest terms.

64. Change $6\frac{5}{9}$ to an improper fraction.

65. Write $\frac{66}{9}$ as a mixed number in simplest form.

4.5 Adding Fractions

SECTION GOAL

- To find the sum of two or more fractions

We can add two or more fractions only if they have the same denominator. If this is not the case, then we must find their least common denominator and then change the fractions into equivalent fractions with that denominator.

We can then add the fractions by adding their numerators and placing this sum over the denominator. In symbolic form,

To add fractions

If b is not equal to zero, then

$$\frac{a}{b} + \frac{c}{b} = \frac{a+c}{b}$$

The commutative and associative properties of addition (Chapter 1) hold for fractions as well as for whole numbers. And 0 is also the additive identity for fractions.

Identity Element for Addition

If b is not equal to zero, then

$$\frac{a}{b} + 0 = 0 + \frac{a}{b} = \frac{a}{b}$$

Let's look at an example of addition of two fractions with the same denominator.

EXAMPLE 1

Add: $\frac{1}{8} + \frac{3}{8}$

SOLUTION

In this problem, we are adding eighths: *one* eighth plus *three* eighths. The figure below indicates what the sum should be.

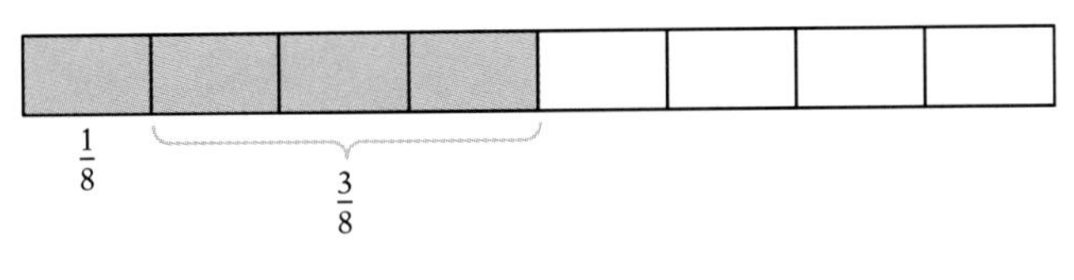

$$\frac{1}{8} + \frac{3}{8} = \frac{1+3}{8} = \frac{4}{8}$$ Add numerators. Keep common denominator.

1a. Add: $\frac{1}{6} + \frac{5}{6}$

1b. Find the sum of $\frac{5}{12}$ and $\frac{1}{12}$.

2a. Add: $\frac{9}{14} + \frac{6}{7}$

2b. What is the total of $\frac{7}{20}$ and $\frac{9}{10}$?

We can reduce this result to

$$\frac{4 \div 4}{8 \div 4} = \frac{1}{2}$$

So $\frac{1}{8} + \frac{3}{8} = \frac{1}{2}$

■ *Now do margin exercises 1a and 1b.*

We need to do a little more work when the denominators are not the same.

EXAMPLE 2

Add: $\frac{13}{18} + \frac{5}{9}$

SOLUTION

We can only add fractions with the same denominator. Therefore, we must first find the least common denominator and express the fractions as equivalent fractions with this denominator. Then we can add them.

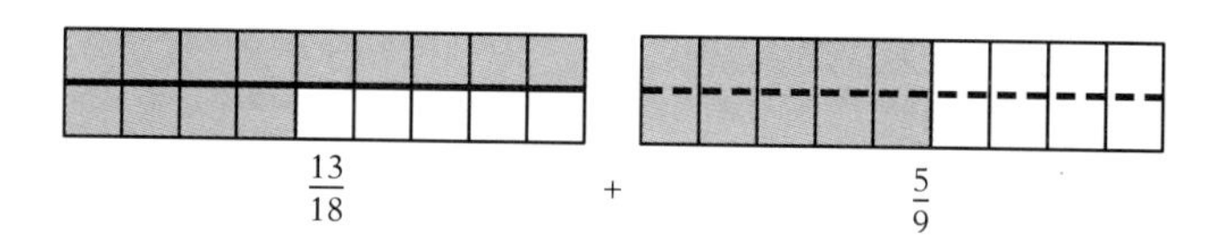

The LCM of 9 and 18 is 18, so the denominator we want is 18.

Expressing these two fractions as equivalent fractions with the denominator 18, we have

$$\frac{13}{18} = \frac{13}{18}$$

$$\frac{5}{9} = \frac{5 \times 2}{9 \times 2} = \frac{10}{18}$$

Now we can add the two fractions.

$$\frac{13}{18} + \frac{10}{18} = \frac{13 + 10}{18} = \frac{23}{18}$$ Add numerators. Keep denominator.

Thus $\frac{13}{18} + \frac{5}{9} = \frac{23}{18}$.

Note that this sum is an improper fraction. To change it to a mixed number, we divide 23 by 18 and write the remainder over 18:

$$\begin{array}{r} 1 \\ 18\overline{)23} \\ \underline{18} \\ 5 \end{array}$$

So $\frac{23}{18} = 1 + \frac{5}{18}$ or $1\frac{5}{18}$.

■ *Now do margin exercises 2a and 2b.*

In Example 2 we expressed the sum of $\frac{13}{18}$ and $\frac{5}{9}$ as $1\frac{5}{18}$. This is a more appropriate way to express a result than $\frac{23}{18}$. In fact, from here on we shall always express results in *simplest form*. That is, we will reduce to lowest terms and eliminate improper fractions as well.

Study Hint

Remember, when we add two fractions they *must* have a common denominator.

$\frac{1}{3} + \frac{4}{5}$ is *not* equal to $\frac{5}{8}$

The next example shows an alternative way to add fractions. Instead of writing the addition problem horizontally, we shall write it vertically. The process remains the same. All we change is the format.

EXAMPLE 3 Add:

$$\begin{array}{r} \frac{5}{8} \\ \frac{6}{7} \\ \frac{1}{2} \\ \hline \end{array}$$

SOLUTION

The LCM of 8, 7, and 2 is 8×7, or 56. We change all three fractions to equivalent fractions with 56 as the denominator.

$$\frac{5}{8} = \frac{5 \times 7}{8 \times 7} = \frac{35}{56}$$

$$\frac{6}{7} = \frac{6 \times 8}{7 \times 8} = \frac{48}{56}$$

$$\frac{1}{2} = \frac{1 \times 28}{2 \times 28} = \frac{28}{56}$$

Now we can add. To add vertically, we line the fractions up vertically, as we did in whole-number addition. We then add the numerators and express this result, as always, over the denominator.

$$\begin{array}{rl} \frac{35}{56} & \\ \frac{48}{56} & \\ +\frac{28}{56} & \\ \hline \frac{111}{56} & \begin{array}{l} 35 + 48 + 28 = 111 \\ \text{Write 111 over 56.} \end{array} \end{array}$$

Remember: Do not add the denominators! The result here is an improper fraction. Dividing 56 into 111 gives 1 with a remainder of 55. Our sum, then, is $1\frac{55}{56}$.

So $\frac{5}{8} + \frac{6}{7} + \frac{1}{2} = 1\frac{55}{56}$.

■ *Now do margin exercises 3a and 3b.*

3a. Find the sum of $\frac{1}{2}$ and $\frac{4}{9}$ and $\frac{2}{3}$.

3b. Add:

$$\begin{array}{r} \frac{3}{7} \\ \frac{5}{14} \\ +\frac{3}{28} \\ \hline \end{array}$$

4a. Add mentally:

$$\begin{array}{r} \frac{1}{19} \\ \frac{11}{19} \\ +\frac{5}{19} \\ \hline \end{array}$$

4b. Add: $\frac{1}{4} + \frac{1}{5} + \frac{1}{6}$

Both the horizontal method and the vertical method are acceptable for adding fractions. You should use whichever is more comfortable for you. We will show both methods in our final example.

EXAMPLE 4

Add: $\frac{4}{7} + \frac{3}{5} + \frac{1}{2}$

SOLUTION

First we find the least common denominator. Because the LCM of 7, 5, and 2 is 70, the LCD is 70.

Expressing these three fractions as equivalent fractions with the same denominator, 70, we have

$$\frac{4}{7} = \frac{4 \times 10}{7 \times 10} = \frac{40}{70}$$

$$\frac{3}{5} = \frac{3 \times 14}{5 \times 14} = \frac{42}{70}$$

$$\frac{1}{2} = \frac{1 \times 35}{2 \times 35} = \frac{35}{70}$$

Adding the three fractions horizontally gives us

$$\frac{40}{70} + \frac{42}{70} + \frac{35}{70} = \frac{117}{70}$$

Because this sum is an improper fraction, we change it to a mixed number.

$$\frac{117}{70} = 1\frac{47}{70}$$

So $\frac{4}{7} + \frac{3}{5} + \frac{1}{2} = 1\frac{47}{70}$.

To add vertically, we again must change all three fractions to equivalent fractions with denominator 70. But now we can add immediately:

$$\begin{array}{rl} \left.\begin{array}{r} \frac{4}{7} = \frac{4 \times 10}{7 \times 10} = \frac{40}{70} \\ \frac{3}{5} = \frac{3 \times 14}{5 \times 14} = \frac{42}{70} \\ +\frac{1}{2} = \frac{1 \times 35}{2 \times 35} = \frac{35}{70} \end{array}\right\} & \text{Equivalent fractions} \\ \hline \frac{117}{70} & \begin{array}{l} 40 + 42 + 35 = 117 \\ \text{Write 117 over 70.} \end{array} \end{array}$$

Dividing 117 by 70 gives us 1 with a remainder of 47.

So $\frac{4}{7} + \frac{3}{5} + \frac{1}{2} = 1\frac{47}{70}$.

We get the same answer with both methods, as we should.

■ *Now do margin exercises 4a and 4b.*

Work the exercises that follow.

4.5 EXERCISES

In Exercises 1 through 36, add the fractions. Express all results in simplest form.

1. $\frac{11}{16} + \frac{7}{16} + \frac{3}{16}$

2. $\frac{5}{12} + \frac{7}{12} + \frac{11}{12}$

3. $\frac{4}{9} + \frac{5}{9} + \frac{7}{9}$

4. $\frac{3}{14} + \frac{11}{14} + \frac{5}{14}$

5. $\frac{2}{15} + \frac{8}{15} + \frac{11}{15}$

6. $\frac{7}{18} + \frac{17}{18} + \frac{2}{18}$

7. $\frac{5}{21} + \frac{10}{21} + \frac{13}{21}$

8. $\frac{7}{20} + \frac{9}{20} + \frac{19}{20}$

9. $\frac{3}{5} + \frac{4}{7}$

10. $\frac{7}{11} + \frac{5}{6}$

11. $\frac{7}{8} + \frac{2}{3}$

12. $\frac{3}{4} + \frac{2}{9}$

13. $\frac{5}{9} + \frac{13}{18}$

14. $\frac{6}{7} + \frac{23}{42}$

15. $\frac{3}{8} + \frac{27}{40}$

16. $\frac{4}{5} + \frac{28}{65}$

17. $\frac{5}{16} + \frac{9}{20}$

18. $\frac{7}{12} + \frac{7}{18}$

19. $\frac{1}{4} + \frac{5}{6}$

20. $\frac{3}{10} + \frac{11}{16}$

21. $\frac{3}{4} + \frac{5}{8} + \frac{7}{16}$

22. $\frac{2}{3} + \frac{5}{6} + \frac{13}{18}$

23. $\frac{4}{5} + \frac{7}{10} + \frac{13}{20}$

24. $\frac{5}{6} + \frac{7}{12} + \frac{11}{18}$

25. $\frac{3}{4} + \frac{5}{6} + \frac{2}{3}$

26. $\frac{2}{5} + \frac{1}{3} + \frac{17}{30}$

27. $\frac{1}{2} + \frac{7}{8} + \frac{5}{12}$

28. $\frac{7}{9} + \frac{3}{4} + \frac{11}{12}$

29. $\frac{8}{11} + \frac{8}{9}$

30. $\frac{2}{47} + \frac{1}{3}$

31. $\frac{3}{17} + \frac{7}{10}$

32. $\frac{6}{13} + \frac{1}{11}$

33. $\frac{1}{5} + \frac{1}{6} + \frac{1}{7}$

34. $\frac{1}{6} + \frac{1}{7} + \frac{1}{8}$

35. $\frac{10}{11} + \frac{11}{12}$

36. $\frac{18}{19} + \frac{1}{10}$

4.5 MIXED PRACTICE

By doing these exercises, you will practice all the topics up to this point in the chapter. Express all answers in simplest form.

37. Add: $\frac{5}{12} + \frac{1}{4} + \frac{1}{2}$

38. Change $\frac{92}{12}$ to a mixed number.

39. Determine which is larger, $\frac{2}{3}$ or $\frac{5}{7}$.

40. Add: $\frac{4}{5} + \frac{3}{4} + \frac{1}{8}$

41. Reduce $\frac{250}{1000}$ to lowest terms.

42. What fraction of a leap year represents the number of days in June? What fraction represents the number of days of the leap year that are not in June?

43. Arrange $\frac{1}{4}$, $\frac{2}{9}$, and $\frac{2}{3}$ in order from smallest to largest.

44. Insert the proper symbol, $<$ or $>$, between $\frac{5}{9}$ and $\frac{4}{15}$.

4.6 Adding Fractions and Mixed Numbers

SECTION GOAL

- To find the sum of two or more fractions or mixed numbers

Recall that a mixed number is a whole number combined with a fraction. You have already learned how to add whole numbers, and you have just learned how to add fractions.

To add two or more mixed numbers

1. Add the fractions.
2. Add the whole numbers.
3. Combine the sums.
4. Reduce to lowest terms.

It is a good idea to estimate the sum before you start, so that you can tell if your answer is reasonable. Let's look at some examples.

EXAMPLE 1

Add: $64\frac{1}{8} + 23\frac{2}{7}$

SOLUTION

Rounding these mixed numbers to the nearest whole numbers and adding, we can see that the sum should be at least 87.

Sums of mixed numbers are most easily found in a vertical format, so we write

$$\begin{array}{r} 64\frac{1}{8} \\ +23\frac{2}{7} \\ \hline \end{array}$$

The fractions have different denominators, so we must find their LCD. The least common multiple of 8 and 7 is 56, and so the LCD must be 56.

$$\frac{1}{8} = \frac{1 \times 7}{8 \times 7} = \frac{7}{56}$$

$$\frac{2}{7} = \frac{2 \times 8}{7 \times 8} = \frac{16}{56}$$

The problem is, then,

$$\begin{array}{r} 64\frac{7}{56} \\ +23\frac{16}{56} \\ \hline \end{array}$$

We can now add.

$$\begin{array}{r} 64\frac{7}{56} \\ +23\frac{16}{56} \\ \hline 87\frac{23}{56} \end{array}$$

$64 + 23 = 87$ $7 + 16 = 23$ Write 23 over 56.

Because the fraction is in lowest terms, we have $64\frac{1}{8} + 23\frac{2}{7} = 87\frac{23}{56}$.

1a. Add: $28\frac{1}{3} + 14\frac{3}{7}$

1b. Find the sum: $13\frac{7}{9} + 6\frac{1}{5}$

2a. Add: $12\frac{5}{22} + 7\frac{3}{4} + \frac{7}{11}$

2b. Find the total of $8\frac{3}{8}$ and $19\frac{7}{12}$ and $\frac{5}{6}$.

■ *Now do margin exercises 1a and 1b.*

Don't forget to add whole numbers separately from fractions. Let's look at one more example, the addition of three numbers.

EXAMPLE 2

Add: $5\frac{3}{4} + 6\frac{2}{7} + \frac{9}{14}$

SOLUTION

The sum should be about 13, because rounding gives us 6 + 6 + 1 = 13.

We rewrite our problem in vertical form:

$$\begin{array}{r} 5\frac{3}{4} \\ 6\frac{2}{7} \\ +\ \frac{9}{14} \\ \hline \end{array}$$

Because the fractions have different denominators, we need their least common denominator. The LCM of the denominators 4, 7, and 14 is 28, so our LCD is 28.

After we rewrite each fraction as an equivalent fraction with a denominator of 28, our problem becomes

$$\begin{array}{r} 5\frac{21}{28} \\ 6\frac{8}{28} \\ +\ \frac{18}{28} \\ \hline \end{array}$$

We then add the fractions and the whole numbers.

$$\begin{array}{rrl} & 5\frac{21}{28} & \\ & 6\frac{8}{28} & \\ & +\ \frac{18}{28} & \\ \hline 5 + 6 = 11 & 11\frac{47}{28} & 21 + 8 + 18 = 47 \\ & & \text{Write 47 over 28.} \end{array}$$

We still must change the improper fraction to a mixed number.

$$\frac{47}{28} = 28\overline{)47}\ \text{(quotient 1; } 47 - 28 = 19\text{)} = 1\frac{19}{28}$$

Then $11\frac{47}{28} = 11 + 1\frac{19}{28} = 12\frac{19}{28}$. The answer is reasonable because it is near our estimate.

So $5\frac{3}{4} + 6\frac{2}{7} + \frac{9}{14} = 12\frac{19}{28}$

■ *Now do margin exercises 2a and 2b.*

Work the exercises that follow.

4.6 EXERCISES

In Exercises 1 through 48 add the fractions and mixed numbers. Express all answers in simplest form.

1. $\frac{3}{8} + 3\frac{1}{2}$

2. $\frac{5}{6} + 6\frac{2}{3}$

3. $4\frac{1}{14} + \frac{2}{7}$

4. $\frac{1}{4} + 11\frac{1}{3}$

5. $7\frac{4}{5} + 1\frac{7}{10}$

6. $6\frac{3}{8} + 4\frac{5}{16}$

7. $3\frac{5}{11} + 5\frac{9}{22}$

8. $8\frac{5}{6} + 9\frac{7}{12}$

9. $3\frac{1}{3} + 2\frac{7}{12} + 5\frac{5}{6}$

10. $4\frac{3}{4} + 1\frac{5}{8} + 7\frac{9}{16}$

11. $6\frac{3}{5} + 2\frac{3}{10} + 8\frac{13}{20}$

12. $5\frac{4}{9} + 3\frac{5}{18} + 2\frac{2}{3}$

13. $4\frac{6}{7} + 11\frac{3}{5}$

14. $5\frac{2}{3} + 6\frac{9}{13}$

15. $13\frac{3}{8} + 9\frac{6}{11}$

16. $8\frac{5}{9} + 12\frac{4}{5}$

17.
$$\begin{array}{r} 11\frac{1}{2} \\ 6\frac{4}{9} \\ +10\frac{3}{5} \\ \hline \end{array}$$

18.
$$\begin{array}{r} 12\frac{5}{7} \\ 17\frac{2}{3} \\ +\ 4\frac{7}{10} \\ \hline \end{array}$$

19.
$$\begin{array}{r} 5\frac{4}{7} \\ 2\frac{1}{3} \\ +15\frac{3}{4} \\ \hline \end{array}$$

20.
$$\begin{array}{r} 9\frac{1}{2} \\ 14\frac{4}{5} \\ +\ 8\frac{2}{7} \\ \hline \end{array}$$

21. $5\frac{3}{8} + 26\frac{7}{12}$

22. $13\frac{11}{20} + 8\frac{5}{6}$

23. $9\frac{3}{4} + 11\frac{7}{10}$

24. $4\frac{5}{6} + 10\frac{3}{8}$

25. $14\frac{1}{2} + 6\frac{3}{4} + 9\frac{1}{6}$

26. $12\frac{2}{3} + 5\frac{1}{6} + 2\frac{5}{9}$

27. $9\frac{7}{12} + 7\frac{5}{6} + 6\frac{5}{8}$

28. $5\frac{11}{12} + 8\frac{4}{5} + 3\frac{19}{30}$

29. $11\frac{4}{5} + 5\frac{1}{4}$

30. $4\frac{4}{7} + 13\frac{5}{6}$

31. $15\frac{3}{8} + 6\frac{2}{3}$

32. $9\frac{4}{9} + 17\frac{1}{5}$

33. $\frac{3}{5} + 8\frac{2}{3} + 3\frac{1}{2}$

34. $\frac{8}{15} + 8\frac{2}{3} + 13\frac{1}{5}$

35. $4\frac{2}{5} + 3\frac{3}{8} + \frac{1}{3}$

36. $8\frac{5}{6} + 3\frac{7}{12}$

37. $8\frac{5}{6} + 5\frac{11}{12} + 11\frac{23}{24}$

38. $3\frac{13}{18} + 13\frac{7}{9} + 7\frac{1}{6}$

39. $4\frac{17}{30} + 9\frac{14}{15} + 6\frac{4}{5}$

40. $6\frac{5}{44} + 7\frac{6}{11} + 12\frac{9}{22}$

41. $7\frac{3}{10} + 6\frac{5}{24}$

42. $15\frac{5}{6} + 36\frac{13}{14}$

43. $21\frac{3}{8} + 13\frac{7}{12}$

44. $8\frac{7}{12} + 19\frac{19}{20}$

45. $18\frac{7}{12} + 14\frac{11}{18} + 13\frac{1}{2}$

46. $12\frac{9}{10} + 5\frac{1}{2} + 7\frac{3}{5}$

47. $20\frac{3}{7} + 9\frac{1}{2} + 3\frac{23}{28}$

48. $10\frac{8}{15} + 5\frac{4}{5} + 16\frac{5}{9}$

4.6 MIXED PRACTICE

By doing these exercises, you will practice all the topics up to this point in the chapter. Express all results in simplest form.

49. Add: $6\frac{3}{8} + 5\frac{7}{12} + \frac{1}{6}$

50. Change $\frac{451}{23}$ to a mixed number.

51. Add: $\frac{2}{3} + \frac{3}{4} + \frac{5}{6}$

52. Reduce $\frac{75}{250}$ to lowest terms.

53. In a certain city there are 35 office buildings. Of these, 20 are owned by banks. What fraction of the office buildings is not owned by banks?

54. Which is smaller, $14\frac{11}{12}$ or $14\frac{14}{15}$?

55. Reduce $\frac{190}{570}$ to lowest terms.

56. Rewrite $\frac{33}{64}$, $\frac{7}{8}$, and $\frac{2}{4}$ as equivalent fractions with the least common denominator.

57. Add: $11\frac{1}{3} + 6\frac{2}{5}$.

58. Change $\frac{109}{12}$ to a mixed number.

59. Arrange $\frac{2}{7}$, $\frac{4}{5}$, and $\frac{3}{4}$ in order from largest to smallest.

4.7 Subtracting Fractions

SECTION GOAL

- To find the difference of two fractions

As in addition, one fraction can be subtracted from another only if the two fractions have the same denominator. If the denominators are different, then we must find their least common denominator and rewrite the fractions as equivalent fractions with that denominator. Then we can subtract by subtracting the numerators and placing the result over the denominator. In symbolic form,

To Subtract Fractions

If b is not equal to zero, then

$$\frac{a}{b} - \frac{c}{b} = \frac{a-c}{b}$$

Let's look at a few examples.

EXAMPLE 1

Subtract: $\frac{5}{7} - \frac{2}{7}$

SOLUTION

Using the procedure just introduced, we have

$$\frac{5}{7} - \frac{2}{7} = \frac{5-2}{7} = \frac{3}{7}$$

The answer is already in simplest form.

Thus $\frac{5}{7} - \frac{2}{7} = \frac{3}{7}$. The following figure shows this pictorially.

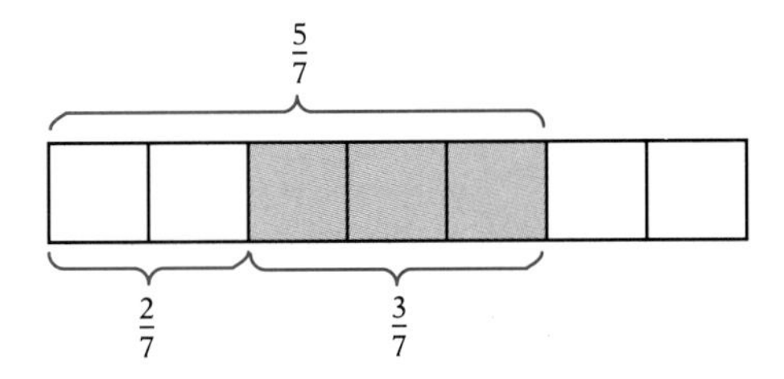

■ *Now do margin exercises 1a and 1b.*

1a. Subtract: $\frac{5}{9} - \frac{2}{9}$

1b. What is the difference between $\frac{11}{17}$ and $\frac{6}{17}$?

In the next example, we will first have to find a common denominator.

EXAMPLE 2

Subtract: $\frac{11}{16} - \frac{2}{3}$

SOLUTION

Because the two fractions have different denominators, we must find their least common denominator before we can subtract. The LCM of the denominators 16 and 3 is 48, so the least common denominator of the two fractions is 48.

2a. Subtract: $\frac{21}{25} - \frac{4}{5}$

2b. How much larger is $\frac{13}{15}$ than $\frac{3}{5}$?

3a. Subtract: $\begin{array}{r} \frac{7}{8} \\ -\frac{1}{6} \\ \hline \end{array}$

3b. Subtract: $\frac{9}{42} - \frac{12}{63}$

Expressing the fractions with a denominator of 48, we get

$$\frac{11}{16} = \frac{11 \times 3}{16 \times 3} = \frac{33}{48}$$

$$\frac{2}{3} = \frac{2 \times 16}{3 \times 16} = \frac{32}{48}$$

Now we subtract.

$$\frac{33}{48} - \frac{32}{48} = \frac{33 - 32}{48} = \frac{1}{48}$$

Because the answer is in simplest form, we are finished.

So $\frac{11}{16} - \frac{2}{3} = \frac{1}{48}$.

■ *Now do margin exercises 2a and 2b.*

Let's look at one last example.

EXAMPLE 3

Subtract: $\begin{array}{r} \frac{5}{18} \\ -\frac{1}{24} \\ \hline \end{array}$

SOLUTION

The denominators are different, so we must find a least common denominator.

$$\text{Prime factors of } 18 = 2 \times 3 \times 3$$
$$\text{Prime factors of } 24 = 2 \times 2 \times 2 \times 3$$

The least common multiple of 18 and 24 is $2 \times 2 \times 2 \times 3 \times 3$, which is 72. So we need to find equivalent fractions with a denominator of 72.

$$\frac{5}{18} = \frac{5 \times 4}{18 \times 4} = \frac{20}{72}$$

$$\frac{1}{24} = \frac{1 \times 3}{24 \times 3} = \frac{3}{72}$$

We can now subtract the numerators and place the result over the denominator.

$$\frac{20}{72} - \frac{3}{72} = \frac{20 - 3}{72} = \frac{17}{72}$$

So $\frac{5}{18} - \frac{1}{24} = \frac{17}{72}$.

■ *Now do margin exercises 3a and 3b.*

Work the exercises that follow.

4.7 EXERCISES

In Exercises 1 through 32, subtract as indicated. Express all answers in simplest form.

1. $\frac{5}{6} - \frac{1}{6}$

2. $\frac{5}{7} - \frac{2}{7}$

3. $\frac{11}{12} - \frac{7}{12}$

4. $\frac{13}{18} - \frac{5}{18}$

5. $\frac{4}{5} - \frac{2}{5}$

6. $\frac{8}{9} - \frac{4}{9}$

7. $\frac{9}{10} - \frac{2}{10}$

8. $\frac{9}{14} - \frac{3}{14}$

9. $\frac{8}{9} - \frac{2}{7}$

10. $\frac{5}{6} - \frac{3}{5}$

11. $\frac{3}{4} - \frac{5}{9}$

12. $\frac{5}{8} - \frac{3}{16}$

13. $\frac{3}{5} - \frac{9}{20}$

14. $\frac{3}{13} - \frac{2}{39}$

15. $\frac{21}{25} - \frac{4}{5}$

16. $\frac{7}{12} - \frac{3}{10}$

17. $\frac{13}{14} - \frac{5}{6}$

18. $\frac{7}{8} - \frac{5}{12}$

19. $\frac{17}{18} - \frac{23}{36}$

20. $\frac{23}{30} - \frac{11}{20}$

21. $\frac{3}{4} - \frac{13}{20}$

22. $\frac{15}{16} - \frac{1}{10}$

23. $\frac{17}{18} - \frac{7}{12}$

24. $\frac{2}{5} - \frac{1}{4}$

25. $\frac{13}{18} - \frac{3}{5}$

26. $\frac{10}{11} - \frac{3}{7}$

27. $\frac{2}{33} - \frac{1}{99}$

28. $\frac{1}{11} - \frac{1}{132}$

29. $\frac{4}{123} - \frac{1}{41}$

30. $\frac{5}{95} - \frac{1}{19}$

31. $\frac{5}{13} - \frac{1}{17}$

32. $\frac{6}{19} - \frac{2}{15}$

4.7 MIXED PRACTICE

By doing these exercises, you will practice all the topics up to this point in the chapter. Express all answers in simplest form.

33. For the 1988 summer Olympics, only 131,000 of the expected 250,000 foreign spectators showed up. What fraction of the expected foreign spectators did not show up?

34. Which is greater, $26\frac{3}{4}$ or $26\frac{2}{3}$?

35. Subtract: $\frac{4}{7} - \frac{1}{12}$

36. Add: $52\frac{1}{3} + 71\frac{3}{5}$

37. Add: $39\frac{7}{8} + 15\frac{5}{6}$

38. Subtract: $\frac{11}{12} - \frac{5}{9}$

39. Rewrite $\frac{4}{9}$ and $\frac{3}{7}$ as equivalent fractions with their least common denominator.

40. Subtract: $\frac{5}{6} - \frac{1}{2}$

41. Reduce $\frac{185}{200}$ to lowest terms.

42. Arrange $\frac{2}{3}$, $\frac{3}{4}$, and $\frac{4}{9}$ in order so that you can use the symbol >.

4.8 Subtracting Fractions and Mixed Numbers

Section Goal

- To find the difference of two fractions or mixed numbers

When we subtract mixed numbers, we subtract the whole-number parts separately from the fractions.

To subtract two mixed numbers

1. Subtract the fractions.
2. Subtract the whole numbers.
3. Combine the two differences.
4. Reduce to lowest terms.

Example 1

Subtract: $4\frac{5}{6} - 3\frac{2}{7}$

SOLUTION

Rounding to the nearest whole number, we estimate the difference as $5 - 3 = 2$.

As in addition of mixed numbers, we rewrite the problem in vertical form, lining up the fractions and the whole numbers.

$$\begin{array}{r} 4\frac{5}{6} \\ -3\frac{2}{7} \\ \hline \end{array}$$

The least common denominator of the fractions is 42, so we have

$$\frac{5}{6} = \frac{5 \times 7}{6 \times 7} = \frac{35}{42}$$

$$\frac{2}{7} = \frac{2 \times 6}{7 \times 6} = \frac{12}{42}$$

The problem is, then,

$$\begin{array}{r} 4\frac{35}{42} \\ -3\frac{12}{42} \\ \hline \end{array}$$

1a. Subtract: $48\frac{3}{4} - 7\frac{2}{3}$

1b. $6\frac{4}{7}$ is how much less than $34\frac{7}{8}$?

We can now subtract.

$$\begin{array}{lrl} & 4\frac{35}{42} & \\ & -3\frac{12}{42} & \\ \hline 4 - 3 = 1 & 1\frac{23}{42} & 35 - 12 = 23 \text{ Write 23 over 42.} \end{array}$$

So $4\frac{5}{6} - 3\frac{2}{7} = 1\frac{23}{42}$.

■ *Now do margin exercises 1a and 1b.*

Don't forget to subtract whole numbers separately from fractions.

Recall that we can rewrite a whole number as a fraction. For example, the whole number 1 is equal to the unit fraction $\frac{6}{6}$, $\frac{5}{5}$, $\frac{4}{4}$, and so on. We need this idea to work the next example, in which we subtract a mixed number from a whole number.

EXAMPLE 2

Subtract: $6 - 4\frac{3}{5}$

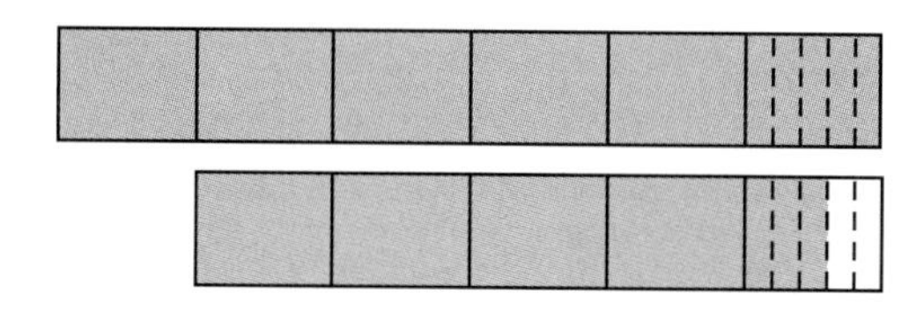

SOLUTION

We estimate the answer to be about $6 - 5 = 1$.

We must first change the whole number 6 to $5 + \frac{5}{5}$. We chose $\frac{5}{5}$ instead of $\frac{4}{4}$ or $\frac{3}{3}$, because the fractional part of the number we are going to subtract has 5 as its denominator.

This gives us

$$\begin{array}{rl} 5\frac{5}{5} & \text{Because } 6 = 5 + \frac{5}{5} \\ -4\frac{3}{5} & \\ \hline \end{array}$$

Because there is already a common denominator, we can subtract immediately.

$$\begin{array}{lcl} & 5\frac{5}{5} & \\ & -4\frac{3}{5} & \\ \hline 5-4=1 & 1\frac{2}{5} & 5-3=2 \\ & & \text{Write 2 over 5.} \end{array}$$

The estimate of about 1 shows that $1\frac{2}{5}$ is a reasonable answer.

So $6 - 4\frac{3}{5} = 1\frac{2}{5}$.

■ *Now do margin exercises 2a and 2b.*

In the next example, we take the idea of renaming whole numbers and fractions one step further.

EXAMPLE 3

Subtract: $7\frac{2}{9} - 4\frac{1}{3}$

SOLUTION

We estimate our answer as about $7 - 4 = 3$.

We rewrite the problem in vertical form.

$$\begin{array}{r} 7\frac{2}{9} \\ -4\frac{1}{3} \\ \hline \end{array}$$

The least common denominator of the two fractions is 9. Rewriting them as equivalent fractions with the common denominator gives us

$$\begin{array}{r} 7\frac{2}{9} \\ -4\frac{3}{9} \\ \hline \end{array}$$

The next step would normally be to subtract. However, we cannot subtract this as written, because we cannot take 3 ninths from 2 ninths.

2a. Find the difference between 18 and $14\frac{3}{11}$.

2b. Subtract: $27 - 15\frac{2}{5}$

3a. Subtract: $33\frac{1}{14} - 14\frac{2}{7}$

3b. Subtract $21\frac{7}{16}$ from $96\frac{3}{8}$.

Recall what we did in Example 2. We must take one whole unit from the whole number in $7\frac{2}{9}$, express it as $\frac{9}{9}$, and then add it to the fraction.

$$7\frac{2}{9} = 6 + \frac{9}{9} + \frac{2}{9} = 6\frac{11}{9}$$

We can now subtract.

$$\begin{array}{rcl} & 6\frac{11}{9} & \\ & \cancel{7\frac{2}{9}} & \\ & -4\frac{3}{9} & \\ \hline 6 - 4 = 2 & 2\frac{8}{9} & 11 - 3 = 8 \\ & & \text{Write 8 over 9.} \end{array}$$

The difference of $2\frac{8}{9}$ is very close to the estimate of 3, so our answer is reasonable.

Thus $7\frac{2}{9} - 4\frac{3}{9} = 2\frac{8}{9}$.

■ *Now do margin exercises 3a and 3b.*

Work the exercises that follow.

4.8 EXERCISES

In Exercises 1 through 52, subtract as indicated. Be sure all answers are expressed in simplest form.

1. $2\frac{5}{6} - 1\frac{1}{6}$

2. $4\frac{13}{18} - 3\frac{11}{18}$

3. $3\frac{17}{21} - 1\frac{5}{21}$

4. $2\frac{19}{30} - 1\frac{11}{30}$

5. $44 - \frac{3}{8}$

6. $10 - \frac{2}{5}$

7. $36 - \frac{11}{15}$

8. $62 - \frac{4}{9}$

9. $27\frac{7}{8} - 19\frac{1}{4}$

10. $28\frac{11}{12} - 2\frac{5}{6}$

11. $35\frac{23}{27} - 26\frac{4}{9}$

12. $41\frac{37}{45} - 38\frac{3}{5}$

13. $25\frac{1}{2} - 11\frac{3}{7}$

14. $34\frac{3}{5} - 15\frac{2}{11}$

15. $17\frac{2}{3} - 13\frac{1}{16}$

16. $29\frac{3}{4} - 7\frac{2}{9}$

17. $50\frac{3}{4} - 32\frac{2}{3}$

18. $61\frac{14}{15} - 23\frac{5}{6}$

19. $46\frac{7}{8} - 24\frac{1}{6}$

20. $55\frac{13}{14} - 20\frac{3}{8}$

21. $30\frac{3}{5} - 3\frac{4}{5}$

22. $4\frac{8}{13} - 1\frac{11}{13}$

23. $16\frac{7}{20} - 8\frac{9}{20}$

24. $31\frac{11}{15} - 21\frac{14}{15}$

25. $40\frac{4}{11} - 7\frac{3}{5}$

26. $64\frac{10}{23} - 29\frac{1}{2}$

27. $37\frac{1}{6} - 13\frac{4}{5}$

28. $73\frac{7}{20} - 42\frac{2}{3}$

29. $81\frac{15}{26} - 47\frac{12}{13}$

30. $78\frac{9}{16} - 51\frac{3}{4}$

31. $69\frac{1}{2} - 48\frac{13}{18}$

32. $80\frac{3}{14} - 60\frac{15}{28}$

33. $79 - 63\frac{8}{9}$

34. $66 - 59\frac{3}{7}$

35. $77 - 45\frac{4}{11}$

36. $68 - 56\frac{6}{13}$

37. $72 - 53\frac{23}{27}$

38. $76 - 67\frac{17}{19}$

39. $83 - 71\frac{14}{15}$

40. $75 - 58\frac{4}{9}$

41. $28\frac{3}{4} - 11\frac{15}{16}$

42. $45\frac{7}{12} - 21\frac{2}{3}$

43. $59\frac{5}{6} - 15\frac{11}{18}$

44. $34\frac{7}{30} - 33\frac{4}{15}$

45. $27\frac{1}{8} - 3\frac{1}{6}$

46. $30\frac{3}{8} - 14\frac{1}{10}$

47. $41\frac{5}{9} - 13\frac{5}{12}$

48. $36\frac{5}{6} - 5\frac{7}{9}$

49. $29\frac{7}{16} - 7\frac{5}{12}$

50. $32\frac{5}{14} - 27\frac{1}{4}$

51. $18\frac{23}{25} - 9\frac{11}{15}$

52. $42\frac{9}{16} - 17\frac{5}{6}$

4.8 MIXED PRACTICE

By doing these exercises, you will practice all the topics up to this point in the chapter.

53. Subtract: $17 - 15\frac{4}{5}$

54. Add: $42\frac{1}{4} + 15\frac{1}{6}$

55. Change $\frac{67}{9}$ to a mixed number.

56. Subtract: $15\frac{7}{12} - 10\frac{5}{6}$

57. Add: $17\frac{2}{3} + 24\frac{3}{5} + 9\frac{3}{4}$

58. Subtract: $9 - 6\frac{4}{9}$

59. Find the sum: $11\frac{3}{7} + 19\frac{5}{12} + 5\frac{1}{2}$

60. Which is smaller, $2\frac{2}{7}$ or $2\frac{3}{8}$?

4.9 Applying Addition and Subtraction of Fractions and Mixed Numbers: Word Problems

SECTION GOAL

- To solve applications of the addition and subtraction of fractions and mixed numbers

Word problems involving addition and subtraction of fractions and mixed numbers are very much like those we worked in Chapters 1 and 2. We use the same models; the main difference is that we need to perform operations on *mixed numbers and fractions.*

To do the problems in this section, you will first have to decide which operation is needed—addition, subtraction, or perhaps both. As a general rule, before you start your computations, think first about what is being asked.

Let's look at some examples.

EXAMPLE 1

Mount Everest in Asia is $5\frac{1}{2}$ miles high. Mount Koskiusko, in Australia, is $1\frac{1}{3}$ miles high. How much taller is Mount Everest?

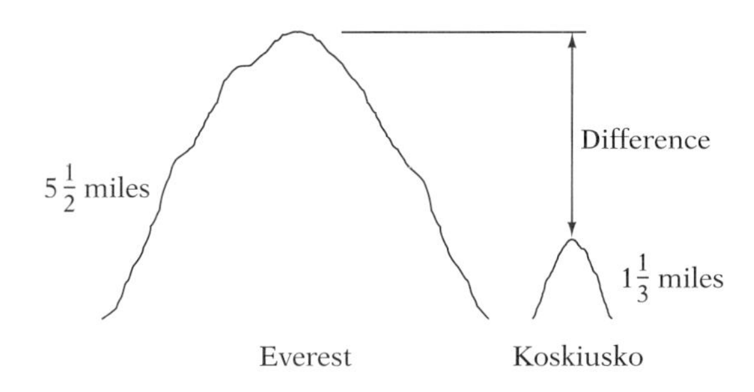

SOLUTION

The question asks for a comparison of heights: "How much taller . . . ." The operation called for is subtraction. We need to subtract the height of Mount Koskiusko from that of Mount Everest.

$$5\frac{1}{2} = 5\frac{3}{6} \qquad \text{The LCD is 6.}$$
$$-1\frac{1}{3} = -1\frac{2}{6}$$
$$4\frac{1}{6} \qquad \text{Subtracting}$$

Mount Everest is $4\frac{1}{6}$ miles higher.

■ *Now do margin exercises 1a and 1b.*

1a. A speed skater can skate $30\frac{11}{50}$ miles per hour or can skateboard $53\frac{9}{20}$ miles per hour. How much faster can he skateboard?

1b. Your time for one mile in last year's New York Marathon was $8\frac{3}{10}$ minutes. Your time for one mile in this year's New York Marathon was $7\frac{1}{2}$ minutes. How much faster was your speed this year?

EXAMPLE 2

A child measured $32\frac{3}{4}$ inches last year. This year she grew $2\frac{1}{2}$ inches. How tall is she now?

2a. The Cuyaguatega cave system in Cuba is $32\frac{7}{10}$ miles long. The Flintridge cave system in the United States is $148\frac{7}{10}$ miles longer. How long is the Flintridge system?

2b. The monthly growth rate of a wild grass is $5\frac{9}{10}$ inches. The monthly growth rate of bamboo is $29\frac{1}{2}$ inches more. How much does bamboo grow in a month?

3a. A certain recipe calls for $1\frac{1}{4}$ cups of flour, $2\frac{2}{3}$ cups of sugar and $1\frac{3}{4}$ cups of oat bran. After these are mixed together, $1\frac{1}{2}$ cups of the mixture are held for a second batch. About how much of the recipe is used for the first batch?

3b. Steven's weight record for the last three months is as follows. In the first month, he gained $4\frac{1}{4}$ pounds, in the second he lost $5\frac{1}{2}$ pounds, and in the third he gained $3\frac{3}{4}$ pounds. If his starting weight was 165 pounds, what is his new weight?

SOLUTION

The question gives you a starting height and an amount of growth and asks for a new, greater height. The operation is addition.

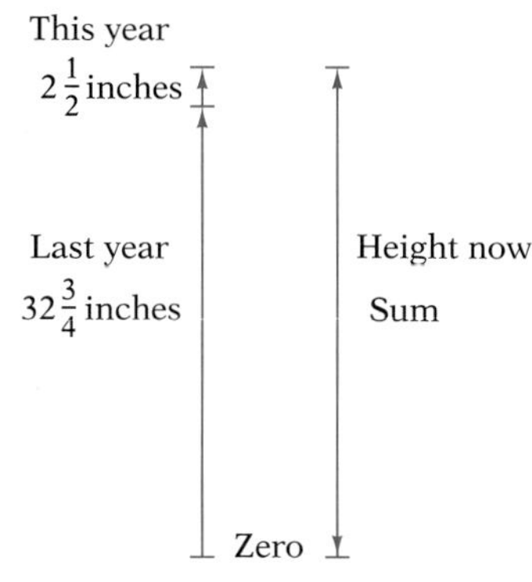

$$\begin{array}{rcl} 32\frac{3}{4} & = & 32\frac{3}{4} \\ +\ 2\frac{1}{2} & = & +\ 2\frac{2}{4} \\ \hline & & 34\frac{5}{4} = 35\frac{1}{4} \end{array}$$

The LCD is 4. In simplest form

The child's height this year is $35\frac{1}{4}$ inches.

■ *Now do margin exercises 2a and 2b.*

Let's look at one last example.

EXAMPLE 3

A certain stock opened the day at $14\frac{3}{4}$ points but dropped $2\frac{1}{2}$ points during the day. The second day it rose $1\frac{5}{8}$ points. What was the closing price the second day?

SOLUTION

The question gives you a starting price and then gives you two different actions that occurred: a drop (subtraction) and a rise (addition). Two operations are required: first subtraction, then addition. Subtracting the decrease of $2\frac{1}{2}$ points yields

$$\begin{array}{rcl} 14\frac{3}{4} & = & 14\frac{3}{4} \\ -\ 2\frac{1}{2} & = & -\ 2\frac{2}{4} \\ \hline & & 12\frac{1}{4} \end{array}$$

The LCD is 4. Subtracting

Then adding the increase of $1\frac{5}{8}$ points gives

$$\begin{array}{rcl} 12\frac{1}{4} & = & 12\frac{2}{8} \\ +\ 1\frac{5}{8} & = & +\ 1\frac{5}{8} \\ \hline & & 13\frac{7}{8} \end{array}$$

The LCD is 8. Adding

The stock closed at $13\frac{7}{8}$ points.

■ *Now do margin exercises 3a and 3b.*

Work the exercises that follow.

4.9 EXERCISES

In Exercises 1 through 28, work the word problems. Make sure your answer is reasonable.

1. A wooly mammoth's shoulder height was about 14 feet $4\frac{1}{2}$ inches. How much taller was a mammoth than an African elephant, which has a shoulder height of 10 feet 6 inches?

2. Randy Barnes, a U.S. shot putter, threw 73 feet $5\frac{1}{2}$ inches in the summer Olympics but was beaten by an East German throwing 73 feet $8\frac{3}{4}$ inches. By how much did he lose?

3. There are $29\frac{1}{2}$ days from one new moon to the next new moon. How many days are there from the first new moon in a year to the third new moon?

4. Nguyen measured $45\frac{3}{8}$ inches when he was measured this week. This was $2\frac{1}{4}$ inches more than he measured last year. How tall was he last year?

5. A species of dragonfly from Borneo has a wing span of $7\frac{1}{2}$ inches and a body length of $4\frac{1}{4}$ inches. How much longer than its body is the wing span of this insect?

6. The wing span of a Marabou stork is $11\frac{1}{2}$ feet, and that of a wandering albatross is $11\frac{5}{6}$ feet. How much longer is the albatross's wing span?

7. The Antarctic glacier flows $84\frac{3}{5}$ yards in a week; the Greenland glacier flows $236\frac{9}{10}$ yards in the same time. How much farther does the Greenland glacier flow in a week?

8. The estimated land area of Pittsburgh, Pennsylvania, is $55\frac{1}{2}$ square miles. The estimated land area of Omaha, Nebraska, is $45\frac{4}{10}$ square miles more than that of Pittsburgh. What is the estimated land area of Omaha?

Pittsburgh, PA
Omaha, NB

9. Marie's cat weighs $13\frac{1}{2}$ pounds. When she got on the scale with her cat, the scale read $165\frac{3}{8}$ pounds. How much does Marie weigh?

10. The driving distance from Queens, New York, to Ramsey, New Jersey, is $34\frac{3}{4}$ miles. If I drive this distance and then an additional $13\frac{1}{2}$ miles from Ramsey to Tuxedo, New York, how far have I driven?

11. A measuring cup holds 8 ounces when full. If you have already poured $4\frac{3}{8}$ ounces of water into the empty cup, how many ounces of milk can you add on top of this before your cup overflows?

12. The average American male is 5 feet 9 inches tall; the average female is 5 feet $3\frac{3}{4}$ inches tall. How much taller is the average male?

13. The average size of the adult male Helena's hummingbird from Cuba is $2\frac{1}{4}$ inches from bill tip to tail. The North American ostrich is sometimes nine feet tall. How much longer is a 1-foot ruler than the hummingbird?

14. In the summer Olympics in Seoul, Korea, the American who won the heptathlon, Jackie Joyner-Kersee, had come to the high jump with a previous personal-best jump of 6 feet 4 inches. In the Olympic event, however, she jumped 6 feet $1\frac{1}{4}$ inches. By how much did she miss her personal best?

15. There are two ways home from school to your house. The first, by train, takes $1\frac{1}{2}$ hours. The second, by car, takes $\frac{5}{6}$ of an hour. How much faster is the car trip?

16. The span of a starfish is 4 feet 6 inches. The claw span of a giant spider crab is 12 feet $1\frac{1}{2}$ inches. How much longer is the spider crab's span?

17. A large can of corn contains $15\frac{2}{3}$ ounces of corn. A regular size can contains $7\frac{1}{2}$ ounces. How many more ounces are there in the larger can?

18. The difference between the diameters of a luncheon plate and a larger dinner plate is $2\frac{1}{2}$ inches. If a dinner plate is $11\frac{1}{4}$ inches in diameter, what is the diameter of the luncheon plate?

19. You have lengths of wood that measure $14\frac{2}{3}$ feet, $26\frac{1}{4}$ feet, and $18\frac{1}{2}$ feet. What would they measure in total if you lined them up end to end?

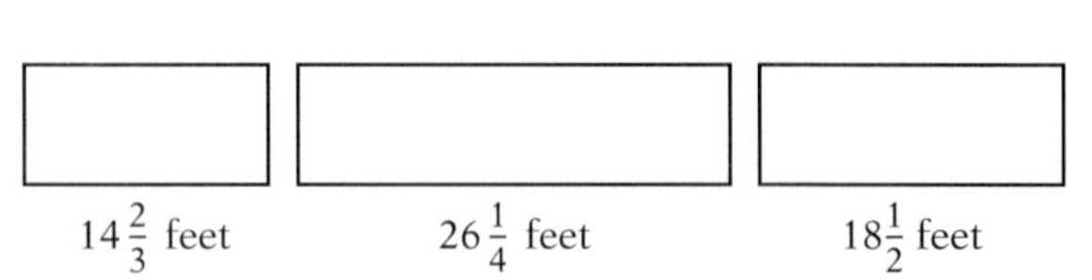

20. A recipe calls for $1\frac{1}{2}$ cups of flour, $2\frac{1}{3}$ cups of sugar, and $\frac{1}{8}$ cup of vanilla. How many cups of ingredients are there in this recipe?

21. A runner ran 10 miles on Monday, $13\frac{1}{3}$ miles on Wednesday, and $18\frac{1}{4}$ miles on Saturday. What total distance did she run?

22. During the day, a person sleeps about $\frac{1}{3}$ of the time, goes to school about $\frac{1}{4}$ of the time, takes about $\frac{1}{8}$ of the time to prepare food and eat, and may do homework about $\frac{1}{12}$ of the day. What fraction of a day does this leave for other activities?

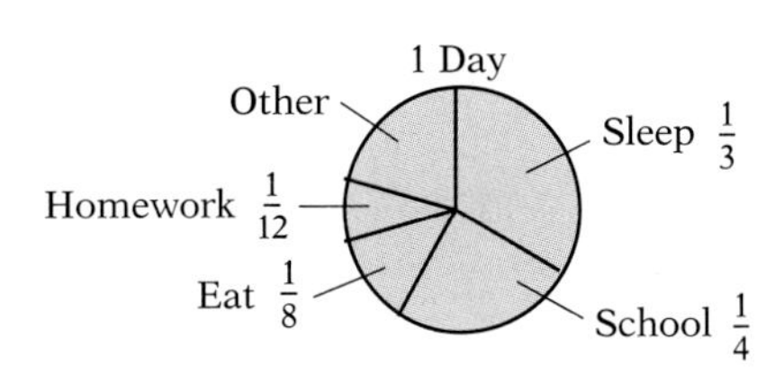

23. Stephanie is $3\frac{11}{12}$ years older than her brother Shepherd. If Shepherd is $16\frac{1}{2}$ years old, how old is Stephanie? How old will they be in $3\frac{1}{4}$ years?

24. During a recent work slowdown, supervisors were required to work additional hours. The foreman usually works a 160-hour month. During the slowdown, however, in addition to his normal hours, he worked $12\frac{3}{4}$ extra hours the first week, $20\frac{1}{8}$ the second week, and $13\frac{5}{6}$ the third week. How many hours did he work altogether that particular month?

25. In making a budget, a person should spend no more than about $\frac{1}{3}$ of his or her salary on rent or housing, should save about $\frac{1}{10}$ regularly, and should spend no more than about $\frac{1}{3}$ on taxes. What fraction of the person's salary is then left for food and other necessities?

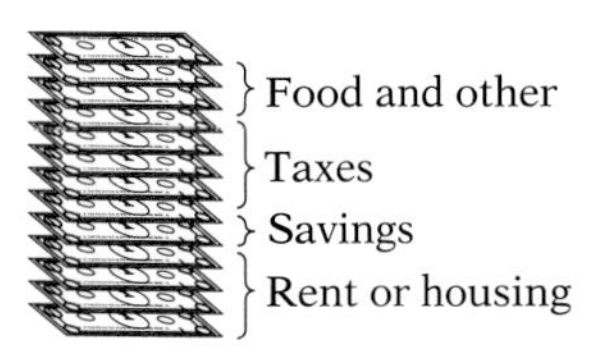

26. A Turkish weight lifter shattered an Olympic record by $115\frac{3}{4}$ pounds for total lift when he lifted a total of 755 pounds. What was the Olympic record prior to his lift?

27. The distance between El Paso and Indianapolis is 1418 miles. If I started in El Paso and drove $759\frac{7}{10}$ miles, how much more would I have to drive to get to Indianapolis?

28. The total east–west length of your property is 160 feet. If a house is built there with an east–west width of $97\frac{2}{3}$ feet, how much east–west footage remains?

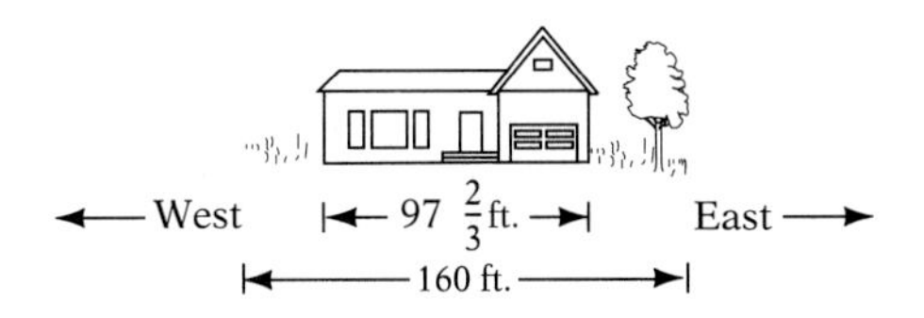

4.9 MIXED PRACTICE

By doing these exercises, you will practice all the topics up to this point in the chapter. Express all answers in simplest form.

29. The Eiffel Tower in Paris was $985\frac{11}{12}$ feet tall when a mast was added. The tower is now $1052\frac{1}{3}$ feet tall. How tall is the mast?

30. The Suez Canal is $100\frac{3}{5}$ miles long. The Panama Canal is $50\frac{7}{10}$ miles long. How much longer is the Suez Canal?

31. Find the sum: $26\frac{3}{10} + 41\frac{5}{16}$

32. Add: $\frac{37}{48} + \frac{11}{24}$

33. The Northern Line Subway tunnel is $7\frac{1}{2}$ miles longer than the Henderson rail line tunnel. The Henderson rail line tunnel is $9\frac{4}{5}$ miles long. How long is the Northern Line Subway tunnel?

34. Add: $18\frac{5}{7} + 39\frac{9}{14}$

35. Find the difference: $27\frac{1}{4} - 3\frac{3}{8}$

36. Arrange $\frac{4}{9}$, $\frac{3}{7}$, and $\frac{2}{5}$ in order from smallest to largest.

37. Find the difference: $76\frac{2}{7} - 64\frac{5}{12}$

38. Find the difference: $135 - 42\frac{3}{5}$

An Application to Statistics: Using Probability Information

Sometimes in mathematics, we are given information that can be used to find unstated information. This is particularly true when we are working with "chance" problems. The chance of any event happening is a number between 0 and 1. The chance of an event happening *plus* the chance of that event *not* happening always equals 1.

Chance of event happening + chance of event not happening = 1

In these problems the words *chance* and *probability* mean the same thing.

EXAMPLE 1

Ten balls with numbers from 0 to 9 are placed in a container. The chance of choosing the ball with the number 6 on it, if you draw just one ball, is 1 out of 10, or $\frac{1}{10}$. Under these same conditions, what are the chances of *not* drawing the ball with the number 6 on it?

SOLUTION

The number required by the problem is the chance that the number 6 will *not* be drawn. This number is equal to 1 minus the chance that it will be drawn. So we write 1 as $\frac{10}{10}$ and subtract $\frac{1}{10}$.

$$\frac{10}{10} - \frac{1}{10} = \frac{9}{10}$$

Thus there is a $\frac{9}{10}$ probability of *not* drawing the ball with a 6 on it.

■ *Now do margin exercises 1a and 1b.*

Work the exercises that follow. Make sure your answers are reasonable.

1a. The chance of choosing a ball with the number 6, 7, or 8 on it, if you draw just one ball, is 3 out of 10, or $\frac{3}{10}$. Under these same conditions, what is the probability of *not* drawing a ball with the number 6, 7, or 8 on it?

1b. The chance of winning a prize in a drawing is $\frac{3}{1000}$. What is the chance that you will lose?

EXERCISES

1. You have 1 chance in 4 of getting exactly two heads when you flip two coins. What fraction represents the probability that you will not get exactly 2 heads?

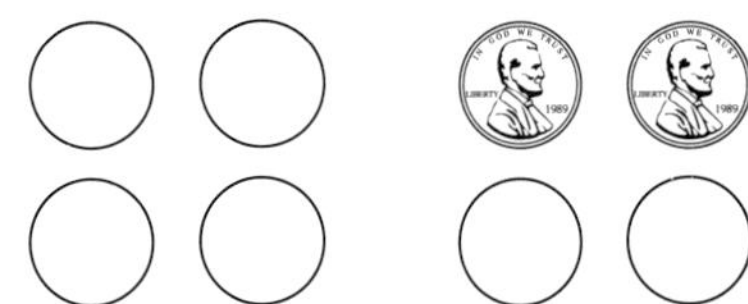

2. There is a $\frac{45}{100}$ chance of rain. What is the chance that it will not rain?

3. When one is tossing two dice, the probability of getting a sum of 10 or more is $\frac{1}{6}$. What is the probability of getting a sum of less than 10?

4. The probability of winning the lottery for a piece of soft-sided luggage is 4 in 500. What fraction represents the probability of *not* winning the lottery?

5. You have probably heard the expression "You're one in a million." What is the probability that you are *not* one in a million?

6. The chances of winning a certain scratch-off game are 3 out of 35. What is the probability of losing?

7. The chance of rolling two dice and getting a sum greater than 10 is $\frac{1}{12}$. What is the probability of getting a sum less than or equal to 10?

8. Three out of every four doctors recommend a certain product. How many doctors do not recommend the product?

9. You passed 70 out of 90 items on a test. How many items did you not pass?

10. You have taken 21 out of 125 credits needed for graduation. How many have you not taken?

11. Half of the students in a room are female. What fraction are male?

12. Only one person in 365,000 wins the state lottery. How many lose?

Chapter 4 Review

ERROR ANALYSIS

These problems have been worked incorrectly. Tell what the error is and then write the correct solution.

1. Reduce $\frac{18}{54}$ to lowest terms.

Incorrect Solution *Correct Solution*

$18\overline{)54}$ with quotient 3, so $\frac{18}{54} = 3$

Error ______________________

2. Which is larger, $\frac{3}{8}$ or $\frac{2}{3}$?

Incorrect Solution *Correct Solution*

$\frac{3}{8}$ is larger because its numerator and denominator are larger.

Error ______________________

3. Add: $\frac{1}{3} + \frac{2}{5}$

Incorrect Solution *Correct Solution*

$$\frac{1}{3} + \frac{2}{5} = \frac{3}{8}$$

Error ______________________

4. Add: $1\frac{5}{8} + \frac{3}{5}$

Incorrect Solution *Correct Solution*

$$\begin{array}{rcr} 1\frac{5}{8} & = & 1\frac{5}{40} \\ +\frac{3}{5} & = & +\frac{3}{40} \\ \hline & & 1\frac{8}{40} \end{array}$$

Error ______________________

5. Subtract: $7 - \frac{3}{4}$

Incorrect Solution *Correct Solution*

$$7 - \frac{3}{4} = 6\frac{3}{4}$$

Error ______________________

6. Subtract: $\begin{array}{r} 3\frac{1}{4} \\ -1\frac{5}{8} \\ \hline \end{array}$

Incorrect Solution

$$\begin{array}{rr} 3\frac{1}{4} = & 3\frac{2}{8} \\ -\frac{5}{8} = & -1\frac{5}{8} \\ \hline & 2\frac{3}{8} \end{array}$$

Error ______________________________

Correct Solution

INTERPRETING MATHEMATICS

By working these exercises, you will test and strengthen your mathematics vocabulary.

Use the fraction $\frac{2}{3}$ and identify and define the words underlined in Exercises 1 and 2.

1. the numerator

2. the denominator

3. Compare and contrast proper and improper fractions.

4. In your own words, tell why the mixed number $3\frac{1}{4}$ is the same as $\frac{13}{4}$.

5. Give two sets of equivalent fractions. Explain what the name means.

6. What is meant by a fraction written in simplest form?

7. What is the difference between a common denominator and a least common denominator? Can we use both in computations?

8. Explain in your own words why, when we add and subtract fractions, we use common denominators. Also explain why we add or subtract only the numerators.

REVIEW PROBLEMS

The exercises that follow will give you a good review of the material presented in this chapter. Work through them and check your answers at the back of the book. Express in simplest form wherever you can.

Section 4.1

1. There are 24 chairs with 20 people sitting in them. What fraction of the chairs is empty?

2. Write $27\frac{3}{5}$ as an improper fraction.

3. Change $\frac{56}{3}$ to a mixed number.

4. A water bottle holds 32 ounces. It currently has 11 ounces in it. What fraction of the bottle is filled?

Change each of the following mixed numbers to an improper fraction.

5. $74\frac{3}{7}$

6. $34\frac{4}{9}$

Section 4.2

7. Rewrite $\frac{1}{15}$ and $\frac{3}{4}$ with the least common denominator.

8. Express $\frac{25}{36}$ with a denominator of 72.

9. Which fraction or fractions are equivalent to $\frac{4}{9}$?
   a. $\frac{24}{54}$ b. $\frac{4}{72}$ c. $\frac{60}{135}$

10. Rewrite $\frac{17}{25}$ with a denominator of 125.

Section 4.3

Reduce each of the following to lowest terms.

11. $\frac{185}{315}$

12. $\frac{200}{1000}$

13. $\frac{36}{120}$

Section 4.4

Arrange these numbers in order using <.

14. $\frac{27}{34}$ and $\frac{15}{17}$

15. $17\frac{3}{7}$, $17\frac{4}{9}$, and $17\frac{1}{3}$

16. $11\frac{3}{4}$, $11\frac{1}{8}$, and $11\frac{2}{5}$

17. $1\frac{2}{45}$, $1\frac{3}{48}$, and $1\frac{20}{120}$

Round each of the following to the nearest whole number.

18. $56\frac{12}{51}$

19. $32\frac{17}{35}$

20. $\frac{29}{40}$

Section 4.5

Find the sum of each of the following.

21. $\frac{7}{9} + \frac{2}{3}$

22. $\frac{12}{13} + \frac{3}{4}$

23. $\frac{11}{33} + \frac{1}{3} + \frac{3}{4}$

24. $\frac{1}{2} + \frac{2}{3} + \frac{4}{7}$

Section 4.6

Find the sum of each of the following.

25. $13\frac{2}{9} + 7\frac{1}{3}$

26. $6\frac{3}{5} + 9\frac{1}{8}$

27. $8\frac{3}{4} + 12\frac{4}{9} + 3\frac{1}{6}$

28. $5\frac{2}{3} + 3\frac{11}{12}$

29. $17\frac{9}{11} + 9\frac{5}{9}$

30. $1\frac{1}{3} + 2\frac{1}{4} + 3\frac{1}{7}$

Section 4.7

Find the difference of each of the following.

31. $\frac{13}{16} - \frac{5}{7}$

32. $\frac{5}{12} - \frac{1}{3}$

33. $\frac{1}{12} - \frac{1}{30}$

34. $\frac{2}{5} - \frac{1}{250}$

Section 4.8

Subtract:

35. $11 - 4\frac{5}{6}$

36. $9 - 3\frac{4}{5}$

37. $7\frac{8}{9} - 5$

38. $11\frac{2}{3} - 4\frac{7}{8}$

39. $5\frac{1}{500} - 2\frac{1}{250}$

40. $23 - 3\frac{1}{32}$

Section 4.9

41. A skirt is $27\frac{5}{12}$ inches long with an additional $2\frac{2}{3}$ inches of lace edging. How long is the skirt with the lace?

42. The interest rate went up $\frac{3}{8}$ percent from $10\frac{1}{4}$ percent. What is the new interest rate?

43. The temperature went to $80\frac{2}{3}$ degrees, which was up $2\frac{1}{8}$ degrees from the original temperature. What was the original temperature?

44. Neal Jr.'s height is $1\frac{5}{34}$ times his mother Floride's height. His father, Neal Sr., is $1\frac{3}{17}$ times Floride's height. Which Neal is taller, Junior or Senior?

45. Majid, Shareefa, and Barrock have planted 120 flowers. Majid planted $\frac{2}{3}$ of the flowers and Shareefa planted $\frac{1}{4}$ of them. What fraction of the flowers did Barrock plant?

46. In changing margins, we added $\frac{1}{4}$ inch to each side of a column that was $6\frac{2}{3}$ inches wide to start with. What is the new column size?

Mixed Review

47. Write $\frac{227}{23}$ as an improper fraction with a denominator of 690.

48. What fraction of a year represents the number of days in December, March, and April? What fraction represents the number of days of the year (not a leap year) that are not in those three months?

49. Rewrite $\frac{4}{190}$ and $\frac{3}{170}$ as equivalent fractions with the least common denominator.

50. Arrange $1\frac{2}{3}$, $1\frac{3}{4}$, and $1\frac{4}{9}$ in order so that you can use the symbol >.

52. In a collection of 24 glasses, 8 are cracked and 7 are taller than the others. What fraction of the glasses are not cracked? What fraction of the glasses are shorter than the 7? Which of these fractions is larger?

53. Add: $426\frac{1}{9} + 163\frac{2}{7} + 828\frac{2}{21}$

54. Find the difference: $201\frac{1}{2} - 3\frac{2}{3}$

55. There is a kite on a $195\frac{2}{13}$-foot string and one on a $75\frac{1}{3}$-foot string. If the kites are the same size, how much higher can one kite fly than the other?

CHAPTER 4 TEST

This exam tests your knowledge of the topics in Chapter 4. Simplify results wherever possible.

1. a. Write as a mixed number: $\frac{37}{7}$

   b. Write $6\frac{7}{9}$ as an improper fraction.

   c. Write $\frac{2}{5}$, $\frac{9}{11}$, and $\frac{17}{22}$ as fractions with the least common denominator.

2. a. Write $\frac{100}{250}$ as a fraction with 5 as a denominator.

   b. Which is larger, $45\frac{6}{17}$ or $45\frac{7}{18}$?

   c. Round $19\frac{5}{9}$ to the nearest whole number.

3. a. Add: $\frac{6}{7}$ and $\frac{8}{49}$

   b. Add: $\frac{2}{5} + \frac{1}{10} + \frac{4}{15}$

   c. Add: $\frac{1}{34} + \frac{16}{17}$

4. a. Find the sum of $5\frac{6}{18} + 17\frac{2}{9}$.

   b. Add: $3\frac{1}{5} + 3\frac{5}{7} + 1\frac{1}{2}$

   c. Find the sum of $18\frac{4}{5}$ and $9\frac{5}{9}$.

5. **a.** How much greater than $\frac{4}{15}$ is $\frac{16}{25}$?

b. Subtract $\frac{6}{7}$ from $\frac{13}{14}$.

c. Take $\frac{1}{1000}$ from $\frac{1}{100}$.

6. **a.** Subtract: $2\frac{7}{8} - \frac{9}{10}$

b. Subtract: $16\frac{5}{7} - 3\frac{5}{6}$

c. Subtract $112\frac{2}{5}$ from $113\frac{5}{9}$.

7. Solve the following application questions.

a. The average male child is 2 feet $10\frac{1}{2}$ inches tall at age 2 and is 3 feet $4\frac{1}{2}$ inches tall at 4. How much does he grow altogether during that time?

b. You have $3\frac{3}{4}$ quarts of water and you add $2\frac{7}{8}$ quarts and then $1\frac{1}{2}$ quarts more. You need 16 quarts altogether. How much more do you need?

c. From the sum of $2\frac{2}{7}$ and $3\frac{1}{2}$, take away $4\frac{6}{7}$. The result is how much less than 10?

Cumulative Review

CHAPTERS 1–4

The exercises that follow will help you maintain the skills you have developed. Write all answers in simplest form.

1. Add: $1095 + 267$

2. Subtract: $4000 - 299$

3. $\begin{array}{r} 904 \\ \underline{\times 370} \end{array}$

4. Divide: $4807 \div 23$

5. $\begin{array}{r} 7 \text{ feet } 3 \text{ inches} \\ \underline{-4 \text{ feet } 8 \text{ inches}} \end{array}$

6. Which is larger, 11,001 or 10,100?

7. Subtract: $50 - 26\frac{2}{3}$

8. Write $3^4 \times 4^2$ as a whole number.

9. Find the perimeter of a rectangle with a length of $12\frac{3}{4}$ inches and a width of $10\frac{1}{4}$ inches.

10. Add: $16\frac{4}{7} + 9\frac{3}{8}$

11. Simplify using the standard order of operations.

$$2(13 - 6)^2 - 2 \div 3$$

12. Simplify: $2^3(5^2)$

13. Write in numerals: one million, six thousand, nine

14. Find the perimeter of a square with a side of $23\frac{1}{2}$ inches.

15. Subtract: $27\frac{2}{7} - 15\frac{3}{4}$

16. Marjorie has planks of the following lengths: $24\frac{3}{4}$ feet, $28\frac{1}{4}$ feet, and $35\frac{1}{2}$ feet. She needs 175 feet of wood to successfully complete her project. How many more feet of wood will she need?

17. Multiply: 120×832

18. Divide: $2100 \div 15$

19. Multiply: 5 pounds 9 ounces
$\underline{\times 3}$

20. Which is smaller, $60\frac{2}{3}$ or $60\frac{3}{5}$?

5 Multiplying and Dividing Fractions and Mixed Numbers

A company that sells cereal or cookies or any kind of food must be concerned with the food's packaging. The package needs to take up minimum shelf space but must look as large as packages made by competitors. The relationship between the surface area and the volume is often critical.

- *Which of these containers do you think holds the most cereal? How can you test your answer?*

Skills Check

*T*ake this short quiz to see how well prepared you are for Chapter 5. The answers are given at the bottom of the page. So are the sections to review if you get any answers wrong.

1. Multiply: $30 \times 24 \times 5$

2. Reduce $\frac{18}{90}$ to lowest terms.

3. Change $\frac{36}{5}$ to a mixed number.

4. Simplify: $21\frac{100}{31}$

5. Change $9\frac{14}{15}$ to an improper fraction.

6. Reduce $\frac{36}{64}$ to lowest terms.

7. Add: $\frac{1}{8} + \frac{5}{12}$

8. Subtract: $2\frac{1}{8} - \frac{15}{16}$

9. Divide: $128 \div 16$

10. What is the GCF of 63 and 54?

ANSWERS: 1. 3600 [Section 2.2] 2. $\frac{1}{5}$ [Section 4.3] 3. $7\frac{1}{5}$ [Section 4.1] 4. $24\frac{7}{31}$ [Section 4.1] 5. $\frac{149}{15}$ [Section 4.1] 6. $\frac{9}{16}$ [Section 4.3] 7. $\frac{13}{24}$ [Section 4.5] 8. $1\frac{3}{16}$ [Section 4.8] 9. 8 [Section 2.4] 10. 9 [Section 3.7]

5.1 Multiplying Fractions

We can multiply fractions together by using the following rule:

To Multiply Fractions

If b and d are not equal to zero, then

$$\frac{a}{b} \times \frac{c}{d} = \frac{a \times c}{b \times d}$$

This tells us that to multiply fractions, we multiply the numerators together and multiply the denominators together. We can also use this procedure to multiply a whole number by a fraction, because a whole number n can always be written as the fraction $\frac{n}{1}$.

EXAMPLE 1

Multiply: $\frac{1}{2} \times 8$

SOLUTION

First let's look at a diagram and try to determine what the answer should be.

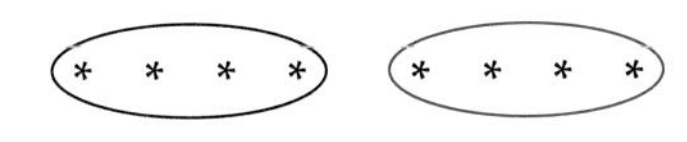

This figure shows eight stars, separated into two equal groups. One of the two groups ($\frac{1}{2}$) has four stars, so we should get a result of 4 when we multiply $\frac{1}{2} \times 8$.

We first rewrite

$$\frac{1}{2} \times 8 \quad \text{as} \quad \frac{1}{2} \times \frac{8}{1}$$

Then we multiply numerators and denominators.

$$\frac{1}{2} \times \frac{8}{1} = \frac{1 \times 8}{2 \times 1} = \frac{8}{2}$$

Simplified, the product is 4, as expected.

■ *Now do margin exercises 1a and 1b. The correct answers are given at the back of the book.*

The commutative and associative properties of multiplication hold for fractions. So do the following properties:

Identity Element for Multiplication

If b is not equal to zero, then

$$\frac{a}{b} \times 1 = 1 \times \frac{a}{b} = \frac{a}{b}$$

SECTION GOAL

■ To find the product of two or more fractions

1a. Multiply: $\frac{1}{4} \times 8$

1b. Find the product of $\frac{1}{3}$ and 9.

2a. Multiply: $\frac{2}{3} \times \frac{1}{4}$

2b. Multiply: $\frac{5}{8} \times \frac{1}{5}$

3a. Multiply: $\frac{1}{9} \times \frac{3}{5} \times \frac{4}{7}$

3b. Multiply: $\frac{2}{5} \times \frac{8}{9} \times \frac{7}{8} \times \frac{1}{2}$

Multiplication Property of Zero

If b is not equal to zero, then

$$\frac{a}{b} \times 0 = 0 \times \frac{a}{b} = 0$$

Let's work another example and use the commutative property for multiplication of fractions to check the answer.

EXAMPLE 2

Find the product of $\frac{1}{3}$ and $\frac{3}{5}$.

SOLUTION

Multiplying the numerators together and the denominators together gives us

$$\frac{1}{3} \times \frac{3}{5} = \frac{1 \times 3}{3 \times 5} = \frac{3}{15}$$

Reducing to lowest terms, we have

$$\frac{3}{15} = \frac{3 \div 3}{15 \div 3} = \frac{1}{5}$$

The commutative property tells us we can reverse the factors, multiply together, and expect the same product.

$$\frac{3}{5} \times \frac{1}{3} = \frac{3 \times 1}{5 \times 3} = \frac{3}{15} = \frac{1}{5}$$

The products are the same, so the work is probably correct.

When we multiply whole numbers together, the product is always larger than either of our numbers. But, as you can see in Example 2, when we multiply proper fractions, the result is smaller than either original fraction.

■ *Now do margin exercises 2a and 2b.*

Sometimes you may need to multiply three (or more) fractions together. Then, you may multiply the fractions together two by two. Or you may simply do the work as in the next example.

EXAMPLE 3

Multiply: $\frac{1}{2} \times \frac{3}{8} \times \frac{4}{7}$

SOLUTION

Multiplying the numerators together and the denominators together gives

$$\frac{1}{2} \times \frac{3}{8} \times \frac{4}{7} = \frac{(1 \times 3) \times 4}{(2 \times 8) \times 7} = \frac{3 \times 4}{16 \times 7} = \frac{12}{112}$$

Reducing to lowest terms, we have

$$\frac{12}{112} = \frac{12 \div 4}{112 \div 4} = \frac{3}{28}$$

So $\frac{1}{2} \times \frac{3}{8} \times \frac{4}{7} = \frac{3}{28}$.

■ *Now do margin exercises 3a and 3b.* Work the exercises that follow.

Name Section Date

5.1 EXERCISES

In Exercises 1 through 36, multiply and express your answers in simplest form.

1. $18 \times \frac{1}{2}$
2. $16 \times \frac{1}{8}$
3. $20 \times \frac{1}{5}$
4. $14 \times \frac{1}{7}$
5. $\frac{2}{3} \times 9$
6. $\frac{3}{4} \times 12$
7. $\frac{3}{8} \times 16$
8. $\frac{4}{5} \times 25$
9. $\frac{1}{9} \times \frac{8}{9}$
10. $\frac{1}{10} \times \frac{1}{8}$
11. $\frac{1}{5} \times \frac{5}{9}$
12. $\frac{3}{10} \times \frac{2}{9}$
13. $\frac{5}{6} \times \frac{6}{7}$
14. $\frac{1}{4} \times \frac{4}{7}$
15. $\frac{1}{2} \times \frac{5}{8}$
16. $\frac{2}{5} \times \frac{1}{8}$
17. $\frac{1}{7} \times 7$
18. $4 \times \frac{1}{4}$
19. $8 \times \frac{1}{8}$
20. $18 \times \frac{1}{9}$
21. $\frac{3}{5} \times \frac{5}{7}$
22. $\frac{2}{9} \times \frac{1}{2}$
23. $\frac{3}{7} \times \frac{1}{3}$
24. $\frac{1}{2} \times 2$
25. $\frac{3}{4} \times \frac{1}{4}$
26. $\frac{6}{7} \times \frac{1}{9}$
27. $\frac{4}{7} \times \frac{7}{10}$
28. $\frac{4}{5} \times \frac{7}{9}$
29. $\frac{21}{25} \times \frac{5}{7} \times \frac{2}{3}$
30. $\frac{1}{8} \times \frac{3}{5} \times \frac{12}{19}$
31. $\frac{8}{9} \times \frac{5}{7} \times \frac{4}{7}$
32. $\frac{33}{35} \times \frac{2}{7} \times \frac{2}{3}$
33. $\frac{3}{11} \times \frac{2}{7} \times \frac{2}{5} \times \frac{1}{2}$
34. $\frac{1}{12} \times \frac{3}{4} \times \frac{5}{6} \times \frac{2}{5}$
35. $\frac{1}{4} \times \frac{2}{7} \times \frac{1}{8} \times \frac{4}{5}$
36. $\frac{2}{9} \times \frac{6}{7} \times \frac{7}{6} \times \frac{3}{4}$

EXTRA

Using Cancelation to Multiply and Simplify

There is a short-cut way to multiply and simplify in the same step. Follow along in this example.

EXAMPLE Multiply and simplify as you go:

$$\frac{2}{3} \times \frac{9}{10} \times \frac{1}{18}$$

SOLUTION Begin by setting up the problem.

$$\frac{2}{3} \times \frac{9}{10} \times \frac{1}{18} = \frac{2 \times 9 \times 1}{3 \times 10 \times 18}$$

Next look and find which numbers have common factors and remove those factors. Here, we can divide the 2 in the numerator by 2 and the 10 in the denominator by 2. We cross out the numbers that have been divided, and we place the quotients above the numbers.

$$\frac{2 \times 9 \times 1}{3 \times 10 \times 18} = \frac{\overset{1}{\cancel{2}} \times 9 \times 1}{3 \times \underset{5}{\cancel{10}} \times 18}$$

This indicates the division of both numerator and denominator by 2. Next we can divide the 9 in the numerator and the 18 in the denominator by 3, yielding 3 in the numerator and 6 in the denominator; then we can divide the 3 and the 6 by 3 again, yielding 1 and 2. Note that we continue to cross out dividends and place the resulting quotient above the numbers.

$$\frac{\overset{1}{\cancel{2}} \times 9 \times 1}{3 \times \underset{5}{\cancel{10}} \times 18} = \frac{\overset{1}{\cancel{2}} \times \overset{3}{\cancel{9}} \times 1}{3 \times \underset{5}{\cancel{10}} \times \underset{6}{\cancel{18}}} = \frac{\overset{1}{\cancel{2}} \times \overset{\overset{1}{\cancel{3}}}{\cancel{9}} \times 1}{3 \times \underset{5}{\cancel{10}} \times \underset{\underset{2}{\cancel{6}}}{\cancel{18}}} = \frac{1 \times 1 \times 1}{3 \times 5 \times 2} = \frac{1}{30}$$

We could have shortened the process by dividing the numerator 9 and denominator 18 by 9 immediately.

Canceling like this reduces the possibility of errors because it gives smaller multipliers. You may use this process on any multiplication or division problem in this chapter.

5.2 Multiplying Fractions and Mixed Numbers

SECTION GOAL

- To find the product of two or more fractions or mixed numbers

It is a short step from the multiplication of fractions to the multiplication of mixed numbers such as $3\frac{1}{3}$. All we need to do is express the mixed numbers as improper fractions; then we can multiply their numerators, multiply their denominators, and simplify the product.

To Multiply Two or More Mixed Numbers

1. Change the mixed numbers to improper fractions.
2. Multiply the numerators.
3. Multiply the denominators.
4. Simplify the answer.

EXAMPLE 1

Multiply: $8\frac{1}{4} \times 13\frac{2}{3}$

SOLUTION

Here, we estimate the product by multiplying 8×14. Our answer should be about 112.

Before we multiply, we need to rewrite each mixed number as an improper fraction.

To rewrite $8\frac{1}{4}$, we multiply 8 by the denominator 4 and add the numerator 1. Then we write the result over the denominator 4.

$$8\frac{1}{4} = \frac{8 \times 4 + 1}{4} = \frac{33}{4}$$

To rewrite $13\frac{2}{3}$, we multiply 13 by 3 and add 2. We then write this result over 3, the denominator.

$$13\frac{2}{3} = \frac{13 \times 3 + 2}{3} = \frac{41}{3}$$

Now we can multiply, simplifying as we go.

$$\frac{33}{4} \times \frac{41}{3} = \frac{\overset{11}{\cancel{33}} \times 41}{4 \times \underset{1}{\cancel{3}}} = \frac{451}{4}$$

1a. Find the product: $12\frac{1}{6} \times 8\frac{3}{5}$

1b. Find the product: $8\frac{2}{5} \times 10\frac{5}{8}$

2a. Multiply: $\frac{1}{9} \times 4\frac{1}{5} \times 5\frac{4}{7}$

2b. Multiply: $\frac{2}{5} \times 10\frac{8}{9} \times 6\frac{7}{8}$

To rewrite $\frac{451}{4}$ as a mixed number, we divide 451 by 4 and place the remainder over 4.

$$\begin{array}{r} 112 \\ 4\overline{)451} \\ \underline{4} \\ 5 \\ \underline{4} \\ 11 \\ \underline{8} \\ 3 \end{array} \qquad \frac{451}{4} = 112\frac{3}{4}$$

This answer is close to 112, as we predicted, and so is reasonable.

Thus $8\frac{1}{4} \times 13\frac{2}{3}$ is $112\frac{3}{4}$.

■ *Now do margin exercises 1a and 1b.*

As when we multiplied proper fractions, we can multiply more than two mixed numbers and fractions together.

EXAMPLE 2

Find: $2\frac{1}{3} \times \frac{5}{7} \times 8\frac{2}{5}$

SOLUTION

To estimate the product, we note that because $\frac{5}{7}$ is close to 1, the product will be close to $2 \times 1 \times 8 = 16$.

For the actual multiplication, we first rewrite the mixed numbers as improper fractions.

$$2\frac{1}{3} = \frac{2 \times 3 + 1}{3} = \frac{7}{3} \quad \text{and} \quad 8\frac{2}{5} = \frac{8 \times 5 + 2}{5} = \frac{42}{5}$$

Our problem can then be rewritten as

$$\frac{7}{3} \times \frac{5}{7} \times \frac{42}{5}$$

Multiplying and simplifying yield:

$$\frac{7}{3} \times \frac{5}{7} \times \frac{42}{5} = \frac{\overset{1}{\cancel{7}} \times \overset{1}{\cancel{5}} \times \overset{14}{\cancel{42}}}{\underset{1}{\cancel{3}} \times \underset{1}{\cancel{7}} \times \underset{1}{\cancel{5}}} = \frac{14}{1}$$

Our answer, 14, is close to our estimate and so is reasonable.

■ *Now do margin exercises 2a and 2b.*

Reciprocals

Reciprocal

The **reciprocal** of any non-zero fraction $\frac{a}{b}$, is $\frac{b}{a}$.

Therefore, the reciprocal of $\frac{5}{12}$ is $\frac{12}{5}$. And, of course, the reciprocal of $\frac{12}{5}$ is $\frac{5}{12}$.

Because we can write any non-zero whole number a as $\frac{a}{1}$, the reciprocal of a is $\frac{1}{a}$. So the reciprocal of 8 is $\frac{1}{8}$. The reciprocal of 1 is simply 1, because $\frac{1}{1}$ is the same as 1. *Zero does not have a reciprocal.*

Let's see what happens when we multiply a number by its reciprocal.

EXAMPLE 3

Multiply $\frac{4}{5}$ by its reciprocal.

SOLUTION

To find the reciprocal of $\frac{4}{5}$ we reverse numerator and denominator, obtaining $\frac{5}{4}$. Then we multiply.

$$\frac{4}{5} \times \frac{5}{4} = \frac{4 \times 5}{5 \times 4} = \frac{20}{20} = 1$$

So $\frac{4}{5}$ times its reciprocal $\frac{5}{4}$ is 1.

■ *Now do margin exercises 3a and 3b.*

The results of Example 3 and margin exercises 3a and 3b illustrate an important property of reciprocals: When we multiply a number by its reciprocal, we always get 1 as the product. In symbols,

Multiplying by the Reciprocal

For any non-zero number a,

$$a \times \frac{1}{a} = \frac{1}{a} \times a = 1$$

For a number written as a fraction $\frac{a}{b}$,

$$\frac{a}{b} \times \frac{b}{a} = \frac{ab}{ba} = 1$$

3a. Find the product of $\frac{3}{8}$ and $\frac{8}{3}$.

3b. Multiply $\frac{45}{99}$ by its reciprocal.

This property holds for mixed numbers as well. We will use this fact in our final example.

EXAMPLE 4

Multiply $121\frac{1}{3}$ times its reciprocal.

SOLUTION

To find the reciprocal of a mixed number, we must first rewrite the mixed number as an improper fraction.

$$121\frac{1}{3} = \frac{121 \times 3 + 1}{3} = \frac{364}{3}$$

Now we find the reciprocal in the same way as before: We reverse the fraction. The reciprocal of $\frac{364}{3}$ is $\frac{3}{364}$.

Our problem is then

$$\frac{364}{3} \times \frac{3}{364} = \frac{364 \times 3}{3 \times 364} = 1$$

Because $364 \times 3 = 3 \times 364$

So the product of $121\frac{1}{3}$ and its reciprocal is 1.

■ *Now do margin exercises 4a and 4b.*

Work the exercises that follow.

4a. Find the reciprocal of $396\frac{1}{100}$.

4b. Multiply $973\frac{9}{11}$ by its reciprocal. Show all steps.

5.2 EXERCISES

In Exercises 1 through 4, change each mixed number to an improper fraction.

1. $6\frac{5}{8}$
2. $4\frac{4}{5}$
3. $7\frac{1}{9}$
4. $10\frac{7}{10}$

In Exercises 5 through 24, find each product. Make sure your answer is in simplest form.

5. $8\frac{3}{7} \times 9\frac{1}{3}$
6. $3\frac{3}{10} \times 1\frac{9}{11}$
7. $13\frac{1}{8} \times \frac{2}{5}$
8. $10\frac{1}{8} \times \frac{1}{9}$
9. $1\frac{1}{3} \times 10\frac{5}{7}$
10. $\frac{2}{5} \times 18\frac{1}{3}$
11. $16\frac{3}{5} \times \frac{1}{2}$
12. $\frac{6}{7} \times 15\frac{1}{9}$
13. $\frac{2}{7} \times 1\frac{2}{5}$
14. $\frac{5}{9} \times 40\frac{3}{5}$
15. $\frac{7}{9} \times 1\frac{9}{10} \times \frac{1}{7}$
16. $4\frac{2}{9} \times \frac{9}{19} \times \frac{1}{2}$
17. $1\frac{7}{9} \times 1\frac{1}{8} \times \frac{1}{3}$
18. $6\frac{1}{2} \times \frac{2}{9} \times 6\frac{2}{3}$
19. $1\frac{1}{6} \times 1\frac{1}{5} \times \frac{1}{7}$
20. $4\frac{3}{8} \times 5\frac{1}{7} \times \frac{1}{2}$
21. $3\frac{3}{10} \times \frac{10}{11} \times \frac{1}{12}$
22. $2\frac{1}{7} \times 5\frac{3}{5} \times 8\frac{1}{3}$
23. $4\frac{3}{5} \times 3\frac{5}{7}$
24. $6\frac{4}{9} \times 1\frac{1}{4}$

In Exercises 25 through 32, write the reciprocal of each number.

25. $\frac{7}{15}$
26. $\frac{5}{7}$
27. 5
28. 11
29. $\frac{19}{3}$
30. $\frac{21}{8}$
31. $2\frac{1}{2}$
32. $8\frac{3}{7}$

In Exercises 33 through 56, perform the following multiplications. Make sure your answer is in simplest form.

33. $3\frac{1}{3} \times 2\frac{7}{10} \times 5\frac{5}{9}$

34. $3\frac{3}{4} \times \frac{1}{53} \times 7\frac{1}{15}$

35. $6\frac{3}{5} \times 2\frac{1}{3} \times 3\frac{1}{2}$

36. $5\frac{4}{9} \times 3\frac{5}{7} \times 2\frac{2}{13}$

37. $4\frac{6}{7} \times 1\frac{3}{17}$

38. $4\frac{2}{3} \times 6\frac{5}{7}$

39. $3\frac{3}{8} \times 9\frac{7}{9}$

40. $8\frac{5}{9} \times 1\frac{4}{11}$

41. $12\frac{6}{7} \times 4\frac{9}{10}$

42. $3\frac{1}{33} \times 3\frac{1}{3}$

43. $8\frac{1}{3} \times 1\frac{7}{15}$

44. $4\frac{1}{8} \times 6\frac{2}{3} \times 9\frac{1}{7}$

45. $9\frac{1}{11} \times \frac{7}{47} \times 6\frac{5}{7}$

46. $5\frac{1}{12} \times 8\frac{4}{5} \times 3\frac{3}{11}$

47. $7\frac{5}{6} \times 2\frac{1}{47} \times 11\frac{1}{5}$

48. $3\frac{1}{18} \times \frac{9}{11} \times 1\frac{1}{3}$

49. $2\frac{1}{17} \times 1\frac{4}{13} \times 1\frac{3}{10}$

50. $1\frac{4}{5} \times 7\frac{6}{7} \times \frac{5}{9}$

51. $8\frac{7}{11} \times 4\frac{9}{10} \times \frac{11}{14}$

52. $12\frac{1}{10} \times 5\frac{1}{11} \times 7\frac{1}{7}$

53. $20\frac{4}{7} \times 9\frac{1}{3} \times \frac{7}{144}$

54. $\frac{13}{18} \times 5\frac{1}{5} \times 1\frac{5}{13}$

55. $1\frac{1}{4} \times 5\frac{4}{7} \times 2\frac{1}{3} \times \frac{4}{5} \times \frac{3}{7}$

56. $5\frac{12}{13} \times 3\frac{6}{7} \times 1\frac{5}{9} \times \frac{7}{27} \times \frac{9}{14}$

57. What is the reciprocal of the reciprocal of a non-zero fraction a/b?

58. What is the recriprocal of a non-zero whole number c?

5.2 MIXED PRACTICE

By doing these exercises, you will practice all the topics up to this point in the chapter.

In Exercises 59 through 64, multiply and write in simplest form.

59. $\frac{7}{9} \times \frac{3}{4} \times \frac{8}{35}$

60. $\frac{3}{4} \times 8$

61. $2\frac{1}{5} \times 1\frac{1}{8}$

62. $\frac{2}{3} \times 3\frac{4}{7} \times 1\frac{1}{2}$

63. Multiply $6\frac{2}{3}$ by its reciprocal.

64. Multiply $\frac{1}{3}$ by 13,765,222.

5.3 Dividing Fractions

We use our knowledge of multiplication to divide. The division of fractions is related to multiplication in the following way:

To Divide Fractions

If b and d are not equal to zero, then

$$\frac{a}{b} \div \frac{c}{d} = \frac{a}{b} \times \frac{d}{c} = \frac{a \times d}{b \times c}$$

In other words, to do the division $\frac{a}{b} \div \frac{c}{d}$, multiply the dividend $\frac{a}{b}$ by the reciprocal of the divisor, $\frac{d}{c}$. Then simplify the resulting fraction.

Example 1

Divide: $\frac{1}{2} \div \frac{3}{4}$

SOLUTION

First we find the reciprocal of the divisor. The reciprocal of $\frac{3}{4}$ is $\frac{4}{3}$.

Next we set up the problem as a multiplication by this reciprocal.

$$\frac{1}{2} \div \frac{3}{4} \quad \text{becomes} \quad \frac{1}{2} \times \frac{4}{3}$$

Then we multiply the numerators and the denominators.

$$\frac{1}{2} \times \frac{4}{3} = \frac{1 \times 4}{2 \times 3} = \frac{4}{6}$$

Finally, simplifying gives us

$$\frac{4}{6} = \frac{4 \div 2}{6 \div 2} = \frac{2}{3}$$

■ *Now do margin exercises 1a and 1b.*

To see why this "division" procedure works, note first that

$$\frac{1}{2} \div \frac{3}{4} \quad \text{can be written as} \quad \frac{\frac{1}{2}}{\frac{3}{4}}$$

Recall that the value of a fraction does not change when we multiply the numerator and the denominator by the same number. Recall also that a number times its reciprocal is equal to 1. Accordingly,

$$\frac{\frac{1}{2}}{\frac{3}{4}} = \frac{\frac{1}{2} \times \frac{4}{3}}{\frac{3}{4} \times \frac{4}{3}} = \frac{\frac{1}{2} \times \frac{4}{3}}{1} = \frac{1}{2} \times \frac{4}{3}$$

This is exactly the multiplication we performed to find $\frac{1}{2} \div \frac{3}{4}$ in Example 1.

The next example illustrates division by a whole number.

SECTION GOAL

■ To find the quotient of two fractions

1a. Divide: $\frac{5}{8} \div \frac{1}{5}$

1b. Divide: $\frac{3}{7} \div \frac{4}{7}$

2a. Divide: $\frac{3}{10} \div 3$

2b. Divide: $\frac{2}{7} \div 5$

EXAMPLE 2

Divide: $\frac{2}{5} \div 2$

SOLUTION

First, we rewrite the problem as multiplication by the reciprocal of the divisor.

$$\frac{2}{5} \div 2 = \frac{2}{5} \div \frac{2}{1} = \frac{2}{5} \times \frac{1}{2}$$

Then we perform this multiplication.

$$\frac{2}{5} \times \frac{1}{2} = \frac{2 \times 1}{5 \times 2} = \frac{2}{10} = \frac{1}{5}$$

So $\frac{2}{5} \div 2$ is $\frac{1}{5}$.

■ *Now do margin exercises 2a and 2b.*

EXAMPLE 3

Divide: $2 \div \frac{1}{2}$

SOLUTION

First we rewrite the problem as multiplication by the reciprocal of the divisor. We also write 2 as $\frac{2}{1}$.

$$\frac{2}{1} \div \frac{1}{2} = \frac{2}{1} \times \frac{2}{1}$$

Then we perform the multiplication.

$$\frac{2}{1} \times \frac{2}{1} = \frac{2 \times 2}{1 \times 1} = \frac{4}{1} = 4$$

So $2 \div \frac{1}{2}$ is 4.

3a. Divide: $17 \div \frac{1}{5}$

■ *Now do margin exercises 3a and 3b.*

Example 3 is so simple that we can almost "see" the answer before we get it. But, fortunately, the procedures of mathematics have been designed to work for both simple and complicated numbers, whether or not we can "see" the answer.

3b. Divide: $3 \div \frac{1}{7}$

Let's try another example. Note that in this case, because we are dividing by a proper fraction, we get a quotient that is larger than the dividend.

EXAMPLE 4

Divide: $\frac{3}{16} \div \frac{7}{8}$

SOLUTION

4a. Divide: $\frac{5}{7} \div \frac{3}{14}$

First we rewrite our problem as multiplication by the reciprocal of $\frac{7}{8}$. The problem becomes

$$\frac{3}{16} \times \frac{8}{7}$$

Then we multiply and simplify.

$$\frac{3}{16} \times \frac{8}{7} = \frac{3 \times 8}{16 \times 7} = \frac{3 \times \overset{1}{\cancel{8}}}{\underset{2}{\cancel{16}} \times 7} = \frac{3}{14}$$

4b. Divide: $\frac{4}{9} \div \frac{3}{7}$

■ *Now do margin exercises 4a and 4b.* Work the exercises that follow.

5.3 EXERCISES

In Exercises 1 through 4, find the reciprocal of each fraction.

1. $\frac{17}{27}$

2. $\frac{18}{23}$

3. $\frac{37}{49}$

4. $\frac{120}{121}$

In Exercises 5 through 40, divide. Make sure your answer is in simplest form.

5. $4 \div \frac{1}{4}$

6. $\frac{5}{12} \div \frac{1}{12}$

7. $7 \div \frac{1}{7}$

8. $\frac{3}{8} \div \frac{1}{8}$

9. $\frac{2}{3} \div 2$

10. $3 \div \frac{1}{3}$

11. $\frac{5}{7} \div 5$

12. $\frac{6}{11} \div 6$

13. $\frac{5}{6} \div \frac{1}{6}$

14. $5 \div \frac{1}{5}$

15. $\frac{7}{10} \div \frac{1}{10}$

16. $\frac{3}{4} \div 3$

17. $\frac{1}{3} \div \frac{1}{6}$

18. $\frac{1}{2} \div \frac{1}{8}$

19. $\frac{7}{90} \div \frac{3}{27}$

20. $\frac{8}{9} \div \frac{15}{90}$

21. $\frac{2}{35} \div \frac{2}{25}$

22. $\frac{4}{45} \div \frac{3}{5}$

23. $\frac{3}{10} \div \frac{7}{90}$

24. $\frac{6}{75} \div \frac{5}{15}$

25. $\frac{23}{75} \div \frac{7}{25}$

26. $\frac{21}{91} \div \frac{11}{13}$

27. $\frac{27}{38} \div \frac{15}{19}$

28. $\frac{9}{10} \div \frac{9}{11}$

29. $\frac{35}{65} \div \frac{8}{13}$

30. $\frac{17}{19} \div \frac{14}{95}$

31. $\frac{14}{27} \div \frac{34}{54}$

32. $\frac{5}{11} \div \frac{95}{121}$

33. $\frac{51}{53} \div \frac{1}{10}$

34. $\frac{21}{56} \div \frac{11}{12}$

35. $\frac{3}{55} \div \frac{4}{15}$

36. $\frac{2}{14} \div \frac{2}{35}$

37. Divide $\frac{120}{121}$ by $\frac{10}{11}$.

38. Divide $\frac{320}{450}$ by $\frac{20}{63}$.

39. Divide $\frac{710}{891}$ by $\frac{10}{11}$.

40. Divide $\frac{525}{600}$ by $\frac{25}{30}$.

In Exercises 41 through 48, find the reciprocal of each mixed number.

41. $2\frac{17}{19}$

42. $15\frac{6}{7}$

43. $19\frac{2}{3}$

44. $304\frac{7}{10}$

45. $28\frac{11}{20}$

46. $15\frac{13}{15}$

47. $63\frac{21}{25}$

48. $100\frac{16}{17}$

5.3 MIXED PRACTICE

By doing these exercises, you will practice all the topics up to this point in the chapter. Write all answers in simplest form.

49. Multiply: $\frac{1}{3} \times \frac{2}{9} \times \frac{5}{12}$

50. Find the product of $7\frac{1}{3}$ and $2\frac{5}{11}$.

51. Multiply: $\frac{2}{7} \times \frac{3}{5}$

52. Divide: $\frac{11}{13}$ by $\frac{18}{39}$

53. Divide: $\frac{3}{4} \div \frac{9}{26}$

54. Find the product: $\frac{11}{13} \times \frac{1}{8}$

55. Find the quotient: $\frac{2}{9} \div \frac{14}{81}$

56. Find the product: $2 \times \frac{13}{16}$

57. Multiply: $3\frac{2}{5} \times 8$

58. Find the quotient: $\frac{7}{15} \div \frac{14}{17}$

5.4 Dividing Fractions and Mixed Numbers

SECTION GOAL

- To find the quotient of two fractions or mixed numbers

If a division problem includes mixed numbers, we first change them to improper fractions. Then we use the procedure for dividing fractions.

To Divide Two or More Mixed Numbers

1. Change the mixed numbers to improper fractions.
2. Proceed as in the division of fractions.

EXAMPLE 1

Divide: $2\frac{1}{3} \div \frac{1}{3}$

SOLUTION

First we change the mixed number to an improper fraction.

$$2\frac{1}{3} = \frac{(2 \times 3 + 1)}{3} = \frac{7}{3}$$

Then we rewrite the problem as multiplication by the reciprocal of the divisor.

$$\frac{7}{3} \div \frac{1}{3} = \frac{7}{3} \times \frac{3}{1}$$

Then we multiply and simplify.

$$\frac{7}{3} \times \frac{3}{1} = \frac{7 \times 3}{3 \times 1} = \frac{21}{3} = 7$$

We can check the result by multiplying the quotient (7) by the divisor $(\frac{1}{3})$.

$$7 \times \frac{1}{3} = \frac{7}{1} \times \frac{1}{3} = \frac{7}{3} = 2\frac{1}{3}$$

■ *Now do margin exercises 1a and 1b.*

1a. Divide: $3\frac{1}{5} \div \frac{1}{2}$

1b. Divide: $13\frac{1}{8} \div \frac{3}{4}$

Now we will divide a proper fraction by a mixed number.

EXAMPLE 2

Divide: $\frac{1}{2} \div 2\frac{1}{2}$

SOLUTION

First we rewrite the mixed number as a fraction.

$$2\frac{1}{2} = \frac{(2 \times 2 + 1)}{2} = \frac{5}{2}$$

Then we rewrite the problem as multiplication by the reciprocal of the divisor.

$$\frac{1}{2} \div \frac{5}{2} = \frac{1}{2} \times \frac{2}{5}$$

Now we multiply and simplify.

2a. Divide: $\frac{1}{8} \div 2\frac{1}{4}$

2b. Divide: $\frac{2}{5} \div 3\frac{1}{10}$

$$\frac{1}{2} \times \frac{2}{5} = \frac{1 \times 2}{2 \times 5} = \frac{2}{10} = \frac{1}{5}$$

We can check by multiplying the quotient by the divisor.

$$\frac{1}{5} \times 2\frac{1}{2} = \frac{1}{5} \times \frac{5}{2} = \frac{1}{2}$$

So $\frac{1}{2} \div 2\frac{1}{2} = \frac{1}{5}$.

■ *Now do margin exercises 2a and 2b.*

EXAMPLE 3

Divide: $5\frac{5}{7} \div 6\frac{2}{3}$

3a. Divide: $5\frac{3}{8} \div 6\frac{1}{4}$

3b. Divide: $7\frac{7}{10} \div 8\frac{2}{5}$

SOLUTION

Note that the divisor is larger than the dividend. Therefore, the quotient for this division will be smaller than 1.

First we change both mixed numbers to improper fractions.

$$5\frac{5}{7} = \frac{(5 \times 7 + 5)}{7} = \frac{40}{7}$$

$$6\frac{2}{3} = \frac{(6 \times 3 + 2)}{3} = \frac{20}{3}$$

Then we rewrite the problem as multiplication by the reciprocal of the divisor.

$$\frac{40}{7} \div \frac{20}{3} = \frac{40}{7} \times \frac{3}{20}$$

Now we multiply and simplify.

$$\frac{40}{7} \times \frac{3}{20} = \frac{40 \times 3}{7 \times 20} = \frac{\overset{2}{\cancel{40}} \times 3}{7 \times \underset{1}{\cancel{20}}} = \frac{6}{7}$$

We can check by multiplying.

$$\frac{6}{7} \times 6\frac{2}{3} = \frac{6}{7} \times \frac{20}{3} = \frac{\overset{2}{\cancel{6}} \times 20}{7 \times \underset{1}{\cancel{3}}} = \frac{40}{7} = 5\frac{5}{7}$$

So $5\frac{5}{7} \div 6\frac{2}{3} = \frac{6}{7}$.

■ *Now do margin exercises 3a and 3b.*

Work the exercises that follow.

5.4 EXERCISES

In Exercises 1 through 52, divide and simplify your answers.

1. $4\frac{1}{2} \div \frac{1}{2}$
2. $\frac{1}{4} \div 3\frac{1}{4}$
3. $5\frac{1}{8} \div \frac{1}{8}$
4. $\frac{1}{8} \div 2\frac{1}{8}$
5. $3\frac{1}{6} \div 2\frac{2}{3}$
6. $6\frac{1}{3} \div 2\frac{5}{6}$
7. $8\frac{1}{4} \div 1\frac{1}{8}$
8. $4\frac{2}{5} \div 2\frac{1}{10}$
9. $10\frac{1}{5} \div 2\frac{1}{10}$
10. $7\frac{1}{3} \div 8\frac{1}{6}$
11. $1\frac{1}{7} \div 2\frac{3}{14}$
12. $3\frac{1}{9} \div 1\frac{5}{18}$
13. $3\frac{1}{5} \div 6\frac{2}{7}$
14. $4\frac{1}{2} \div 2\frac{1}{4}$
15. $10\frac{2}{3} \div 4\frac{2}{9}$
16. $4\frac{2}{5} \div 1\frac{1}{10}$
17. $1\frac{2}{5} \div 5\frac{7}{10}$
18. $7\frac{3}{8} \div 2\frac{1}{4}$
19. $2\frac{3}{16} \div 17\frac{1}{2}$
20. $5\frac{4}{7} \div 1\frac{5}{14}$
21. $1\frac{1}{2} \div 9\frac{5}{6}$
22. $5\frac{1}{15} \div 38$
23. $3\frac{1}{7} \div 6\frac{2}{7}$
24. $6\frac{1}{9} \div 11$
25. $8\frac{2}{5} \div 4\frac{2}{3}$
26. $2\frac{2}{7} \div 32$
27. $4\frac{3}{8} \div 14\frac{1}{8}$
28. $7\frac{3}{5} \div 5\frac{3}{7}$
29. $\frac{3}{8} \div 3\frac{1}{2}$
30. $\frac{5}{6} \div 6\frac{2}{3}$
31. $4\frac{1}{14} \div \frac{2}{7}$
32. $\frac{1}{4} \div 11\frac{1}{3}$
33. $9\frac{3}{4} \div 11\frac{7}{10}$
34. $4\frac{5}{6} \div 3\frac{5}{8}$
35. $11\frac{1}{5} \div 3\frac{1}{5}$
36. $4\frac{1}{12} \div 1\frac{1}{6}$
37. $1\frac{3}{8} \div 3\frac{2}{3}$
38. $17\frac{1}{9} \div 17\frac{1}{9}$
39. $1\frac{17}{21} \div 2\frac{5}{7}$
40. $2\frac{19}{30} \div 7\frac{9}{10}$
41. $6\frac{1}{20} \div 5\frac{1}{2}$
42. $3\frac{11}{15} \div 11\frac{1}{5}$
43. $9\frac{1}{6} \div 3\frac{2}{36}$
44. $4\frac{1}{5} \div 3\frac{4}{15}$
45. $8\frac{3}{40} \div 3\frac{1}{20}$
46. $4\frac{7}{12} \div 3\frac{23}{36}$
47. $\frac{1}{25} \div 3\frac{4}{5}$
48. $4\frac{4}{5} \div 4\frac{6}{25}$

49. $1\frac{1}{48} \div \frac{1}{36}$

50. $10\frac{1}{8} \div 24\frac{3}{4}$

51. $1\frac{5}{39} \div 4\frac{12}{13}$

52. $\frac{9}{26} \div 1\frac{47}{52}$

5.4 MIXED PRACTICE

By doing these exercises, you will practice all the topics up to this point in the chapter.

53. Multiply: $\frac{5}{6} \times \frac{4}{7} \times \frac{12}{25}$

54. Divide: $\frac{3}{5} \div 4$

55. Find the quotient: $5\frac{3}{7} \div 8\frac{1}{2}$

56. Find the product: $9\frac{3}{7} \times 8\frac{1}{6}$

57. Divide: $\frac{4}{7} \div \frac{11}{49}$

58. Multiply: $\frac{2}{5} \times 4\frac{3}{7} \times 2\frac{1}{2}$

59. Divide: $8\frac{2}{5} \div 9\frac{5}{8}$

60. Find the product: $15 \times \frac{4}{5}$

61. Find the quotient: $4\frac{1}{16} \div 2$

62. Find the product: $\frac{3}{10} \times \frac{5}{9}$

63. Find the quotient: $\frac{2}{7} \div 6$

5.5 Applying Multiplication and Division of Fractions and Mixed Numbers: Word Problems

In Chapter 2, you learned how to work word problems that require multiplication or division of whole numbers for their solution. Let's look at some problems that require those same operations on fractions. Our concern, as always, is to read and interpret the problem, decide which method to use, apply the method, and then review the problem to determine if the answer is reasonable.

SECTION GOAL

- To solve applications of multiplication and division of fractions and mixed numbers

EXAMPLE 1

In a certain pharmacy, there are 29 bottles of Cophex cold medicine on a shelf. If each of these bottles holds $\frac{3}{4}$ ounce, how many ounces of cold medicine are available altogether?

SOLUTION

Here we use multiplication as a substitute for repeated addition. There are 29 bottles and each bottle holds $\frac{3}{4}$ ounce. We solve this problem by multiplying the number of bottles by the number of ounces per bottle.

$$29 \times \frac{3}{4} = \frac{87}{4} = 21\frac{3}{4}$$

The 29 bottles contain $21\frac{3}{4}$ ounces of cold medicine.

■ *Now do margin exercises 1a and 1b.*

1a. In a soda bottling plant, $8\frac{1}{2}$ ounces of soda are being put into each bottle. There are 5900 bottles filled. How much soda do they contain altogether?

1b. Wood chips are being packed with $5\frac{1}{2}$ pounds per box. There are 150 boxes. How many pounds of wood chips are there altogether?

EXAMPLE 2

For a formal dinner, $12\frac{1}{2}$ cantaloupes are to be cut into portions that are each the size of $\frac{1}{4}$ cantaloupe. How many people can be served cantaloupe?

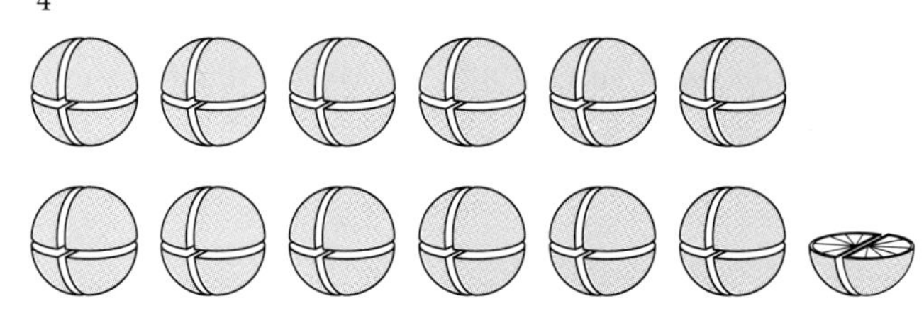

SOLUTION

We have $12\frac{1}{2}$ cantaloupes and want to cut (or divide) each into pieces that are $\frac{1}{4}$ of a cantaloupe. We therefore need to divide $12\frac{1}{2}$ by $\frac{1}{4}$ to find the number of pieces.

$$12\frac{1}{2} \div \frac{1}{4} = \frac{25}{2} \times \frac{4}{1} = \frac{100}{2} = 50$$

Thus 50 people can be served cantaloupe.

2a. You have $6\frac{2}{3}$ boards that need to be cut into lengths that are each $10\frac{1}{3}$ feet long. How many lengths can you cut if each board is 22 feet long?

2b. Out of a piece of wire, there were 29 pieces cut that were each $1\frac{3}{4}$ feet in length. There were $1\frac{1}{4}$ feet left over. What was the length of the original wire?

3a. A runner was jogging around Washington Square Park, moving at $7\frac{1}{2}$ miles per hour. Another runner was running $1\frac{1}{5}$ times that speed. How fast was this second runner moving?

3b. A car was driving up the Blue Ridge Parkway, moving at $30\frac{1}{2}$ miles per hour. Another car on a similar roadway was moving about $2\frac{1}{2}$ times that speed. How fast was this second car moving?

■ *Now do margin exercises 2a and 2b.*

In Example 2, we could also reason that we obtain 4 pieces from each cantaloupe. Then we would get $4 \times 12\frac{1}{2} = 50$ pieces from all $12\frac{1}{2}$ cantaloupes.

EXAMPLE 3

A ship, sailing up the Hudson River, was clocked at $5\frac{1}{2}$ knots (nautical miles per hour). On open water, the ship can move about $3\frac{1}{2}$ times that speed. How fast can it move on open water?

SOLUTION

This problem involves a scaled increase; therefore, we multiply.

To find the speed on open water, we must multiply its river speed ($5\frac{1}{2}$ knots) by the scale factor ($3\frac{1}{2}$ times that speed).

$$5\frac{1}{2} \times 3\frac{1}{2}$$

We first change the mixed numbers to improper fractions.

$$5\frac{1}{2} = \frac{(5 \times 2 + 1)}{2} = \frac{11}{2}$$
$$3\frac{1}{2} = \frac{(3 \times 2 + 1)}{2} = \frac{7}{2}$$

Then we multiply.

$$\frac{11}{2} \times \frac{7}{2} = \frac{77}{4}$$

The unit here is knots. After simplifying $\frac{77}{4}$, we see that the ship can move at $19\frac{1}{4}$ knots on open water.

■ *Now do margin exercises 3a and 3b.*

The next example requires division, because we are "breaking down" an amount into several smaller amounts.

EXAMPLE 4

You have $3\frac{1}{2}$ pounds of sunflower seeds that you want to sell in 14 equal, smaller packages. How much will each package weigh?

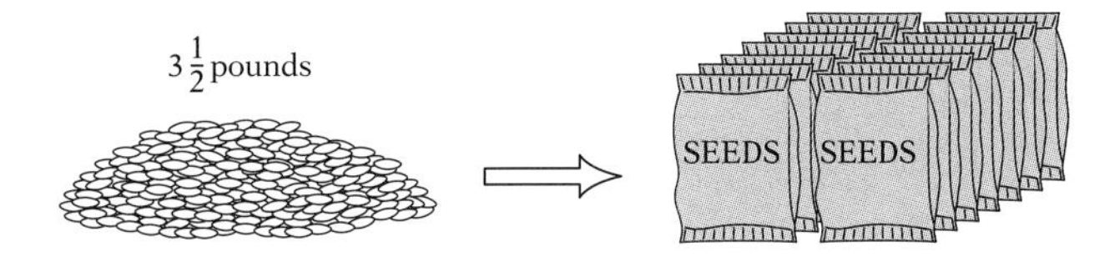

SOLUTION

Here we want to divide the $3\frac{1}{2}$ pounds among 14 separate packages.

$$3\frac{1}{2} \div 14$$

We first rewrite both $3\frac{1}{2}$ and 14 as improper fractions.

$$3\frac{1}{2} = \frac{7}{2} \qquad 14 = \frac{14}{1}$$

Then we rewrite the problem as a multiplication, multiply, and simplify.

$$\frac{7}{2} \div \frac{14}{1} = \frac{7}{2} \times \frac{1}{14} = \frac{7}{28} = \frac{1}{4}$$

So the 14 packages will weigh $\frac{1}{4}$ pound each.

■ *Now do margin exercises 4a and 4b.*

Work the exercises that follow.

4a. You have $60\frac{2}{3}$ pounds of birdseed. You want to break this up into 91 packets that can be opened to fill bird feeders throughout the winter. How much will each packet weigh?

4b. A roll of ribbon is 200 feet long. How many pieces of ribbon can be cut that are each $1\frac{1}{3}$ feet long? How much ribbon will be left over?

EXTRA

Using Fractions to Identify Fingerprints

Fractions have many uses. One of the most fascinating is in classifying fingerprints. Fingerprints, which are unique to each individual and remain unchanged throughout a person's lifetime (except for scarring), are made up of unique patterns of arches, loops, and whorls on each finger.

A fingerprint identification method that is still in use in North American and most English speaking countries was proposed in 1900 by Edward Richard Henry. In one part of this system, the values are assigned as follows:

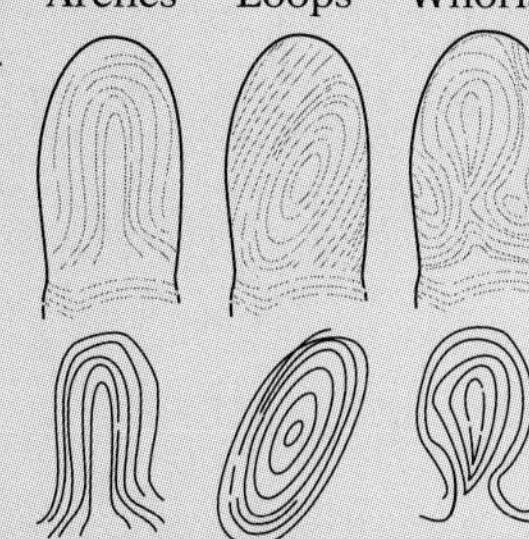

$$\text{loop} = 0$$
$$\text{arch} = 0$$
$$\text{whorl} = 1$$

These values are then substituted into the following identification formula:

$$\frac{16 \times (R\text{ Index}) + 8 \times (R\text{ Ring}) + 4 \times (L\text{ Thumb}) + 2 \times (L\text{ Middle}) + 1 \times (L\text{ Little}) + 1}{16 \times (R\text{ Thumb}) + 8 \times (R\text{ Middle}) + 4 \times (R\text{ Little}) + 2 \times (L\text{ Index}) + 1 \times (L\text{ Ring}) + 1}$$

where L and R stand for left and right, respectively. Thus, if a person has a whorl on the left and right index fingers and on the left middle finger and loops and arches on all others, her or his classification number is

$$\frac{16\,(1) + 8\,(0) + 4\,(0) + 2\,(1) + 1\,(0) + 1}{16\,(0) + 8\,(0) + 4\,(0) + 2\,(1) + 1\,(0) + 1} = \frac{16 + 2 + 1}{2 + 1} = \frac{19}{3} = 6\frac{1}{3}$$

This formula divides all fingerprints into 1024 different classes. Other information is then used to further classify a person's fingerprints.

1. What classification number would you have if your fingerprints contained only loops and arches?
2. Say if every finger and both thumbs showed a whorl. What would be the classification number then?
3. What is your classification number?

*From George Knill, "Fingerprints and Fractions," The Mathematics Teacher, November 1980, p. 608.

5.5 EXERCISES

Solve each of the following word problems. Be sure to read each problem carefully before starting and review the problem when finished to see if the result is reasonable.

1. The human body produces $1\frac{1}{2}$ million red blood cells every second. How many does it produce per minute?

2. Five million, eight hundred forty-three thousand, thirty-three square miles of land are cultivated. This represents approximately $\frac{1}{10}$ of the total land area of the world. What is the approximate land area of the world?

3. A certain glass holds $5\frac{1}{3}$ ounces. How many glasses could be filled from a 32-ounce pitcher of water? How much water would be left in the pitcher?

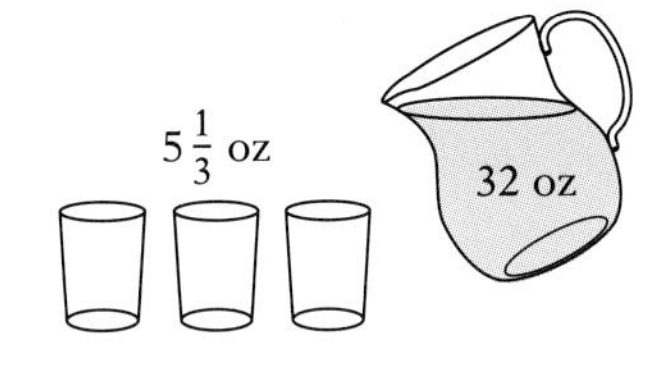

4. The highest point of elevation in the Netherlands is about $\frac{1}{4}$ the height of the Empire State Building. If the building is 1472 feet tall, about how high is the highest point in the Netherlands?

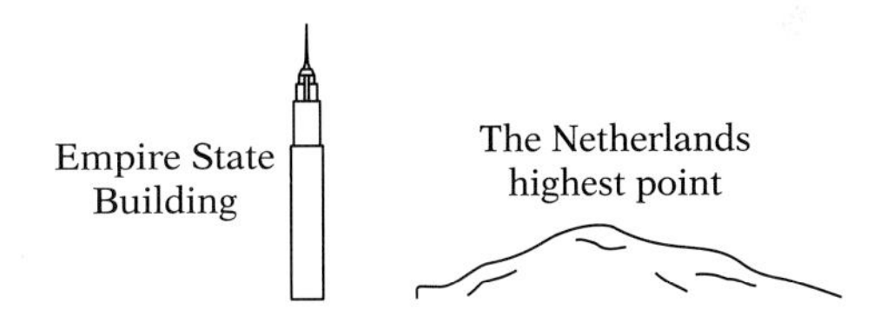

5. There is a Leonid meteor shower every $33\frac{1}{4}$ years. How many showers will have occurred in the next 200 years?

6. A barrel holds $22\frac{3}{4}$ gallons of fluid. How many gallons are there in the barrel when it is half full?

7. A scout troop is making stuffed animals for the children's ward. If each bear requires $2\frac{3}{4}$ square feet of material, how much material will the scouts need for 84 bears?

8. A penny is $\frac{3}{4}$ inch wide. How many pennies, placed in a line, would it take to reach 12 inches?

9. A trained frog can jump 17 feet $6\frac{3}{4}$ inches. If a man can jump about $\frac{3}{4}$ that distance, how far can he jump?

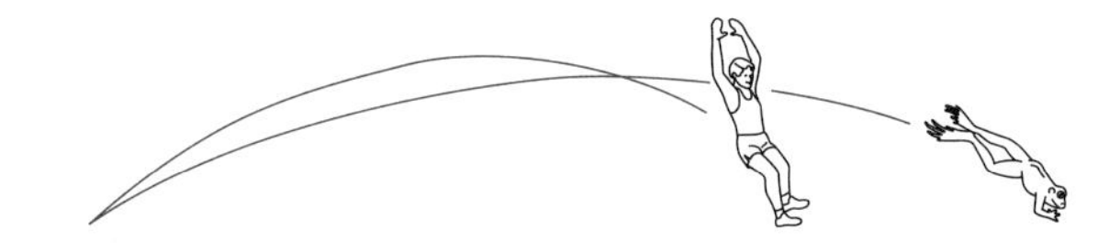

10. Two-thirds of the human body weight is water. If I weigh 230 pounds, how much of my weight is water?

11. Cans of turnip greens are on sale for 49 cents per can. If each can holds $21\frac{3}{16}$ ounces, how many ounces of greens are there in 144 cans?

12. A horse has jumped 8 feet $1\frac{3}{4}$ inches high. The pole vault record is twice that plus $11\frac{1}{2}$ inches. About how high is the pole vault record jump?

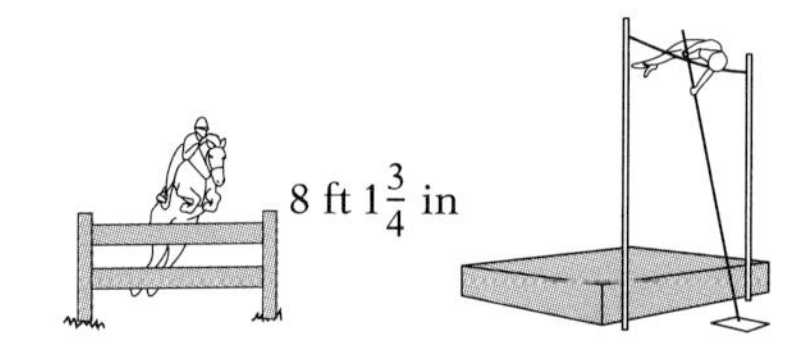

13. A Saturn year is equivalent to $29\frac{1}{2}$ Earth years. How many Saturn years are there in 1000 Earth years?

14. The population of London in the mid-1970s was about $4\frac{4}{5}$ times that of the population of Warsaw. About how large was London's population if Warsaw's was 1,500,000?

15. Of the London area's estimated seven million inhabitants, only about 8 of every 10,000 live in the actual "City" of London. About how many people live in the "City"?

16. A Pluto day is equivalent to $6\frac{1}{2}$ Earth days. How many Pluto days are there in 390 Earth days?

17. A table tennis ball weighs about $\frac{9}{100}$ ounce. It would take $2868\frac{1}{2}$ table tennis balls to equal the weight of a ten-pin bowling ball. How much does such a bowling ball weigh?

18. An elephant weighs about $6\frac{5}{6}$ tons. A tractor weighs about $1\frac{1}{5}$ times as much as an elephant and consumes about 3590 gallons of diesel fuel each year. How much does a tractor weigh?

19. The distance from the pitcher's mound to the home plate in baseball is $60\frac{1}{2}$ feet. If a player's stride is $2\frac{1}{2}$ feet long, how many steps would he take to walk from the mound to home plate?

20. At one time, the world triple jump record was 58 feet $8\frac{1}{2}$ inches. How long was each of the three parts of the jump if they were all the same length?

21. Neptune's diameter is $17\frac{1}{4}$ times that of Earth. What is the diameter of Neptune if Earth's diameter is 7926 miles?

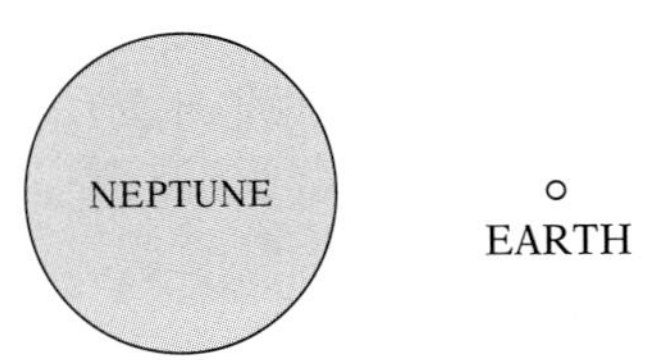

22. A flea can jump $1\frac{1}{12}$ feet. A kangaroo can jump $10\frac{1}{2}$ feet farther. An elephant does not jump. How many times farther can a kangaroo jump than a flea?

23. Transportation costs $\frac{1}{8}$ of your weekly allowance and food costs $\frac{2}{5}$ of it. If you receive $40 per week, what do you have left after paying for transportation and food?

24. A paper clip is $1\frac{1}{10}$ inches long. How many paper clips would it take to reach from one end of a piece of notebook paper to the other if the paper is 11 inches long?

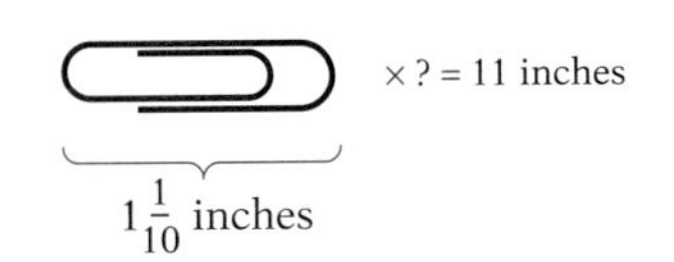

25. There are 500 sheets of paper in one ream. If I have $4\frac{1}{5}$ reams of paper, how many sheets do I have?

26. A Greyhound bus weighs about $1\frac{9}{10}$ times as much as an elephant. An elephant weighs about $6\frac{5}{6}$ tons. How much do the elephant and the bus weigh together?

27. Letters can be typed in elite print or pica print. Pica print allows about 10 letters an inch. If my paper is $8\frac{1}{2}$ inches wide, how many pica letters can I fit on one line?

28. The approximate land area of the world is 58,433,000 square miles. One-third of this land is arid. How many square miles are arid?

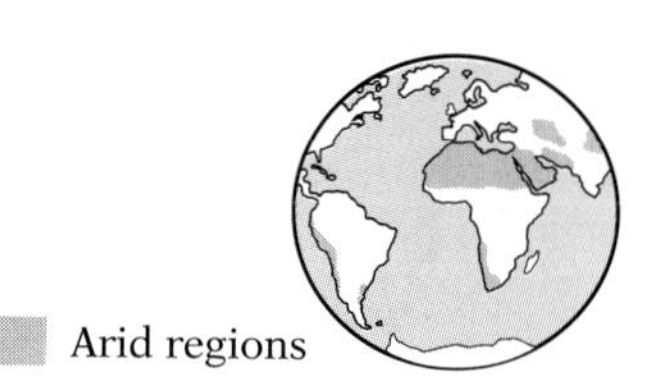

29. In 1981, the median wage of full-time female workers age 15 and over was approximately $\frac{2}{3}$ that of male workers. If the average male earned approximately $20,260 that year, what did the average female earn?

30. The road distance (in miles) between Memphis and Pittsburgh is approximately $1\frac{1}{4}$ times the air distance (in statute miles) between Pittsburgh and Memphis. If the air distance is approximately six hundred sixty statute miles, how many miles is the road distance?

31. The land area of the former Soviet Union is approximately three-quarters of the land area of Africa. If the former Soviet Union has a land area of approximately 9,000,000 square miles, what is the approximate area of Africa?

32. Use estimation to answer the following: The fraction of women finishing the New York City marathon in 1986 was 3323 of 19,689 total finishers, or $\frac{3323}{19{,}689}$. In 1987, it was $\frac{3689}{21{,}244}$.

Which fraction is larger, and what does this mean? That is, did the fraction of women finishers increase or decrease from 1986 to 1987?

5.5 MIXED PRACTICE

By doing these exercises, you will practice all the topics up to this point in the chapter. Write all answers in simplest form.

33. Multiply: $\frac{7}{12} \times \frac{6}{11} \times \frac{4}{21}$

34. A certain hamburger chain advertises square hamburgers that are $2\frac{1}{2}$ inches on each side. If the hamburgers are put side to side, how many can be lined up on one side of a pan with a length of 15 inches?

35. Multiply: $3\frac{4}{7} \times 5\frac{3}{5}$

36. Divide: $7 \div 6\frac{1}{2}$

37. A certain package is $1\frac{1}{2}$ times the length of a second. How big is the first package if the second is 2 feet long?

38. Divide: $9\frac{2}{7} \div 4\frac{1}{3}$

39. Multiply: $\frac{4}{5} \times \frac{8}{9}$

40. Divide: $7\frac{3}{5} \div 2$

41. Multiply: $4\frac{2}{3} \times 3\frac{1}{2}$

42. Your younger sister weighs 150 pounds. This is $\frac{2}{3}$ your weight. How much do you weigh?

43. Divide: $2\frac{4}{5} \div 6\frac{1}{8}$

44. A runner jogs $4\frac{2}{3}$ miles per day. How many miles does she jog in 6 days?

APRIL 1992 $4\frac{2}{3}$ miles per day

S	M	T	W	Th	F	S
			1	2	3	4
5	6	7	8	9	10	11
12	13	14	15	16	17	18
19	20	21	22	23	24	25
26	27	28	29	30		

5.6 Other Operations on Fractions and Mixed Numbers

SECTION GOALS

- To raise fractions and mixed numbers to a power
- To evaluate an expression using the standard order of operations

The operations that we performed on whole numbers also apply to fractions and mixed numbers—so does the standard order of operations.

Raising Fractions to a Power

Fractions can be squared or raised to any power, just as whole numbers can. The only difference in this case is that both the numerator and the denominator must be raised to the appropriate power.

In the expression $\left(\frac{a}{b}\right)^n$, $\frac{a}{b}$ is the *base* and n is the *exponent.* The exponent tells how many times the base is used as a factor.

As an example,

$$\left(\frac{2}{3}\right)^5 = \underbrace{\frac{2}{3} \times \frac{2}{3} \times \frac{2}{3} \times \frac{2}{3} \times \frac{2}{3}}_{\text{5 times}} = \frac{32}{243}$$

(exponent: 5; base: $\frac{2}{3}$)

The expression $\left(\frac{2}{3}\right)^5$ is read "two-thirds to the fifth power." For fractions, as for whole numbers, $\left(\frac{a}{b}\right)^0 = 1$.

EXAMPLE 1

Rewrite $\left(2\frac{1}{3}\right)^4$ without an exponent.

SOLUTION

First we rewrite $2\frac{1}{3}$ as an improper fraction.

$$2\frac{1}{3} = \frac{2 \times 3 + 1}{3} = \frac{7}{3}$$

Then we use $\frac{7}{3}$ as a factor 4 times.

$$\left(\frac{7}{3}\right) \times \left(\frac{7}{3}\right) \times \left(\frac{7}{3}\right) \times \left(\frac{7}{3}\right) = \frac{7 \times 7 \times 7 \times 7}{3 \times 3 \times 3 \times 3} = \frac{2401}{81} = 29\frac{52}{81}$$

■ *Now do margin exercises 1a and 1b.*

1a. Rewrite $\left(1\frac{1}{6}\right)^5$ without using an exponent.

1b. Rewrite $\left(1\frac{2}{5}\right)^4$ without using an exponent.

Standard Order of Operations

The standard order of operations is given in Section 3.4. If you don't remember it well, review it now—before we apply it to fractions and mixed numbers.

EXAMPLE 2

Simplify: $42 - \frac{1}{2}\left(5 + \frac{5}{7}\right) \times 7$

SOLUTION

Using the standard order of operations, we simplify within the parentheses first.

$$42 - \frac{1}{2}\left(5 + \frac{5}{7}\right) \times 7$$

$$42 - \frac{1}{2}\left(5\frac{5}{7}\right) \times 7$$

We then multiply.

$$42 - \frac{1}{2}\left(\frac{40}{7}\right) \times 7 \qquad \left(5\frac{5}{7} = \frac{40}{7}\right)$$

$$42 - \frac{20}{7} \times 7$$

$$42 - 20$$

Finally we subtract.

$$42 - 20 = 22$$

So $42 - \frac{1}{2}\left(5 + \frac{5}{7}\right) \times 7 = 22$.

■ *Now do margin exercises 2a and 2b.*

2a. Simplify: $5 \times \left(\frac{1}{2}\right) + \left(4 + \frac{2}{3}\right) - 1$

2b. Simplify: $5 - \frac{2}{3} \div \frac{1}{3}$

The example on page 321 includes an exponent as well as parentheses. When you work with a combination of parentheses and exponents, be careful to work one step at a time.

EXAMPLE 3

Simplify: $\left[\left(\frac{1}{2}\right)^2\left(3\frac{1}{6} - 2\frac{2}{3}\right)\right]^2 + 1\frac{3}{8} - 1\frac{1}{4}$

SOLUTION

Using the standard order of operations, we work inside the parentheses first.

$$\left[\left(\frac{1}{2}\right)^2\left(3\frac{1}{6} - 2\frac{2}{3}\right)\right]^2 + 1\frac{3}{8} - 1\frac{1}{4}$$

$$\left[\left(\frac{1}{2}\right)^2\left(3\frac{1}{6} - 2\frac{4}{6}\right)\right]^2 + 1\frac{3}{8} - 1\frac{1}{4} \quad \text{Rewriting with LCD}$$

$$\left[\left(\frac{1}{2}\right)^2\left(2\frac{7}{6} - 2\frac{4}{6}\right)\right]^2 + 1\frac{3}{8} - 1\frac{1}{4} \quad \text{Borrowing } \tfrac{6}{6} \text{ from 3}$$

$$\left[\left(\frac{1}{2}\right)^2\left(\frac{3}{6}\right)\right]^2 + 1\frac{3}{8} - 1\frac{1}{4} \quad \text{Subtracting}$$

Next we eliminate the exponent inside the brackets and simplify where we can.

$$\left[\left(\frac{1}{2}\right)^2\left(\frac{1}{2}\right)\right]^2 + 1\frac{3}{8} - 1\frac{1}{4} \quad \text{Simplifying}$$

$$\left[\left(\frac{1}{4}\right)\left(\frac{1}{2}\right)\right]^2 + 1\frac{3}{8} - 1\frac{1}{4} \quad \text{Squaring } \tfrac{1}{2}$$

3a. Simplify:
$\left[11-(6-3)\right]^3+\left(\frac{2}{3}\right)^2(2)(3^2)$

3b. Simplify:
$2^2-\left[(45-44)^2-\left(\frac{6}{7}\right)\right]^2-1$

We continue, multiplying inside the brackets.

$$\left[\left(\frac{1}{4}\right)\left(\frac{1}{2}\right)\right]^2+1\frac{3}{8}-1\frac{1}{4}$$

$$\left[\frac{1}{8}\right]^2+1\frac{3}{8}-1\frac{1}{4} \quad \text{Multiplying}$$

We square $\frac{1}{8}$ as indicated, obtaining

$$\frac{1}{64}+1\frac{3}{8}-1\frac{1}{4}$$

Then we do additions and subtractions from left to right.

$$\frac{1}{64}+1\frac{24}{64}-1\frac{16}{64} \quad \text{Rewriting with LCD}$$

$$1\frac{25}{64}-1\frac{16}{64} \quad \text{Adding}$$

$$\frac{9}{64} \quad \text{Subtracting}$$

Therefore,

$$\left[\left(\frac{1}{2}\right)^2\left(3\frac{1}{6}-2\frac{2}{3}\right)\right]^2+1\frac{3}{8}-1\frac{1}{4}=\frac{9}{64}$$

■ *Now do margin exercises 3a and 3b.*

Work the exercises that follow.

5.6 EXERCISES

In Exercises 1 through 16, rewrite the following without exponents. In Exercises 9 through 16, use your calculator to help with exponents.

1. $\left(\frac{1}{2}\right)^2$

2. $\left(\frac{1}{2}\right)^3$

3. $\left(\frac{1}{3}\right)^3$

4. $\left(\frac{1}{3}\right)^2$

5. $\left(7\frac{2}{5}\right)^2$

6. $\left(9\frac{3}{8}\right)^2$

7. $\left(5\frac{3}{7}\right)^2$

8. $\left(2\frac{2}{3}\right)^2$

9. $\left(1\frac{4}{9}\right)^3\left(\frac{3}{8}\right)^2$

10. $\left(2\frac{3}{5}\right)^2\left(\frac{3}{7}\right)^2$

11. $\left(\frac{2}{11}\right)^2\left(\frac{4}{7}\right)^2$

12. $\left(\frac{5}{7}\right)^3\left(\frac{1}{2}\right)^2$

13. $\left(\frac{6}{7}\right)^4\left(\frac{2}{5}\right)^2$

14. $\left(\frac{4}{5}\right)^3\left(\frac{2}{5}\right)^2$

15. $\left(1\frac{3}{8}\right)^3\left(\frac{1}{3}\right)^2$

16. $\left(\frac{7}{10}\right)^4\left(\frac{3}{7}\right)^2$

In Exercises 17 through 54, simplify each of the following expressions.

17. $26 - 14 + \frac{1}{9}$

18. $(12 \times 4) + 16 \div \frac{1}{4}$

19. $15 - 12 \times \frac{1}{4}$

20. $4 + 7 \times 8 + \left(\frac{1}{6}\right)^2$

21. $8 + \frac{1}{4} \times 3$

22. $26 + \frac{1}{8} \times 6$

23. $35 + \frac{1}{7} \times 5$

24. $9 + \frac{1}{6} \times 2$

25. $(11 + 5) \times \frac{1}{4} - 2$

26. $\left(29 + \frac{1}{4}\right) \times 2 - 5$

27. $\left(33 + \frac{1}{3}\right) \times 2 - 17$

28. $(15 + 3) \times \frac{1}{3} \div 11$

29. $(3 \times 5) + 4 \div \frac{1}{2}$

30. $47 - 10 + \frac{2}{7}$

31. $(51 \times 2) + 27 \div \frac{1}{3}$

32. $8\frac{1}{4} + (8 - 7)^5 \times 5$

33. $\left(\frac{1}{3}\right)^2 + 9 \times 3 + 7$

34. $22 + 3 \times \left(\frac{1}{2}\right)^3 + 5$

35. $17 + 4 \times 9 + \left(\frac{1}{4}\right)^2$

36. $98 - 12 \div \frac{1}{6}$

37. $2\frac{1}{2} + (6 - 2)^2 \times 3$

38. $7\frac{2}{3} + (5 - 2)^3 \times 4$

39. $5\frac{3}{4} + (8 - 5)^3 \times 2$

40. $(11 \times 6) + 15 \div \frac{1}{5}$

41. $8\left(\frac{7}{12}-\frac{1}{3}\right)^2+5$

42. $15\left(\frac{2}{3}-\frac{4}{9}\right)-\frac{2}{3}$

43. $\left(\frac{6}{11}-\frac{1}{8}\right)\div\frac{3}{8}$

44. $\left(\frac{1}{7}+\frac{3}{5}\right)\div 3\frac{1}{4}$

45. $2\frac{1}{3}+\left(6\frac{1}{5}-\frac{1}{5}\right)^2$

46. $27\div\left(\frac{1}{3}\right)^2\times\frac{1}{4}+\frac{5}{12}$

47. $5\frac{2}{5}+2\frac{1}{2}\times 6\frac{3}{5}\div\left(\frac{1}{5}\right)^2$

48. $16\div\left(4^3+\frac{3}{8}\right)\div\frac{4}{5}$

49. $[(2\div 7)^2-1\div 49]\times 7^2+3\times\frac{1}{6}$

50. $\left[3-\left(\frac{1}{18}\div\frac{2}{9}+1\right)^2\right]\div 23$

51. $\left[\left(\left(4-\frac{1}{9}\right)\times 3\right)^2\div\left(7\div 18\right)\right]^2$

52. $\left[4-\left(\frac{2}{3}\div\frac{1}{2}\right)^2\right]+5\div\frac{2}{3}$

53. $18+\left(\frac{2}{3}\times 60-3\div\frac{1}{2}\right)^2$

54. $10-\left(\frac{3}{8}\times 2\div\frac{1}{2}\right)^2\div 4$

5.6 MIXED PRACTICE

By doing these exercises, you will practice all the topics up to this point in the chapter. Express all answers in simplest form.

55. Multiply: $21\frac{1}{3}\times 6\frac{3}{4}$

56. Divide: $3\frac{3}{5}\div 2\frac{4}{5}$

57. A new office is being furnished with bookshelves. If each shelf is $15\frac{5}{12}$ feet long, how long are 10 such shelves?

58. Simplify: $\left[2\frac{1}{6}+3\frac{1}{4}\right]^2 \div 65+2$

59. Rewrite $\left(\frac{3}{4}\right)^2$ without exponents.

60. Multiply: $\frac{3}{7}\times\frac{16}{21}\times 1\frac{5}{9}$

61. Simplify: $\left(\frac{1}{2}\right)^3+2\left[\left(6-3\frac{1}{2}\right)^2+1\frac{1}{4}\right]$

62. Rewrite $\left(\frac{1}{5}\right)^3$ without exponents.

63. A roast is cut into slices that are $\frac{1}{4}$ inch thick. If the roast is 18 inches long after it is cooked, how many slices will there be?

64. Multiply: $7\frac{2}{5}\times 6\frac{1}{4}$

65. A tablespoon is equivalent to $\frac{1}{16}$ of a cup. How many cups do I have if I have 32 tablespoons of liquid?

66. Divide: $11\frac{2}{5}\div 4\frac{1}{3}$

67. Simplify: $3\left(7-3\frac{1}{3}\right)^2-5\div\frac{1}{4}$

5.7A Geometry and Measurement: Applying Operations on Fractions and Mixed Numbers

Section Goals

- To find the area of a triangle
- To find the area of a trapezoid

In this section, we will be finding the area of a triangle and the area of a trapezoid.

Area of a Triangle

To find the area of a triangle, we must first identify the *base* and the *height* of a triangle. Then:

To Find the Area of a Triangle

If b is the length of the base and h is the height, then

$$\text{Area} = \frac{1}{2} \times b \times h, \text{ or } A = \frac{1}{2}bh$$

Any side can be chosen as the base of a triangle. The height is then the shortest distance from the opposite angle to that base. If the triangle is a right triangle, then the height can be any side except the hypotenuse. Usually, though, the height is not the same as the length of a side. But the height always makes a **right (90°) angle** with the base.

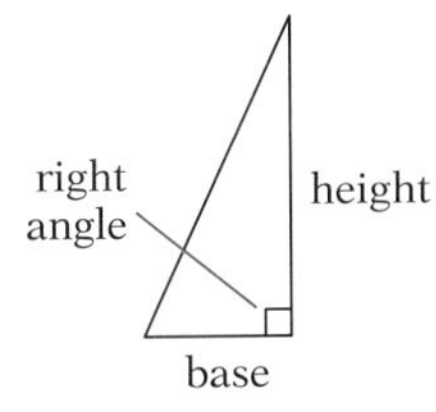

Right triangle

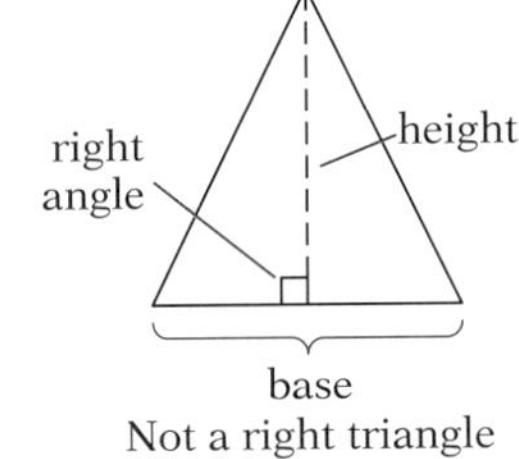

Not a right triangle

Let's look at an example.

Example 1

Find the area of this triangle.

The unit of measure is feet.

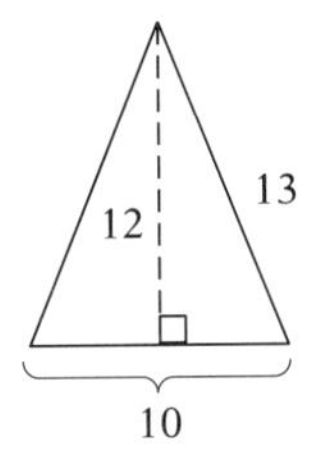

SOLUTION

We choose the 10-foot side as the base; then the height is equal to 12 feet, not 13 feet. Using the formula, we substitute the known values for b and h.

$$\begin{aligned} A &= \frac{1}{2}bh \\ &= \frac{1}{2} \times 10 \times 12 \\ &= \frac{120}{2} \\ &= 60 \end{aligned}$$

So the area of the triangle is 60 square feet. (Remember, area is expressed in *square* units.)

1a. Find the area of the following triangle.

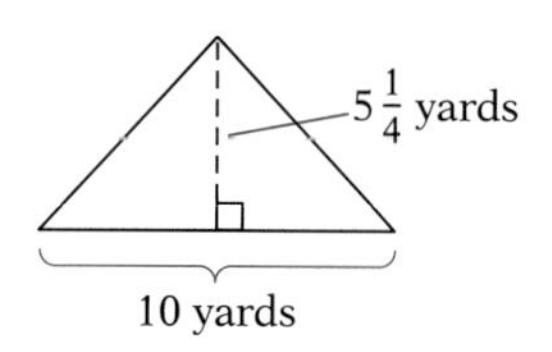

1b. Find the area of a triangle that has a height of 6 meters and a base of $7\frac{1}{2}$ meters.

2a. Find the area of a trapezoid with bases of $12\frac{1}{2}$ and $15\frac{1}{2}$ inches and height of 10 inches.

2b. Find the area of a trapezoid that has bases of 5 and $3\frac{1}{2}$ units and a height of $\frac{3}{4}$ unit.

■ *Now do margin exercises 1a and 1b.*

Area of a Trapezoid

Trapezoid *ABCD* in this figure has been divided into two triangles.

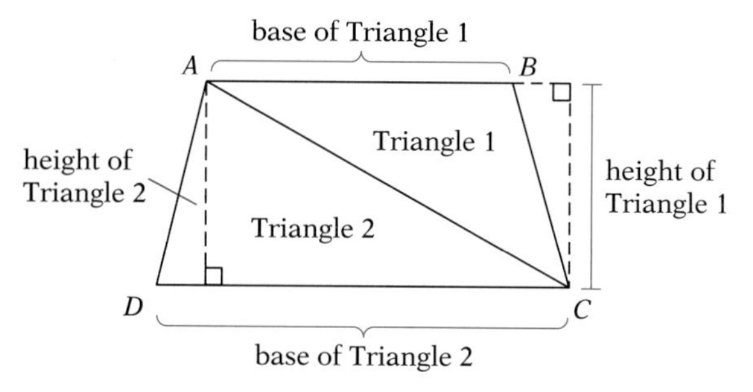

To find the area of the trapezoid, we can add the areas of the triangles. That is,

$$\text{Area of trapezoid} = \text{area ①} + \text{area ②}$$
$$= \frac{1}{2}\text{ base ①} \times \text{height ①} + \frac{1}{2}\text{ base ②} \times \text{height ②}$$

The heights of the triangles are equal, so we can write this as

$$\text{Area of trapezoid} = \frac{1}{2}\text{ height (base ①} + \text{base ②)}.$$

In symbolic form,

To Find the Area of a Trapezoid

If h is the height, and b_1 and b_2 are the lengths of the bases, then

$$A = \frac{1}{2}h(b_1 + b_2)$$

EXAMPLE 2

Find the area of a trapezoid with bases of 12 and 36 units and height 10 units.

SOLUTION

We substitute the values for b_1, b_2, and h into the formula.

$$A = \frac{1}{2}h(b_1 + b_2)$$
$$= \frac{1}{2}(10)(12 + 36)$$
$$= \frac{1}{2}(10)(48)$$
$$= 5(48)$$
$$= 240$$

So the area is 240 square units.

■ *Now do margin exercises 2a and 2b.*

Work the exercises that follow.

5.7A EXERCISES

1. Find the area of a trapezoid with bases of $12\frac{2}{3}$ and $16\frac{1}{2}$ inches and height 10 inches.

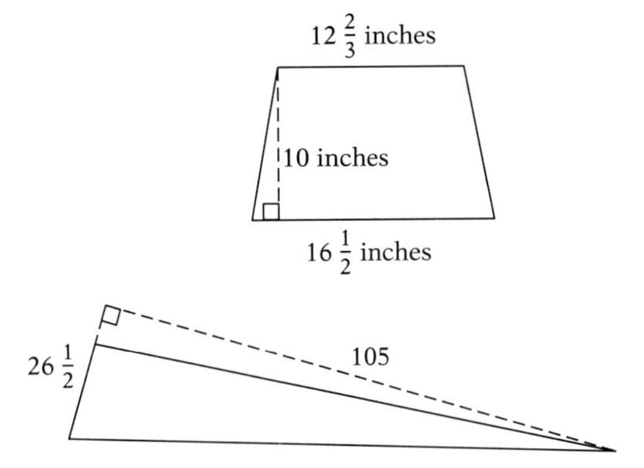

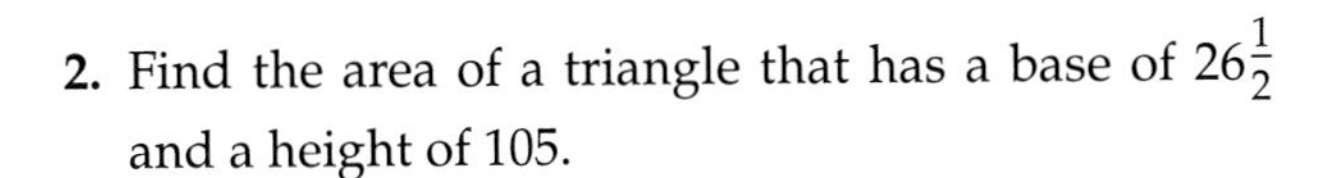

2. Find the area of a triangle that has a base of $26\frac{1}{2}$ and a height of 105.

3. Given that $h = 2A/(b_1 + b_2)$, find h when A is 48 square inches, b_1 is 15 inches, and b_2 is 20 inches.

4. Find the area of a trapezoid whose height is 9 feet and whose bases are 11 feet and 13 feet, respectively.

5. Which has greater area, a triangle with a base of $18\frac{1}{2}$ inches and a height of $3\frac{1}{2}$ inches, or a trapezoid with bases of $18\frac{1}{2}$ inches and $10\frac{1}{2}$ inches and a height of $2\frac{1}{2}$ inches?

6. Which has the greater area, a triangular sail whose base is $3\frac{1}{5}$ yards and whose height is $7\frac{3}{5}$ yards, or a $5\frac{1}{2}$-yard square sail?

7. Two triangles have equal bases and equal heights, but one is a right triangle and one is a scalene triangle. Which has the larger area? Justify your answer.

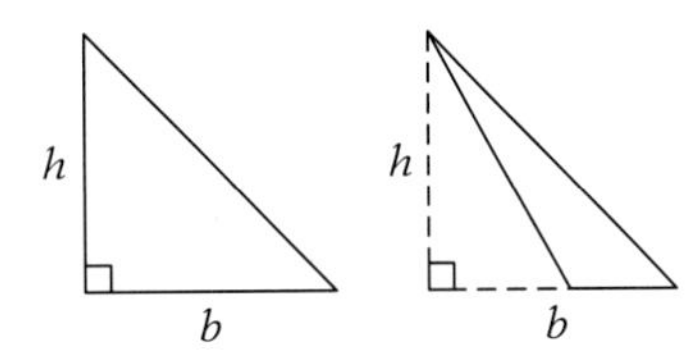

8. Trapezoids with bases of $12\frac{1}{2}$ and $15\frac{1}{2}$ inches and heights of $1\frac{1}{2}$ inches are fitted together around a square tile to form a larger square tile (see accompanying figure). What is the area of the larger square tile, and what is the area of the smaller one?

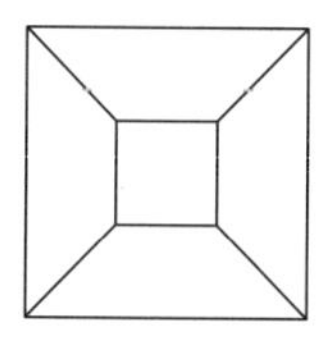

9. A farmer wants to fence off a triangular piece of his land that stretches 12 yards by 13 yards by 5 yards. Show that this piece of land is a right triangle and then find the area.

10. The kite shown in the accompanying figure is made up of two equal cloth triangles, with two braces holding the corners in place. The lengthwise brace has a length of $26\frac{1}{2}$ inches, and the crosswise brace has a length of $10\frac{1}{2}$ inches. What is the area of the kite?

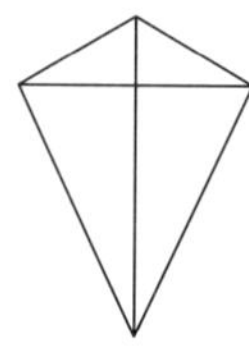

11. A neon sign is being erected along the road. The sign will be in the shape of a trapezoid with bases of $5\frac{1}{4}$ feet and $8\frac{1}{2}$ feet and a height of $4\frac{1}{3}$ feet. How much space will there be on the sign to write the message?

12. A field shaped like a trapezoid is fenced off diagonally from the southwest corner to the northeast corner. The area of the southernmost triangle is $20\frac{1}{4}$ square yards, and the area of the northernmost triangle is $30\frac{1}{6}$ yards. If the height of both triangles is $5\frac{1}{2}$ yards, what are the lengths of the bases? Which base is located farthest north? Which is located farthest south?

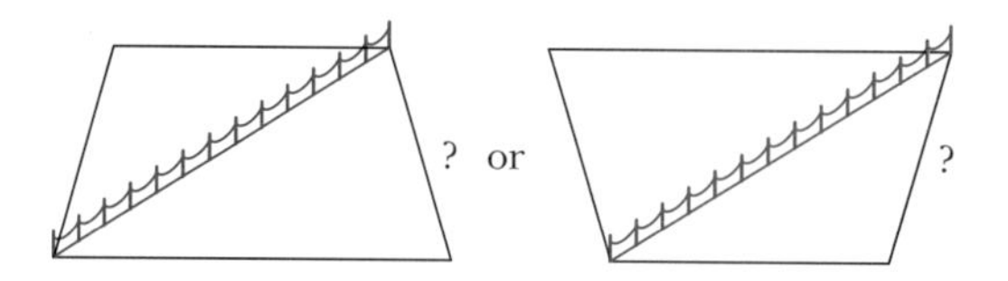

13. A square tile with an area of $36\frac{1}{8}$ square inches has diagonals with a length of $25\frac{1}{2}$ inches. What is the area of each of the four equal triangles formed by the diagonals?

14. A quilt is made entirely of triangular pieces of cloth like those shown in the accompanying figure. Pairs of triangles are sewn into small squares, each with an area of 25 square centimeters as shown in the middle figure. This makes up a quilt with a length of 160 centimeters and a width of 120 centimeters, as shown at the right. Use this information to answer the following questions.

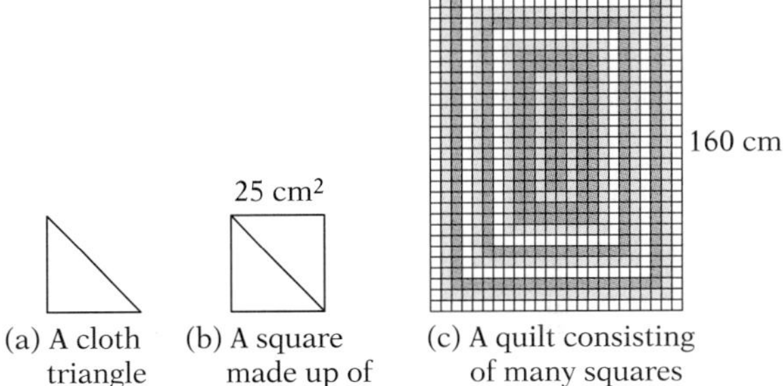

(a) A cloth triangle (b) A square made up of two triangles (c) A quilt consisting of many squares

a. What is the area of each small triangular piece of cloth?

b. How many squares altogether are in this quilt?

c. If the quilt is made up of only 3 colors, equally distributed, what area of cloth is needed in each color?

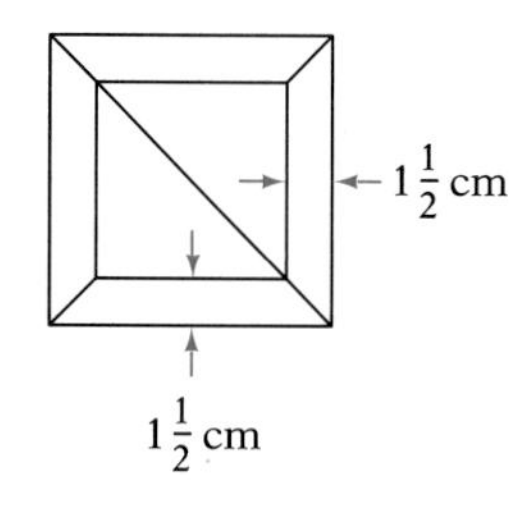

15. Another quilter used the squares shown in Exercise 14 but put a $1\frac{1}{2}$-centimeter border around each, as the accompanying figure shows.

a. If this quilter used the same number of squares as the quilter in Exercise 14, what would be the dimensions of this quilt?

b. If this quilt had the same outside dimensions as the quilt in Exercise 14, how many of these new squares would be in this quilt?

c. The border around each square in this quilt is made up of 4 trapezoids. What are the dimensions of each trapezoid? What is the area of each border trapezoid?

5.7B Geometry and Measurement: Additional Applications

SECTION GOALS

- To find the volume of a cube
- To find the volume of a rectangular solid
- To find the surface area of a cube
- To find the surface area of a rectangular prism

Calculating Volume

Recall from Chapter 1 that the volume of a three-dimensional object is equal to the amount of space it encloses. The units of volume are cubic units: cubic inches, cubic feet, and so on. A cubic inch is a cube with its length, width, and height all equal to 1 inch; a cubic foot is a cube with its length, width, and height all equal to 1 foot.

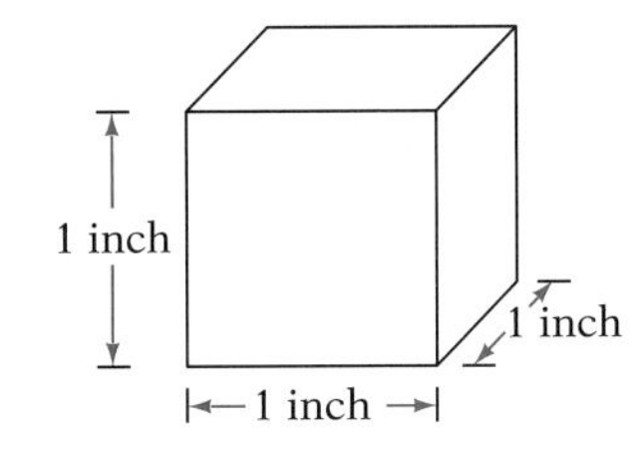

Cube

The geometric figure called a **cube** is a solid whose length, width, and height are equal in measure. Each side of a cube is called an **edge.**

To find the volume of any cube, we multiply its length, width, and height together. Because they all have the same measure:

To Find the Volume of a Cube

If s is the length of an edge, then

Volume $V = s^3$

EXAMPLE 1

Find the volume of a cube with a side of $4\frac{1}{4}$ feet.

SOLUTION

We use the formula $V = s^3$.

$V = s^3$	The formula
$= \left(4\frac{1}{4}\right)^3$	Substituting $s = 4\frac{1}{4}$
$= \frac{17}{4} \times \frac{17}{4} \times \frac{17}{4}$	Multiplying out
$= \frac{4913}{64}$ or $76\frac{49}{64}$	Simplifying

So the volume is $76\frac{49}{64}$ cubic feet.

■ *Now do margin exercises 1a and 1b.*

1a. Find the volume of a cube with edge $3\frac{1}{4}$ feet.

1b. Find the volume of this cube:

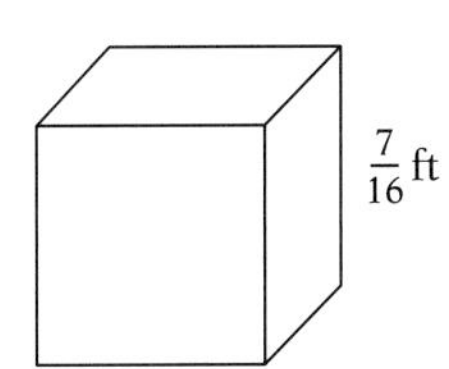

2a. Find the volume of a rectangular solid with length $3\frac{1}{3}$ feet, width 8 feet, and height $6\frac{1}{4}$ feet.

2b. Find the volume of the rectangular solid with sides of $2\frac{1}{10}$, $5\frac{2}{3}$, and 5 inches.

A cube has six square surfaces that are called its **faces.** A solid whose faces are rectangles is called a **rectangular solid.** A rectangular solid looks somewhat like a cube, but its dimensions—length, width, and height—are not necessarily all equal.

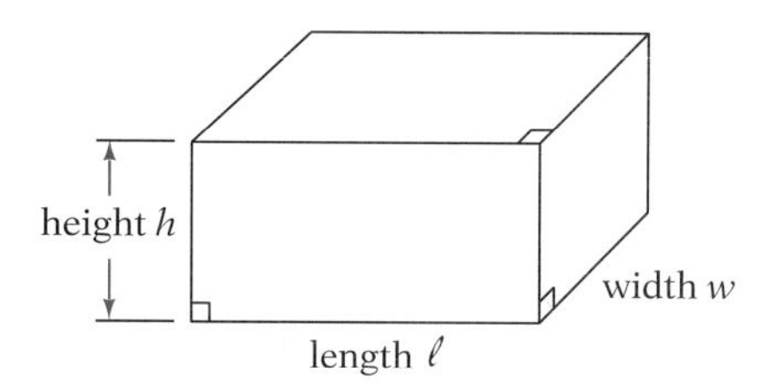

To find the volume of a rectangular solid, we multiply its length, width, and height. In symbols,

To Find the Volume of a Rectangular Solid

If ℓ is the length, w is the width, and h is the height, then

$$\text{Volume } V = \ell \times w \times h$$

EXAMPLE 2

Find the volume of a rectangular solid with length 14 inches, width 13 inches, and height 8 inches.

SOLUTION

We substitute the values for ℓ, w, and h in the formula and solve.

$V = \ell \times w \times h$	The formula
$= 14 \times 13 \times 8$	Substituting
$= 1456$	Multiplying

So the volume is 1456 cubic inches.

■ *Now do margin exercises 2a and 2b.*

Calculating Surface Area

Next, we will find the surface area of a cube and a rectangular solid.

A cube has six faces, each in the shape of a square.

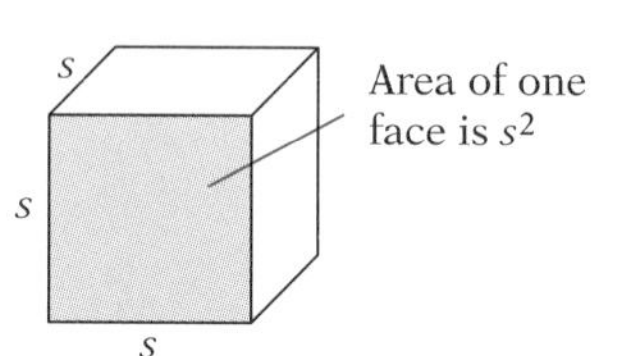

To find the surface area of a cube, we simply find the area of one of the square faces and multiply it by 6.

To Find the Surface Area of a Cube

If s is the side of a cube, then

$$\text{Surface area} = SA = 6s^2$$

3a. Find the surface area of a cube with a side equal to $2\frac{1}{3}$ inches.

3b. Find the surface area of a cube with a side equal to $8\frac{1}{2}$ inches.

EXAMPLE 3

Find the surface area of a cube whose side is $3\frac{1}{2}$ inches long.

SOLUTION

Again, we substitute into the formula and solve.

$SA = 6s^2$	The formula
$= 6\left(3\frac{1}{2}\right)^2$	Substituting
$= 6\left(\frac{7}{2}\right)^2$	Changing to improper fraction
$= 6\left(\frac{49}{4}\right)$	Removing the exponent
$= 73\frac{1}{2}$	Multiplying and simplifying

The surface area of the cube with a side of $3\frac{1}{2}$ inches is $73\frac{1}{2}$ square inches.

■ *Now do margin exercises 3a and 3b.*

Notice that surface areas are given in square units.

To find the surface area of a rectangular solid (sometimes called a **rectangular prism**), we find the areas of the sides and the area of the ends and add these areas together. The prism shown in the following figure has two faces of area ℓh, two faces of area wh, and two faces of area $w\ell$.

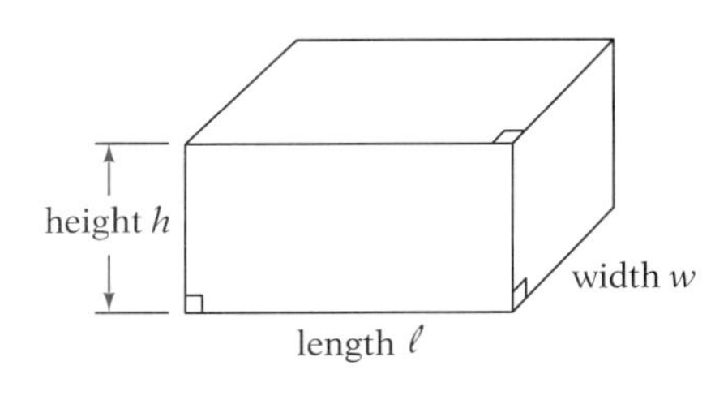

To Find the Surface Area of a Rectangular Prism

If ℓ is the length, w is the width, and h is the height, then

$$\text{Surface area } SA = 2(\ell h + wh + w\ell)$$

4a. Find the surface area of a rectangular prism with a length of 12 inches, a width of 10 inches, and a height of $21\frac{1}{2}$ inches.

4b. Find the surface area of a rectangular prism with a length of $1\frac{2}{5}$ inches, a width of $2\frac{1}{6}$ inches, and a height of $\frac{1}{5}$ inch.

EXAMPLE 4

Find the surface area of a rectangular prism with a length of $2\frac{1}{2}$ inches, a width of 1 inch, and a height of $1\frac{1}{4}$ inches.

SOLUTION

We substitute in the formula and solve.

$SA = 2(\ell h + wh + w\ell)$ The formula

$SA = 2\left[\left(2\frac{1}{2}\right)\left(1\frac{1}{4}\right) + (1)\left(1\frac{1}{4}\right) + (1)\left(2\frac{1}{2}\right)\right]$ Substituting

$= 2\left[\left(\frac{5}{2}\right)\left(\frac{5}{4}\right) + \frac{5}{4} + \frac{5}{2}\right]$ Changing to improper fractions

$= 2\left[\left(\frac{25}{8}\right) + \frac{5}{4} + \frac{5}{2}\right]$ Multiplying

$= 2\left[\frac{25}{8} + \frac{10}{8} + \frac{20}{8}\right]$ Rewriting with LCD

$= 2\left(\frac{55}{8}\right)$ Adding

$= \frac{110}{8} = 13\frac{3}{4}$ Multiplying and simplifying

Thus the surface area of the rectangular prism is $13\frac{3}{4}$ square inches.

■ *Now do margin exercises 4a and 4b.*

Work the exercises that follow. Make sure your answers are reasonable.

5.7B EXERCISES

Use your calculator to assist you in solving these problems.

1. Find the volume *and* surface area of the rectangular solid to the right.

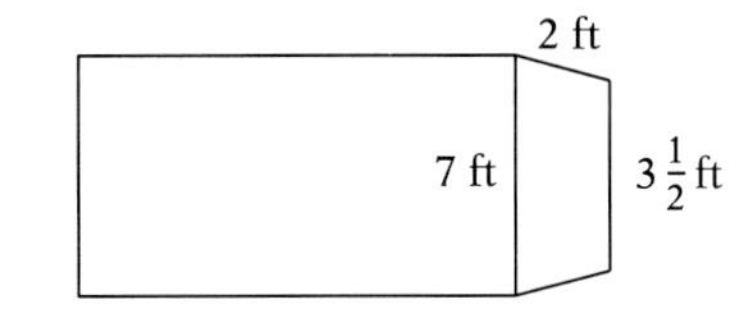

2. Find the number of cubic feet in a freezer that has a height of $3\frac{1}{2}$ feet, a length of 8 feet, and a width of $2\frac{1}{4}$ feet.

3. To find the required capacity of an air conditioner (in BTUs), we multiply the volume of the room it is to cool by 3. How much capacity is needed for a room that measures $17\frac{1}{3}$ feet long, $11\frac{1}{2}$ feet wide, and 9 feet high?

4. A mini-storage facility advertises bins that are 15 feet wide, $10\frac{1}{2}$ feet long, and $8\frac{1}{4}$ feet wide. What is the volume of each bin?

5. Find the volume of a cube that has an edge of $1\frac{1}{2}$ inches. What is its surface area?

6. A refrigerator is said to have a volume of 32 cubic feet. If the length of the interior of the refrigerator is 3 feet, and the interior width is 2 feet, how high must the interior of the refrigerator be? Use the formula $h = V/w\ell$.

7. A toy box is advertised as holding 40 cubic feet of toys. If the length of the toy box is 5 feet, and its height is 3 feet, how wide must the toy box be? Use the formula $w = V/(\ell\ h)$. If you paint the toy box on the outside, what surface area must the paint cover? Include the top.

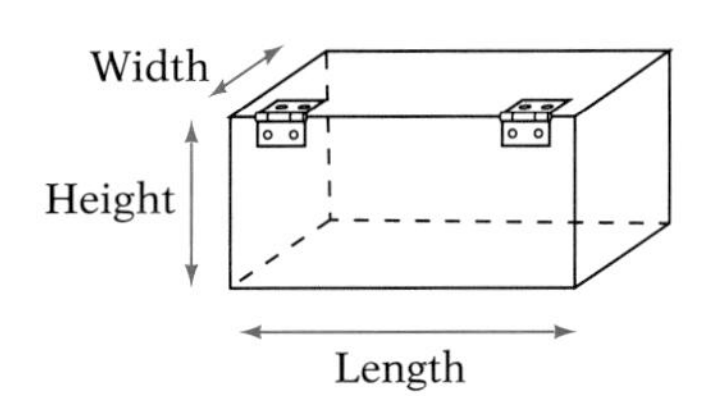

8. A fish tank is $14\frac{1}{2}$ inches long, 8 inches wide, and 11 inches high.

a. How many cubic inches of volume are there in the fish tank?

b. What is the surface area of the tank? (The tank does not have a top.)

9. Which has the larger surface area: a prism with length $2\frac{1}{2}$ inches, width $3\frac{1}{4}$ inches, and height 1 inch, or a cube with one side equal to $2\frac{5}{8}$ inches?

10. The gas tank in your car is the shape of a rectangular solid. It measures 2 feet by $3\frac{1}{2}$ feet by $\frac{7}{8}$ foot. What is its capacity (volume) in cubic feet?

11. The trunk of your car has a height of 24 inches, a width of 60 inches, and a depth of 36 inches. Find its volume.

12. A cube measures $\frac{1}{2}$ inch on a side. What is its volume? What is its surface area?

13. You are replacing your sidewalk and need to figure out how much concrete you will need. The area to be filled measures 120 feet by $2\frac{1}{2}$ feet and will be $\frac{1}{4}$ foot deep. What volume of concrete should you buy?

5.7 MIXED PRACTICE

By doing these exercises, you will practice all the topics up to this point in the chapter.

14. Divide: $8\frac{3}{5} \div 5$

15. Rewrite $\left(\frac{4}{5}\right)^4$ without exponents.

16. The sea horse moves $10\frac{1}{2}$ inches per minute. How long would it take one to move 34 inches?

17. A train takes $13\frac{1}{4}$ minutes to get from one destination to another. A commuter bus takes two and one-half times as long. How long does the bus take?

18. If I can drive $23\frac{1}{2}$ miles per gallon of gas, how many miles can I drive on $15\frac{1}{4}$ gallons?

19. Simplify: $11 - \left(\frac{3}{5}\right)^2 + 2(9 - 8\frac{1}{4})$

20. Multiply: $6\frac{2}{5} \times 3\frac{7}{16}$

21. Elite print on a typewriter will type 12 characters to the inch. How many characters can be typed on 6 lines that each measure $8\frac{1}{2}$ inches?

22. Simplify: $\left(\frac{1}{7}\right)^2 + 15 - 36^0 \div \frac{1}{9}$

23. Find the volume of a rectangular solid that has a length of 10 inches, a width of $4\frac{1}{4}$ inches, and a height of 9 inches.

24. Find the area of a trapezoid with bases $5\frac{2}{3}$ inches and $7\frac{1}{3}$ inches and height $4\frac{1}{2}$ inches.

25. Divide: $\frac{15}{17} \div \frac{3}{8}$

26. Find the surface area of a cube with an edge of $3\frac{3}{4}$ feet.

27. Find the area of a triangle that has a height of $4\frac{1}{4}$ feet and a base of $3\frac{1}{2}$ feet.

CHAPTER 5 REVIEW

ERROR ANALYSIS

These problems have been worked incorrectly. Tell what the error is and then write the correct solution.

1. Multiply: $2\frac{2}{3} \times 3\frac{1}{7}$

Incorrect Solution

$$2\frac{2}{3} \times 3\frac{1}{7} = 6\frac{2}{21}$$

Correct Solution

Error ______________________________

____

2. Divide: $4\frac{2}{7} \div 2\frac{1}{6}$

Incorrect Solution

$$4\frac{2}{7} \div 2\frac{1}{6} = \frac{30}{7} \div \frac{13}{6}$$
$$= \frac{7}{30} \times \frac{13}{6}$$
$$= \frac{91}{180}$$

Correct Solution

Error ______________________________

____

3. Multiply: $\frac{2}{3} \times \frac{3}{4}$

Incorrect Solution

$$\frac{2}{3} \times \frac{3}{4} = \frac{8}{9}$$

Correct Solution

Error ______________________________

____

4. Multiply: $2\frac{1}{4} \times \frac{3}{8}$

Incorrect Solution

$$2\frac{1}{4} \times \frac{3}{8} = \frac{9}{4} \times \frac{3}{8}$$
$$= \frac{27}{32} = 1\frac{5}{32}$$

Correct Solution

Error ______________________________

____

5. Divide and reduce: $3\frac{1}{4} \div 2\frac{2}{3}$

Incorrect Solution

$$3\frac{1}{4} \div 2\frac{2}{3} = \frac{13}{\not{4}_{1}} \div \frac{\not{8}^{2}}{3}$$
$$= \frac{26}{3} = 8\frac{2}{3}$$

Error ____

____

Correct Solution

INTERPRETING MATHEMATICS

By working these exercises, you will test and strengthen your mathematics vocabulary.

1. Explain, in your own words, how to multiply fractions.

2. Describe the operation of division of fractions. Use the word reciprocal.

3. Draw a triangle. Identify its base and height. Tell how to find the area.

4. Describe a trapezoid. Use the word parallel.

5. Describe a cube. Use the words edge and face.

6. Explain how to find the surface area of a rectangular prism.

REVIEW PROBLEMS

The exercises that follow will give you a good review of the material presented in this chapter. Work through them and check your answers at the back of the book. Express your answers in simplest form.

Section 5.1

Multiply:

1. $\frac{1}{5} \times \frac{3}{4}$

2. $\frac{2}{7} \times \frac{4}{5}$

3. $\frac{1}{6} \times \frac{2}{3} \times \frac{1}{2}$

4. $\frac{1}{5} \times \frac{5}{10} \times \frac{1}{2}$

5. $\frac{8}{9} \times \frac{5}{7} \times \frac{4}{7}$

6. $\frac{33}{35} \times \frac{5}{7} \times \frac{2}{3}$

Section 5.2

Multiply:

7. $2\frac{1}{4} \times 3\frac{1}{2}$

8. $9\frac{3}{5} \times 2\frac{2}{3}$

9. $15\frac{1}{2} \times \frac{9}{62}$

10. $3\frac{1}{2} \times 8\frac{2}{5}$

11. $8\frac{4}{9} \times 4 \times 4\frac{1}{2}$

12. $2\frac{1}{6} \times 12 \times \frac{1}{5}$

Section 5.3

Divide:

13. $\frac{1}{4} \div \frac{3}{8}$

14. $\frac{5}{16} \div \frac{4}{7}$

15. $\frac{3}{8} \div \frac{9}{32}$

16. $\frac{4}{9} \div 16$

17. $9 \div \frac{3}{14}$

18. $10 \div \frac{5}{18}$

Section 5.4

Divide:

19. $8\frac{1}{2} \div \frac{3}{8}$

20. $\frac{1}{18} \div 2\frac{2}{3}$

21. $6\frac{1}{8} \div 2\frac{1}{3}$

22. $4\frac{1}{4} \div 6\frac{4}{5}$

23. $3\frac{2}{7} \div 1\frac{4}{7}$

24. $1\frac{1}{7} \div \frac{8}{17}$

Section 5.5

Solve:

25. A flight of stairs contains 12 steps that are each $\frac{3}{4}$ foot high. How many feet have you climbed up if you have walked up 6 flights?

26. A telephone pole contains steps that are $1\frac{1}{2}$ feet apart. How many steps are there in a pole that is 22 feet high if the first step is at a height of $1\frac{1}{2}$ feet?

27. You are planting flowers in a garden. If you have $32\frac{3}{4}$ square feet of space, and you are planting flowers that must each have $\frac{1}{2}$ square foot of space to grow in, how many flowers can you plant in the entire garden?

28. Your score in a board game is $2\frac{1}{2}$, $3\frac{1}{4}$, $2\frac{1}{3}$, and $\frac{2}{3}$. Your friend's score is $5\frac{1}{2}$, $\frac{3}{4}$, $\frac{2}{3}$, and $1\frac{1}{3}$. Who has the larger score, and how much larger is it?

29. An orchestra is made up of about 100 instruments. If approximately $\frac{2}{3}$ of these instruments are in the string family, how many instruments are string instruments? Round your answer, if necessary.

30. It is said that kidney beans have 4 grams or more of fiber per $\frac{1}{2}$-cup serving. Approximately how much fiber is there in 8 cups of kidney beans?

Section 5.6

Simplify:

31. $\left(\frac{4}{9}\right)^3$

32. $\left(\frac{7}{9}\right)^4$

33. $\left(\frac{5}{6}\right)^3 \times \left(\frac{2}{3}\right)^0$

34. $\left(\frac{1}{2}\right)^4 + \left(\frac{3}{4}\right)^2$

35. $3\left(6 - \frac{1}{3}\right)^2 + 8\frac{3}{4}$

36. $\left(\frac{3}{7}\right)^2 - \frac{5}{7} \times \frac{1}{5} + 6\frac{2}{7}$

Section 5.7

37. Find the area of a triangle with a base of 24 units and a height of $5\frac{1}{2}$ units.

38. Find the area of a trapezoid with a height of $10\frac{2}{3}$ inches and bases of $20\frac{1}{2}$ and $10\frac{3}{4}$ inches.

39. Find the volume and surface area of a rectangular solid that is $15\frac{1}{2}$ inches high, $2\frac{2}{3}$ inches wide, and $2\frac{1}{4}$ inches long.

40. Find the surface area and volume of a cube-shaped die with an edge of $\frac{1}{2}$ inch.

41. A triangular sail is $5\frac{1}{3}$ yards tall and 10 yards wide. What is its area?

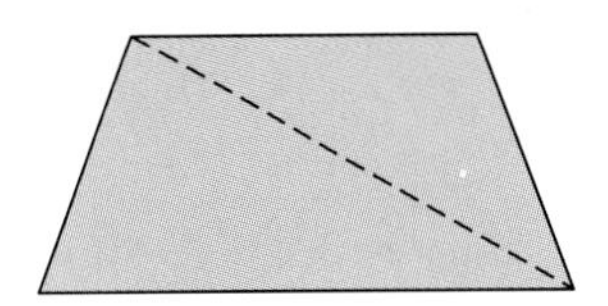

42. A trapezoid is cut from one corner diagonally to the other into two triangles. One triangle has a height of $6\frac{1}{3}$ inches and a base of $12\frac{1}{2}$ inches, and the other triangle has a height of $6\frac{1}{3}$ inches and a base of 10 inches. What was the area of the original trapezoid?

Mixed Review

43. Multiply: $3\frac{2}{5} \times 1\frac{1}{2}$

44. Multiply: $\frac{19}{48} \times \frac{3}{5} \times \frac{12}{19}$

45. In an odds-and-ends store, pencils sell for $2\frac{1}{2}$ cents apiece. How much will a dozen cost?

46. A good student spends $1\frac{1}{2}$ hours of study time outside of class for each hour of actual class. If Martin has 15 hours of actual class time, how much time should he study each week, if he is a good student?

47. Multiply: $7 \times 3\frac{1}{2}$

48. Multiply: $3\frac{1}{5} \times \frac{5}{16}$

49. The human eye weighs approximately $1\frac{1}{2}$ ounces. How much would both eyes weigh?

50. Your yearly salary is $24,000. You spend $\frac{1}{5}$ of it on food. How much do you spend per month on food?

51. Divide: $3 \div \frac{1}{2}$

52. Divide: $7\frac{1}{3} \div 8$

53. Multiply: $2\frac{1}{6} \times 12 \times \frac{1}{5}$

54. Divide: $\frac{7}{8} \div \frac{17}{24}$

Name Section Date

Chapter 5 Test

This exam tests your knowledge of the topics in Chapter 5. Express answers in simplest form wherever possible.

1. Multiply:

a. $\frac{2}{3} \times \frac{6}{7}$

b. $\frac{7}{9} \times \frac{3}{5} \times \frac{1}{14}$

c. $3\frac{3}{5} \times 5\frac{1}{3}$

2. Divide:

a. $\frac{11}{12} \div \frac{3}{8}$

b. $\frac{7}{12} \div \frac{4}{9}$

c. $7\frac{3}{5} \div 1\frac{4}{15}$

3. Solve:

a. A book is $3\frac{3}{4}$ inches thick. Five of these books stacked together would be how thick?

b. Thirteen and three-eighths ounces of water must be divided into three equal portions. How much will be in each portion?

c. In a recent Olympic games, over 120 countries participated. This was $9\frac{3}{13}$ times the number of countries who participated in the first Olympic games. How many countries participated in the first games?

4. a. Rewrite $\left(\frac{3}{4}\right)^3$ without an exponent.

b. Use the standard order of operations to simplify: $\left[\left(\frac{2}{3}\right)^2 + \left(1\frac{1}{3} \div \frac{2}{5}\right)\right] - \left(\frac{2}{3}\right)^3$

c. Simplify: $1 - \left[\left(\frac{3}{5}\right)^2 - \left(\frac{1}{5}\right)^2\right] - \frac{2}{5}$

5. Solve:

a. Find the area of a triangle that has a base of 15 inches and a height of 12 inches.

b. Find the volume of a rectangular solid that is 16 feet high, 3 feet wide, and 4 feet long.

c. Find the surface area of a cube with one edge equal to 10 inches.

Cumulative Review

CHAPTERS 1–5

The questions that follow will help you maintain the skills you have developed. Write all fractions in simplest form.

1. Add: 4 hr 55 min
   +3 hr 12 min

2. Subtract: 3090 − 64

3. Multiply: 21×106

4. Divide: $18\overline{)3960}$

5. Divide: 6 gallons 1 quart by 5

6. Find the area of a triangle with a base of $1\frac{1}{2}$ inches and a height of $2\frac{1}{2}$ inches.

7. Find the area of a trapezoid with bases of 20 feet and $6\frac{1}{2}$ feet and a height of $16\frac{2}{3}$ feet.

8. Determine whether 575 is a perfect square.

9. Marcel is conducting an experiment with four glasses of water, each one being $\frac{3}{4}$ full. He pours $\frac{1}{2}$ of the water out of the first glass and $\frac{1}{4}$ of the water out of the second glass. How much water does he now have altogether?

10. Add: $27\frac{1}{2} + 16\frac{2}{3} + 9\frac{1}{4}$

11. Find the greatest common factor of 136 and 48.

12. Subtract: $11\frac{2}{3} - 4\frac{7}{8}$

13. Multiply: $2\frac{1}{2} \times 3\frac{2}{5}$

14. Divide: $8\frac{6}{7} \div 2\frac{2}{5}$

15. Find the perimeter of a regular dodecagon with a side of $3\frac{3}{4}$ cm.

16. Find the volume and surface area of a rectangular prism with dimensions of $3 \times 5\frac{1}{2} \times 6\frac{1}{4}$ units.

17. Evaluate: $2^5 \times 3^3$

18. Find the LCM of 18 and 4.

19. Simplify: $3^4 - (2^2 + 5) \div 3$

20. Compare $\frac{2}{3}$ and $\frac{5}{6}$ and determine which is larger.

6 Operations with Decimals

Have you ever watched Olympic gymnastics? The gymnasts receive decimal scores for their performances. For example, at the 1992 Summer Olympics, six women gymnasts from the United States won the bronze medal in the team competition. Their score of 394.704 was calculated by adding scores received by the six gymnasts in various events. The difference between their score and the score of the women who won the silver medal was only 0.375, less than four tenths of a point.

- *Describe a situation in which you have received a decimal score.*

Skills Check

Take this short quiz to see how well prepared you are for Chapter 6. The answers are given at the bottom of the page. So are the sections to review if you get any answers wrong.

1. Add: $1306 + 84 + 128$

2. Subtract: $6307 - 198$

3. Multiply: 230×804

4. Divide: $2035 \div 198$

5. Round 3456 to the nearest ten.

6. Which is greater, 120698 or 12095?

7. Multiply: $2\frac{1}{10} \times \frac{3}{100}$

8. Add $\frac{3}{10} + \frac{21}{100} + \frac{3}{1000}$

9. Which is greater, $\frac{3}{10}$ or $\frac{29}{100}$?

10. Divide $3\frac{1}{10}$ by $\frac{31}{100}$.

ANSWERS: 1. 1518 [Section 1.4] 2. 6109 [Section 1.5] 3. 184,920 [Section 2.2] 4. $10\frac{55}{198}$ [Section 2.4] 5. 3460 [Section 1.3] 6. 120,698 [Section 1.2] 7. $\frac{63}{1000}$ [Section 5.2] 8. $\frac{513}{1000}$ [Section 4.5] 9. $\frac{3}{10}$ [Section 4.4] 10. 10 [Section 5.4]

6.1 Place Value: Naming Decimals

Identifying the Place Values of Decimal Numbers

Every time you write a check or get a store receipt, you are exposed to decimal numbers. All decimal numbers can be written using the ten *digits* 0, 1, 2, 3, 4, 5, 6, 7, 8, 9 and a decimal point that looks like a period. This point designates the end of the **whole-number part** of the decimal number and the beginning of the **fractional part.**

decimal point
↓
$\underbrace{2\ 3}_{\text{whole-number part}} . \underbrace{6\ 0\ 9}_{\text{fractional part}}$

Note that decimal numbers are like mixed numbers in that they both have a whole-number part and a fractional part. The fractional part of a decimal number is understood to have 10 or a power of 10 as the denominator; it is also called the **decimal part.**

Each digit in a decimal number has a meaning based on its face value (the name of the digit) and its place value (its place in the number). For example, the digit 5 has a different meaning in each of the following numbers:

12.5 Here 5 means five tenths, or $\frac{5}{10}$, because 5 is in the *tenths* place.

123.05 Here 5 means five hundredths, or $\frac{5}{100}$, because 5 is in the *hundredths* place.

224.095 Here 5 means five thousandths, or $\frac{5}{1000}$, because 5 is in the *thousandths* place.

The following figure shows the decimal place values to millionths.

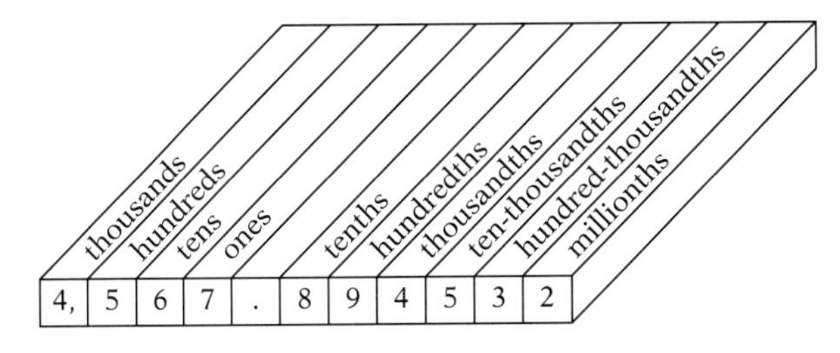

Section Goals

- To identify decimal place values up to billionths
- To write decimal numbers in standard form, in expanded notation, and in words.
- To convert decimal numbers to fractions

Example 1

What does the digit 7 mean in 2.7, in 1.37, and in 98.940007?

SOLUTION

We use the preceding chart to find each place value. The face value is seven.

In 2.7, the digit 7 is in the tenths place and means *seven tenths*, or $\frac{7}{10}$.

In 1.37, the digit 7 is in the hundredths place and means *seven hundredths*, or $\frac{7}{100}$.

1a. What does the digit 3 mean in 1.73, in 2.7003, and in 78.347?

1b. What does the digit 1 mean in 2.0701, in 4.7223001, and in 78.000002147?

2a. Write 4106.53007 in words. (Name the number.)

2b. Write 30.030040001 in words. (Name the number.)

In 98.940007, the digit 7 is in the millionths place and means *seven millionths*, or $\frac{7}{1{,}000{,}000}$.

■ *Now do margin exercises 1a and 1b. The correct answers are at the back of the book.*

Reading and Writing Decimal Numbers

The last number in Example 1 is read as ninety-eight and nine hundred forty thousand, seven millionths. Here's how we get that.

To read or write (in words) any decimal number that is written with digits

1. Read the number to the left of the decimal point.
2. Say *and.*
3. Read the number to the right of the decimal point as though it were a whole number.
4. Name the place value of the last digit on the right.

EXAMPLE 2

Write 400600.30040 in words.

SOLUTION

Grouped, the whole-number portion of the number is 400,600. We write this portion as four hundred thousand, six hundred.

Grouped, the decimal portion of the number is 30,040. We write this as thirty thousand, forty.

The place value of the last digit, 0, is hundred-thousandths.

So the number 400600.30040 is read as four hundred thousand, six hundred and thirty thousand, forty hundred-thousandths.

■ *Now do margin exercises 2a and 2b.*

Next we will "translate" a decimal number from words to digits.

EXAMPLE 3

Write five hundred thousand, six and two hundred six ten-thousandths using digits in standard notation.

SOLUTION

The word *and* separates the whole-number part from the decimal part of any decimal number. We write the parts separately and join them with a decimal point.

five hundred thousand, six:	500,006
and	
two hundred six:	206

Because the decimal part is ten-thousandths, we need four decimal places.

Thus five hundred thousand, six and two hundred six ten-thousandths is written

500,006.0206

■ *Now do margin exercises 3a and 3b.*

3a. Write two million, five thousand, six and two thousand, six hundred-thousandths as a number.

3b. Write fifty thousand, four and seven hundred sixteen hundred-thousandths as a number.

Writing Decimal Numbers Using Expanded Notation

Decimal numbers can be written with digits in two ways. *Standard notation* makes use of the place values and decimal point, as in 17.493. This is the decimal notation we have been using so far. *Expanded notation* makes use of the meanings of the individual digits, with plus signs connecting the parts in a sum. The number 17.493 in expanded notation is

$$10 + 7 + \frac{4}{10} + \frac{9}{100} + \frac{3}{1000}$$

EXAMPLE 4

Write the number 3.00894 in expanded notation.

SOLUTION

3 means three ones, or 3

8 means eight thousandths, or $\frac{8}{1000}$

9 means nine ten-thousandths, or $\frac{9}{10{,}000}$

4 means four hundred-thousandths, or $\frac{4}{100{,}000}$

So in expanded notation, this number is written

$$3 + \frac{8}{1000} + \frac{9}{10{,}000} + \frac{4}{100{,}000}.$$

■ *Now do margin exercises 4a and 4b.*

4a. Write the number 41.0382 in expanded notation.

4b. Write the number 603.080005 in expanded notation.

Converting Decimal Numbers to Fractional Form

Sometimes, it is more convenient to write numbers as proper fractions or mixed numbers than as decimal numbers. Decimal numbers can be converted quickly and easily to fractions or mixed numbers.

To change a decimal number to a mixed number or fraction

1. Write the whole-number part of the decimal (if there is one) as a whole number.
2. Write a fraction with the decimal part as the numerator and the place value of the last digit in the decimal part as the denominator.

5a. Rewrite 0.0103 as a fraction.

5b. Rewrite 1.20307 as a mixed number.

In the last example, we will use this procedure to rewrite a decimal number as a mixed number.

EXAMPLE 5

Rewrite 5.01009 as a mixed number.

SOLUTION

The whole-number part is 5.

The decimal part is 01009, so the numerator of the fraction will be 1009.

The place value of the last digit in the decimal part is hundred-thousandths, so the denominator will be 100,000.

So 5.01009 is the same as $5\frac{1009}{100,000}$.

■ *Now do margin exercises 5a and 5b.*

Note: You generally will see fractions such as $\frac{3}{100}$ written as 0.03 or .03. By including the whole number zero, you make it very clear that the decimal number is less than one.

Work the exercises that follow.

Name Section Date

6.1 EXERCISES

In Exercises 1 through 8, identify the place value of each of the underlined digits.

1. $\underline{3}\underline{3}.4\underline{2}05\underline{7}8$

2. $2\underline{7}3.\underline{6}1\underline{8}2$

3. $\underline{9}.\underline{8}8\underline{3}\underline{7}05$

4. $\underline{9}0.\underline{1}3\underline{8}$

5. $\underline{9}75.\underline{3}8\underline{2}1\underline{6}$

6. $\underline{5}.9\underline{2}2\underline{7}$

7. $\underline{9}03\underline{1}.28\underline{4}$

8. $\underline{9}.1\underline{3}08\underline{4}7$

In Exercises 9 through 16, rewrite each number in expanded notation.

9. 107.5894

10. 59.017345

11. 244.998

12. 2.0314

13. 69104.536

14. 825.39

15. 61.730546

16. 63.0479

In Exercises 17 through 40, rewrite each of the underlined numbers using words if it is written with digits, and vice versa.

17. Size of a DNA molecule: 0.000002 millimeter.

18. Size of a queen honeybee: 1.523 centimeters.

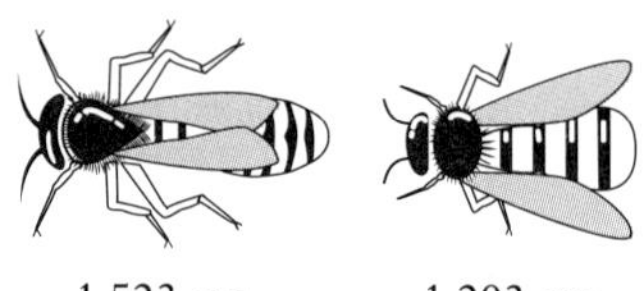

19. Size of a drone bee: 1.203 centimeters.

20. Length of a tobacco mosaic virus: 0.0000171 meter.

21. Length of a honeybee parasite: eighty-three hundredths millimeter.

22. Length of a crab louse: one and four tenths centimeters.

23. Length of a Chloroplast: eighty-four ten-thousandths millimeter.

24. Size of a plant mite: fifty-two hundred-thousandths meter.

25. Width of a tobacco mosaic virus: 0.000044 millimeter.

26. Diameter of a house spider: 0.00103 millimeter.

27. Width of an influenza virus: 0.00000111 dekameter.

28. Average body cell size: 0.00005093 meter.

29. Diameter of a bacterium called *Escherichia coli:* thirteen ten-thousandths millimeter.

30. Wavelength of red light: seven hundred-thousandths centimeter.

31. Weight of a chicken: three and one hundred fifty-three thousandths kilograms.

32. Thickness of a coat of paint: twelve hundred-thousandths meter.

33. Weight of a panel pin: 0.00010687 kilogram.

34. The length of time it takes for a nerve impulse to cross a synapse: 0.00001498 minute.

35. Area of a rugby football pitch: 10.038777 square meters.

36. Energy to split an atom: 0.00009867 erg.

37. Weight of a United Kingdom fifty-pence piece: fifteen thousandths kilogram.

38. Time it takes for a bullet to explode: one millionth second.

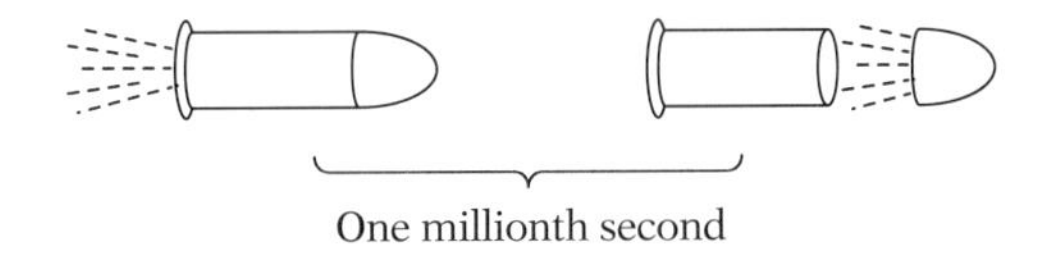

39. Length of time of a bee's wing beat: thirty-three thousandths second.

40. Length of a flea: one and four hundred ninety-two thousandths millimeters.

In Exercises 41 through 48, write the decimals in fractional form.

41. 18.063

42. 0.0008

43. 16.60407

44. 0.69

45. 0.0009

46. 24.076078

47. 0.07032102

48. 18.2051111

In Exercises 49 through 53, write each underlined number as a fraction or mixed number. Then indicate which is greater, the U.S. or British measure.

49. One U.S. pint is twenty-eight and eight hundred seventy-five thousandths cubic inches.

One British pint is thirty four and six hundred seventy-seven thousandths cubic inches.

50. One U.S. quart is 57.75 cubic inches.

One British quart is 69.355 cubic inches.

51. One U.S. gallon is 231.00 cubic inches.

One British gallon is 277.42 cubic inches.

52. One British peck is five hundred fifty-four and eighty-four hundredths cubic inches.

One U.S. peck is five hundred thirty-seven and six tenths cubic inches.

53. One U.S. bushel is 3150.4 cubic inches.

One British bushel is 2219.4 cubic inches.

6.2 Place Value: Rounding and Comparing Decimal Numbers

SECTION GOALS

- To compare and order decimal numbers
- To round decimal numbers to the nearest place up to a billionth

Ordering Decimal Numbers

When we compared whole numbers, we lined up the numbers and compared digits. We will do the same with decimal numbers.

To compare any two decimal numbers:

1. Write the numbers, one underneath the other, with the decimal points aligned.
2. Compare the whole-number parts. If one whole number is greater than another, the entire decimal number is greater. If not, go on to step 3.
3. Compare the digits to the right of the decimal point, place by place. When the digits differ in one place, the decimal number with the greater digit is the greater number.

Before we use this procedure to order decimal numbers, let's consider a property of decimal parts: If we add zeros to the end of a decimal number, to the right of the decimal point, we do not change the value of the decimal number. This means, for example, that 0.01 and 0.010 and 0.0100 all represent the same fraction. Let's look more closely at these numbers.

By definition,

$$0.01 \text{ is } \frac{1}{100}$$

but

$$0.010 \text{ is } \frac{10}{1000} = \frac{10 \div 10}{1000 \div 10} = \frac{1}{100}$$

and

$$0.0100 \text{ is } \frac{100}{10{,}000} = \frac{100 \div 100}{10{,}000 \div 100} = \frac{1}{100}$$

Because we can reduce all these numbers to the same fraction, $\frac{1}{100}$, they are equivalent.

This property makes comparisons easier. We shall use it in our first example.

EXAMPLE 1

Arrange 67.00098, 67.009, and 67.0009 in order from largest to smallest.

6	7	.	0	0	0	9	8
6	7	.	0	0	9	0	0
6	7	.	0	0	0	9	0

1a. Arrange 98.404, 98.04, and 98.003 in order from smallest to largest.

1b. Arrange 75.214, 75.204, and 75.024 in order from largest to smallest.

SOLUTION

We begin by comparing. First we write the numbers, lining up the decimal points. Then we add zeros at the right, as necessary, to give all three the same number of decimal places. That is what makes comparison easier.

67.00098
67.00900
67.00090

Now we compare the numbers, place by place, until they differ. The whole-number parts are equal, so we begin at the decimal point.

67.00098
↕↕*
67.00900
↕↕*
67.00090

These numbers differ in the thousandths place, where 9 is greater than zero. This makes 67.009 the greatest of our three numbers.

Continuing the comparison with the remaining two numbers, we have

67.00098
↕*
67.00090

These two numbers differ in the hundred-thousandths place, where 8 is greater than zero. This makes 67.00098 the next greatest of our three numbers and 6.0009 the least.

So 67.009 > 67.00098 > 67.0009.

■ *Now do margin exercises 1a and 1b.*

Rounding Decimal Numbers

We use the same procedure to round decimal numbers that we used when we rounded whole numbers. Here's an example:

EXAMPLE 2

Round 12,496.798 to the nearest thousand.

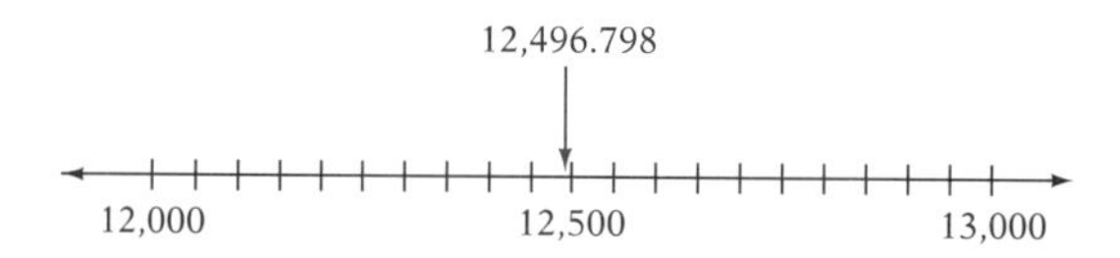

SOLUTION

To round to the nearest thousand, we first underline the digits from the left, through the thousands place. Then we mark the digit to the right of the last underlined digit.

$\underline{12}$,496.798
↑

Because 4 is less than 5, we leave the underlined number as 12 and replace all digits to the right with zeros, leaving 12,000 as our rounded number. Note that we do not need to extend the rounded whole number past the decimal point.

So 12,496.798 rounded to the nearest thousand is 12,000.

■ *Now do margin exercises 2a and 2b.*

2a. Round 2798.456 to the nearest thousand.

2b. Round 15,672.498 to the nearest hundred.

If we are rounding the decimal part of the number only, the procedure differs slightly.

To round a decimal number to a decimal place

1. Write the number, underlining all the digits from the left through the place you are rounding to.
2. If the digit just to the right of the last underlined digit is 5 or greater, increase the underlined number by 1; disregard the decimal point.
3. If the digit just to the right of the last underlined digit is less than 5, leave the underlined number as is.
4. Delete any digits to the right.

Follow along in this example to see what to do with the rounded digits.

3a. Round 0.746785 to the nearest hundred-thousandth.

3b. Round 8.14165 to the nearest ten-thousandth.

EXAMPLE 3

Round 4.76873 to the nearest thousandth.

SOLUTION

First we underline the number through the thousandths place and look at the digit to the right of the last underlined digit.

$$\underline{4.768}73$$
$$\uparrow$$

Because 7 is greater than 5, we increase the 4.768 to 4.769.

$$\underline{4.76\overset{9}{\not{8}}}73$$
$$\uparrow$$

Then we drop the digits to the right, leaving 4.769 as our rounded number.

■ *Now do margin exercises 3a and 3b.*

Do you see why we drop the digits to the right of the place we rounded to? They should be changed to zeros, and zeros at the right can be dropped.

What happens when the digit we are rounding is a nine? Follow this example carefully.

EXAMPLE 4

Round 12.539995 to the nearest hundred-thousandth.

SOLUTION

First we underline the number through the hundred-thousandths place and look at the digit to the right of the last underlined digit.

$$\underline{12.53999}5$$
$$\uparrow$$

4a. Round 12.539995 to the nearest millionth.

4b. Round 132.59899995 to the nearest millionth.

5a. Rewrite each of the following to the closest tenth of a thousand dollars.

$4567
$6477
$8190
$8566

5b. Rewrite each of the following to the closest tenth of a million.

1,499,999
2,500,001
5,740,000
3,622,144

Because the number to the right is 5, we increase 12.53999 by 1 (disregarding the decimal point) to 12.54000. We drop all digits to the right of the hundred-thousandths place, so our rounded number is 12.54000. We keep the three zeros to indicate how this number was rounded.

■ *Now do margin exercises 4a and 4b.*

Sometimes we use decimal numbers to express very large numbers! This might sound impossible, but the next example will illustrate it.

EXAMPLE 5

The following is a list of the oil-carrying capacities, in barrels, of four supertankers. Express each amount to the closest tenth of a million barrels.

Bellamya	3,304,205
Pierre Guillaumat	3,307,317
Globtik London	2,820,658
Globtik Tokyo	2,796,883

SOLUTION

We are interested in rounding to the nearest tenth of a million. To do so, we first rewrite the numbers as millions by placing a decimal point immediately after the millions place. Then we can round them exactly as we rounded decimal numbers in the last few examples. We first get

Bellamya	3.304205 million
Pierre Guillaumat	3.307317 million
Globtik London	2.820658 million
Globtik Tokyo	2.796883 million

Rounding each to the nearest tenth, we obtain

3.304205 rounded is 3.3 million
3.307317 rounded is 3.3 million
2.820658 rounded is 2.8 million
2.796883 rounded is 2.8 million

Our list with rounded values looks like this:

Bellamya	3.3 million
Pierre Guillaumat	3.3 million
Globtik London	2.8 million
Globtik Tokyo	2.8 million

■ *Now do margin exercises 5a and 5b.*

Work the exercises that follow.

Name Section Date

6.2 EXERCISES

In Exercises 1 through 4, write a statement about the three decimal numbers using the symbol <.

1. 27.41 36.52 17.83

2. 35.29 43.61 81.54

3. 44.13 19.74 35.18

4. 86.55 52.86 76.53

In Exercises 5 through 8, insert < or > to make each statement true.

5. 43.126 ____ 36.532

6. 65.18 ____ 78.417

7. 27.36 ____ 49.364

8. 18.14 ____ 54.128

In Exercises 9 through 12, identify the greater decimal number. Use the correct symbol to rewrite each pair with the lesser number first.

9. 0.8643 0.08643

10. 0.09531 0.9531

11. 0.042017 0.04217

12. 0.06572 0.065721

In Exercises 13 through 16, write a statement about the decimal numbers using the symbol >.

13. 0.0004 0.0007

14. 0.005 0.006

15. 0.009 0.001

16. 0.005 0.003

In Exercises 17 through 20, insert < or > to make each statement true.

17. 0.0002 ____ 0.001

18. 0.0031 ____ 0.02

19. 0.0075 ____ 0.005999

20. 0.01006 ____ 0.0104

In Exercises 21 through 32, use what you know about decimal numbers to answer the questions.

21. The average life expectancy of a male age 10 in the United States is 72.0982 years. The average life expectancy of a male age 10 in Canada is 72.921 years. Which has the longer life expectancy?

22. Compare the area of these two cities and determine which has the smaller area.

Monaco	0.7297006 square mile
Vatican City	0.16999 square mile

23. Here are the densities of some substances, in grams per cubic centimeter(g/cm^3). Arrange them in order from greatest to least.

Alcohol	0.789332 g/cm^3
Rubber	0.9301152 g/cm^3
Petroleum	0.87865 g/cm^3
Beechwood	0.72 g/cm^3

24. Arrange the following average weights in order from greatest to least.

Chinchilla	1.49873 pounds
Guinea pig	1.537221 pounds
Ferret	2.03560001 pounds

25. The surface gravity of the Moon is 0.1553298. The surface gravity of Mercury is 0.28499. Which has the greater gravitational pull?

26. Roberto Clemente's batting averages for his four best years were

1961	.3508
1964	.3387
1965	.329
1967	.356701

Arrange the years in order from his highest batting average to his lowest.

27. There are three different ways of measuring a year. Each results in a measurement of approximately 365 days, but they all differ in the decimal approximations. Arrange these three years in order from greatest number of days to least.

Sidereal year	365.2563656 days
Anomalistic year	365.2596 days
Trophical year	365.242198781 days

28. The first eight places of the decimal equivalents for $\frac{1}{12}$, $\frac{1}{13}$, $\frac{1}{14}$, and $\frac{1}{15}$ (not necessarily in order) are

0.07142857
0.07692307
0.06666666
0.08333333

Arrange these decimal numbers in order from greatest to least.

29. At 20 degrees Celsius, gold has a density of 19.28793 g/cm^3 and another metal has a density of 19.26 g/cm^3. Which has the higher density?

30. Compared to the gravity of the Earth, the gravity of Uranus is 37.68991 and the gravity of Saturn is 37.039. Which has the lower gravity?

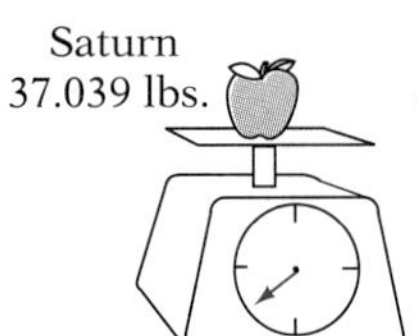

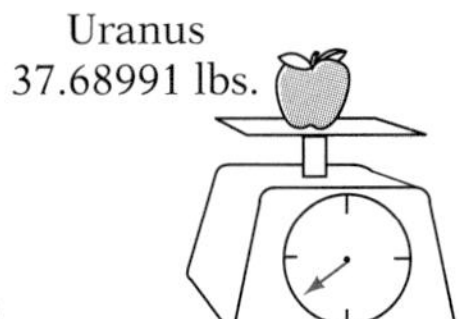

31. Wade Boggs was the Batting Champion for four years with the following batting averages. Arrange the years in order from his best year to his worst.

1987	0.363
1986	0.357
1985	0.368
1983	0.361

32. The finishing times for four runners in a race were

Runner A	6 min 23.09 sec
Runner B	6 min 23.1 sec
Runner C	6 min 25.52 sec
Runner D	5 min 18.05 sec

In what order did the runners finish the race?

In Exercises 33 through 36, use the following table to classify chickens with the given weights.

Chicken Type	Weights
capons	from 5.0 to 8.0 pounds
roasters	from 3.5 to 5 pounds
fryers	from 2.5 to 3.5 pounds
broilers	2.5 pounds or less

33. 3.75 pounds?

34. 2.50983 pounds?

35. 3.481102 pounds?

36. 5.0487612 pounds?

In Exercises 37 through 46, the average personal income per capita in 1984 is given for several major cities. Round each number to the given place.

37. New York–New Jersey metropolitan area: $15, 957 (nearest tenth of a thousand dollars)

38. Los Angeles metropolitan area: $14,591 (nearest ten thousand dollars)

39. Chicago metropolitan area: $14,456 (nearest hundred dollars)

40. San Francisco metropolitan area: $17,171 (nearest tenth of a thousand dollars)

41. Philadelphia metropolitan area: $13,785 (nearest tenth of a thousand dollars)

42. Detroit metropolitan area: $13,984 (nearest hundred dollars)

43. Boston metropolitan area: $15,932 (nearest tenth of a thousand dollars)

44. Houston metropolitan area: $14,374 (nearest tenth of a thousand dollars)

45. Washington, D.C. metropolitan area: $17,724 (nearest ten thousand dollars)

46. Dallas metropolitan area: $15,272 (nearest tenth of a thousand dollars)

In Exercises 47 through 52, the number on the left has been rounded. Of the other three numbers, circle those that it could have been before rounding.

47. 6.9 6.85 6.94999 6.9049

48. 7.23 7.239 7.23 7.2349875

49. 2.0 2.093 1.99 1.957867

50. 3.04 3.0399876 3.04499 3.045

51. 2.19 2.1949 2.18997 2.1854

52. 3.611 3.61107 3.61195 3.610997

6.2 MIXED PRACTICE

By doing these exercises, you will practice all the topics up to this point in this chapter.

53. Round 67.98 to the nearest tenth.

54. Write 87.603 in words.

55. Arrange 1.08, 1.008, and 1.88 in order from largest to smallest.

56. Write $60{,}000 + 500 + \frac{3}{1000}$ in standard notation.

57. Round 42.9012 to the nearest thousandth.

58. Write 245.0802 in expanded notation.

59. Arrange 96.536, 96.563, 96.057, and 96.0563 in order from largest to smallest.

60. Write sixty and two hundred forty ten-thousandths as a numeral.

61. Write 9.067 as a mixed number.

6.3 Adding Decimal Numbers

The addition of decimal numbers is similar to the addition of whole numbers, except that we must be careful of the decimal point. The numbers we add are the *addends;* the result is the *sum*. The same addition properties apply to decimal numbers as to whole numbers.

As always, before adding you should estimate the sum. To do so, simply round to the nearest whole number and add.

EXAMPLE 1

Add: 6.8 + 7.5

SOLUTION

First we estimate. Rounding each decimal number to the nearest whole number gives us 7 + 8, or 15, as an estimate.

To add, we write the problem in vertical form, lining up the decimal points.

$$\begin{array}{r} 6.8 \\ +7.5 \\ \hline \end{array}$$

↑ Align.

Next we begin the addition, adding the rightmost column as usual.

$$\begin{array}{rl} ^{1} & \\ 6.8 & \\ +7.5 & 8+5=13 \\ \hline 3 & \text{Write 3 and carry 1.} \end{array}$$

We continue adding, ignoring the decimal point until the addition is completed.

$$\begin{array}{rl} ^{1} & \\ 6.8 & \\ +7.5 & 1+6+7=14 \\ \hline 14\ 3 & \text{Write 14.} \end{array}$$

We now insert the decimal point in our answer directly beneath the decimal points in the problem.

$$\begin{array}{r} 6.8 \\ +7.5 \\ \hline 14.3 \end{array}$$

↑ Align.

We estimated our answer to be about 15, so the sum 14.3 is reasonable.

■ *Now do margin exercises 1a and 1b.*

EXAMPLE 2

Add: 15.6 + 9 + 11.893

SOLUTION

First we estimate. Rounding and adding, we get 16 + 9 + 12, or 37.

Next we rewrite the problem in vertical form, lining up the decimal points. Remember, *a whole number has an understood decimal point after the number!*

SECTION GOAL

■ To find the sum of two or more decimal numbers

1a. Add: 5.4 + 6.7

1b. Find the sum of 4.81 and 1.92.

2a. Add: 7435.3 + 358 + 12.4

2b. Add: 214.8 + 638 + 7331.7

3a. Add: 1445.8 + 15 + 376.23

3b. Find the sum: 37,234.8 + 89 + 1852.07 + 1.3

4a. Add on your calculator: 0.0008 + 67.492 + 1093.13

4b. Find the sum 960.0007 + 8330.0495 + 23.174 on your calculator.

$$\begin{array}{r} 15.6 \\ 9. \\ +11.893 \\ \hline \end{array}$$ Add the decimal point.

We add zeros to the right of the decimal points to give all our addends the same number of decimal places. This makes it easier to add the columns.

$$\begin{array}{r} 15.600 \\ 9.000 \\ +11.893 \\ \hline \end{array}$$

Now we add the columns and insert the decimal point in the answer.

$$\begin{array}{r} 15.600 \\ 9.000 \\ +11.893 \\ \hline 36.493 \end{array}$$

Our estimate of the sum was about 37, so the result 36.493 is reasonable.

■ *Now do margin exercises 2a and 2b.*

Let's try another example.

EXAMPLE 3

Add: 13.5 + 1000.76 + 12.000713

SOLUTION

As an estimate, we get 14 + 1001 + 12, or 1027. We then rewrite the problem in vertical form, line up the decimal points, and add zeros, as necessary, to the right of the decimal points.

$$\begin{array}{r} 13.500000 \\ 1000.760000 \\ +\ \ 12.000713 \\ \hline \end{array}$$

We add the columns and insert the decimal point in the answer.

$$\begin{array}{r} 13.500000 \\ 1000.760000 \\ +\ \ 12.000713 \\ \hline 1026.260713 \end{array}$$

Our estimate of 1027 indicates that 1026.260713 is reasonable.

■ *Now do margin exercises 3a and 3b.*

To add decimal numbers using a calculator, follow the same procedure as for whole numbers. But don't forget to key in the decimal points.

EXAMPLE 4

Add: 456.789 + 23.456 + 0.00003

SOLUTION

Key in 456.789 [+] 23.456 [+] .00003

The display should read 480.24503, which is the required sum.

■ *Now do margin exercises 4a and 4b.*

Work the exercises that follow. Remember to estimate sums.

6.3 EXERCISES

In Exercises 1 through 43, add.

1. $\begin{array}{r} 20.5376 \\ +20.2423 \\ \hline \end{array}$

2. $\begin{array}{r} 71.25 \\ 3.48 \\ +\ 0.85 \\ \hline \end{array}$

3. $\begin{array}{r} 11.91 \\ +95.85 \\ \hline \end{array}$

4. $\begin{array}{r} 40.812 \\ +40.165 \\ \hline \end{array}$

5. $\begin{array}{r} 14 \\ +33.35 \\ \hline \end{array}$

6. $\begin{array}{r} 711 \\ +\ 0.073 \\ \hline \end{array}$

7. $\begin{array}{r} 31 \\ +12.83 \\ \hline \end{array}$

8. $\begin{array}{r} 64 \\ +84.62 \\ \hline \end{array}$

9. 28 + 0.006

10. 36 + 0.080

11. 84 + 0.009

12. 20 + 0.9040

13. 7 + 0.63 + 18.04

14. 83 + 0.08 + 39.01

15. 94.31 + 8.26 + 0.005

16. 95.85 + 51.26 + 29

17. 0.85 + 0.7 + 5.2009

18. 0.671 + 0.0911 + 1.3

19. 0.29 + 0.3076 + 0.2

20. 41 + 74.031 + 8 + 6.9

21. 9.775 + 82.77211 + 6.11

22. 7.338 + 50.59

23. 7.338 + 505.9

24. 62 + 28.3

25. 27.63 + 137.4

26. 85.41 + 68.9598

27. 16.59 + 49.63211 + 5172

28. 8725.16 + 66.19

29. 18,354.3 + 511.18

30. 856,267.9 + 213.28

31. 26.7 + 430.4968 + 6793

32. 4162 + 44.09 + 35.6 + 0.3

33. 76.18 + 185.37 + 8000.007 + 624.375

34. 88,923 + 13.0402 + 128.3

35. 29,873 + 56.068 + 6879 + 2.5

36. 990.34 + 9820.05 + 100,019.00023

37. 1537.21 + 658.7 + 102,937.483

38. 6024.5 + 92.010 + 123.845708

39. 96986.4 + 7.635 + 3827.3098

40. 18,222.36 + 7.15 + 0.102

41. 87,256.34 + 9245.676 + 82,254.9388

42. 872,861 + 80.005 + 9357.06

43. 8219.2 + 667,185.6376 + 9.99923 + 92,373

In Exercises 44 through 51, fill in the missing digits.

44.
```
   13.005
    2.9__
 +  0.002
 --------
   15.987
```

45.
```
   28.8_
 +  3._7
 -------
   32.47
```

46.
```
   27.6_8
    _._16
 +  0.490
 --------
   31.64_
```

47.
```
   1_.0_
 +  4._6
 -------
   _8.84
```

48.
```
   __8.0_
     6._0
 +  1_.38
 --------
   639.39
```

49.
```
     2_.073
    1_8._40
 +    0.0__
 ----------
    136.621
```

50.
```
   1_.638
   _2._7_
 +  8.0_6
 --------
   95.618
```

51.
```
   _29.0_4
    31_._30
 +    0.71_
 ----------
    848.393
```

6.3 MIXED PRACTICE

By doing these exercises, you will practice all the topics up to this point in this chapter.

52. Add and round your answer to the nearest tenth: $18 + 62.08 + 56.24$

53. Write 41.396 in words.

54. Convert 0.0067 to a fraction.

55. Arrange 18.202, 18.024, and 18.204 in order from smallest to largest.

56. Add: $52 + 194.03 + 0.0009$

57. Write sixty-three thousandths as a numeral.

58. Round to the nearest tenth and add: $40 + 0.007 + 0.25$

59. Convert 12.418 to a fraction.

60. Round 17.0992 to the nearest thousandth.

61. Write in standard notation: $\frac{4}{10{,}000} + \frac{8}{1{,}000} + \frac{6}{100} + \frac{8}{10}$

6.4 Subtracting Decimal Numbers

SECTION GOAL

- To find the difference of two decimal numbers

The subtraction of decimal numbers is similar to the subtraction of whole numbers, except that we must be careful of the decimal point. The number we subtract from is the *minuend*. The number subtracted is the *subtrahend;* the result is the *difference.*

As in addition, it is useful to estimate the answer. Then we subtract, after first aligning the decimal points. Finally, we check the result by using addition.

EXAMPLE 1

Subtract: 3.2 – 1.7

SOLUTION

We estimate by rounding to the nearest whole number and subtracting. The number 3.2 rounds to 3, and 1.7 rounds to 2. Our answer will be about 3 – 2, or 1.

Now we rewrite the problem vertically, making sure to line up the decimal points.

$$\begin{array}{r} 3.2 \\ \underline{-1.7} \\ \uparrow \end{array} \quad \text{Align.}$$

Next we subtract, regrouping as needed and disregarding the decimal points for the moment. We get

$$\begin{array}{r} \overset{2}{\cancel{3}}.2 \\ \underline{-1.7} \\ 1\ 5 \end{array}$$

Then we place the decimal point in the answer, directly below the decimal points in the problem.

$$\begin{array}{r} \overset{2}{\cancel{3}}.2 \\ \underline{-1.7} \\ 1.5 \\ \uparrow \end{array} \quad \text{Align.}$$

1a. Subtract: 22.1 – 4.6

1b. Find the difference between 15.6 and 7.9

2a. Subtract: 200.004 – 34.39

2b. What is the largest number that can be subtracted from 425.89 and still yield an answer greater than 422? Answer to the nearest thousandth.

Our estimate of about 1 shows that our decimal point is properly placed. We check by adding the result (1.5) to the subtrahend (1.7) to get the minuend (3.2).

$$\begin{array}{r} 1.5 \\ +1.7 \\ \hline 3.2 \end{array} \quad \text{Correct}$$

■ *Now do margin exercises 1a and 1b.*

In the next example, we subtract hundredths from thousandths.

EXAMPLE 2

Subtract: 100.005 – 7.06

SOLUTION

We estimate the answer by rounding and subtracting. We get 100 – 7, or about 93.

We next write the problem vertically, lining up the decimal points. Before we actually subtract, however, we add zeros to the right end of the "shorter" decimal number to make the decimal parts the same "length." Remember, this does not change the value of the decimal number.

$$\begin{array}{r} 100.005 \\ -\ \ 7.060 \\ \hline \end{array}$$

↑ Align.

Then we subtract, placing the decimal point only *after* we subtract.

$$\begin{array}{r} {\scriptstyle 9\,9\,9\,1} \\ \not{1}\not{0}\not{0}.\not{0}05 \\ -\ \ 7.060 \\ \hline 92.945 \end{array}$$

↑ Place decimal point.

Comparison with our estimate of approximately 93 shows that this answer is reasonable. However, we can also check by addition.

$$\begin{array}{r} 92.945 \\ +\ 7.060 \\ \hline 100.005 \end{array} \quad \text{Correct}$$

■ *Now do margin exercises 2a and 2b.*

In the next example, we subtract a decimal number from a whole number.

EXAMPLE 3

Subtract: 9854 – 0.36

SOLUTION

We estimate that the answer will be slightly smaller than 9854 – 0, or 9854.

We then rewrite the problem in vertical form, lining up the decimal points. A whole number has an understood decimal point after that number, so we can write 9854 as 9854.00. We subtract as usual, disregarding the decimal point until the end. We get

$$\begin{array}{r} 985\overset{3}{\cancel{4}}.\overset{9}{\cancel{0}}{}^{1}0 \\ -\quad 0.36 \\ \hline 9853.64 \end{array}$$

The estimate tells us that our result is reasonable.

■ *Now do margin exercises 3a and 3b.*

3a. Subtract: 918 – 7.985

3b. Find the difference between 830.008 and 70.0016.

4a. Use your calculator to find the difference between 82.18 and 45.

4b. From 298.38, subtract 273.4009.380.

EXAMPLE 4

Subtract 1932.896 – 212.3846 using your calculator.

SOLUTION

Just as in whole-number computations, the minuend must be keyed in first.

Key in

1932.896 [–] 212.3846 [=]

The display should show 1720.5114.

■ *Now work margin exercises 4a and 4b.*

Work the exercises that follow.

6.4 EXERCISES

In Exercises 1 through 39, subtract as indicated.

1. $\begin{array}{r} 18.37 \\ -\ 8.05 \\ \hline \end{array}$

2. $\begin{array}{r} 9.89 \\ -5.73 \\ \hline \end{array}$

3. $\begin{array}{r} 7.26 \\ -2.14 \\ \hline \end{array}$

4. 0.145 – 0.026

5. 0.179 – 0.115

6. 336.72 – 15.61

7. 654.81 – 27.53

8. 492.27 – 16.18

9. 536.85 – 28.37

10. $\begin{array}{r} 316.53 \\ -\ 87.16 \\ \hline \end{array}$

11. $\begin{array}{r} 8294.28 \\ -\ 53.13 \\ \hline \end{array}$

12. $\begin{array}{r} 7185.73 \\ -\ 29.51 \\ \hline \end{array}$

13. $\begin{array}{r} 472.64 \\ -\ 86.18 \\ \hline \end{array}$

14. $\begin{array}{r} 115.004 \\ -\ 1.385 \\ \hline \end{array}$

15. $\begin{array}{r} 212.007 \\ -\ 3.628 \\ \hline \end{array}$

16. From 95.008, subtract 82.4313

17. From 73.002, subtract 16.7191

18. From 681.003, subtract 9.26526

19. Subtract 0.8 from 10.676

20. Subtract 0.37 from 0.405

21. Subtract 0.29 from 310.319

22. From 770.83, subtract 6.90352

23. From 56.14, subtract 3.700423

24. Find the difference between 620.1 and 0.37

25. Find the difference between 920.7 and 0.41

26. 980.4 – 0.36

27. 850.6 – 0.24

28. 78,257 – 153.8638

29. 92,903 – 447.282

30. Subtract 0.326 from 127

31. From 632, take away 0.435

32. Subtract 0.961 from 3185

33. Subtract 10.823 from 5435

34. 8297 – 0.41898

35. 11.864 – 0.90767

36. 2751 – 830.70412

37. 1806 – 160.56243

38. 6893 – 14.6932

39. 3164 – 89.0154

6.4 MIXED PRACTICE

By doing these exercises, you will practice all the topics up to this point in this chapter.

40. Add 537; 108.5; and 0.0009 and round to the nearest thousandth.

41. Write $37\frac{26}{1000}$ in words.

42. Write 0.6008 as a fraction.

43. Subtract: 43.805 – 29.362

44. Find the sum of twenty-seven and thirty-six hundredths, and eight hundred four and ninety-eight hundredths, and seven ten-thousandths.

45. Find the difference: 104.375 – 65.89

46. Arrange 10.076, 11.088, and 10.78803 in order from smallest to largest.

47. Add: 136.85 + 23.762 + 830

6.5 Multiplying Decimal Numbers

In Chapter 2, you learned how to multiply whole numbers. The same procedure is used to multiply decimal numbers. The terminology is also the same. We multiply two *factors* (or a *multiplier* and a *multiplicand*), to get their *product*.

The Decimal Place Rule

The properties of whole-number multiplication also hold for decimal numbers. The major difference is that we must place the decimal point correctly in the product. To do so, we use the following rule:

> The number of decimal places in a product is equal to the sum of the numbers of decimal places in the factors.

In our first example, you will see how to use this rule.

EXAMPLE 1

Multiply: 9.05×6

SOLUTION

First we round to estimate the size of the product. Because 9.05 rounds to 9, and the multiplier 6 is whole number, the product will be close to 9×6, or 54.

Next we rewrite the problem in vertical form and multiply, ignoring the decimal point until the end of the problem.

$$\begin{array}{r} 9.05 \\ \times \quad 6 \\ \hline 54\ 30 \end{array}$$

Here, 9.05 has two decimal places and 6 has zero decimal places. Thus we locate the decimal point $2 + 0$ or two places from the right in our answer.

$$\begin{array}{rl} 9.05 & \text{Two decimal places} \\ \times \quad 6 & \text{Zero decimal places} \\ \hline 54.30 & \text{Two decimal places} \end{array}$$

Recall that we estimated the product as being close to 54. Placing the decimal point differently would give answers that are not close to 54, so the decimal point is in the correct place.

Therefore, 9.05×6 is 54.3.

■ *Now do margin exercises 1a and 1b.*

SECTION GOAL

■ To find the product of two or more decimal numbers

1a. Multiply: 4.08×9

1b. Multiply: 652×0.6

Multiplying by Tenths and Tens

Before we continue, let's look at what happens when a large whole number is multiplied by tenths, hundredths, thousandths, and so on.

$$3{,}000{,}000 \times 0.1 = 300000.0$$
$$3{,}000{,}000 \times 0.01 = 30000.00$$
$$3{,}000{,}000 \times 0.001 = 3000.000$$
$$3{,}000{,}000 \times 0.0001 = 300.0000$$
$$3{,}000{,}000 \times 0.00001 = 30.00000$$
$$3{,}000{,}000 \times 0.000001 = 3.000000$$
$$3{,}000{,}000 \times 0.0000001 = 0.3000000$$

As you can see, each decimal place in the multiplier removes a ten from the multiplicand. This pattern suggests a short cut.

To multiply any multiplicand by 0.1, 0.01, 0.001, and so on

1. Write down the multiplicand.
2. Move the decimal point in the multiplicand to the *left* the same number of decimal places as are in the multiplier.

For example, to multiply by 0.0001, move the decimal point four places to the left in the multiplicand.

Similarly, note what happens when a decimal number is multiplied by 10, 100, 1000, and so on.

$$0.00002 \times 10 = 0.0002$$
$$0.00002 \times 100 = 0.002$$
$$0.00002 \times 1000 = 0.02$$
$$0.00002 \times 10000 = 0.2$$
$$0.00002 \times 100000 = 2$$
$$0.00002 \times 1000000 = 20$$
$$0.00002 \times 10000000 = 200$$

Each ten in the multiplier removes a decimal place from the multiplicand. This pattern also suggests a short cut.

To multiply any multiplicand by 10, 100, 1000, and so on

1. Write down the multiplicand.
2. Move the decimal point in the multiplicand to the *right* one place for each zero in the multiplier.

To multiply by 10,000, then, move the decimal point four places to the right.

These short cuts can be particularly helpful in estimating answers.

More Multiplication of Decimal Numbers

EXAMPLE 2

Multiply: $\begin{array}{r} 0.03 \\ \times 0.02 \\ \hline \end{array}$

SOLUTION

Here it is not easy to estimate the answer. However, each of these numbers is close to 0.01, so a rough estimate of the answer would be 0.01 × 0.01, or 0.0001.

Next we multiply, ignoring the decimal points until we find a product. We get

$$\begin{array}{r} 0.03 \\ \times 0.02 \\ \hline 6 \end{array}$$

To determine the placement of the decimal point, we add the numbers of decimal places in the two factors:

0.03 has	*2 decimal places*
0.02 has	*2 decimal places*
Product has	*4 decimal places*

We count off four decimal places from the far right, moving left. We have to insert three zeros to place the decimal point correctly:

$$.\underset{4}{0}\,\underset{3}{0}\,\underset{2}{0}\,\underset{1}{6}$$

Recall that our estimate was about 0.0001. Therefore, the answer we obtained is reasonable.

So 0.03 × 0.02 = 0.0006.

■ *Now do margin exercises 2a and 2b.*

Let's do another example where both numbers have a decimal portion.

EXAMPLE 3

Multiply: 26.003 × 0.0009

SOLUTION

First we estimate. To do so, we round 0.0009 to 0.001. Then the product will be very close to 26 × 0.001, or 0.026.

Next we rewrite the problem vertically and multiply, ignoring the decimal point.

$$\begin{array}{r} 26.003 \\ \times 0.0009 \\ \hline 234027 \end{array}$$

2a. Find the product: 2.014 × 0.05

2b. Multiply: 6.08 × 0.0901

3a. What is the product of 2.8603 and 0.009?

3b. Multiply: 60.773×0.6206

4a. Multiply: 17.03302×5.00014

4b. Find the product: 25.06×18.08903

5a. Use your calculator to multiply: 0.99999×357.8

5b. Multiply $0.28 \times 0.996 \times 9.32$ on your calculator.

Because 26.003 has 3 decimal places and 0.0009 has 4 decimal places, the product must have $3 + 4 = 7$ decimal places. So we count off seven places from the right in the product 234027, preceding it with a zero, and obtain

$$.0\,2\,3\,4\,0\,2\,7$$
$$7\;6\;5\;4\;3\;2\;1$$

So $26.003 \times 0.0009 = 0.0234027$.

■ *Now do margin exercises 3a and 3b.*

EXAMPLE 4

Multiply: 19.002×1.00034

SOLUTION

The factors here round to 19 and 1, so our estimate of the product is 19×1, or 19.

We write the problem in vertical form and multiply, ignoring the decimal points.

$$\begin{array}{r} 19.002 \\ \times\, 1.00034 \\ \hline 76008 \\ 570060 \\ 1900200000 \\ \hline 1900846068 \end{array}$$

Counting off the decimal places in the two factors, we find that there should be a total of 8 decimal places in the answer, giving 19.00846068. Our estimate of 19 tells us that this is a reasonable answer.

■ *Now do margin exercises 4a and 4b.*

When you multiply decimal numbers with a calculator, it keeps track of the decimal places. You must, however, remember to key in the decimal points in the proper places.

EXAMPLE 5

Using your calculator, multiply

$$246.8 \times 37.9$$

SOLUTION

Key in the following:

246.8 [×] 37.9 [=]

The display will show the product, 9353.72. Note that there are a total of two decimal places in the factors and two decimal places in the product.

■ *Now do margin exercises 5a and 5b.*

Work the exercises that follow. Make sure your answers are reasonable.

6.5 EXERCISES

In Exercises 1 through 60, multiply as indicated. Estimate to check your answers.

1. 9×1.4

2. 6×3.6

3. 3×8.5

4. 5×7.05

5. 0.3×0.55

6. 0.09×0.3

7. 2.10×0.4

8. 11.6×4.09

9. 146×0.004

10. 83×0.09

11. $\begin{array}{r} 1.5 \\ \underline{\times 4.0} \end{array}$

12. $\begin{array}{r} 6.008 \\ \underline{\times \quad 5.9} \end{array}$

13. $\begin{array}{r} 3.4 \\ \underline{\times 6.0} \end{array}$

14. $\begin{array}{r} 7.2 \\ \underline{\times 3.0} \end{array}$

15. $\begin{array}{r} 34.3 \\ \underline{\times \ 5.8} \end{array}$

16. $\begin{array}{r} 49.1 \\ \underline{\times \ 9.8} \end{array}$

17. 9.5×0.0201

18. 11.7×0.4001

19. $\begin{array}{r} 9.02 \\ \underline{\times \ 3.8} \end{array}$

20. $\begin{array}{r} 2.6 \\ \underline{\times 8.0} \end{array}$

21. 21.6×1.03

22. 3.25×20.6

23. 7.21×80.5

24. 3.4×0.16

25. $\begin{array}{r} 23.8 \\ \underline{\times 2.02} \end{array}$

26. $\begin{array}{r} 2.6 \\ \underline{\times 5.105} \end{array}$

27. $\begin{array}{r} 1.8 \\ \underline{\times 20.25} \end{array}$

28. 2882.4×93.008

29. 31.5×0.17

30. 68.8×0.24

31. 62.9×11.01

32. 113.07×8.205

33. $82{,}320.7 \times 60.26$

34. $91{,}230.9 \times 30.12$

35. $56{,}440.6 \times 80.35$

36. 6816.5×2.46

37. 11132.6×0.035

38. 3410.5×0.00046

39. 6782.3×0.0028

40. 258.970×0.0009

41. 648.7×7.013

42. 183.9×6.025

43. 3701.5×17.001

44. 2.7×70.36

45. 1.0035×2.04

46. 3.02006×7.008

47. 2.035×3.04009

48. 58×9.7001

49. 31.078×0.04

50. 12.003×2.06

51. 340.081×7.683

52. 6854.9×0.0097

53. 315.017×20.4

54. 728.223×41.5

55. 16.702×30.9

56. 6111.52×87.08

57. 123.202×40.04

58. 2604.333×68.15

59. 81.007×3.25

60. 718.4×17.08

In Exercises 61 through 64, use your calculator to multiply each pair of numbers in turn. Stop multiplying when you discover a pattern to the answers. Indicate where you stopped multiplying, give the remaining answers, and describe in your own words the pattern you see.

61. $1.01 \times 11 =$ ________
$1.01 \times 111 =$ ________
$1.01 \times 1111 =$ ________
$1.01 \times 11111 =$ ________
$1.01 \times 111111 =$ ________
$1.01 \times 1111111 =$ ________

62. $9.9 \times 123 =$ ________
$9.9 \times 1234 =$ ________
$9.9 \times 12345 =$ ________
$9.9 \times 123456 =$ ________
$9.9 \times 1234567 =$ ________
$9.9 \times 12345678 =$ ________

63. $1.23 \times 555 =$ ________
$1.23 \times 5555 =$ ________
$1.23 \times 55555 =$ ________
$1.23 \times 555555 =$ ________
$1.23 \times 5555555 =$ ________
$1.23 \times 55555555 =$ ________
$1.23 \times 555555555 =$ ________

64. $7777 \times 1.05 =$ ________
$77777 \times 1.05 =$ ________
$777777 \times 1.05 =$ ________
$7777777 \times 1.05 =$ ________
$77777777 \times 1.05 =$ ________
$777777777 \times 1.05 =$ ________
$7777777777 \times 1.05 =$ ________

6.5 MIXED PRACTICE

By doing these exercises, you will practice all the topics up to this point in this chapter.

65. Multiply: 34.56×12.9

66. Add: $29 + 36.4 + 1.8$

67. Write 84.005 in expanded notation.

68. Multiply: 345×0.23

69. Multiply 1.01×12 and round your answer to the nearest tenth.

70. Subtract three and ninety-eight thousandths from sixty-three.

71. Subtract: $179.6 - 85.53$

72. The following list gives the comparative masses of the planets (with Earth given mass 1). Put them in order from least mass to greatest.

Venus	0.81
Mercury	0.06
Pluto	0.17
Mars	0.11
Earth	1
Neptune	17.23
Uranus	14.54
Saturn	95.15
Jupiter	317.83

SUN MERCURY VENUS EARTH MARS JUPITER SATURN URANUS NEPTUNE PLUTO

73. Add 38.7 + 40 + 0.076 and round to the nearest hundredth.

74. Subtract: 100.007 – 65.439

75. Add: 98.34 + 9 + 203.4

76. Write 0.0986 as a fraction.

6.6 Dividing Decimal Numbers

SECTION GOAL

- To find the quotient of two decimal numbers

Dividing decimal numbers is an everyday experience. Suppose you and two friends go out to dinner, and the check comes to \$34.05. To share the cost you have to divide by 3. This is a problem in dividing decimal numbers. The quotient is \$11.35.

Meal Check			
# In Party *3*	Waiter *RM*	Table *6*	Date *2/14/92*
			\$29.61
		Tip 15%	\$4.44
		Total	\$34.05

Whole-Number Divisors

Division of decimal numbers is very similar to division of whole numbers. A *dividend* is divided by a *divisor*, and the result is a *quotient*. The main difference involves placing the decimal point in the quotient.

When only the dividend has a decimal point, you can use this rule:

> When the divisor has no decimal point, the quotient's decimal point should be placed directly over the decimal point in the dividend.

In our dinner check example, the decimal points would be placed like this:

$$3\overline{)34.05}\quad 11.35$$

EXAMPLE 1

Divide 1190.01111 by 30.

SOLUTION

First we estimate to determine the approximate size of the answer. The dividend is slightly less than 1200, and 1200 divided by 30 is 40. Thus the quotient will be slightly less than 40.

Next we rewrite the division and place the decimal point.

$$30\overline{)1190.01111}$$

1a. Divide 27.9 by 3

1b. What is the quotient of 35.5 and 5?

Then we do the division as usual.

```
         39.667037
   30)1190.011110
       90            30 × 3
       290
       270           30 × 9
        200
        180          30 × 6
         201
         180         30 × 6
          211
          210        30 × 7
            11
            00       30 × 0
            111
             90      30 × 3
             210
             210     30 × 7
               0
```

A quotient slightly less than 40 (as we predicted) seems reasonable.

■ *Now do margin exercises 1a and 1b.*

Dividing by Tens and Tenths

Before we do another division, we can find two more useful short cuts. First note what happens when a whole number is divided by a tenth, a hundredth, and so on.

$$3 \div 0.1 = 30$$
$$3 \div 0.01 = 300$$
$$3 \div 0.001 = 3000$$
$$3 \div 0.0001 = 30000$$
$$3 \div 0.00001 = 300000$$
$$3 \div 0.000001 = 3000000$$
$$3 \div 0.0000001 = 30000000$$

Each decimal place in the divisor multiplies the dividend by ten. This gives us one short cut:

To divide by 0.1, 0.01, 0.001, and so on

1. Write down the dividend.
2. Move the decimal point in the dividend *to the right* the same number of decimal places as there are in the divisor.

Thus, to divide 630 by 0.0001, we move the decimal point to the right four places in the dividend:

630 0 0 0 0.
1 2 3 4

So 630 ÷ 0.0001 is 6,300,000.

Now note what happens when a whole number is divided by 10, 100, 1000, and so on.

$$2 \div 10 = 0.2$$
$$2 \div 100 = 0.02$$
$$2 \div 1000 = 0.002$$
$$2 \div 10000 = 0.0002$$
$$2 \div 100000 = 0.00002$$
$$2 \div 1000000 = 0.000002$$
$$2 \div 10000000 = 0.0000002$$

Each ten in the divisor adds a decimal place to the dividend. The pattern here provides the second short cut.

To divide by 10, 100, 1000, and so on

1. Write down the dividend.
2. Move the decimal point in the dividend *to the left* one place for each zero in the divisor.

To divide 927 by 10,000, we move the decimal point to the left four places in the dividend:

.0 9 2 7

So 927 ÷ 10,000 is 0.0927.

Decimal-Number Divisors

Most often in division problems, there will be decimal points in both the dividend and the divisor.

EXAMPLE 2

Divide: $2.1\overline{)18.942}$

SOLUTION

The quotient here will be close to 9, because 18 divided by 2 is 9.

Before we divide, we must place the quotient's decimal point. We need to somehow remove the decimal point from the divisor. We can do this by multiplying both the divisor and the dividend by 10 to get

$$21\overline{)189.42}$$

2a. Divide: $3.2\overline{)19.2}$

2b. Divide 21.3 by 7.1

Here's why it works: First write the original problem in fractional form.

$$\frac{18.942}{2.1}$$

Next multiply the numerator and denominator of the fraction by 10 (remember, this does not change the value of the fraction).

$$\frac{18.942}{2.1} = \frac{18.942 \times 10}{2.1 \times 10} = \frac{189.42}{21}$$

This is equivalent to the division $21\overline{)189.42}$ that we obtained above.

Dividing, we get

$$\begin{array}{rl} & \text{Align.} \\ 9.02 & \\ 21\overline{)189.42} & \\ \underline{189} & 9 \times 21 \\ 0\,4 & \\ \underline{0} & 0 \times 21 \\ 42 & \\ \underline{42} & 2 \times 21 \\ 0 & \end{array}$$

We can check by multiplying.

$2.1 \times 9.02 = 18.942$ Correct

■ *Now do margin exercises 2a and 2b.*

Our short cut for multiplying by 10, 100, and so on gives us the following procedure for removing the decimal point from a divisor.

To remove a decimal point from the divisor

1. Move the decimal point to the far right of the divisor. (This essentially eliminates it.)
2. Move the decimal point in the dividend to the right the same number of places as it was moved in the divisor.
3. Add zeros to the dividend as necessary.

Let's look at another example and use the rule.

EXAMPLE 3

Divide: $0.035\overline{)1.470}$

SOLUTION

We must move the divisor's decimal point three places to the right (0.035 becomes 35.). Then we have to move the dividend's decimal point to the right three places (1.470 becomes 1470).

The problem we must now solve is $35\overline{)1470}$. Because the dividend is close to 1500, and the divisor is near 30, the quotient will be near (less than) 50. We place the decimal point in the quotient *first*, and then the solution looks like this:

```
          Place the
    42. ← decimal point.
35)1470.
   140    35 × 4
    70
    70    35 × 2
```

This quotient is reasonable.

■ *Now do margin exercises 3a and 3b.*

Let's divide a decimal number into a whole number.

EXAMPLE 4

Divide: $0.008\overline{)10}$

SOLUTION

To move the decimal point out of the divisor, we must move the decimal point three places to the right. To do the same in the dividend, we first have to add zeros to the right of its decimal point. We get

```
0.0 0 8.)10.0 0 0.
  1 2 3     1 2 3
```

We remove unneeded zeros, place the decimal point in the quotient directly over its new place in the dividend, and then divide.

```
   1250.
8)10000.
  8
  20
  16
   40
   40
    00
```

We can check by multiplying the original divisor (0.008) times the quotient (1250) to get the dividend (10).

■ *Now do margin exercises 4a and 4b.*

3a. Divide: $0.002\overline{)0.1470}$

3b. Divide: 1.650/0.05

4a. Divide: $9 \div 0.003$

4b. Divide: $20 \div 0.0004$

5a. Use your calculator to divide 456.45 by 3.4.

5b. Use your calculator to divide 3847.9 by 0.0009. Round your answer to the nearest tenth.

EXAMPLE 5

Divide 12.5 by 800 using a calculator, and round to the nearest thousandth.

SOLUTION

We use the division key [÷]. Key in

12.5 [÷] 800 [=]

The display should read 0.015625. Because the answer is to be rounded to the nearest thousandth, we round to 0.016.

■ *Now do margin exercises 5a and 5b*

Work the exercises that follow. Make sure your answers are reasonable.

6.6 EXERCISES

In Exercises 1 through 32, divide as indicated. Use your calculator to check your answers.

1. $13\overline{)300.3}$
2. $9\overline{)104.4}$
3. $11\overline{)237.6}$
4. $21\overline{)346.5}$
5. $\frac{0.24}{6}$
6. $\frac{0.48}{4}$
7. $\frac{0.96}{3}$
8. $\frac{6.4}{0.8}$
9. $1615.1 \div 3.1$
10. $344.0 \div 1.6$
11. $152.4 \div 1.2$
12. $10.72 \div 0.16$
13. $0.15\overline{)0.45}$
14. $0.12\overline{)0.96}$
15. $0.06\overline{)0.84}$
16. $14 \div 0.7$
17. $\frac{83.8}{0.4}$
18. $\frac{5.3}{0.5}$
19. $\frac{4.9}{0.4}$
20. $852.0 \div 0.4$
21. Divide 15.12 by 0.2
22. Divide 0.65 by 5
23. Divide 0.196 by 0.14
24. Divide 0.198 by 0.11
25. Divide 0.28 by 0.08
26. Divide 0.0063 by 0.0007

27. Divide 0.729 by 0.9

28. Divide 0.156 by 0.3

29. Divide 0.184 by 0.2

30. Divide 6256 by 1.7

31. Divide 8 by 0.2

32. Divide 20 by 0.4

In Exercises 33 through 36, divide. Round your answer to the nearest hundredth.

33. $16 \div 0.83$

34. $0.081\overline{)0.88}$

35. $1.8\overline{)26463}$

36. $2.2\overline{)112645}$

In Exercises 37 through 40, divide as indicated.

37. $\frac{0.637}{0.7}$

38. $\frac{2.40}{0.006}$

39. $\frac{0.00021}{0.700}$

40. $\frac{0.00027}{0.009}$

6.6 MIXED PRACTICE

By doing these exercises, you will practice all the topics up to this point in this chapter.

41. Divide: $2545.2 \div 4.04$

42. Subtract: $506.84 - 13.007$

43. Add $3.86 + 29.358$ and round your answer to the nearest hundredth.

44. Multiply: 10.986×2.308

45. Add: $34.67 + 85.201 + 55$

46. Divide: $98.004 \div 0.04$

47. Multiply: 34.892×0.001

48. Subtract $26.85 - 15.883$ and round your answer to the nearest hundredth.

49. Arrange 106.35, 106.035, and 106.0305 in order from smallest to largest.

50. Subtract: $106.004 - 23.8$

6.7 Applying Addition, Subtraction, Multiplication, and Division of Decimal Numbers: Word Problems

Section Goal

- To solve applications of addition, subtraction, multiplication, and division of decimal numbers

We have already solved word problems using whole numbers and fractions. Word problems involving decimal numbers, such as dollar amounts or metric measurements, are solved in exactly the same way. The first example requires comparison; we use subtraction to find out by how much two things differ.

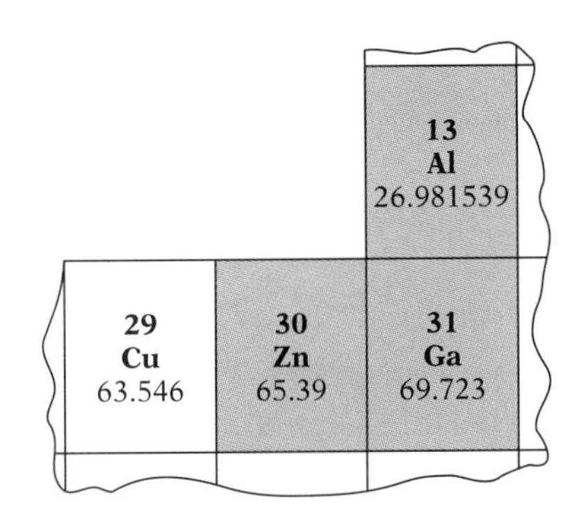

Example 1

The atomic weight of aluminum is 26.981539. The atomic weight of copper is 63.546. How much heavier than aluminum is copper?

SOLUTION

To find the *difference* in the two weights, we subtract.

$$\begin{array}{r} 63.546000 \\ \underline{-26.981539} \\ 36.564461 \end{array}$$

The atomic weight of copper is 36.564461 units more than that of aluminum.

- *Now do margin exercises 1a and 1b.*

1a. The melting point of aluminum is 660.37 degrees. The melting point of iron is 1535 degrees. What is the difference in the melting points of these two substances?

1b. The distance from the planet Pluto to the Sun is 3675.27 million miles. The planet Neptune is 879.81 million miles closer. How far from the Sun is Neptune?

The next example involves scaled increase.

Example 2

Frosty, the Siberian husky, always runs with his owner, William. William runs about 7.06 miles per hour (an 8.5-minute mile). Frosty runs about 1.5 times faster than that. How fast does this dog run?

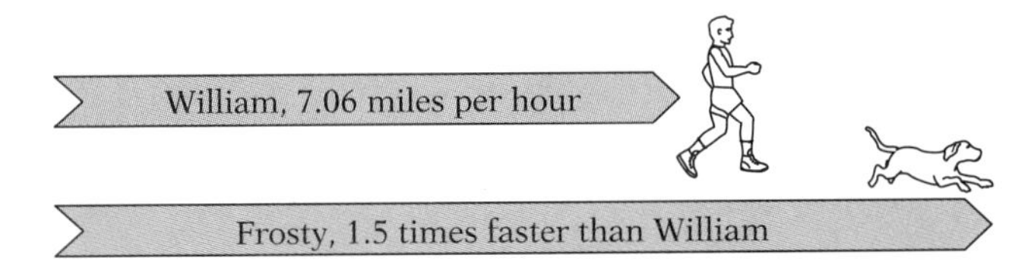

SOLUTION

Here we are given a speed (7.06 miles per hour) and told that a second speed (Frosty's) is a number of times faster (1.5). We are looking for a larger number, and we must multiply 7.06×1.5.

2a. The estimated 1987 population of the U.S.S.R. was 0.284 billion. India's population was 0.5163 billion more than the population of the U.S.S.R. What was the estimated 1987 population of India?

2b. The sea horse moves 10.5 inches per minute. How far will it travel in 30 minutes?

3a. How much net income does the firm in Example 3 make on a shipment of 137,000 yards of silk thread?

3b. Granny Smith apples are sold for 99 cents a pound. They also come packed in a container of 4 apples for $3. If each apple weighs 0.81 pound, how much do you save by buying them in the package?

$$\begin{array}{r} 7.06 \\ \times\ 1.5 \\ \hline 3530 \\ 706 \\ \hline 10.590 \end{array}$$

Frosty's speed is 10.59 miles per hour.

■ *Now do margin exercises 2a and 2b.*

Our final example requires more than one step for its solution.

EXAMPLE 3

A firm has imported 560,000 yards of silk thread to be cut into lengths (called skeins) of 91 yards each and sold to upholsterers. The firm makes a profit of $3.98 per skein. However, it loses 15 cents per yard for any thread that is left over. How much net income will the firm have from this shipment?

SOLUTION

We need to find the number of skeins sold and the number of yards left over. We must divide 560,000 by 91. Then the quotient is the number of skeins, and the remainder is the left-over yards. The division is

$$\begin{array}{rl} 6\,153. & \text{Quotient} \\ 91\overline{)560{,}000.} & \\ 546 & \\ \hline 140 & \\ 91 & \\ \hline 490 & \\ 455 & \\ \hline 350 & \\ 273 & \\ \hline 77 & \text{Remainder} \end{array}$$

So the firm cuts 6153 skeins and has 77 yards left over. The profit for each skein is $3.98, so

$$\text{Profit} = \$3.98 \times 6153 = \$24{,}488.94$$

The loss for each left-over yard of thread is 15 cents, or $0.15. So the total loss is

$$\text{Loss} = \$0.15 \times 77 = \$11.55$$

The net income, then, is the difference between the profit and the loss.

$$\text{Net} = \$24{,}488.94 - \$11.55 = \$24{,}477.39$$

■ *Now do margin exercises 3a and 3b.*

Work the exercises that follow. Remember to see if your answers are reasonable.

6.7 EXERCISES

1. The odometer on your car reads 7098.3 miles. Then you travel 685.8 miles. What is the new odometer reading?

2. It takes 224.7 Earth days for Venus to revolve around the Sun. How many Earth weeks is this?

3. A small wheel rotates 2.25 times more than a larger wheel during the same period. If the larger wheel rotates 45 times, how many times has the smaller wheel rotated during the same period?

4. Lucy and Billy had just walked from their home in Hamilton into town. Their walk was 2.5 miles. This was half (0.5 times) their usual distance. How far do they usually walk?

5. The Lake Ponchartrain Bridge in Louisiana is 23.87 miles long from end to end. How far will I go if I make 6 round trips across the bridge in one week?

6. The gap between electrodes on a spark plug is generally 0.04 millimeters (mm). When you take one from a box, it measures 0.032 mm. How far does it have to be adjusted?

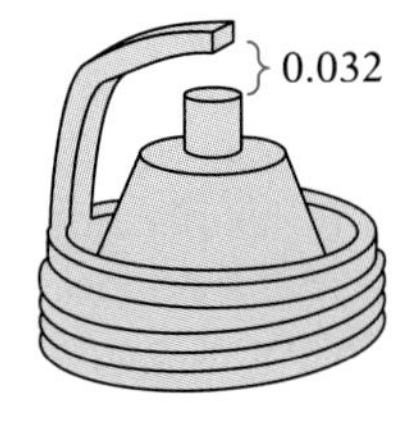

7. In Shanghai, 2.58 million bike trips are taken per day. How many million are taken in a week?

8. Frosty eats dog food that costs about $3.95 a pound. If he eats 3.5 pounds a day, how much does it cost per week?

9. Two of the NBA Lifetime Leaders in scoring averages are Wilt Chamberlain with 30.1 points per game and Elgin Baylor with 27.4. What is the difference in their scoring averages?

10. The estimated numbers of people who consider themselves Moslem, Buddhist, Protestant, Catholic, and Jewish are as follows:

Moslem	935 million
Buddhist	303 million
Protestant	73.5 million
Catholic	50.5 million
Jewish	5.9 million

What is the total number of people (to the nearest tenth of a billion) who consider themselves part of the five major religions?

Use the following information to answer the Exercises 11 and 12.

According to John Finley in his 1926 essay "The Wisdoms of Leisure," the following list indicates the amount of time a 70-year-old person has spent in each activity over a lifetime. We rewrote the numbers in terms of years.

Activity	Years	Activity	Years
Sleep	23	Walking	2.25
Sickness	1.5	Idling	2.5
Washing, shaving, etc.	2.5	Sundries	7
School	1.25	Waiting for a train	0.04
Reading	7	"Sweethearting"	0.16
Play	1.75	Entertaining	0.5

11. How many years are accounted for altogether in the activities listed?

12. If we add 18.0 years of work to the foregoing list, how many years of a person's life are unaccounted for?

Use the following chart to estimate the answer to Exercises 13 through 16. This chart gives the rotational velocity, at the equator, of each of the planets.

Venus	4.05 miles per hour
Mercury	6.73 miles per hour
Pluto	76.56 miles per hour
Mars	538 miles per hour
Earth	1040 miles per hour
Neptune	6039 miles per hour
Uranus	9193 miles per hour
Saturn	22,892 miles per hour
Jupiter	28,325 miles per hour

13. About how many times faster is Pluto's rotation than that of Venus? Is it closer to:

20 times? 2 times? 200 times? 0.2 times?

14. About how many times faster is Earth's rotation than that of Pluto? Is it closer to:

13 times? 1.3 times? 130 times? 0.13 times?

15. About how many times faster is Jupiter's rotation than that of Mercury? Is it closer to:

400 times? 4000 times? 4 times? 0.4 times?

16. About how many times faster is Neptune's rotation than that of Mars? Is it closer to:

1.2 times? 12 times? 120 times? 0.12 times?

Use the paycheck stub shown in the following table to answer Exercises 17 through 20.

Taxes		**Other Deductions**	
Federal	$231.78	Social Security	$116.54
State	$ 97.08	Pension	$ 37.72
City	$ 6.98	Insurance	$ 4.13
		Union dues	$ 15.75
PAY PERIOD:	9/13/92 to 9/26/92		
GROSS PAY:	$1551.83		

17. How much did you pay in taxes?

18. How much did you pay in other deductions?

19. What would your check have been if there had been only taxes taken out and not any "other deductions"?

20. What is your net pay (your pay minus all deductions)?

21. In a car repair book, the directions say that a certain valve should be open 0.28 millimeter (mm). At a tune-up, you discover that the valve is open 0.235 mm. How much more must the valve be opened?

22. The diameter of a bass drum is 4 times larger than that of a snare drum. If the snare drum is 1.317 feet wide, how wide is the bass drum?

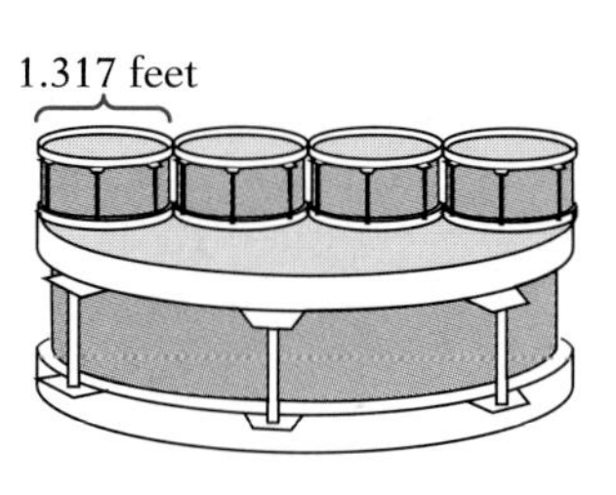

23. The lowest frequency your FM radio can receive is 87.9 megahertz. Its highest frequency is 20.107 megahertz greater than the lowest frequency. What is the highest frequency receivable on your radio?

24. In the 1988 summer Olympics, the winner of the gold medal in women's diving won with a score of 580.23. This was 45.9 points ahead of the silver medalist and the silver medalist was 47.04 points ahead of the bronze medalist. What was the score of the bronze medalist?

25. A counter top costs $84.95 per running foot. How much will it cost me if I need to cover 15 feet of counter space?

26. Death Valley has received about 1.6 inches of rain a year for the last 12 years. How much rain has it received in that time altogether?

27. A softball has a mass of 198.4 grams. A baseball has a mass of 148.8 grams. How much more mass does a softball have?

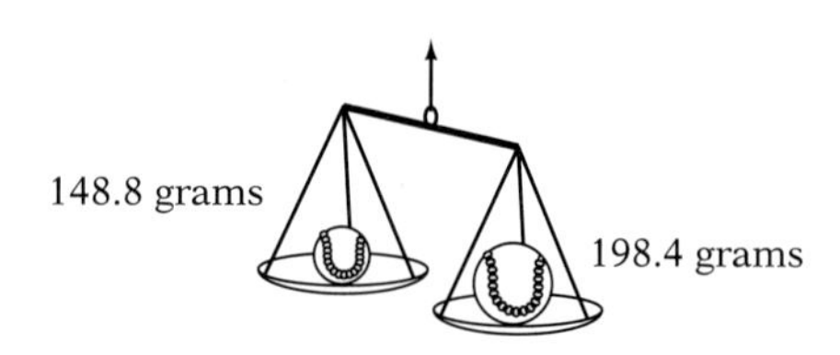

28. Six Flags Over Georgia, an amusement park, boasts the world's largest and tallest wooden roller coaster, the Scream Machine. It is 105 feet tall, and the cars travel 57 miles per hour (0.95 miles per minute). It takes about 1.5 minutes to complete one trip. How many miles long is the trip?

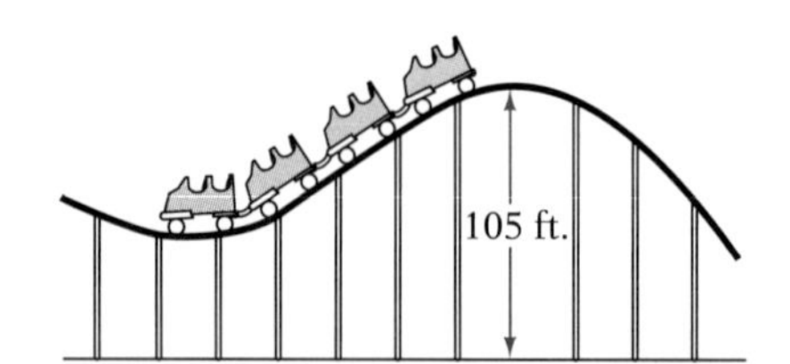

29. A tennis racket has a mass of 396.9 grams, and a badminton racket has a mass of 127.6 grams. How much mass do a dozen tennis rackets have?

30. An ice hockey player's stick is 134.6 centimeters long from the head of the shaft to the heel. A field hockey stick is 81.33 centimeters long. How much longer is the ice hockey stick?

31. During the Civil War, it was estimated that nearly one-third of all the currency in circulation was counterfeit. Approximately 1600 state banks were designing and printing their own money, and there were about 7000 different varieties of genuine bills. If there were 2 million bills in circulation, about how many were counterfeit? (*Hint:* Beware of unneeded information.) Use 0.333 as an estimate for $\frac{1}{3}$.

32. If a baseball player's run from home around the bases and back again is 109.7 meters and he runs 4 equal "legs" from base to base, how far is it between bases?

33. A horseshoe weighs 1.13 kilograms. How much does a set of four horseshoes weigh?

Use the following table to answer Exercises 34 and 35.

Animal	Weight
hedgehog	1.88 pounds
chinchilla	1.5 pounds
ferret	2.04 pounds
beaver	58.5 pounds
guinea pig	1.54 pounds
wild boar	302.5 pounds

34. What is the difference in weight between the heaviest animal and the lightest?

35. What is the total weight of the three lightest animals?

36. About 0.6 of your body weight is water. If you weigh 149 pounds, how many pounds are "water weight"?

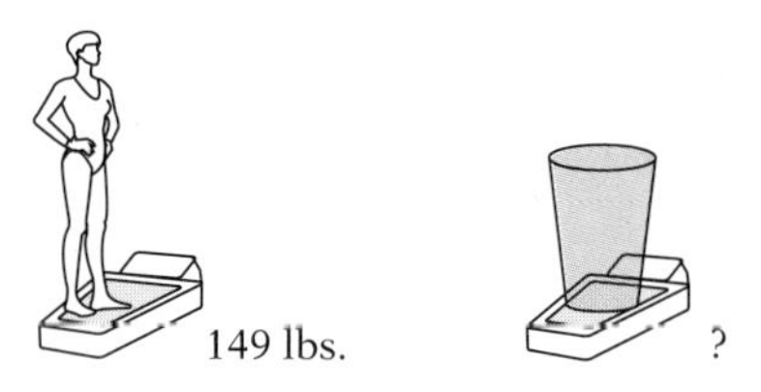

37. An ostrich has a normal body temperature of 102.58 degrees, and an owl has a normal body temperature of 104.4 degrees. How much hotter is the body temperature of the owl?

38. A certain Cadillac that was specially built in 1976 is 29.5 feet long. How many of these cars, parked end to end, would it take to span the length of the Humber Estuary suspension bridge, which is 4626 feet long?

Use the following information to answer Exercises 39 and 40.

In Kentucky, there are a lot of camping sites with hiking trails. Some, along with the miles of hiking trails available, are listed here.

Camping Site	Length of Trails
Levi Jackson State Park	8.5 miles
Jenny Wiley State Resort Park	7.2 miles
John James Audubon State Park	5.7 miles
Pine Mountain State Resort Park	9 miles

39. Two of the parks listed are State Resort Parks. Find the two that are *not* and determine their total number of miles of hiking trails.

40. Find the park with the greatest length of hiking trails and that with the least. Then compute the difference in their trail length.

6.7 MIXED PRACTICE

By doing these exercises, you will practice all the topics up to this point in this chapter.

41. The speed of a garden snail is approximately 0.03 mph. The speed of a spider is approximately 39 times as fast. What is the speed of the spider?

42. Subtract: $82.537 - 0.0888$

43. The three fastest times for the indoor mile were 3 minutes 49.78 seconds, 3 minutes 50.6 seconds, and 3 minutes 51.2 seconds. What is the difference between the fastest time and the third-fastest time?

44. Divide: $305.94 \div 0.003$

45. Find the total land area of these five cities in the United States.

Columbus, Ohio — 189.272 square miles
Honolulu, Hawaii — 617 square miles
Washington, D.C. — 68.25 square miles
Seattle, Washington — 144.6 square miles
Cleveland, Ohio — 79 square miles

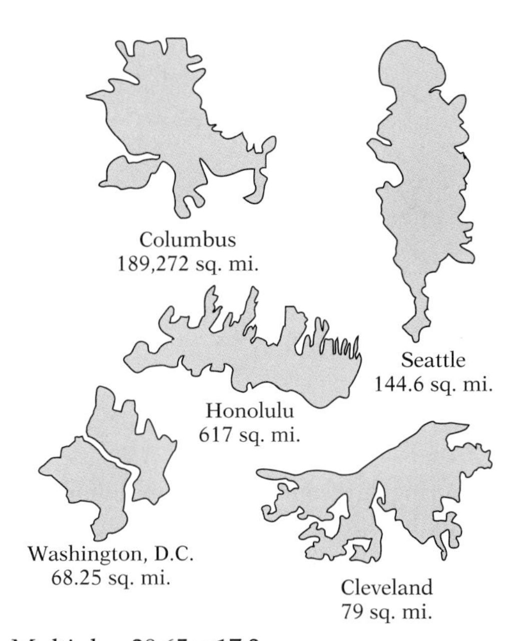

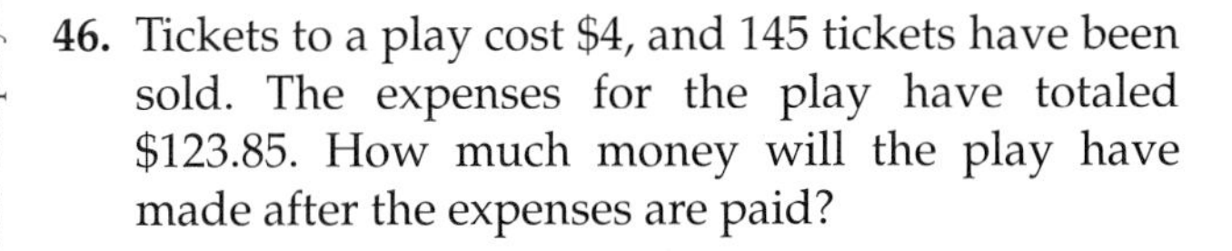

46. Tickets to a play cost \$4, and 145 tickets have been sold. The expenses for the play have totaled \$123.85. How much money will the play have made after the expenses are paid?

47. Multiply: 29.65×17.3

48. During a recent work stoppage, you were required to work the following hours of overtime:

Week 1	3.75 hours
Week 2	4.5 hours
Week 3	6.33 hours
Week 4	5 hours

How many hours did you work overtime that particular month?

49. Multiply: 0.005×34.76

50. The length of a football cannot be less than 10.875 inches nor greater than 11.4375 inches. What is the difference between the two limits?

51. Write in standard notation: $100 + 50 + 4 + \frac{6}{100} + \frac{3}{1000}$

52. Round 86.9499 to the nearest thousandth.

53. Multiply: 52.8×17.02

54. A measuring cup contains 360.5 milliliters (ml) of water to be used to dilute some paint for a preschool activity. If you need to pour some water into 40 cups, approximately how many milliliters of water should you pour in each cup? Round your answer to the nearest tenth of a milliliter.

6.8 Other Operations on Decimal Numbers

Section Goals

- To raise a decimal number to a power
- To use the standard order of operations on decimal numbers
- To convert fractions to decimal numbers

Raising to a Power

Decimal numbers can be raised to powers just as whole numbers and fractions can. Recall that in the exponential notation a^n, a is called the *base* and n is called the *exponent*. The exponent indicates how many times the base is used as a factor.

Example 1

Evaluate: $(0.3)^2$

SOLUTION

To raise 0.3 to the second power, we multiply it by itself.

$$0.3 \times 0.3 = 0.09$$

■ *Now do margin exercises 1a and 1b.*

1a. Rewrite $(1.05)^2$ without using exponents.

1b. Rewrite $(2.031)^2$ without using exponents.

Here's another definition that extends to decimal numbers.

> For a decimal number a not equal to zero, a^0 is defined as 1.

So, for example, $(0.5)^0$ is equal to 1.

Standard Order of Operations

The standard order of operations is another concept that applies to decimal numbers as well as whole numbers and fractions. Here it is, repeated briefly:

> **Standard Order of Operations**
>
> 1. Simplify within parentheses.
> 2. Remove exponents.
> 3. Multiply and divide, from left to right.
> 4. Add and subtract, from left to right.

2a. Simplify:
$[(11)^2 - (6.2 - 3)] + (2.3)^2 \times 6$

2b. Use your calculator to simplify:
$\{3[4.5 - 1.8 + (0.6)]^2 - 8.07\}^2$

Let's use the standard order of operations to simplify an expression involving decimal numbers.

EXAMPLE 2

Simplify:

$$3 \times \{[(1.2 - 0.7)^2 - (0.2)^3] \div 6^0\} - 0.2$$

SOLUTION

Using the standard order of operations, we work inside the parentheses first.

$$3 \times \{[(1.2 - 0.7)^2 - (0.2)^3] \div 6^0\} - 0.2$$

$$3 \times \{[(0.5)^2 - (0.2)^3] \div 6^0\} - 0.2 \qquad \text{Subtracting}$$

Next we simplify the exponents and remove the parentheses.

$$3 \times \{[(0.5)^2 - (0.2)^3] \div 6^0\} - 0.2$$

$$3 \times \{[0.25 - 0.008] \div 1\} - 0.2 \qquad \text{Clearing exponents}$$

$$3 \times \{0.242 \div 1\} - 0.2 \qquad \text{Subtracting}$$

$$3 \times 0.242 - 0.2 \qquad \text{Dividing}$$

$$0.726 - 0.2 \qquad \text{Multiplying}$$

$$0.526 \qquad \text{Subtracting}$$

So $3 \times \{[(1.2 - 0.7)^2 - (0.2)^3] \div 6^0\} - 0.2$ is 0.526.

■ *Now do margin exercises 2a and 2b.*

Converting Fractions to Decimal Numbers

We know that $\frac{1}{2}$ and $\frac{5}{10}$ are equivalent fractions. But 0.5 is also a name for the same fraction—in decimal form. We use division to convert fractions to decimal numbers.

3a. Write the decimal name for $\frac{3}{8}$.

3b. Write the decimal name for $8\frac{1}{4}$.

EXAMPLE 3

Write $\frac{1}{8}$ in decimal form.

SOLUTION

First, we write the fraction in long-division form, placing a decimal point after the 1 and adding several zeros.

$$8\overline{)1.000}$$

Then we divide.

$$\begin{array}{r} .125 \\ 8\overline{)1.000} \\ \underline{8} \\ 20 \\ \underline{16} \\ 40 \\ \underline{40} \end{array}$$

So $\frac{1}{8}$ is equivalent to 0.125.

■ *Now do margin exercises 3a and 3b.*

4a. Rewrite $\frac{5}{6}$ as a decimal number to three decimal places.

4b. Rewrite $\frac{9}{11}$ as a decimal number to 4 decimal places.

Sometimes, the result of the division is a repeating decimal—that is, a decimal part whose digits repeat no matter how many places we carry it to. Examples include $3\overline{)1.000} = 0.33333\ldots$ and $11\overline{)12.00} = 1.090909\ldots$.

EXAMPLE 4

Rewrite $\frac{2}{3}$ as a decimal number to three decimal places.

SOLUTION

Following the procedure of Example 3, we divide to four decimal places:

$$
\begin{array}{r}
.6666 \\
3\overline{)2.0000} \\
\underline{1\,8} \\
20 \\
\underline{18} \\
20 \\
\underline{18} \\
20 \\
\underline{18} \\
2
\end{array}
$$

The quotient is a repeating decimal that we can write either as $0.666\ldots$ or as $0.66\overline{6}$. The bar indicates that the digit or digits beneath the bar repeat.

Rounded to three places, the quotient is 0.667.

■ *Now do margin exercises 4a and 4b.*

Work the exercises that follow.

Name Section Date

6.8 EXERCISES

In Exercises 1 through 12, rewrite each expression without exponents.

1. $(1.2)^2$
2. $(1.2)^3$
3. $(0.00071)^0$
4. $(0.25)^2$
5. $(0.38)^0$
6. $(0.057)^3$
7. $(0.014)^3$
8. $(0.035)^2$
9. $(0.23)^2$
10. $(0.0007)^4$
11. $(0.0005)^3$
12. $(1.3)^2$

In Exercises 13 through 28, simplify each expression.

13. $(8 + 1.4 \times 3)^2$
14. $26 + 1.8 \times 6$
15. $(3.5 + 1.7) - 0.007 \times 5$
16. $[(4 + 0.8)^2 + (1.6)]^2 - 15^0$
17. $\{[(1.1 + 5) \times 1.4]^2 - 2\}5$
18. $[(1.5 + 3)^2 \times 1.3 - 11]^2$
19. $4[(3.3 + 1.3 \times 2)^2 - 1.7]$
20. $(1.1^2 \times 6 + 15 \div 1.5)^2$
21. $\{[(3 \times 0.5)^2 + 4] \div 1.2 - 5\}^0$
22. $7[(1.2 \times 4) + 1.6 \div (0.4)^2]$
23. $(5.1 \times 2 + 2.7^2) \div 0.1^3$
24. $2.9 + 1.4 \times 2^5 - 0.5 \times 3^4$
25. $8[(1.3)^2 + 0.9 \times (3.7 + 16)^0]$
26. $\{[2.2 + 3 \times (1.2)^3 + 5] \div 5\}^2$
27. $\{0.17 + [0.4 \times 9 + (1.4)^2]\}^2$
28. $[(0.9 + 1.6 \times 2)^2 \times 0.1^5]^2$

In Exercises 29 through 44, convert each fraction to decimal numbers. For repeating decimal parts, round to the nearest ten-thousandth.

29. $\frac{1}{2}$
30. $\frac{3}{4}$
31. $\frac{1}{9}$
32. $\frac{4}{9}$
33. $\frac{1}{6}$
34. $\frac{1}{5}$
35. $\frac{3}{5}$
36. $\frac{1}{7}$
37. $3\frac{3}{5}$
38. $5\frac{5}{7}$
39. $16\frac{1}{3}$
40. $38\frac{1}{2}$
41. $7\frac{9}{11}$
42. $19\frac{3}{100}$
43. $102\frac{3}{4}$
44. $178\frac{3}{8}$

6.8 MIXED PRACTICE

By doing these exercises, you will practice all the topics up to this point in this chapter.

45. Your annual salary is \$25,603.67. Your overtime rate is computed by dividing your annual salary by 1500. What is your overtime rate to the nearest cent?

46. Simplify: $(3.4^2 - 0.006) + 2.57 \times (3.1)^2$

47. The price on a sweater is \$45. The tax is found by multiplying the cost of the sweater by 0.0825. What is the tax on the sweater?

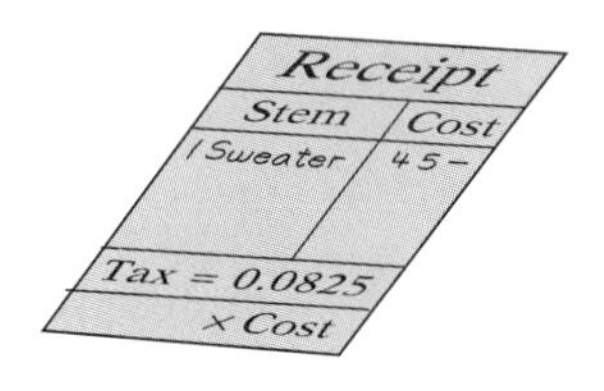

48. Write in standard notation: $30{,}000 + 600 + \frac{5}{10{,}000}$

49. Rewrite $(0.3)^4$ without exponents.

50. Add $0.0008 + 37.12 + 46.805$ and round your answer to the nearest thousandth.

51. Divide: $6.39 \div 0.006$

52. Convert $\frac{2}{11}$ to a decimal rounded to the nearest hundredth.

53. Simplify: $9.05 \times 2.1 - 3.6 + 4.9$

54. Multiply: 36.8×0.125

55. Simplify: $[(11.1)^2 - 2.4 \times 1.03 + 5] \div 2.5$

56. Five friends split a \$25 dinner bill evenly, so that each paid \$5 (plus tax and tip). The costs of four of the dinners were \$3.50, \$6, \$4.90, and \$7. What was the cost of the fifth?

"The Diner" Bill	
5 in party	
1 platter..............	\$7.00
1 salad................	\$3.50
1 daily special....	\$6.00
1 fish..................	\$4.90
1 hummus/pita..	
Total	\$25.00

6.9 Geometry and Measurement: Applying Operations on Decimal Numbers

SECTION GOAL

- To find the area and the circumference of a circle

Up until this point in the geometry sections, we have been dealing with polygons and their corresponding solids. One characteristic of a polygon is that it has only straight sides. The figure we will discuss in this section—the circle—has no straight sides at all.

Circles

Circle

A **circle** is the set of points that are equidistant from a fixed point called the **center.**

This means that any point on a circle is the same distance from the center as any other point. That distance is the measure of the radius.

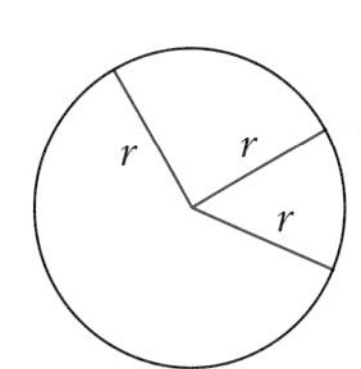

Three radii are shown in the accompanying figure.

Radius

A **radius** (plural: radii) is a line segment drawn from the center of a circle to any point on the circle.

Diameter

A **diameter** is a line segment drawn from one point on a circle through the center to another point on the circle.

In the next figure, we can see that a diameter has measure equivalent to that of two radii of the same circle. Similarly, the radius of a circle is equal in measure to one-half that of the diameter. In symbols,

If a circle has radius of measure r and diameter of measure d, then

$$r = \frac{d}{2} \quad \text{and} \quad d = 2r$$

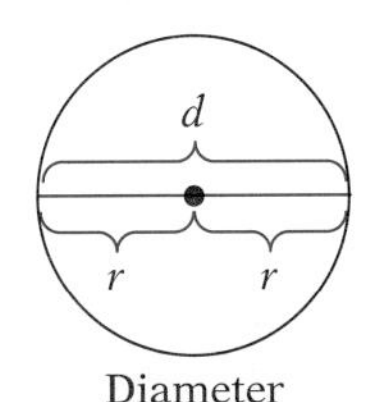

Diameter

1a. Find the circumference of a clock that has a diameter of 11.5 inches. ($\pi \approx 3.14$)

1b. What is the circumference of a circle with radius of 0.901 meter? ($\pi \approx 3.14$)

Circumferences of Circles

Circumference

The **circumference** of a circle is the distance around the circle—its perimeter.

To calculate the circumference C, we use the fact that $C/d = \pi$ (the Greek letter pi) for any circle. This fact gives us the following formulas.

To find the circumference C of a circle

If d is the diameter or r is the radius of the circle, then

$$C = \pi d \text{ or } C = 2\pi r$$

π is a constant—a fixed number like 7 or 342—but its value cannot be determined exactly. To ten decimal places, it is 3.1415926536. This is usually approximated as 3.14 or $\frac{22}{7}$.

Circumferences have units of length—that is, feet, inches, miles, and such.

EXAMPLE 1

A circular vegetable patch is being planted and enclosed by a fence. If the patch is 14 feet in diameter, how much fence will be needed to enclose it? (Use 3.14 as an estimate for π.)

SOLUTION

Here the value of d, the diameter, is known. We first substitute in the formula $C = \pi d$ and then simplify.

$$\begin{aligned} C &= (3.14)(14) \\ &= 43.96 \end{aligned}$$

So 43.96 feet of fence will be needed.

■ *Now do margin exercises 1a and 1b.*

Here are some related formulas that may be useful in solving the problems at the end of this section.

If C is the circumference of a circle, then its diameter is

$$d = \frac{C}{\pi}$$

and its radius is

$$r = \frac{C}{2\pi}$$

Areas of Circles

The formula for the area of a circle also includes the constant π.

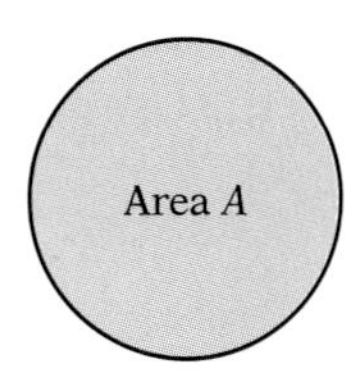

To find the area A of a circle

If r is the radius of the circle, then

$$A = \pi r^2$$

Areas are measured in square units, as we saw in Section 2.7.

EXAMPLE 2

A round cake pan measures $10\frac{1}{2}$ inches in diameter. What is the area of the cake pan? (Use $\frac{22}{7}$ to approximate π.)

SOLUTION

To find the area, we need to use the formula $A = \pi r^2$. We know π, but we don't know the radius r. However, we do know that the diameter d is $10\frac{1}{2}$ and that $r = d/2$. So

$$r = d/2 = (10\tfrac{1}{2})/2$$
$$= \frac{21}{2} \div 2 = \frac{21}{2} \times \frac{1}{2} = \frac{21}{4}$$

We save a little work by leaving the radius in the form of an improper fraction. We can now substitute in the area formula.

$$A = \pi r^2 = \left(\frac{22}{7}\right)\left(\frac{21}{4}\right)^2$$

Using the standard order of operations, we find that

$$A = \left(\frac{22}{7}\right)\left(\frac{441}{16}\right) \quad \text{Squaring}$$

$$= \frac{\overset{11}{\cancel{22}}}{\underset{1}{\cancel{7}}} \times \frac{\overset{63}{\cancel{441}}}{\underset{8}{\cancel{16}}} = \frac{693}{8} = 86\frac{5}{8} \quad \text{Reducing}$$

The area of the cake pan is $86\frac{5}{8}$ square inches.

■ *Now do margin exercises 2a and 2b.*

2a. Find the area of a circle that has a radius of 2.2 inches. ($\pi \approx 3.14$)

2b. A round tablecloth measures 6.5 feet in diameter. What is the area of this cloth? ($\pi \approx 3.14$)

3a. How does the area of a circle change when its radius is tripled?

3b. Can a circle have a radius of 3.4 feet and an area of 28.9 square feet? Explain. (Use $\pi \approx 3.14$)

EXAMPLE 3

A circle has a radius of 1 inch. How does its area change if its radius is doubled? (Use $\pi = 3.14$.)

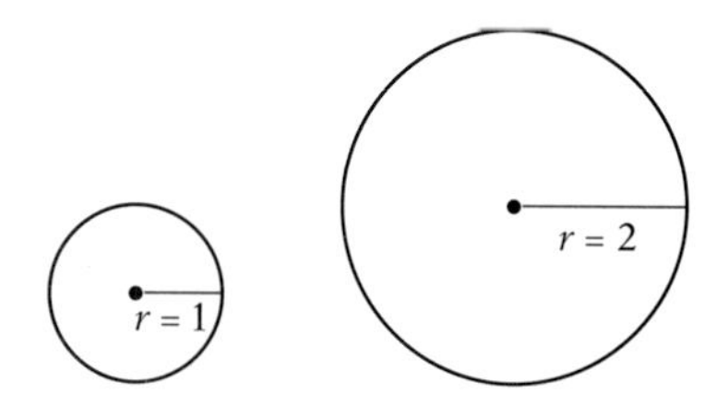

SOLUTION

The area of a circle with radius 1 is

$$A = \pi r^2 = (3.14)(1^2)$$
$$= 3.14 \text{ square inches}$$

The area of a circle with radius 2 is

$$A = \pi r^2 = (3.14)(2^2)$$
$$= (3.14)(4) \text{ square inches}$$

The area of the second circle is 4 times greater. When the radius of a circle is doubled, its area is quadrupled!

■ *Now do margin exercises 3a and 3b.*

Work the exercises that follow.

6.9 EXERCISES

Use the formulas you learned in this section to work the following problems. If you need an estimate for pi, use either 3.14 or $\frac{22}{7}$.

1. A pumpkin decoration for Thanksgiving has a radius of 7.39 inches. What is the approximate circumference of the pumpkin?

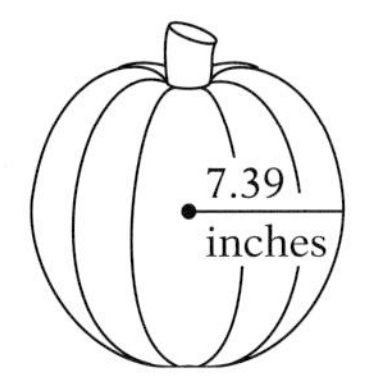

2. When baseball was first played, a regulation baseball had a diameter of $2\frac{3}{84}$ inches. Find the radius of a circle with the same diameter.

3. Find the radius of a quarter if its diameter is $\frac{11}{12}$ inches.

4. Find the largest possible size of a second hand for a clock that has a diameter of 14.75 inches.

5. The diameter of the top of a round lampshade is 22.3 inches. What is its radius?

6. Find the radius of a circle with diameter 13.25 feet.

7. The diameter of the Moon is approximately 2160 miles. If a space vehicle were to move in a straight path all the way around the Moon, how far would it travel?

8. The widest point on a baseball bat can be no larger than 2.75 inches in diameter. What is the largest that the circumference can be?

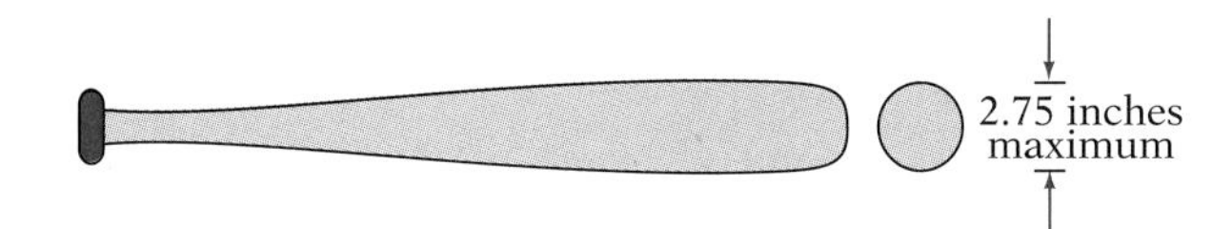

9. The largest sunspot recorded was 124,274 miles across. What was the radius of this sunspot?

10. Find the diameter of a dinner plate that has a radius of 11.7 inches.

11. Find the diameter of a circle if the radius is 34.67 yards.

In Exercises 12 through 17, find the area that each of the following coins would take up lying flat on a table.

12. a quarter: diameter 0.95 inch

13. a nickel: radius 0.42 inch

14. a half-dollar: diameter 1.24 inches

15. a dime: radius 0.35 inch

16. a Diamond Jubilee New York City subway token: diameter 0.90 inch

17. A penny has a diameter of 0.748 inch. What is the area of one side?

18. One of the first bicycles made in 1876, had a front wheel that measured 5.19 feet in diameter. What was the radius of the wheel?

19. A computer diskette has a radius of 2.625 inches, and it is enclosed in a square paper envelope that has a side equivalent to the diameter of the diskette. Find the area of the top or bottom of the square envelope.

20. Find the area of a circular mirror with a 14.5-inch radius.

21. The Earth's equator is a circle with a radius of 3963.49 miles. What is its diameter?

22. A flat Christmas decoration is circular in shape. If the diameter is 14.8 inches, what is the amount of ribbon that is needed to go around the decoration?

23. The actual wrestling area in a wrestling ring has a diameter of 29.5 feet. What does this area measure?

24. A regulation basketball hoop measures 18 inches in diameter. What is the measure of the distance around the rim of the hoop? Use $\frac{22}{7}$ as an estimate for pi in your calculations.

25. Find the radius of an embroidery hoop that has a diameter of 12.675 inches.

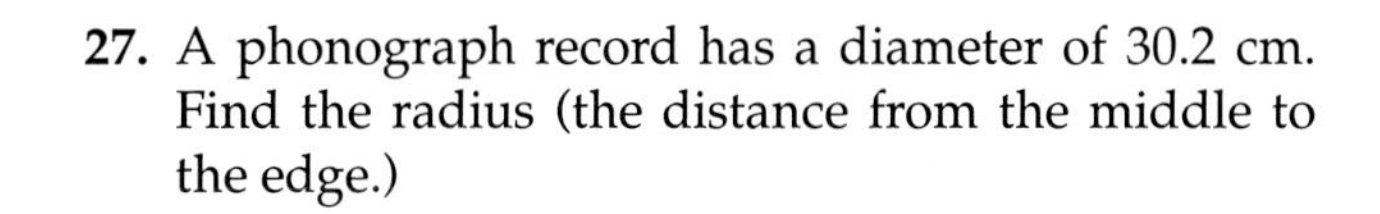

26. The largest magnet on record is in Russia and has a diameter of 196.1 feet. What is the circumference of this magnet?

27. A phonograph record has a diameter of 30.2 cm. Find the radius (the distance from the middle to the edge.)

28. One of the first bicycles, made in 1876, had a front wheel that measured 5.7 feet in diameter. What was the circumference of the wheel?

29. How does the square of the radius of a circle (that is, r^2) change when the area is doubled? (Use the fact that $r^2 = \frac{A}{\pi}$. Consider circles with areas of 31.4 and 62.8 square units.)

30. How does the circumference of a circle change when the diameter is doubled?

31. How does the circumference of a circle change when the radius is doubled?

32. How does the area of a circle change when the diameter is doubled?

6.9 MIXED PRACTICE

By doing these exercises, you will practice all the topics up to this point in this chapter.

33. A piece of lumber that measures 24.7 centimeters (cm) long is being cut from a larger piece that measures 40 cm in length. How much of the piece of lumber will be left?

34. Divide: $13\overline{)26.39}$

35. The radius of a circle is 4.5 cm. Find the area of the circle.

36. Business expenses for one month were as follows:

Heat	$100.56
Rent	$1251.85
Supplies	$98

What were the total expenses for the month?

37. Multiply: 10.4×3.86

38. Convert 9.0208 to a fraction.

39. The smallest that a regulation golf ball can be in diameter is 1.68 inches, and the smallest that a tennis ball can be in diameter is 2.625 inches. Find the difference between these limits for the two balls.

40. Write 236.07 in words.

41. Find the area of a circle with circumference 11.932 inches. (Use $\pi = 3.14$.)

42. Find the area of a plate with radius 4.3 inches.

43. Martin's dog Thor is 1.2 years old. Martin is 22.5 times the age of his dog. How old is Martin?

44. Find the difference: $5235.7 - 1235.671$

45. A book is 0.75 times as wide as it is long. If it is 15.36 cm long, how wide is it?

46. Simplify: $[14.5 - (2.3)^2](6) + 0.7 \div 4 - 2^2$

An Application to Statistics: Finding the Mean, Median, and Mode

Finding the Mean

An important concept that is used in calculating grades, in controlling the quality of products, and in many other situations is the "average," or **mean**, of a set of measurements. The average of a set of numbers is one of three measures that indicate where the "center" of the set is. The others are the median and the mode. We will discuss all three in this section. The mean can be thought of as the balance point in a set of measurements.

To find the mean of a set of *n* numbers

1. Add the numbers together.
2. Divide by *n*.

Let's look at an example.

EXAMPLE 1

A man measures the heights of some shrubs in front of his house. They measure 33.4 inches, 35.2 inches, 42.5 inches, and 38.1 inches. What is the average height?

SOLUTION

Here $n = 4$, because there are four measurements. To find the mean, we add all the measurements and then divide that total by 4.

$$\text{Add: } \begin{array}{r} 33.4 \\ 35.2 \\ 42.5 \\ \underline{38.1} \\ 149.2 \end{array} \qquad \text{Divide by 4: } \begin{array}{r} 37.3 \\ 4\overline{)149.2} \\ \underline{12} \\ 29 \\ \underline{28} \\ 12 \\ \underline{12} \\ 0 \end{array}$$

The average of the four heights is 37.3 inches.

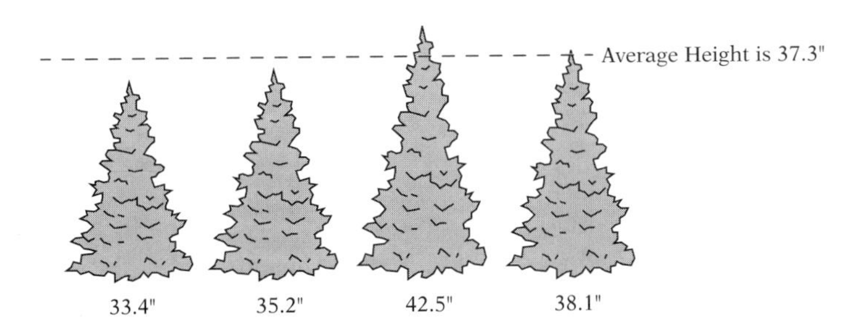

■ *Now do margin exercises 1a and 1b.*

1a. A boy measures the length of some fishing poles he has collected. They measure 73.2 inches, 85.8 inches, 72.6 inches, and 90 inches. What is the mean length?

1b. Rainfall in an area measures 5.02 inches, 1.3 inches, 6.12 inches, 5.01 inches, and 3 inches over a five-month span. What is the mean rainfall during these five months?

2a. What is the median of 1, 87, 32, 46, and 0?

2b. What is the median of 1989, 1943, 1933, 2001, 27, 3, and 1842?

Finding the Median

The **median** of a set of numbers is the "central number" of that set.

To find the median of a set of *n* numbers

1. Arrange the numbers in order.
2. Find the "central number" of those numbers. The "central number" is the number such that there are as many values greater than it as there are values less than it.

EXAMPLE 2

Find the median of 43, 39, 47, 26, and 28.

SOLUTION

Arrange the numbers in order.

26 28 39 43 47

The central number is 39, because there are 2 numbers on either side of it.

So 39 is the median.

■ *Now do margin exercises 2a and 2b.*

Finding the median is not so easy when we have an even number of values.

EXAMPLE 3

A child cuts some string in the following lengths: 43.1 inches, 47.2 inches, 42 inches, and 48.5 inches. What is the median length?

SOLUTION

To find the median, we first write the numbers in order:

$$42 \quad 43.1 \quad 47.2 \quad 48.5$$

There is no "central number" because there are four numbers. To find the median in such a case, we must find the average of the two middle numbers—here 43.1 and 47.2. We add them and divide their sum by 2.

$$\frac{43.1 + 47.2}{2} = \frac{90.3}{2} = 45.15$$

So 45.15 inches is the median length.

■ *Now do margin exercises 3a and 3b.*

3a. The heights of certain fence posts are 73.4 inches, 85.2 inches, 72.5 inches, and 90.1 inches. What is the median height?

3b. Snowfalls in the New York metropolitan area measured 6.02 inches, 6 inches, 4.05 inches, and 3.1 inches. What was the median snowfall?

Finding the Mode

Our third measure, the **mode,** is also relatively easy to find. It is the value that occurs most frequently in a set of numbers. A given set of numbers may have one mode, no mode, or many modes.

To find the mode of a set of *n* numbers

1. Determine how many times each number appears in the set (by tallying)
2. The mode is the number (or numbers) that appears (or appear) most.

4a. Find the mode of the following set of numbers:
1, 3, 2, 6, 4, 7, 7, 4, 8, 2, 3, 2, 2, 4

4b. Find the mode of the following set of numbers:
0, 0, 1, 4, 7, 3, 5, 9, 2, 7, 1, 10, 2, 3, 1, 2, 6

EXAMPLE 4

Find the mode of the following numbers:

1, 3, 2, 4, 5, 3, 7, 2, 3, 7, 3

SOLUTION

First we write down all the numbers in order. Then we tally how many of each there are.

	1	/
→	2	//
	3	////
	4	/
	5	/
	7	//

Because the number 3 occurs four times, and no other number occurs as many times, 3 is the mode of this set of numbers.

■ *Now do margin exercises 4a and 4b.*

Work the exercises that follow.

EXERCISES

1. The lengths of the spans of the five longest steel arch bridges in the world are given in the following table.

New River Gorge	1699.58 feet
Bayonne	1625.4 feet
Sydney Harbour	1649.66 feet
Fremont	1254.9 feet
Port Mann	1199.95 feet

What are the mean and median lengths for these five steel arch bridges?

2. The lengths of three of the longest canals in the world are as follows: White Sea, 141.3 miles; Suez, 100.25 miles; and Volga, 61.75 miles. What is the mean for these canal lengths?

3. The number of electoral votes for Bush in the 1988 Presidential Election are listed here for 10 states. What is the median number of electoral votes for Bush for these 10 states? What is the mode?

47, 21, 12, 24, 20, 7, 11, 25, 29, 11

4. The last six Apollo missions had durations of $245\frac{1}{3}$ hours, 143 hours, $215\frac{3}{4}$ hours, $295\frac{1}{4}$ hours, $266\frac{1}{3}$ hours, and $301\frac{1}{2}$ hours. What was the median duration of these six missions?

5. In a certain six-year period, the following numbers of people have finished the New York City marathon: 13,599, 14,546, 14,492, 15,887, 19,689, and 21,244. What was the median number of finishers during these six years? What was the mean?

6. In the first eight years of the running of the New York City marathon, there were the following numbers of women finishers: none, 4, 2, 5, 9, 36, 63, 184. What was the median number of women that finished in these eight years? What was the mean?

7. On six of the entries in the Tall Ship race in New York City on July 4, 1976, the numbers of crew members were as follows: 104, 189, 99, 162, 236, and 16. What was the median number of crew members on these six ships?

Number of Crew Members on each ship

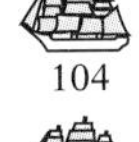

8. The total number of miles Buzzie drove for each of six weeks were as follows: 25.4 miles, 33.2 miles, 28.7 miles, 27.9 miles, 31.8 miles, and 42.2 miles. What was the median number of miles for the 6 weeks? What was the mean?

Use the following information for Exercises 9 through 12.

Between any two fractions on a number line, there is another fraction that is halfway between them. We call this property **betweenness**. Use the idea of the mean to find the fraction that lies halfway between each of the following pairs of fractions.

9. $\frac{1}{5}$ and $\frac{1}{125}$

10. $\frac{1}{7}$ and $\frac{1}{8}$

11. $\frac{1}{100}$ and $\frac{1}{10}$

12. $\frac{11}{13}$ and $\frac{12}{13}$

Use the following information for Exercises 13 through 16.

This same property (betweenness) holds for decimal fractions. By finding the mean of the pair, find the decimal fraction that lies halfway between each of the following pairs of decimal fractions.

13. 0.21 and 0.22

14. 0.004 and 0.005

15. 0.231 and 0.232

16. 0.10097 and 0.10098

CHAPTER 6 REVIEW

ERROR ANALYSIS

These problems have been worked incorrectly. Tell what the error is and then write the correct solution.

1. Add: 0.7 + 0.7

Incorrect Solution *Correct Solution*

$$\begin{array}{r} 0.7 \\ +0.7 \\ \hline 0.14 \end{array}$$

Error ______________________

____

2. Add: 6.3 + 7 + 94.8

Incorrect Solution *Correct Solution*

$$\begin{array}{r} 6.3 \\ .7 \\ +94.8 \\ \hline 101.8 \end{array}$$

Error ______________________

____

3. Multiply: 4.35 × 2.3

Incorrect Solution *Correct Solution*

$$\begin{array}{r} 4.35 \\ \times 2.3 \\ \hline 1305 \\ 870 \\ \hline 100.05 \end{array}$$

Error ______________________

____

4. Write $2\frac{1}{8}$ as a decimal.

Incorrect Solution *Correct Solution*

2.18

Error ______________________

____

5. Subtract: 100 – 6.84

Incorrect Solution

$$\begin{array}{r} 100 \\ -\ \ 6.84 \\ \hline 94.84 \end{array}$$

Correct Solution

Error ______________________________

____

6. Divide: 23.5 ÷ 7

Incorrect Solution

$$\begin{array}{r} 3.34 \\ 7\overline{)23.5} \\ \underline{21} \\ 25 \\ \underline{21} \\ 4 \end{array}$$

Correct Solution

Error ______________________________

____

INTERPRETING MATHEMATICS

By working these exercises, you will test and strengthen your mathematics vocabulary.

1. Give an example of a number that contains a decimal point and a decimal part. Indicate which is which.

2. What is the difference between a decimal part and a fractional part of a number?

3. Define the term circle.

4. In the circles that follow, indicate each radius, diameter, and circumference.

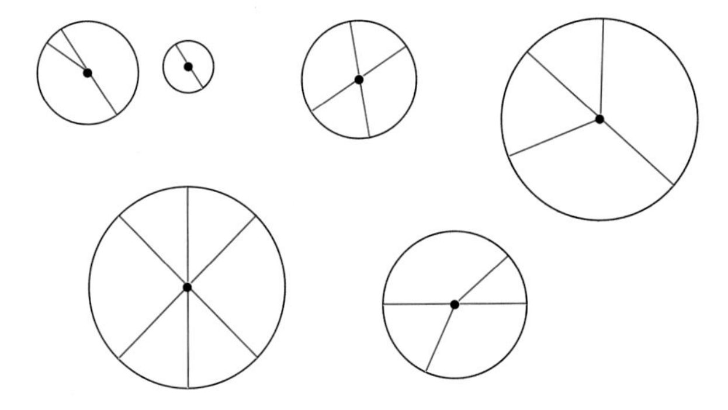

5. How are the circumference of a circle, the circle's diameter, and the value π (pi) related?

6. In words, explain how you find the area of a circle. Also explain why the answer is given in square units even though you are working with a round circle.

REVIEW PROBLEMS

The exercises that follow will give you a good review of the material presented in this chapter. Work through them and check your answers at the back of the book. Remember to write all answers in simplest form.

Section 6.1

Express in expanded notation:

1. 6.003
2. 17.096

Write in numerals:

3. eighteen and seven hundred twenty-five ten-thousandths
4. one hundred thirty-seven and six hundredths

In exercises 5 and 6, give the place value of all the non-zero digits:

5. 20.13050
6. 102.00304

Section 6.2

7. Round 24.5 to the nearest whole number.
8. In 1920 there was $5,698,214,612 in circulation. How many dollars is this rounded to the nearest million?
9. Round 18.4999651 to the nearest hundredth, thousandth, ten-thousandth, and hundred-thousandth.
10. Arrange 9.002, 9.040001, and 9.03 in order from smallest to largest.
11. Round 17.604949498 to the nearest hundredth, thousandth, ten-thousandth, hundred-thousandth, and millionth.
12. Arrange 14.009, 14.0096, and 14.0806 in order from smallest to largest.

Section 6.3

Add:

13. 14.96 + 18 + 5.03

14. 27 + 0.005 + 12.04

15. 4162 + 44.09 + 35.6 + 0.3

16. 88,923 + 13.0402 + 128.3

17. 29 + 56.068 + 6 + 2.5

18. 24.5 + 2.010 + 123.45708

Section 6.4

Subtract:

19. 85.74 – 27.69

20. 230.9 – 11.86

21. 63 – 18.42

22. Subtract 9.83 from one thousand.

23. From 576.85 subtract 108.0709

24. 10.263 – 0.99003

Section 6.5

Multiply:

25. 9.18×0.006

26. 1.04×25.3

27. 0.05×64

28. 0.0009×2.7

29. $11{,}044 \times 1.1$

30. 0.3728×0.99

Section 6.6

31. Divide: $2.232 \div 2.4$

32. Divide: 12.96 by 0.12

33. Find the quotient: $392 \div 1.4$

34. Divide 76.209 by 0.004

35. 1020.45 divided by 0.0625 is what?

36. What is the quotient of 2.5 divided by 6.25?

Section 6.7

37. If Earth has a density of 1, Mars has a density of 0.72, and Saturn has a density of 0.13, how many times more dense is Mars than Saturn? Round to the nearest thousandth.

38. The welcome home ticker-tape parade for the Desert Storm troops attracted 3.5 million spectators. They were spread somewhat evenly along 18 blocks in the financial district of New York City. About how many spectators were there per block? Give your answer to the nearest 100 spectators.

39. Blood flows through the kidneys at a rate of 1.3 liters a minute. At that rate, how many liters will flow through the kidneys in an hour?

40. Lucy and Billy usually walk together. Their walks are about 4 miles long and take about 45 minutes (0.75 hour). How fast do they walk? (Use the formula $r = D/t$, where D is distance in miles and t is time in hours. Your answer will be in miles per hour.)

41. A certain rollercoaster ride lasts 1 minute 50 seconds. If the entire ride is 3250 feet long, what is the average speed in feet per second? (Use the formula $r = D/t$, where D is distance in feet and t is time in seconds. Your answer will be in feet per second.) Round to the nearest tenth.

42. If Kirsten and Najib each wrote 540 problems for mathematics exams, and each problem took about 3.5 minutes to research and compose, how many hours did they spend altogether on these problems?

Section 6.8

43. Simplify: $3.14 + 29.8 \times 2 - (4.1)^2$

44. Change $\frac{5}{16}$ to a decimal.

45. Rewrite 6.2^3 without exponents.

46. Change $\frac{2}{7}$ to a decimal, rounded to the nearest ten-thousandth.

47. Change $\frac{3}{5}$ to a decimal.

48. Simplify: $0.1^3 - 4.5 \times 0.0002 + 2 \div 0.004^2$

Section 6.9

49. Find the area and circumference of a circle with diameter 4.75 feet. Round to the nearest tenth.

50. Find the area and circumference of a circle with radius 4.5 inches.

51. In a regulation wrestling ring, the diameter is 29.5 feet. What is the circumference of the ring?

52. Find the area of a circular platter with a radius of 10 inches.

53. A man runs around a circular track 5 times. The distance across the track through the middle is 1000 feet. How far did he run?

54. One circle has an area of 25 square inches. Another has a diameter of 4.5 inches. Which circle has the larger area? How much larger is it to the nearest tenth of a square inch?

Mixed Review

55. Add: 55.2 + 5.847 + 43.6846

56. Add: 61 + 0.005 + 57.06

57. Write 3062.04 in expanded notation.

58. Your cat weighed 0.25 pounds at birth. She now weighs 9.85 pounds. How much has she gained in her 11 years?

59. The amount of tax on a certain item is found by multiplying the price by 0.0425. Find the total cost of an item marked $35.

60. In a 16-ounce jar of peaches, there are peaches and liquid. If the label on the jar says that there are 5.75 ounces of liquid, how many ounces of peaches are there in the jar?

Peaches =
Liquid = 5.75 oz.
16 oz. total

61. An astronaut can circle the earth in a rocket ship in 1.5 hours. How many times can the astronaut circle the earth in 24 hours?

62. Black Creek in Mississippi ranges in width from 6.67 yards to 33.33 yards. What is the difference between these two widths?

63. Okatoma Creek in Mississippi is from 0.5 to 10 feet deep. What is the difference between these two depths?

64. To score, a baseball player must run 109.7 meters. A softball player runs 73.2 meters. How much farther does a baseball player run?

65. Hotel employees earn $176.90 for 30.5 hours' work. Radio employees earn $374.71 for 37.1 hours' work. Telephone employees earn $501.42 for 41.1 hours' work.

a. Which of these three types of employees earns the greatest salary per hour?

b. How much would each earn for 40 hours' work? Use your results from part (a).

66. It is estimated that the human heart beats once every 0.8 seconds. How many heart beats occur in 24 hours?

67. The following table gives the total annual precipitation of various regions in Mississippi.

Upper Delta	50.9 inches
North Central	52.9 inches
Lower Delta	51.4 inches
East Central	52.4 inches
South Central	57.4 inches
Central	52.3 inches

a. What is the total of these rainfall amounts?

b. What is the difference in rainfall between the region with the highest total and the region with the lowest total?

68. The maximum initial speed of a golf ball is 170.5 miles per hour. How far could it go in a second? (*Hint:* Divide by 3600, the number of seconds in an hour, and round to the nearest hundredth.)

69. The distance between Los Angeles and Quebec, Canada, is 2579 miles. How many hours would it take for a steam locomotive to make the trip if it traveled at 125.5 miles per hour? Round to the nearest tenth of an hour.

71. A spiny anteater has a normal body temperature of 73.9°F, and a goat has a normal body temperature of 103.8°F. How much higher is the goat's temperature?

70. The tallest skyscrapers in the United States are: in Chicago—the Sears Tower (1454 feet tall, 110 stories), the Standard Oil Building (1136 feet tall, 80 stories), and the John Hancock Center (1127 feet tall, 100 stories); and in New York—The World Trade Center (1350 feet tall, 110 stories) and the Empire State Building (1250 feet tall, 102 stories).

a. Which one of these buildings allows the largest number of feet per story? (Use estimation to simplify your work.)

b. What is the average number of stories in the listed buildings?

Chapter 6 Test

This exam tests your knowledge of the topics in Chapter 6. Express answers in simplest form.

1. a. Write 254.098 in expanded notation.

   b. Write the number represented by $300 + 4 + \frac{3}{10} + \frac{8}{10000}$

   c. Write 3.078 as a mixed number.

2. a. Which is smaller, three and fourteen thousandths or 3.0602?

   b. Arrange 8.04, 8.05001, and 8.006 in order from largest to smallest.

   c. Round 103.76 to the nearest tenth.

3. a. Subtract: $23.8 - 12.409$

   b. Add: $0.0009 + 0.6074 + 12.3$

   c. Add: $45.98 + 34.063 + 12.38411$

4. a. Multiply: 14×0.0001

   b. Find the product: 29.7×23.4

   c. Find the quotient: $9.0936 \div 0.4$

5. Solve:

a. The largest salt-water lake is the Caspian Sea, which has an area of 143.2 thousand square miles. This is 111.38 thousand square miles larger than the largest fresh-water lake, Lake Superior. What is the area, to the nearest thousand square miles, of Lake Superior?

b. The estimated population of Denmark was 5.1 million people in 1984. The estimated population of Sweden was 8.35 million people in 1984. How many more people lived in Sweden than in Denmark?

c. Michelle's age is 2.5 times the age of Sarah. If Sarah is 14, how old is Michelle?

6. a. Simplify: $(2.1 - 1.5)^3 + (4.6)^2 \times 3$

b. Rewrite $(3.7)^2\,(6.2)^0$ without exponents.

c. Simplify: $11.1(4.3) - (1.7)^2$

7. Solve:

a. The radius of a circle measures 3.5 inches. Find the area of the circle.

b. Find the circumference of a circle with a radius of 4.2 centimeters (cm).

c. The diameter of a dish is 20.3 centimeters (cm). Find the radius of this dish.

Cumulative Review

CHAPTERS 1–6

The questions that follow will help you maintain the skills you have developed. Write all fractions in simplest form.

1. Add: $236 + 85 + 1079$

2. Subtract: $1000 - 68.3$

3. Multiply: 307×410

4. Divide and show the remainder: $2357 \div 22$

5. Add: $6\frac{2}{3} + 5\frac{1}{5}$

6. Subtract: $8 - 6\frac{2}{9}$

7. Multiply: $7\frac{1}{2} \times 2$

8. Divide: $3\frac{3}{5} \div \frac{9}{13}$

9. Add: $11.57 + 15 + 0.68$

10. Subtract: $74 - 8.39$

11. Multiply: 0.06×3.001

12. Divide: $660 \div 0.3$

13. Round 375.965 to the nearest ten.

14. Round 0.09 to the nearest tenth.

15. Change 0.125 to a fraction, reduced to lowest terms.

16. Simplify: $14 \div [2 + (9 - 4)^2]$. Express your answer as a fraction in simplest form.

17. Compare 13.004, 13.04, and 13.4 and arrange them in order from largest to smallest.

18. Find the volume of a rectangular solid with length 4.5 feet, width 3.6 feet, and height 6.8 feet.

19. Find the surface area of a cube with a side of $\frac{3}{4}$ inches.

20. Subtract: $5\frac{1}{4} - 2\frac{3}{7}$

7 Ratio and Proportion

How fast do you run? Chances are, you give your answer as a number of minutes per mile (like a 10 minute mile) or as a rate of speed (6 miles per hour). In either case, you compared quantities of distance and time and thereby established a ratio. You encounter ratios daily. For example, you buy gas by the gallon or chicken by the pound. Think about how the following questions might be answered: How much did the S&L loan scandal cost each taxpayer? How is your target heart rate measured when you exercise? What is the population density of New York City? The answers to these questions can be expressed as ratios.

■ *Find examples of ratios in the newspaper or magazine.*

Skills Check

Take this short quiz to see how well prepared you are for Chapter 7. The answers are given at the bottom of the page. So are the sections to review if you get any answers wrong.

1. Multiply: 2.13×100

2. Divide: $100\overline{)56.274}$

3. Find the product: $\frac{2}{3} \times \frac{4}{7}$

4. Find the quotient: $\frac{5}{8} \div \frac{7}{10}$

5. Divide: $7.14 \div 3.4$

6. Change $\frac{3}{8}$ to a decimal.

7. Change 0.035 to a fraction. Do not reduce.

8. Reduce $\frac{32}{36}$ to lowest terms.

9. Rewrite $3\frac{1}{4}$ as an improper fraction.

10. Multiply: 32.6×0.835

ANSWERS: 1. 213 [Section 6.5] 2. 0.56274 [Section 6.6] 3. $\frac{8}{21}$ [Section 5.1] 4. $\frac{25}{28}$ [Section 5.3] 5. 2.1 [Section 6.6] 6. 0.375 [Section 6.6] 7. $\frac{35}{1000}$ [Section 6.1] 8. $\frac{8}{9}$ [Section 4.3] 9. $\frac{13}{4}$ [Section 4.1] 10. 27.221 [Section 6.5]

7.1 Writing Ratios and Rates

Section Goals

- To write a ratio to express a comparison of two quantities
- To give the ratio of two different kinds of measures as a rate
- To determine a unit rate

Writing Ratios

How often have you heard statements like "Three out of every four doctors recommend brand X" or "7 of every 10 cars on the road need tune-ups" or "4 out of 5 students are happy"? These are all statements giving comparisons. And they all compare some sort of count.

The units of our comparisons are doctors, cars, and students; others may have more common units, such as feet, inches, or meters. Such comparisons are known as ratios.

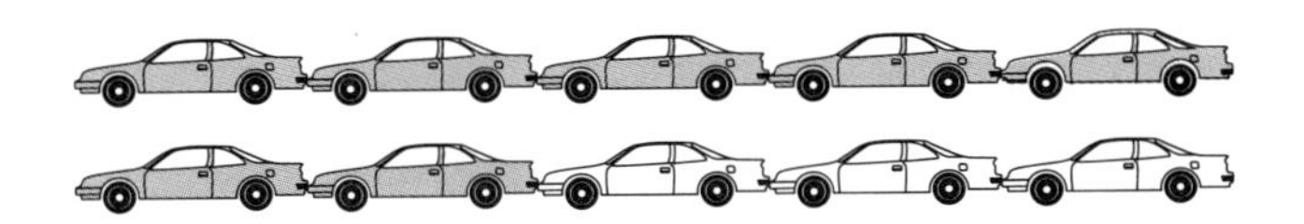

Ratio

A **ratio** is a comparison of two or more counts or measurements with the same units. If a and b are two numbers, the ratio of a to b is written

$$\frac{a}{b} \quad \text{or} \quad a{:}b \quad \text{or} \quad a \text{ to } b$$

So the ratio "three out of every four doctors" can be written

As a fraction: $\frac{3 \text{ doctors}}{4 \text{ doctors}} = \frac{3}{4}$

With a colon: 3 doctors : 4 doctors = 3 : 4

With "to": 3 doctors to 4 doctors = 3 to 4

We usually express a ratio in simplest form—that is, such that the two numbers have no common factor greater than 1. We first form the ratio and then reduce it as necessary.

Example 1

The number of students in a class is 30. The number of male students is 12. Write the ratio of the number of male students to the total number of students in the class. Express this ratio in three different ways, in simplest form.

SOLUTION

To write a ratio, we look for the word "to" in the verbal description and compare the number before "to" with the number after "to." Here we are asked to write the comparison of the number of male students (12) *to* the total number of students in the class (30). Hence 12 is being compared to 30. The ratio of the number of male students to the total number of students is then

$$\frac{12 \text{ students}}{30 \text{ students}} = \frac{12}{30} = \frac{2}{5}$$

1a. Write the ratio of 3 feet to 24 feet. Express it in three different ways, in simplest form.

1b. Write the ratio of 15 pints to 20 pints in simplest terms. Express it in three different ways.

2a. Write the ratio of 1.2 fathoms to 60 fathoms. Express it in three different ways, in simplest form.

2b. Write the ratio of 180 sandspurs to 40 sandspurs. Express it in three different ways, in simplest form.

The other two forms are

$$12:30 = 2:5$$
$$12 \text{ to } 30 = 2 \text{ to } 5$$

■ *Now do margin exercises 1a and 1b. The correct answers are given at the back of the book.*

The ratio in Example 1 simply says that 2 out of every 5 students in the class are males. Some other comparisons that could have been written as the ratio 2 to 5 are 20 students to 50 students, 16 girls to 40 girls, and 10 feet to 25 feet.

In the next example, the ratio involves a fraction. We will simplify to produce a ratio of whole numbers.

EXAMPLE 2

Write a ratio to compare $5\frac{1}{2}$ months to 2 months. Express it in three different ways.

SOLUTION

We first write the ratio of $5\frac{1}{2}$ months to 2 months in fractional form.

$$\frac{5\frac{1}{2}}{2}$$

Then we rewrite the numerator $5\frac{1}{2}$ as an improper fraction. Because $5\frac{1}{2} = \frac{11}{2}$, the ratio becomes

$$\frac{\frac{11}{2}}{2}$$

We simplify by multiplying the numerator and denominator by 2.

$$\frac{\frac{11}{2}}{2} = \frac{\frac{11}{2} \times 2}{2 \times 2} = \frac{11}{4}$$

So our ratio in fractional form is $\frac{11}{4}$. The other two forms are 11:4 and 11 to 4.

In Example 2, we did not rewrite the fractional ratio as a mixed number. Ratios that produce improper fractions are left in that form.

■ *Now do margin exercises 2a and 2b.*

Writing Rates

A rate is a special kind of ratio.

> A **rate** is a ratio in which the numbers (measurements) have different units.

For example, a rate might compare 36 inches to 3 feet, 5 miles to $2\frac{1}{2}$ hours, or 75 failures to 150 tries. Because the numerator and denominator of a rate have different units, we always include the units as part of the rate; we sometimes insert the word "per" between the two units.

We usually rewrite rates so that the numerator or the denominator is a 1; such rates are known as **unit rates.**

To write two measures as a unit rate

1. Write a rate with the first measurement and unit over the second measurement and unit.
2. Simplify the resulting numerical fraction until the denominator is 1. Retain the "units fraction."

We would say that the rate 5 miles to $2\frac{1}{2}$ hours is 2 miles per hour and that the rate 75 failures in 150 tries is 1 failure per 2 tries. In the latter rate, we chose to use a "1" in the numerator; however, we could also have said "0.5 failures per try."

EXAMPLE 3

What is your rate of gasoline usage (in miles per gallon) if you use 5 gallons of gas to drive 150 miles? Write it as a unit rate.

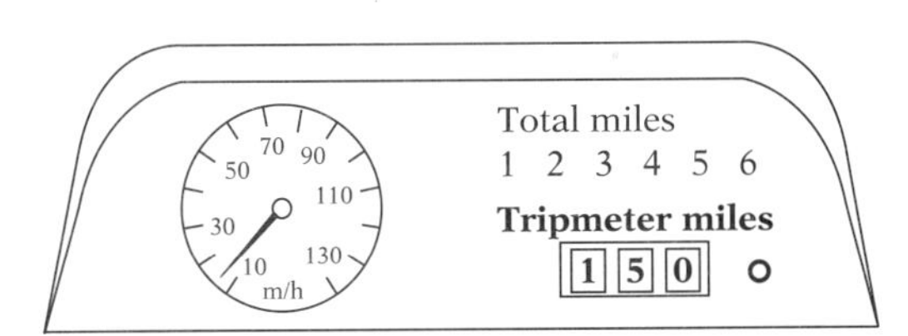

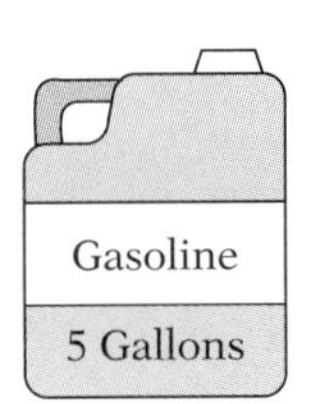

SOLUTION

We need the ratio of 150 miles to 5 gallons, which we write as

$$\frac{150 \text{ miles}}{5 \text{ gallons}}$$

To write this as a unit rate, we must simplify the fraction so that it has the denominator 1 gallon. We do this by dividing both the numerator and the denominator by 5.

$$\frac{150 \text{ miles} \div 5}{5 \text{ gallons} \div 5} = \frac{30 \text{ miles}}{1 \text{ gallon}}$$

We write this as 30 miles/gallon and read it as "30 miles per gallon."

■ *Now do margin exercises 3a and 3b.*

3a. What is your rate of speed in miles per hour if you drive 150 miles in $2\frac{1}{2}$ hours? Write it as a unit rate.

3b. What is your rate of sugar in teaspoons per coffee mug if you add $20\frac{1}{2}$ teaspoons to 15 cups of coffee? Write it as a unit rate.

4a. Write \$4.77 per 9.54 pints as a price per pint.

4b. Write \$25.50 per $8\frac{1}{2}$ pounds as a price per pound.

Our next kind of unit rate is one that you are probably quite familiar with—the unit price. Unit prices are required by law to be posted in grocery stores so that shoppers can easily compare the prices of different-sized packages of the same type of food.

EXAMPLE 4

Write \$2.72 per 8 ounces as a price per ounce.

SOLUTION

The rate is

$$\frac{\$2.72}{8 \text{ ounces}}$$

To find the unit rate, divide numerator and denominator by 8. This gives

$$\frac{2.72 \div 8}{8 \div 8} = \frac{\$.34}{1 \text{ ounce}}$$

We read this as "thirty-four cents per ounce."

■ *Now do margin exercises 4a and 4b.*

Note: Because rates are also ratios, we will often use the word "ratio" in the following sections to refer to both ratios and rates.

Work the exercises that follow. Remember to check to be sure your answers are reasonable.

7.1 EXERCISES

In Exercises 1 through 12, write the ratios in three different ways, each time in simplest form.

1. 23.3 centimeters to 46.6 centimeters

2. 5 meters to 45 meters

3. 120 cars to 50 cars

4. 4500 calories to 90 calories

5. 300.8 miles to 10,000 miles

6. 45 meters to 90 meters

7. $20\frac{1}{3}$ feet to 28 feet

8. $800\frac{1}{2}$ miles to 200 miles

9. Wahid earns $8.50 an hour as a tutor and pays $2 per hour to park his car. Write the ratio of earnings to parking costs per hour.

10. There are 573 public 4-year institutions of higher learning and 960 public 2-year institutions. Write the ratio of the number of 2-year institutions to the number of 4-year institutions.

11. There are 1497 private 4-year institutions of higher learning and 376 private 2-year institutions. Write the ratio of the number of 4-year institutions to the number of 2-year institutions.

12. Out of every 100 students enrolled in college in 1988, 52.9 were women. Write a ratio to compare the number of women to the total number of students.

In Exercises 13 through 18, write each of the rates as a unit rate.

13. 300 miles in 6 hours. Write as miles per hour.

14. 180 words in 4 minutes. Write as words per minute.

15. 300 miles in 6 hours. Write as hours per mile.

16. You spent $400 in 1800 seconds. Write as dollars per second.

17. 26.2 miles in 5 hours. Write as miles per hour.

18. 15.6 miles in 312 minutes. Write as minutes per mile.

In Exercises 19 through 23, write each as a unit price.

19. A ten-foot roll of shelf paper costs $1.89. What is the price per foot?

20. An 8-ounce can of soup costs $.96. It contains $2\frac{1}{2}$ servings per container. What is its approximate cost per serving?

21. A twenty-five-foot roll of aluminum foil costs $2.19. Write the approximate number of feet per dollar. Round to the nearest hundredth.

22. Vegetable oil costs $2 for 25 ounces. What is its unit cost?

23. Ninety tablets of chewable vitamin C at 500 milligrams each sell for \$4.98. What is the approximate price per tablet?

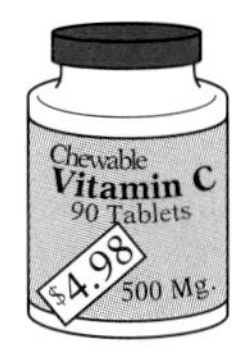

In Exercises 24 through 36, write the following as unit rates with the correct units.

24. Canned tuna fish sells for \$1.40 for $6\frac{1}{2}$ ounces. At this rate, approximately how many ounces could you buy for a dollar? Round to the nearest hundredth.

25. Bill runs 3 miles in 25 minutes. What is his rate in miles per minute?

26. A teacher earns \$55 and teaches 55 minutes. What is her rate in minutes per dollar?

27. A Nissan Pathfinder is driven 10,000 miles in a year. What is this amount in miles per month? Express your answer as a fraction.

28. Lydia runs 5 miles in 50 minutes. How many minutes per mile is this?

29. Decaffeinated coffee is 6 ounces for 90 cents. What is its cost in cents per ounce?

30. A 1000-mile flight costs $280. What is its cost per mile?

31. A car travels 2300 miles on 100 gallons of gas. What is its gasoline useage rate in miles per gallon?

32. A cat eats 6 cans of catfood in a week. What is her consumption in cans per day? Express your answer as a fraction.

33. For every 20 balloons a florist sells, he sells 6 dozen daisies. How many daisies is that per balloon?

34. A football player rushes 500 yards in a season and earns $250,000. How many dollars per yard is that?

35. Suppose the Incredible Hulk weighs 1000 pounds and is seven feet tall. Write a rate to compare his height to his weight.

36. Suppose Popeye weighs 158 pounds and has a forearm that is 20 inches in circumference. Write a rate to compare his forearm circumference to his weight. Express your answer as a fraction.

7.2 Solving Proportions

A **proportion** is an expression that shows that two ratios are equal. We can write proportions in three different ways:

As an equality of fractions

$$\frac{1}{2}=\frac{3}{6}$$

As an equality of ratios that are written with colons

$$1:2=3:6$$

As a comparison of phrases

"1 is to 2 as 3 is to 6"

In a proportion, the numerators have the same unit of measurement and the denominators have the same unit of measurement.

To determine whether a proportion is a true statement, we can reduce the two fractions to lowest terms. If the reduced fractions are equal, then the original fractions are different names for the same fraction, so the proportion is true. Let's look at an example.

EXAMPLE 1

Determine whether the following is a true proportion.

$$1.6:3.6 \stackrel{?}{=} 0.2:0.6$$

SOLUTION

We first rewrite each ratio in fractional form. Then, reducing each fraction to lowest terms, we get

$$\frac{1.6}{3.6}=\frac{1.6\times 10}{3.6\times 10}=\frac{16}{36}=\frac{16\div 4}{36\div 4}=\frac{4}{9}$$

$$\frac{0.2}{0.6}=\frac{0.2\times 10}{0.6\times 10}=\frac{2}{6}=\frac{2\div 2}{6\div 2}=\frac{1}{3}$$

The fractions $\frac{4}{9}$ and $\frac{1}{3}$ are not equal, because $\frac{1}{3}$ is equivalent to $\frac{3}{9}$.

So $1.6:3.6 \stackrel{?}{=} 0.2:0.6$ is not a true proportion.

■ *Now do margin exercises 1a and 1b.*

There is another—sometimes simpler—way to determine whether a proportion is true. Before we discuss it, you need to know two definitions.

> In the proportion
>
> $$\frac{a}{b}=\frac{c}{d}$$
>
> a and d are called the **extremes** of the proportion, and b and c are called the **means** of the proportion.

SECTION GOALS

- To determine whether two ratios are proportional
- To solve any given proportion for its missing term
- To use the cross-products test

1a. Determine whether the following is a true proportion: $\frac{4.8}{20} \stackrel{?}{=} \frac{0.6}{4}$

1b. Is the ratio $2\frac{9}{10}:3.8$ equivalent to $18:20\frac{1}{2}$?

If the proportion is written as

$$\overset{\text{Extremes}}{\overbrace{a:\underset{\text{Means}}{\underbrace{b \;=\; c}}:d}}$$

we can see that a and d are at the extreme outside, whereas b and c are *between* the extremes. (These "means" are not the same as the averages we computed in Chapter 6.)

Now, the second way to determine whether a given proportion is true is to apply the **cross-products test.**

Cross-Products Test

In a true proportion, the product of the means is equal to the product of the extremes.

That is, in the proportion $a:b = c:d$, if

$$ad = bc$$

then the proportion is true.

In fraction form, we can show the cross products as

$$\frac{a}{b} \times \frac{c}{d} \quad \begin{matrix} \rightarrow b \times c \\ \rightarrow a \times d \end{matrix}$$

The process of finding cross products is called **cross-multiplying.**

Let's look at an example.

EXAMPLE 2

Determine whether the following is a true proportion by reducing to lowest terms. Then check with the cross-products test.

$$\frac{16}{8} \stackrel{?}{=} \frac{40}{20}$$

SOLUTION

We first simplify each ratio.

$$\frac{16}{8} = \frac{16 \div 8}{8 \div 8} = \frac{2}{1}$$

$$\frac{40}{20} = \frac{40 \div 20}{20 \div 20} = \frac{2}{1}$$

We see that both fractions can be simplified to $\frac{2}{1}$, so $\frac{16}{8} = \frac{40}{20}$ is a true proportion.

To check, we use the cross-products test.

$$\frac{16}{8} \times \frac{40}{20} \quad \begin{matrix} \rightarrow 8 \times 40 = 320 \\ \rightarrow 16 \times 20 = 320 \end{matrix}$$

Because the product of the extremes is equal to the product of the means, the proportion is true and our work is correct.

■ *Now do margin exercises 2a and 2b.*

Sometimes we are asked to find the value (of a mean or extreme) that will make a proportion true. The value that makes the cross products equal also makes the proportion true.

EXAMPLE 3

Find the value of n that makes the following proportion true.

$$\frac{15}{16} = \frac{n}{9}$$

SOLUTION

We multiply to find the cross products.

Product of extremes: $15 \times 9 = 135$

Product of means: $16 \times n$

In a true proportion, the product of the means is equal to the product of the extremes. In this case,

$$16 \times n = 135$$

We want to find out what n is, but we know only what $16 \times n$ is. However, we do know that division "undoes" multiplication. So if we divide both $16 \times n$ and 135 by 16, we will end up with n alone on both sides of the equation.

$$\frac{16 \times n}{16} = \frac{135}{16} \quad \text{Dividing both sides by 16}$$

$$n = \frac{135}{16} \quad \text{Simplifying the left side}$$

$$n = 8\frac{7}{16} \quad \text{Simplifying the right side}$$

To check, we replace n in the original proportion with $8\frac{7}{16}$ and cross-multiply.

$$\frac{15}{16} \times \frac{8\frac{7}{16}}{9} \quad \rightarrow 16 \times 8\frac{7}{16} = 135 \quad \rightarrow 15 \times 9 = 135$$

The product of the extremes is equal to the product of the means.

Therefore, the value $n = 8\frac{7}{16}$ does make the proportion true.

■ *Now do margin exercises 3a and 3b.*

In the next example, we are asked to "solve for y." This means to find the value of y that makes the proportion true.

EXAMPLE 4

Solve for y in the proportion

2.5 is to y as 2 is to 0.3

2a. Determine whether this is a true proportion by using cross products:
$11.8:12.6 \stackrel{?}{=} 20:23\frac{1}{2}$

2b. Determine whether this is a true proportion by using cross products:
$2.1:3 \stackrel{?}{=} 4.0:6.1$

3a. Find the missing value in the proportion
$\frac{10}{18} = \frac{90}{t}$

3b. Find the missing value in the proportion
$\frac{9\frac{1}{2}}{14} = \frac{r}{70}$

4a. Find the missing value in the proportion

$\frac{9}{12} = \frac{t}{100}$

4b. Find the missing value in the proportion

$6:7 = r:1.4$

5a. Which ratio is larger, $\frac{3}{4.8}$ or $\frac{9}{11.5}$?

5b. Which ratio is smaller, $5:\frac{2}{3}$ or $26:4$?

SOLUTION

First, we rewrite the proportion in fractional form, which is easiest to use.

$$\frac{2.5}{y} = \frac{2}{0.3}$$

We multiply to find the cross products.

Product of extremes: $2.5 \times 0.3 = 0.75$

Product of means: $y \times 2$

Setting the cross products equal, we have

$$y \times 2 = 0.75$$

We have to divide both sides by 2 to get y alone.

$$\frac{y \times 2}{2} = \frac{0.75}{2} = 0.375$$

We check by replacing y in the original problem with 0.375 and cross-multiplying.

The products are equal, so the value $y = 0.375$ makes the proportion true.

$$\frac{2.5}{0.375} = \frac{2}{0.3} \quad \begin{array}{l} 0.375 \times 2 = 0.75 \\ 2.5 \times 0.3 = 0.75 \end{array}$$

■ *Now do margin exercises 4a and 4b.*

We can also use cross products to determine which of two ratios is the larger (or smaller).

For two positive fractions $\frac{a}{b}$ and $\frac{c}{d}$,

If $ad = bc$, then $\frac{a}{b} = \frac{c}{d}$.

If $ad > bc$, then $\frac{a}{b} > \frac{c}{d}$.

If $ad < bc$, then $\frac{a}{b} < \frac{c}{d}$.

EXAMPLE 5

Which is larger: $\frac{15}{22}$ or $\frac{21}{31}$?

SOLUTION

First we write the two fractions as a proportion and find the cross products.

$$\frac{15}{22} \; \frac{21}{31} \quad \begin{array}{l} 22 \times 21 = 462 \\ 15 \times 31 = 465 \end{array}$$

Since the product of the extremes (465) is greater than the product of the means (462), we know that $\frac{15}{22} > \frac{21}{31}$.

■ *Now do margin exercises 5a and 5b.*

Work the exercises that follow.

7.2 EXERCISES

In Exercises 1 through 4, determine whether the proportions are true by reducing the ratios to lowest terms.

1. Is $\frac{2}{3} = \frac{6}{9}$?

2. Is $\frac{4}{9} = \frac{10}{18}$?

3. Is $\frac{20}{36} = \frac{37}{60}$?

4. Is $\frac{12}{14} = \frac{45}{52}$?

In Exercises 5 through 12, determine whether the ratios are proportional by using the cross-products test.

5. $\frac{0.5}{6} \stackrel{?}{=} \frac{4}{18}$

6. $\frac{0.6}{7} \stackrel{?}{=} \frac{3}{35}$

7. $\frac{2.4}{11} \stackrel{?}{=} \frac{4}{15}$

8. $\frac{0.2}{4} \stackrel{?}{=} \frac{4}{80}$

9. $\frac{\frac{2}{3}}{4} \stackrel{?}{=} \frac{3}{8}$

10. $\frac{\frac{1}{2}}{6} \stackrel{?}{=} \frac{5}{60}$

11. $\frac{\frac{5}{6}}{18} \stackrel{?}{=} \frac{\frac{5}{9}}{12}$

12. $\frac{\frac{1}{4}}{14} \stackrel{?}{=} \frac{3}{8}$

In Exercises 13 through 24, find the missing value.

13. $5.1:8 = t:48$

14. $14:15.1 = k:60.4$

15. $11.2:13 = k:169$

16. f is to 32 as $3\frac{1}{2}$ is to 8

17. t is to 126 as $12\frac{1}{3}$ is to 63

18. k is to 93 as 21 is to 3

19. $\frac{1}{2}:f = 3:8$

20. $2:9 = t:\frac{1}{6}$

21. $k:\frac{1}{4} = 3:4$

22. $2:3 = \frac{1}{8}:t$

23. $7:20 = f:60.2$

24. f is to 90 as 7 is to 15.

In Exercises 25 through 32, find the missing value by using the cross-products test.

25. $\frac{7}{9} = \frac{1}{t}$

26. $\frac{3}{15} = \frac{m}{18}$

27. $\frac{t}{5} = \frac{2\frac{1}{4}}{3}$

28. $\frac{3\frac{1}{2}}{8} = \frac{m}{7}$

29. $\frac{0.4}{r} = \frac{5}{7}$

30. $\frac{5}{6} = \frac{2.3}{r}$

31. $\frac{12.5}{1} = \frac{t}{6.25}$

32. $\frac{m}{6} = \frac{8}{1.5}$

In Exercises 33 through 40, place <, >, or = in the blank. Use the cross-products test to determine the order relationship.

33. $\frac{101}{102}$ ___ $\frac{51}{53}$

34. $\frac{123}{350}$ ___ $\frac{369}{800}$

35. 24:47 ___ 73:99

36. 21:67 ___ 38:72

37. $\frac{4.7}{398}$ ___ $\frac{76.3}{3555}$

38. $\frac{55.2}{398}$ ___ $\frac{27.4}{175}$

39. 4.1:99 ___ 8.2:197

40. 9.9:1000 ___ 1.9:113

7.2 MIXED PRACTICE

By doing these exercises, you will practice all the topics in this chapter up to this point. Reduce all answers to lowest terms.

41. Find the missing value in this proportion.
$15:40 = n:8$

42. Which is the smaller ratio? $\frac{3}{4}:\frac{1}{2}$ or $\frac{3}{5}:\frac{2}{3}$?

43. On a used car lot, there are 150 sedans, 200 coupes, and 35 4×4s. Express the ratio of 4×4s to the total number of vehicles on the lot in lowest terms.

44. In 70 minutes, a gear will perform 3000 revolutions. Express the ratio of revolutions per minute as a rate. Round to the nearest hundredth.

45. Determine whether $\frac{1}{2}:6$ and $\frac{3}{4}:9$ are proportional.

46. Which is the larger ratio, $\frac{124}{6}$ or $\frac{136}{8}$?

47. It is estimated that one out of four people owns a CD player. Express the ratio of people who own a CD player to those who do not own a CD player.

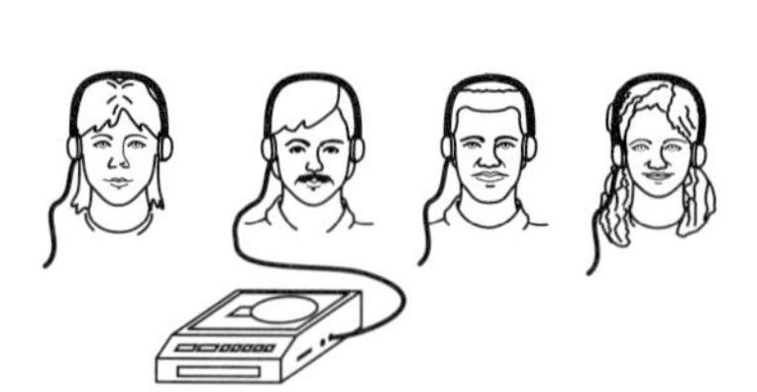

48. Determine whether $\frac{15}{8}$ and $\frac{19}{10}$ are proportional.

49. Find the missing value in this proportion. $\frac{n}{1.2} = \frac{11}{6}$

7.3 Geometry and Measurement: Applying Ratio and Proportion

We use measurements to describe length, area, volume, capacity, mass, time, and temperature. Two systems of measurement are used in the United States, the U.S. Customary system and the metric system. The metric system is used almost exclusively throughout the rest of the world.

In this section, we compare the two systems informally and **convert,** or rename, measurements within the systems and from system to system. The ideas of ratio and proportion are helpful in doing so.

The first measure we will discuss is *length.* The most common units of length are shown in Table 1.

TABLE 1 Units of Length

U.S. customary units
12 inches (in.) = 1 foot (ft)
3 feet = 1 yard (yd)
5280 feet = 1 mile (mi)

Metric units
10 millimeters (mm) = 1 centimeter (cm)
100 centimeters = 1 meter (m)
1000 meters = 1 kilometer (km)

Common conversions

1 in. = 2.54 cm	1 yd = 0.94 m
39.37 in. = 1 m	1 mi = 1.6 km
3.28 ft = 1 m	

The wire in a paper clip is usually about 1 mm thick.
An average-size man's long step is about 1 m. (A meter is a little more than a yard.)

EXAMPLE 1

3245 centimeters is how many millimeters? (3245 cm = ? mm)

SOLUTION

To answer this question, we will set up a proportion involving ratios of centimeters to millimeters. We will use the fact that 10 mm is the same as 1 cm to help us convert this measurement. Then 3245 centimeters is to n millimeters as 1 centimeter is to 10 millimeters.

$$\frac{3245 \text{ cm}}{n \text{ mm}} = \frac{1 \text{ cm}}{10 \text{ mm}}$$

SECTION GOALS

- To convert measures within the U.S. customary system and the metric system
- To solve word problems involving measurement

1a. 18 000 centimeters is how many millimeters?

1b. The distance between two towns is 120 000 kilometers. How many meters is this?

Once we check that the units are the same on both sides, we can drop them.

$$\frac{3245}{n} = \frac{1}{10}$$

Using cross multiplication, we get

$$3245 \times 10 = 1 \times n$$
$$32{,}450 = n$$

So 3245 cm is the same as 32,450 mm.

In the metric system, spaces are sometimes used instead of commas to separate the periods (groups of three digits). Then 32,450 would be written 32 450.

■ *Now do margin exercises 1a and 1b.*

TABLE 2 Units of Area

U.S. customary units
144 square inches (in.^{2}) = 1 square foot (ft^2)
9 square feet = 1 square yard (yd^2)
3,097,600 square yards = 1 square mile (mi^2)

Metric units
100 square millimeters (mm^2) = 1 square centimeter (cm^2)
10 000 square centimeters = 1 square meter (m^2)
1 000 000 square meters = 1 square kilometer (km^2)

Common conversion
1 in.^{2} = 6.45 cm^2

Units of *area* are shown in Table 2. Recall that area is a measure of the extent of a surface. One square unit of area can be thought of as the area of a square whose side is 1 unit long.

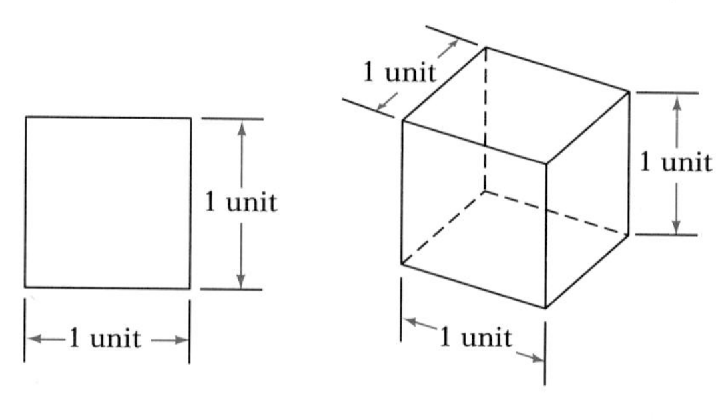

square units
Area

cubic units
Volume

Volume is the measure of the three-dimensional space in a container. Volume units are given in Table 3.

TABLE 3 Units of Volume

U.S. customary units
1 728 cubic inches ($in.^3$) = 1 cubic foot (ft^3)
27 cubic feet = 1 cubic yard (yd^3)

Metric units
1 000 cubic millimeters (mm^3) = 1 cubic centimeter (cm^3)
1 000 000 cubic centimeters = 1 cubic meter (m^3)

Common conversion
1 $in.^3$ is approximately 16.387 cm^3

The measure of liquid volume is usually called *capacity* (Table 4).

TABLE 4 Units of Capacity

U.S. customary units
16 ounces (oz) = 1 pint (pt)
2 pints = 1 quart (qt)
4 quarts = 1 gallon (gal)

Metric units
1000 milliliters (mL) = 1 liter (L)

Common conversions
1 oz = 30 mL
1 qt = 0.95 L

One milliliter is the same as 1 cubic centimeter. A teaspoon holds about 5 mL.
A liter is just a bit larger than a quart.

EXAMPLE 2

1950 ounces is equal to how many quarts?

SOLUTION

We first must look up the relationship between ounces and quarts. Table 4 shows no direct relationship. However, it does show that 16 ounces is 1 pint and that 2 pints is 1 quart. We shall need two proportions.

2a. How many pints are there in $31\frac{1}{4}$ gallons?

2b. 660 oz is equivalent to how many gallons?

Our first proportion is

$$\frac{1950 \text{ oz}}{n \text{ pt}} = \frac{16 \text{ oz}}{1 \text{ pt}} \quad \text{or} \quad \frac{1950}{n} = \frac{16}{1}$$

Then we get

$$1950 \times 1 = 16 \times n \qquad \text{Cross-multiplying}$$

$$\frac{1950}{16} = \frac{16 \times n}{16} \qquad \text{Dividing by 16}$$

$$121\frac{7}{8} = n \qquad \text{Simplifying}$$

So 1950 oz is the same as $121\frac{7}{8}$ pt. We can also write this as 121.875 pt or 121 pt 14 oz.

Now we want to determine the number of quarts equivalent to 121.875 pints. We set up another proportion:

$$\frac{121.875 \text{ pt}}{n \text{ qt}} = \frac{2 \text{ pt}}{1 \text{ qt}} \quad \text{or} \quad \frac{121.875}{n} = \frac{2}{1}$$

Then we have

$$121.875 \times 1 = 2 \times n \qquad \text{Cross-multiplying}$$

$$\frac{121.875}{2} = \frac{2 \times n}{2} \qquad \text{Dividing by 2}$$

$$60.9375 = n \qquad \text{Simplifying}$$

So 121.875 pt is the same as 60.9375 qt and is also equivalent to 60 qts 1 pt 14 oz.

■ *Now do margin exercises 2a and 2b.*

Table 5 lists units of another measurement—**mass,** or the amount of matter in an object. Although many people think that weight and mass mean the same thing, in science, mass and weight are different concepts. Mass is the amount of matter in an object. Weight is a measure of the force of gravity exerted on an object. However, it is common to use the metric units for mass to describe the metric weight of an object.

TABLE 5 Units of Mass

U.S. customary units
16 ounces (oz) = 1 pound (lb)
2000 pounds = 1 ton (t)

Metric units
1000 grams (g) = 1 kilogram (kg)
1000 kilograms = 1 metric ton (t)

Common conversions
1 oz = 28 g
1 lb = 453 g

The mass of one paper clip is about 1 g.
A telephone weighs about 2 kg. (A kilogram is slightly more than 2 pounds.)
The mass of a sub-compact car is about 1 t.

Units of *time* are listed in Table 6. They are used in both the metric and the U.S. customary systems.

TABLE 6 Units of Time

60 seconds (sec) = 1 minute (min)	7 days = 1 week
60 minutes = 1 hour (hr)	365 days = 1 year
24 hours = 1 day	

Let's try one more conversion, this time from one system to the other.

EXAMPLE 3

A courtyard is 10 feet wide and 12 feet long.

What is its area in square meters?

10 ft
12 ft

SOLUTION

We can convert feet to meters and then multiply length × width to find the area, *or* we can multiply to find the area in square feet and then convert. We will convert first.

We see in Table 1 that 1 meter is 3.28 feet. Then, to change 12 feet to meters, we use this proportion:

$$\frac{12 \text{ ft}}{n \text{ m}} = \frac{3.28 \text{ ft}}{1 \text{ m}} \quad \text{or} \quad \frac{12}{n} = \frac{3.28}{1}$$

3a. The net weight of a can of cat food is 6 ounces. How many grams is this?

3b. The base of a triangle measures 5 inches and the height is 4 inches. Express the area of this triangle to the nearest square centimeter.

Then we have

$$12 \times 1 = 3.28 \times n \quad \text{Cross-multiplying}$$
$$\frac{12}{3.28} = \frac{3.28 \times n}{3.28} \quad \text{Dividing by 3.28}$$
$$3.66 = n \quad \text{Simplifying}$$

So 12 feet is equal to 3.66 meters (to two decimal places).

To change 10 feet to meters, we use this proportion:

$$\frac{10 \text{ ft}}{n \text{ m}} = \frac{3.28 \text{ ft}}{1 \text{ m}}$$

Then we have

$$10 \times 1 = 3.28 \times n \quad \text{Cross-multiplying}$$
$$\frac{10}{3.28} = \frac{3.28 \times n}{3.28} \quad \text{Dividing by 3.28}$$
$$3.05 = n \quad \text{Simplifying}$$

Now we have the length and width in meters. We multiply them to find the area of the courtyard.

$$\text{Area} = 3.05 \times 3.66 = 11.163 \text{ m}^2$$

The area of the courtyard is 11.163 m^2.

■ *Now do margin exercises 3a and 3b.*

Work the exercises that follow. You may use a calculator if you wish.

7.3 EXERCISES

The data cited in Exercises 1 through 3 were gathered at the 1988 New York Marathon.

1. 18,000 yards of barricade tape were used. How many miles of tape was this? Express as a fraction.

2. 22,000 cups of coffee were served. If each cup held 6 ounces, how many gallons of coffee were served?

3. The Pulaski Bridge was the half-way point in the marathon. It held the 13.1-mile marker. How many yards was this from the finish?

The heights of four South American volcanoes are given in Exercises 4 through 7. Use a calculator to convert them as requested. In all cases, round to the nearest hundredth of a unit.

4. Aconcagua, 22,834 feet. Convert to miles.

5. Chimborazo, 20,560 feet. Convert to yards.

6. Antisana, 18,713 feet. Convert to inches.

7. Cotopaxi, 19,344 feet. Convert to miles.

The jumping heights actually recorded for certain animals are given in Exercises 8 through 11. Convert these heights into the given units.

8. A German Shepherd dog in K-9 training jumped 11 feet 8 inches. Convert to yards. Express as a fraction.

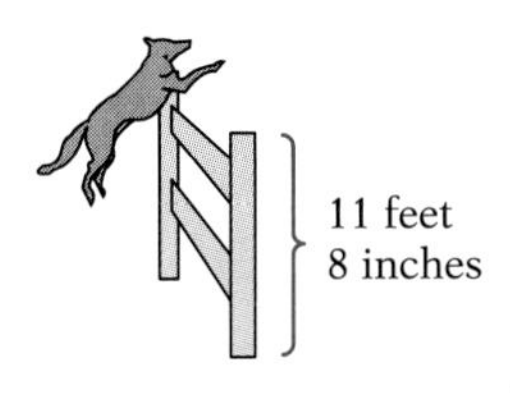

9. A Mako shark can jump 30 feet into the air. Convert to inches.

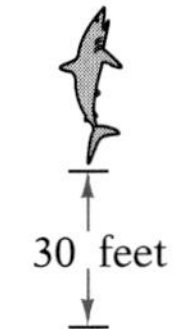

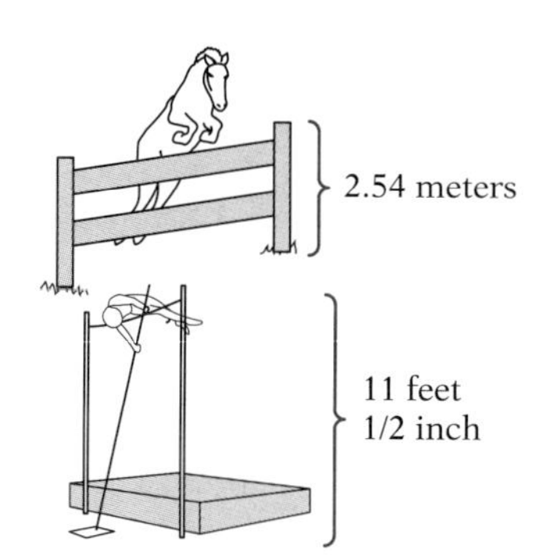

10. An Australian horse has been recorded jumping 2.54 meters. Convert to centimeters.

11. The first recorded pole vault over 11 feet was 11 feet $\frac{1}{2}$ inch. Convert this to inches.

In modern rhythmic gymnastics, certain hand apparatus is used. The sizes of some of this equipment are given in Exercises 12 through 15. Rewrite these measurements in the given units.

12. Hoop, 85 centimeters in diameter. Rewrite in millimeters.

13. Ribbon, 236 inches long. Rewrite in feet expressed as a fraction.

14. Clubs, 17 inches high. Rewrite in centimeters.

15. Ball, 19 centimeters in diameter. Rewrite in inches rounded to the nearest hundredth.

Some animal long-jump records are given in Exercises 16 through 19. Represent them in the requested units.

16. Frog, 17 feet $6\frac{3}{4}$ inches. Represent as inches.

17. Snow leopard, 15.24 meters. Represent this as centimeters.

18. Greyhound, 9.14 meters. Represent this as millimeters.

19. Horse jumping, 32 feet 10 inches. Represent as yards expressed as a fraction.

Animals can travel at very high speeds. In Exercises 20 through 23, convert each of the given speeds as requested.

20. Red kangaroo, 45 miles per hour. Rewrite as kilometers per hour.

21. Cheetah, 98 kilometers per hour. Rewrite as miles per hour.

22. Gazelle, 68 kilometers per hour. Rewrite as miles per hour.

23. Ostrich, 50 miles per hour. Rewrite as kilometers per hour.

The weights of various playing balls are given in Exercises 24 through 27. Rewrite them in the requested units.

24. Table tennis ball, 0.09 ounce. Rewrite in grams.

25. Jai alai ball, 127 grams. Rewrite in ounces expressed as a fraction.

26. Badminton "bird," 5.5 grams. Rewrite in kilograms.

27. Racquetball, 1.4 ounces. Rewrite in pounds.

A marathon runner over 40 years old ran the New York City marathon in 5 hours 51 minutes 53 seconds.

28. How many seconds was this altogether?

29. How many minutes was this altogether? Express your answer as a fraction.

Exercises 30 and 31 give the maximum body weight allowed for weight lifters in the indicated competition categories. Rewrite these weights in pounds rounded to the nearest hundredth.

30. Flyweight, 52 kilograms.

31. Middleweight, 75 kilograms.

7.3 MIXED PRACTICE

By doing these exercises, you will practice all the topics in this chapter up to this point.

32. Motor oil costs $1.19 per quart. You can buy a case of 12 on sale for $10.68. How much do you save on each can if you buy the oil on sale?

33. A 17-ounce can of vegetables has 13 ounces of vegetables and 4 ounces of liquid. How many grams of vegetables does the can hold?

34. A gross (144) of rulers can be purchased for \$150. What is the unit price?

35. Determine whether $\frac{3}{5.25}$ and $\frac{5}{8.75}$ are proportional.

36. A person types 70 words in 1 minute. Write this as a ratio of words per second.

37. Fill in the missing value: $\frac{\frac{1}{6}}{5} = \frac{30}{n}$

38. You ran 26.2 miles in 5.5 hours. Express this as a unit rate of miles per hour rounded to the nearest hundredth.

39. Which is a better buy: Cereal priced at 12 oz for \$1.29 or at 16 oz for \$1.89?

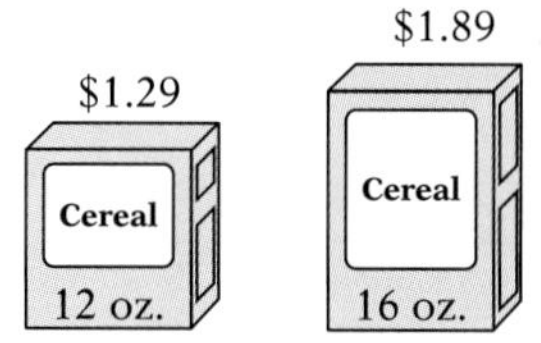

40. Determine whether the following is a true proportion: 12 is to 54 as 4 is to 18

7.4 Applying Ratio and Proportion: Word Problems

In many interesting problems, two related ratios are implied, and one piece of information is missing. To solve them, we have only to equate the two ratios in a proportion so that the units correspond to each other. We then simply solve for the missing value.

EXAMPLE 1

In a certain field of roses and daisies, the ratio of roses to daisies is 2 to 3. If there are 2500 flowers in the field, how many are roses?

SOLUTION

Before we can set up a proportion, we must be able to write two ratios with the same units in the denominator and the same units in the numerator. For one ratio, we have information about the number of roses and the number of daisies. For the second ratio, we have information only about the total number of flowers (2500). We will have to write ratios that use this total number of flowers.

Because the ratio of roses to daisies is 2 to 3, the ratio of roses to total flowers is 2 to 5. (We get the 5 by adding 2 and 3.) This, then, is our first ratio:

$$\frac{2 \text{ roses}}{5 \text{ flowers}}$$

The second ratio must have 2500 flowers in the denominator. We do not know the number of roses. The second ratio is therefore

$$\frac{n \text{ roses}}{2500 \text{ flowers}}$$

We now set these ratios equal to each other, without the units, and cross-multiply.

$$\frac{2}{5} = \frac{n}{2500} \quad \begin{matrix} \rightarrow 5 \times n \\ \rightarrow 2 \times 2500 \end{matrix}$$

Now we set the cross products equal to each other and solve.

$$\begin{aligned} 5 \times n &= 2 \times 2500 & \\ 5 \times n &= 5000 & \text{Multiplying} \\ \frac{5 \times n}{5} &= \frac{5000}{5} & \text{Dividing by 5} \\ n &= 1000 & \text{Simplifying} \end{aligned}$$

Thus, there are 1000 roses.

■ *Now do margin exercises 1a and 1b.*

The next example requires us to find a rate.

SECTION GOAL

■ To apply ratio and proportion in solving word problems

1a. Out of a large class of students, the ratio of male students to female students was four to five. If 120 were female, how many people were in the class?

1b. A man spends one dollar of every four he earns for rent. If he spends \$60 a month on rent, how much does he have left to spend on other things?

2a. The ratio of doctors prescribing a certain antibiotic to those giving a shot to patients is $6\frac{1}{2}$ to 2. If 40 shots were given, how many prescriptions for antibiotics were written?

2b. In a city the ratio of male to female is 4.5 to 6. If the city has a population of 250,000, how many males live there?

EXAMPLE 2

The fastest a greyhound ever ran was 410 yards in 20.1 seconds. What was the dog's speed in feet per minute?

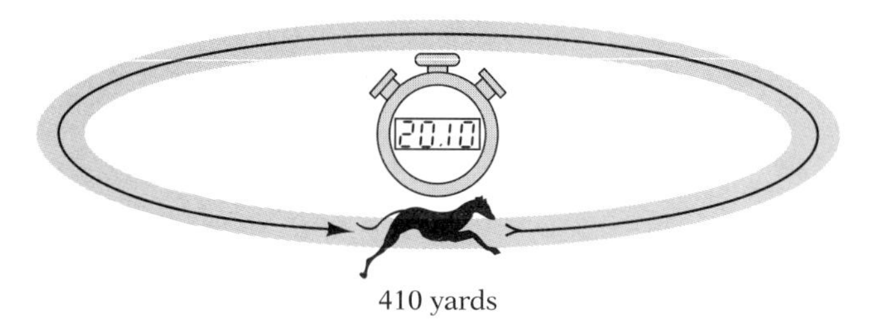

SOLUTION

We must use several steps to solve this problem. Because our information is given in yards and seconds, and we want the answer in feet and minutes, we must first convert our units. We use proportions to do this.

To change 410 yards to a number n of feet, we use the relationship 1 yard = 3 feet and write the proportion.

$$\frac{410 \text{ yards}}{n \text{ feet}} = \frac{1 \text{ yard}}{3 \text{ feet}}$$

$$410 \times 3 = 1 \times n \qquad \text{Equating cross products}$$

$$1230 = n \qquad \text{Multiplying}$$

So 410 yards = 1230 feet.

To change 20.1 seconds to y minutes, we use the relationship 60 seconds = 1 minute.

$$\frac{20.1 \text{ seconds}}{y \text{ minutes}} = \frac{60 \text{ seconds}}{1 \text{ minute}}$$

$$20.1 \times 1 = 60 \times y \qquad \text{Equating cross products}$$

$$20.1 = 60 \times y \qquad \text{Multiplying}$$

$$\frac{20.1}{60} = \frac{60 \times y}{60} \qquad \text{Dividing both sides by 60}$$

$$0.335 = y \qquad \text{Simplifying}$$

So 20.1 seconds = 0.335 minutes.

We want the dog's speed in feet per minute, so we form the ratio

$$\frac{1230 \text{ feet}}{0.335 \text{ minutes}}$$

To find a unit rate, we divide the numerator and denominator by the denominator.

$$\frac{1230 \div 0.335}{0.335 \div 0.335} = \frac{3672}{1}$$

So the greyhound's speed was 3672 feet per minute to the nearest foot.

■ *Now do margin exercises 2a and 2b.*

Work the exercises that follow.

7.4 EXERCISES

In Exercises 1 through 3, use ratio and proportion to solve the problems.

1. Triplets are born once in every 7569 births. Find the number of sets of triplets expected in 1,000,000 births.

2. Quadruplets are born once in every 658,503 births. How many births would there have to be to get 25 quadruplet births?

3. Identical twins are born once in every 270 births. How many sets of identical twins are expected in 1,000,000 births?

In Exercises 4 through 7, use this information to answer the questions.

Every minute, a human breathes 10 pints (6 liters) of air into the lungs.

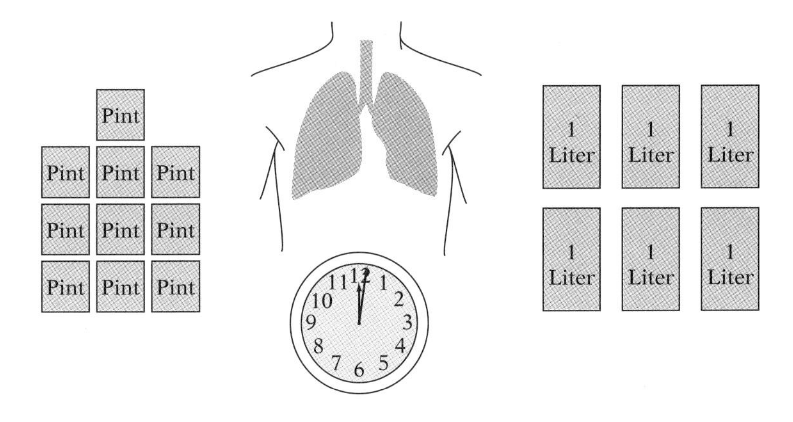

4. How many pints are breathed in over a period of 20 minutes?

5. How many pints are breathed in over a period of 2 hours?

6. How long will it take to breathe in 650 liters of air? Express your answer as a fraction.

7. How many liters of air are breathed in a 24-hour period?

8. Viretta wore out a pair of running shoes by the time she ran 220 miles. At this rate, about how many pairs of running shoes would she have to buy during a year in which she ran 1300 miles?

9. The ratio of the weight of an average adult male to his weight as a year-old baby is 81 to 11.1. What would be the adult weight of a male who weighed 30 pounds at 1 year of age (to the nearest tenth of a pound)?

10. For every 100 people who enter high school in one large metropolitan city, only 54 ever graduate. At this rate, if there are 120,000 high school entrants in that city, how many students can be expected to graduate?

11. The Transit Authority uses buses that hold 63 people but contain only 48 seats. How many people can 400 buses hold?

12. There were 890 students enrolled in a lower-level history course. Of these, 450 finished the course with either an A or a B. Assume that the same ratio held for an introductory English course and find how many students took the course if 1000 received an A or a B.

13. An average 20-question exam takes approximately 45 minutes to complete. At this rate, how long would an 85-question test take?

14. In preparing 140 liters of liquid fertilizer, a company uses 70 liters of water and the rest is chemical concentrate. How much concentrate must be used to make 1000 liters of fertilizer?

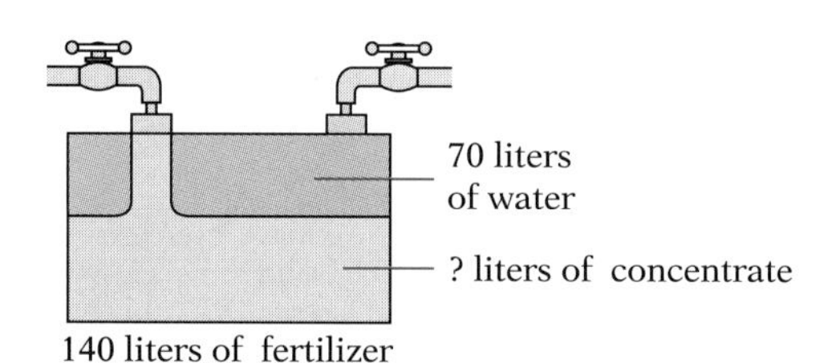

15. Which is more expensive, a 3-lb container of ricotta cheese for $2.99 or three 16-ounce containers for 99 cents each?

16. A major-league outfielder hit 18 times in his first 60 times at bat. If his batting average does not change, how many hits does he have after 100 times at bat?

17. A certain hockey team won 15 of their last 35 games.

a. At this rate, how many games will they win if they play 140 games?

b. If they continue at this rate, how many games will they lose if they win 45 games?

18. A certain baseball player gets 3 hits for every 8 times at bat.

a. At this rate, how many times must he bat to get 66 hits?

b. At this rate, how many hits will he have in 96 times at bat?

19. A million dollars was awarded to the winner of the $1\frac{1}{2}$-mile Breeder's Cup Classic horse race.

a. How far did the winner run per dollar?

b. What was the payoff in dollars per mile?

20. Before the San Francisco "World Series" earthquake, the state sales tax in California was assessed at a rate of 6 cents on the dollar. Then it was temporarily raised to 6.25 cents.

a. If your tax was exactly 33 cents, what was the amount of your purchase (before taxes), both before and after the earthquake?

b. If you paid a tax of $8.25, what were the greatest and least amounts of your purchase (before taxes), after the quake? (Assume the tax is rounded to the nearest cent.)

21. A Kenyan runner ran the fastest mile ever in New Zealand when he ran a mile in $3\frac{1}{2}$ minutes. What was his speed in miles per hour? Express your answer as a fraction.

22. A carousel holds 30 children; a roller coaster holds 12. The rides run in a ratio of 3 carousel rides to every 10 roller coaster rides. How many children can ride the carousel in the time it takes 120 children to ride the roller coaster?

23. In the 1933 movie "King Kong," the gorilla looked 50 feet tall but was really only 18 inches tall. How tall would he have looked if he had been 36 inches tall?

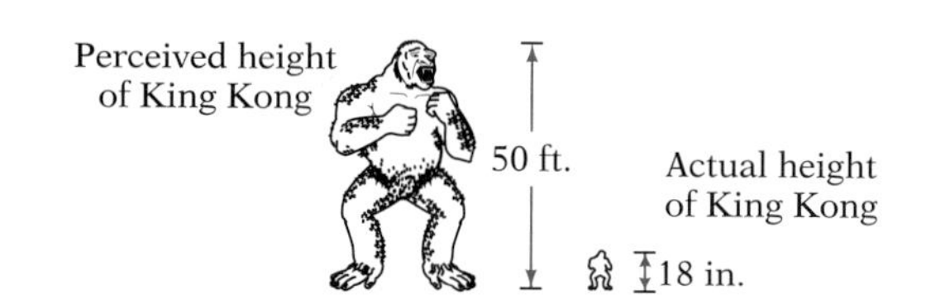

24. A "hat trick" in hockey is three goals in a row in a game. If a hat trick is scored twice in every 5 games, how many hat tricks would you expect to be scored in 255 games?

25. To determine the number of bream in a certain lake, a wildlife ranger tags 425 bream and then releases them into the lake. Later, out of a catch of 300 bream, 18 are found to be tagged. How many bream are there in the lake? You can use a proportion like this:

$$\frac{\text{Number of bream tagged}}{\text{Total number of bream in the lake}} = \frac{\text{Number of tagged bream caught}}{\text{Total number of bream caught}}$$

26. In 1979 a man in Washington completed 5000 successful volleys of a ping pong ball in 45 minutes.

   a. At this rate, how long would it take to complete 1500 successful volleys?

   b. If he had continued at this rate, how many volleys would he have completed in an hour?

7.4 MIXED PRACTICE

By doing these exercises, you will practice all the topics in this chapter up to this point.

27. Candy sells for \$1.49 for 16 ounces. What is the price per ounce?

28. Determine whether $5:\frac{2}{7}$ and $2\frac{1}{2}:\frac{2}{12}$ are proportional.

29. A certain store sold orange juice, tomato juice, and grapefruit juice. On a particular day, the manager sold 15 cans of orange juice, 26 cans of tomato juice, and 11 cans of grapefruit juice. Write the ratio of the amount of grapefruit juice sold to the total amount of the other two juices.

30. Determine whether these are proportional:

$\frac{1}{8}$ is to $\frac{3}{16}$ as 4 is to 6

31. An employee earns \$235.25 for 4 hours of work. What is his rate per hour?

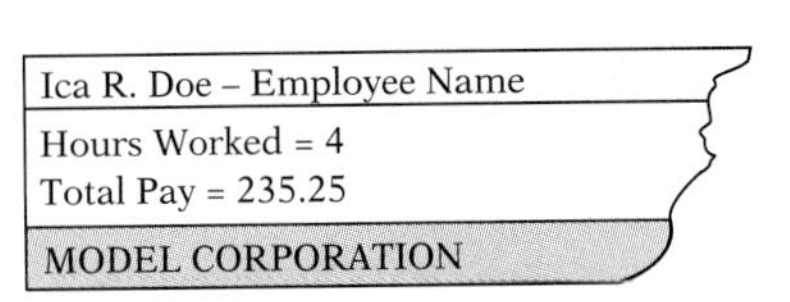
Ica R. Doe – Employee Name
Hours Worked = 4
Total Pay = 235.25
MODEL CORPORATION

32. The speed of light is 186,000 miles per second. Express this as miles per minute.

33. A shoe store sold 4 pairs of high-heeled shoes for every 3 pairs of low-heeled shoes. Write the number of pairs of high-heeled shoes compared to the number of shoes sold as a ratio.

34. Of 20 college students who were surveyed, 7 preferred classical music and 13 preferred rock and roll. At that same ratio, how many prefer classical music if a total of 195 students are found to prefer rock and roll?

35. Every week Andre does laps and sit-ups in a ratio of 15 to 35. If he did 210 situps, how many laps did he run?

36. Find the missing number in this proportion:

$$\frac{2}{5}:n = 2:5$$

37. It costs \$2175.50 to pay an instructor to teach a class. What is the cost per hour if the instructor is paid for 50 hours?

Similar Triangles and Indirect Measurement

To find certain distances, we can sometimes use a method called **indirect measurement,** in which we find the required distance by measuring a second distance and then using ratio and proportion.

Suppose a flagpole is casting a shadow that is 10 feet long at noon; at the same time, a 3-foot-tall fire hydrant is casting a $\frac{1}{2}$-foot shadow. We can use the following proportion to find the height of the flagpole.

$$\frac{\text{Length of shadow}}{\text{Height of object}} = \frac{\frac{1}{2}}{3} = \frac{10}{n}$$

Using the method of Section 7.2, we equate the product of the extremes $\left(\frac{1}{2} \times n\right)$ and the product of the means (30) and solve. Setting these two products equal to each other, we get

$$30 = \frac{1}{2} \times n$$

$$\frac{30}{\frac{1}{2}} = \frac{\frac{1}{2} \times n}{\frac{1}{2}} \qquad \text{Dividing by } \tfrac{1}{2}$$

$$30 \div \frac{1}{2} = n \qquad \text{Simplifying on the right}$$

$$30 \times \frac{2}{1} = n \qquad \text{Simplifying on the left}$$

$$60 = n \qquad \text{Solving}$$

So the flagpole is 60 feet tall.

This procedure works because of a property of triangles. If two triangles have equal angles, then the triangles are said to be **similar,** and their sides are proportional. We were using the fact that the sides are proportional.

Look at the following diagram. The trick to our indirect measurement was that we looked at the shadows of the flagpole and the fire hydrant at the same time of day. Thus the three angles of the flagpole triangle were equal to those of the hydrant triangle.

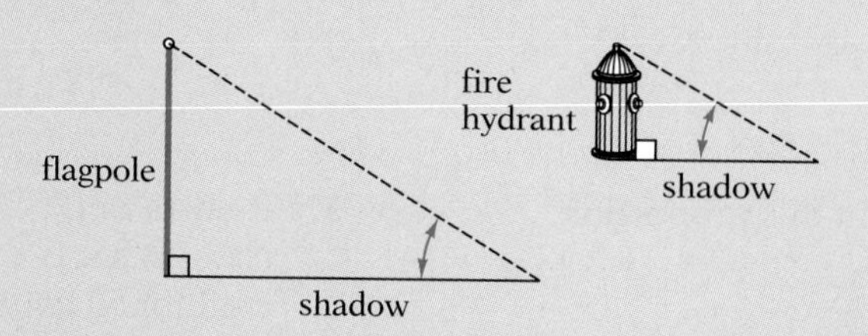

We can use this same property of similar triangles to measure distances across a lake or the height of a building.

7.5 Geometry and Measurement: Finding Volumes and Surface Areas of Spheres and Cylinders

SECTION GOALS

- To find the volume and surface area of a sphere
- To find the volume and surface area of a cylinder

In Chapter 5, we discussed three-dimensional objects whose faces were squares or rectangles. Some three-dimensional figures have curved surfaces. Among them are the two that we will discuss next—the sphere and the cylinder. A globe, a basketball and a marble are familiar examples of spheres. Metal cans are shaped like cylinders.

Volume of a Sphere

A sphere is shown in the following figure; it is a "three-dimensional circle." Every point on the sphere is at a distance r (the radius) from the center.

Sphere

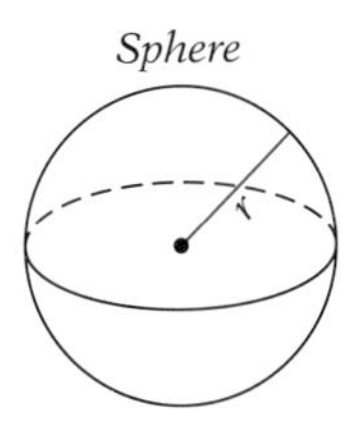

The volume of a sphere is found with a formula:

To find the volume of a sphere

If r is the radius of the sphere, then

$$V = \frac{4}{3}\pi r^3$$

EXAMPLE 1

Find the volume of a sphere with a radius of 6 inches.

SOLUTION

To find the volume, we substitute and simplify.

$$V = \frac{4}{3}\pi r^3$$
$$= \frac{4}{3}(3.14)(6)^3$$
$$= \frac{4}{3}(3.14)(216)$$
$$= \frac{(12.56)\ (216)}{3} = \frac{2712.96}{3}$$
$$= 904.32$$

So the volume of the sphere is 904.32 cubic inches.

■ *Now do margin exercises 1a and 1b.*

Note that volume is always measured in cubic units.

1a. Find the volume of a sphere with a radius of one-eighth inch. $\left(\text{use } \pi \approx \frac{22}{7}\right)$

1b. Find the volume of a sphere with a diameter of one hundred miles. (use $\pi \approx 3.14$)

2a. Find the volume of a canister that has a radius of 10 inches and a height of 12 inches.

2b. Find the volume of a cylindrical candle that has a radius of 5 inches and a height of 11 inches.

Volume of a Cylinder

A right circular cylinder is shown here. (We shall simply call it a cylinder.) Its bases are circles with the same radius, r.

Right Circular Cylinder

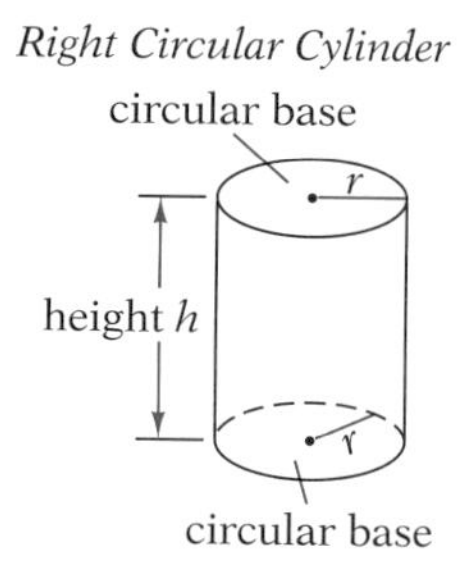

Cylinders are solid figures. (Think of them as being closed on the ends.) The volume of a cylinder is equal to the area of one of its bases times its height. Because the area of a base is πr^2, we have the following formula:

To find the volume of a cylinder

If r is the radius of the base and h is the height, then

$$V = \pi r^2 h$$

EXAMPLE 2

Find the volume of a can that has a radius of 8 inches and a height of 12 inches.

SOLUTION

First we substitute the values into the formula for the volume of a cylinder.

$$\begin{aligned} V &= \pi r^2 h \\ &= (3.14)(8^2)(12) \end{aligned}$$

Then we simplify the resulting expression.

$$V = (3.14)(64)(12) = 2411.52$$

The volume of the cylinder is 2411.52 cubic inches.

■ *Now do margin exercises 2a and 2b.*

Surface Area of a Sphere

If we think about the amount of surface "covering" a sphere or cylinder, we are considering the "surface area." Area is always expressed in square units. The surface area of a sphere is four times the area of a circle with the same radius as the sphere.

To find the surface area of a sphere

If r is the radius of the sphere, then

$$SA = 4\pi r^2$$

EXAMPLE 3

Find the surface area of a sphere with a radius of 2.1.

SOLUTION

To find the surface area, we substitute in the formula and simplify.

$$\begin{aligned} SA &= 4\pi r^2 \\ &= 4(3.14)(2.1)^2 = 12.56(4.41) \\ &= 55.3896 \end{aligned}$$

The surface area of this sphere is 55.3896 square units.

■ *Now do margin exercises 3a and 3b.*

3a. Find the surface area of a sphere with a radius of 5 cm.

3b. A sphere with a diameter of $2\frac{3}{8}$ inches is being covered with special fabric. How much will the covering cost to the nearest dollar if the price of the fabric is $10.90 per square inch?

Surface Area of a Cylinder

The following figure shows that we can think of the cylinder as being made up of three parts: two circular ends and a rectangle rolled up to form the side. The length of the rectangle is the circumference of a base, or $2\pi r$; its width is the height h. The surface area of the cylinder can be found by adding together the areas of the two end circles (πr^2 each) and the area of the rectangular part ($h \times 2\pi r$).

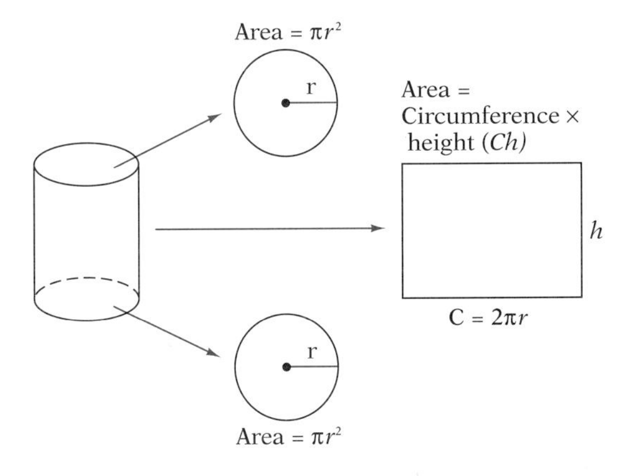

4a. Find the surface area of a cylinder with radius 8 inches and height 12 inches.

4b. A label completely covers the 10 cm height of a can with a diameter of 10 cm. What is the area of the label?

To find the surface area of a cylinder

If r is the radius and h is the height, then

$$SA = \pi r^2 + \pi r^2 + 2\pi rh \text{ or } 2\pi r^2 + 2\pi rh \text{ or } 2\pi r(r + h)$$

EXAMPLE 4

Find the surface area of a cylinder with a radius of 6 inches and a height of 12 inches.

SOLUTION

We substitute into the formula to find the surface area.

$$\begin{aligned} SA &= 2\pi r(r + h) \\ &= 2(3.14)(6)(6 + 12) \\ &= 678.24 \end{aligned}$$

The surface area of the cylinder is 678.24 square inches.

■ *Now do margin exercises 4a and 4b.*

Work the exercises that follow.

7.5 EXERCISES

In Exercises 1 through 27, use 3.14 as an approximation for π unless otherwise directed.

1. Find the surface area of a sphere with a radius of 2.5 inches.

2. Find the volume of a soda can that has a radius of $1\frac{1}{4}$ inches and a height of $4\frac{7}{8}$ inches. Use $\frac{22}{7}$ as an approximation for π.

3. You double the radius of a sphere. How does its surface area change?

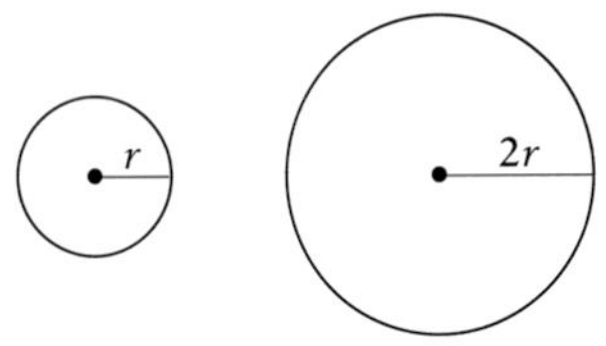

r $2r$

4. You are cutting a label for a can. The can has a diameter of 2.5 inches and is 5.5 inches tall. What is the area of the label that fits exactly on this can?

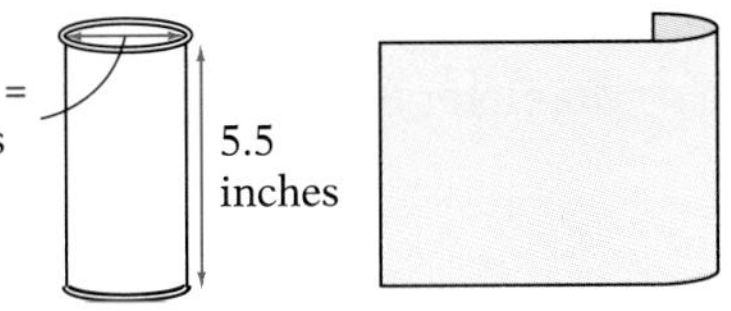

Diameter = 2.5 inches

5.5 inches

5. Find the surface area of a sphere with a radius of 6.25 inches.

6. A beach ball with a radius of 12 inches is filled with air. Another beach ball with a radius of 15 inches is also filled with air. What is the difference in volume between the two balls?

7. A container shaped like a cylinder holds glue for woodworking projects. If the container sits 9.5 inches high and has a diameter of 3 inches, how much glue can it hold?

8. What is the surface area of a cylinder with a radius of 3.5 inches and a height of 2.5 inches?

9. The surface of a sphere with a radius of $\frac{1}{5}$ units is to be painted. How much paint will be needed if 1 gallon of the paint covers exactly 12.5 square units?

10. Modeling clay is shaped into a long cylinder with a diameter of 4 inches. A piece 6 inches long is cut off. What volume of clay is in this piece?

11. What is the surface area of a cylinder with a radius of 0.5 unit and a height of 2.5 units?

12. Find the volume of a pipe that is cylindrical in shape and has a length of 8 feet and a radius of $4\frac{1}{2}$ feet.

13. Find the surface area of (a) a sphere with a radius of 0.001 unit and (b) a sphere whose radius is 4 times greater. (c) How do these compare?

14. A can shaped like a cylinder is filled with water. If it has a radius of 3 inches and a height of 10 inches, how much water can it hold?

Use the following information to answer Exercises 15 through 24. Neither the Sun nor the planets are perfect spheres; they are flattened somewhat at their poles. However, we can use their diameters to determine their approximate surface areas. Do so, using a calculator and the given diameters. (Hint: First round each diameter to the nearest thousand miles.)

15. Sun diameter 865,500 miles

16. Mercury diameter 3032 miles

17. Pluto diameter 3700 miles

18. Mars diameter 4217 miles

19. Venus diameter 7521 miles

20. Earth diameter 7926 miles

21. Neptune diameter 30,800 miles

22. Uranus diameter 32,200 miles

23. Saturn diameter 74,600 miles

24. Jupiter diameter 88,700 miles

25. Water weighs 62.4 pounds per cubic foot. You have a bucket that is shaped like a cylinder with a diameter of 1 foot and a height of 1 foot and it is filled with water. How much does the water weigh?

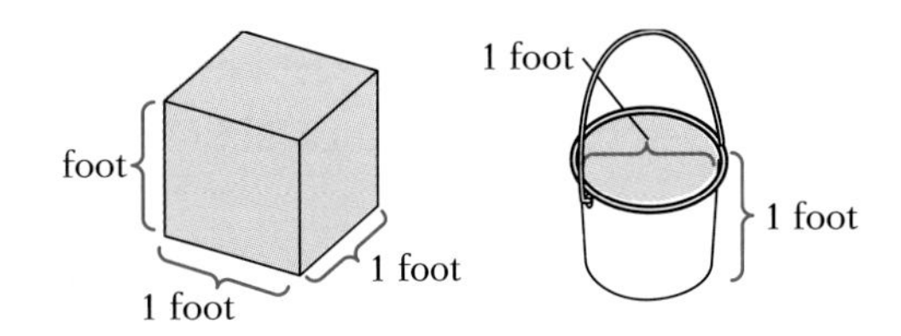

26. What is the surface area of a cylinder with a radius of 0.001 unit and a height of 0.001 unit?

27. A can of soup is emptied into a pan. Both the can and the pan are shaped like cylinders. The soup can is 4.5 inches tall with a diameter of 2.5 inches. The pan is 3 inches tall with a diameter of 5 inches. Is there enough space in the pan for a second can of soup?

In Exercises 28 through 31, find the surface area. Use the relationship that the diameter is equal to the circumference divided by pi (use $\pi \approx 3.14$). Round answers to the nearest hundredth. Use your calculator if you wish.

28. A basketball has a circumference of 78 centimeters.

29. A soccer ball has a circumference of 28 inches.

30. A cricket ball has a circumference of 9 inches.

31. A volleyball has a circumference of 67 centimeters.

7.5 MIXED PRACTICE

By doing these exercises, you will practice all the topics in this chapter up to this point.

32. In the New York City marathon, 284 gallons of blue paint are used to paint the line to direct the runners. How many liters of paint is this?

33. Determine whether $\frac{2}{12}$ and $\frac{3}{18}$ are proportional.

34. What is the surface area of a cylinder that has a height of $4\frac{1}{2}$ inches and a radius of 3 inches? (use $\pi \approx 3.14$)

35. Pens cost $2.24 per dozen. What is the cost per pen?

36. A microwave oven can cook a potato in 7.5 minutes. How many seconds is this?

37. A motor vehicle station can test 40 people in 2 hours. Express the number of people per hour as a unit rate.

38. What is the surface area of a sphere with a radius of 10 mm?

39. Convert the national speed limit, 55 miles per hour, to kilometers per hour.

40. Find the missing value: $n:\frac{2}{5} = \frac{3}{8}:1$

41. A gasoline bill was $15.95 for 13 gallons. Find the unit price.

42. A kitten unraveled a ball of yarn that measured 112.5 yards. How many feet is this?

43. A cylinder has a diameter of 1 mm and a height of 10 mm. What is its volume?

44. How many centimeters long is a carrot that measures 6 inches long?

Chapter 7 Review

ERROR ANALYSIS

These problems have been worked incorrectly. Tell what the error is and then write the correct solution.

1. If 3500 pieces of equipment cost $700,000, how much do 5000 pieces cost?

Incorrect Solution *Correct Solution*

$$\frac{700{,}000}{3500} = \frac{n}{5000}$$

$$\frac{7000}{35} = \frac{n}{5000}$$

$10,000,000

Error ______________________________

____

2. If 7 bowls cost $28, how much do 8 cost?

Incorrect Solution *Correct Solution*

$$\frac{28}{7} = \frac{8}{n}$$

$$28 \times n = 56$$

$$n = 2$$

Error ______________________________

____

3. Twenty-seven boys take art for every 9 girls who take art. This semester, 100 students are taking art. How many are girls?

Incorrect Solution *Correct Solution*

$$\frac{9}{27} = \frac{n}{100}$$

$$27 \times n = 900$$

$n =$ approximately 33 girls

Error ______________________________

____

4. Sixteen cans of cat food cost $4. What is the unit price?

Incorrect Solution

$$4\overline{)16}\quad 4$$

Unit price is $4 per can.

Error ______________________________

____

Correct Solution

INTERPRETING MATHEMATICS

By working these exercises, you will test and strengthen your mathematics vocabulary.

1. Define, in your own words, the terms ratio and rate. How are they alike? How do they differ?

2. What does the word unit mean in unit rate or unit price?

3. The word proportion has a mathematical meaning and a different everyday meaning. Write two sentences, one using proportion as a mathematics word and the other using it as a more common term.

4. The means and the extremes of a proportion are used in a test for equality called the cross-products test. Identify them, using an example, and show how we use them in this test.

5. Next to each of the following units of measurement, write the proper letter to show whether it is a unit of length (L), area (A), volume (V), mass (M) or liquid measure (Q).

a. centimeter
b. cubic millimeter
c. kilometer
d. metric ton
e. square centimeter
f. square millimeter
g. cubic meter
h. kilogram
i. meter
j. milliliter
k. square meter
l. cubic centimeter
m. gram
n. liter
o. millimeter
p. square kilometer

6. Name three familiar items that have the shape of a sphere. Name three that have the shape of a cylinder.

7. A student knows how to find the area of a circle and that of a rectangle. Explain to him, in your own words, how he can use those skills to find the surface area of a cylinder.

REVIEW PROBLEMS

The exercises that follow will give you a good review of the material presented in this chapter. Work through them and check your answers at the back of the book.

Section 7.1

1. A recent purchase of $3848 is to be repaid over a period of 13 months. If the payment is the same each month, write the monthly cost as a ratio of dollars per month.

2. A small college says that it has a high ratio of professors to students. What is the ratio of professors to students if the college has 121 students and 24 professors?

In Exercises 3 and 4, write each of the following as a ratio in simplest form in three different ways.

3. $50 to $45.60

4. $16\frac{1}{4}$ ounces to 8 ounces

5. A school mimeograph machine can duplicate 400 pages in 5 minutes. Determine the unit rate in pages per minute.

6. A person may have a pulse rate of 480 in 600 seconds. Write as beats per second.

Section 7.2

7. Find the missing value: $\frac{69}{92} = \frac{30}{n}$

8. Is 1.2 : 2.4 = 3.6 : 6 a true proportion?

9. Which is the smaller ratio, 1.05 to 3.00 or 2 to 2.5?

10. Does $\frac{13}{39} = \frac{5}{15}$?

11. Find the missing value: $\frac{8}{11} = \frac{n}{77}$

12. Which is the larger ratio, $1\frac{1}{2}$ to 4 or $2\frac{1}{4}$ to 6?

Section 7.3

13. How many meters are there in 480 centimeters?

14. A rope is 45 centimeters long. How many millimeters is this?

15. Convert 1132 centimeters into meters.

16. Two hundred minutes is equivalent to how many hours? (Express as a fraction.)

17. $3\frac{1}{2}$ miles is how many inches?

18. 6.4 hours is how many seconds?

Section 7.4

19. Out of every eight students, three prefer taking business math rather than algebra. If 125 students take algebra, how many students take business math?

20. A worker can assemble 13 machines in 8 hours. At this rate, what is the time needed to assemble one machine? (Express as a fraction.)

21. An assembly line can install 215 car mufflers in 15 minutes. At this rate, how many mufflers can be installed in one hour?

22. Two classes are competing by selling cookies in a local contest. In order to be consistent with last year's results, for every 25 boxes school A sells, school B must sell 40 boxes. If school A sold 1000 boxes, how many boxes must school B sell?

23. A hotel bill for 15 people amounted to $5625. What was the total cost per person if each person paid the same amount?

24. For every 13 tulips he planted, Auris planted 15 irises. If he planted 130 tulips this year, how many irises did he plant?

Section 7.5

In Exercises 25 through 30, use 3.14 as an approximation for π and round your answers to the nearest hundredth where necessary.

25. Find the volume and surface area of a sphere with radius 6.7 inches.

26. A cylinder 6 inches high and $2\frac{1}{2}$ inches in diameter is to be covered with contact paper. How much surface area will have to be covered?

27. Find the surface area of a sphere with a radius of 20 miles.

28. A cylindrical can contains soup. The can is $5\frac{1}{2}$ inches tall and 3 inches in diameter. What volume of soup can it hold?

29. A wheel that has a diameter of 27 inches will have traveled how far after 120 revolutions?

30. There are three wheels on a tricycle. Two have diameters of 8 inches, and one has a diameter of 25 inches. How many revolutions will the smaller wheels have made when the larger one has made 10 complete revolutions?

Mixed Review

31. Write the following as a ratio in three ways. Do not reduce to lowest terms.
$50\frac{1}{2}$ days to 80 days;

32. Write the following as a ratio in three ways. Reduce to lowest terms.
30 pounds to 20 pounds

33. Mary Paul makes $850 per week and spends $200 of that on her rent. Write the ratio of rent per week to earnings per week.

34. Write 160 miles on 10 gallons of gas in terms of gallons per mile.

35. Write 180 words in 240 seconds in terms of seconds per word.

36. Is 11:198 the same proportion as 17:187?

37. A secretarial service required that its employees be able to type 265 words in 5 minutes. What is the number of words per minute that they must type?

38. Which is a smaller ratio: 0.0002:0.0001 or 0.002:0.001?

39. Which is the larger ratio, 2:3 or 7:8?

40. Find the missing value: $\frac{66}{30} = \frac{88}{n}$

41. Determine whether 3:7 = 15:21 is a true proportion.

42. Find the missing value in this proportion.
$2\frac{3}{4}:5\frac{1}{2}$ as $6\frac{1}{8}:n$

43. Write a pulse rate of 480 in 10 minutes in terms of beats per minute.

44. Write 160 miles on 10 gallons of gas in terms of miles per gallon.

45. Which is the better buy, apple juice at 89 cents per $\frac{1}{2}$ gallon or at $1.70 per gallon?

46. Which is the better buy, three 5-ounce cans of cranberry sauce for $1 or one 16-ounce can for $1.29?

47. Baby oil costs $3.29 for 20 ounces. If you buy 240 ounces, how much will you pay?

48. It costs $8.25 to develop 24 photographic prints. What is the cost per print?

Chapter 7 Test

This exam tests your knowledge of the topics in Chapter 7. Express answers in simplest form. Use 3.14 for π.

1. a. The inventory value of coffee is \$0.25 per can, and the inventory value of tea is \$0.15 per bag. What is the ratio of inventory value of coffee to that of tea if the number of bags of tea and the number of cans of coffee are the same?

b. A certain store sold 130 ice cream cones on Tuesday. It sold 60 vanilla cones, and the rest were chocolate. Find the ratio of non-chocolate cones to chocolate cones.

c. A person rides his bike at the rate of 25 miles every 5 hours. Write this as a unit rate.

2. a. Determine which ratio is larger, 1 : 5 or 15 : 25?

b. Find the missing value: $\frac{6}{27} = \frac{42}{n}$

c. Is 11 : 6 = 121 : 66?

3. a. How many grams are in 15 kilograms?

b. How many quarts are there in 5 gallons?

c. How many meters are there in 1500 centimeters?

4. a. A certain machine can reproduce 25 cassettes in 20 minutes. How long would it take to reproduce 135 cassettes?

b. It costs a certain company \$45.36 to manufacture 3 irons. How much does it cost to make 72 irons?

c. A shadow cast by a 32-foot telephone pole is 20 feet long. How tall is a nearby tree that is casting a 23-foot shadow?

5. a. Find the volume of a cylinder with a radius of 4.5 centimeters and a height of 5 centimeters.

b. Find the surface area of a sphere with a diameter of 3 centimeters.

c. A package is wrapped with silver paper to cover a cylinder that has a diameter of 1 foot 3 inches and a height of $6\frac{1}{2}$ inches. What is the surface area to be covered?

Cumulative Review

CHAPTERS 1–7

The questions that follow will help you maintain the skills you have developed. Write all fractions in simplest form.

1. Add: $234 + 708 + 89$

2. Simplify: $11^1 + (9 - 5)^2 - 4(5)$

3. Subtract: $900 - 742$

4. Multiply: 120×506

5. Rewrite 23,509 in expanded notation.

6. Multiply: $2\frac{1}{3} \times 4\frac{5}{7}$

7. Divide $7803 \div 12$; show any remainder using *r*.

8. Find the sum: $56.009 + 12 + 76.9$

9. Divide: $5 \div 3\frac{3}{4}$

10. Cotton swabs cost $1.29 for 90 swabs. What is the unit price? Round to the nearest tenth of a cent.

11. Arrange 2.007, 2.07 and 2.7 in order from largest to smallest.

12. Find the product: 0.008×100

13. Find the difference: $5600 - 76.981$

14. Sales taxes in New York City are assessed at a rate of 8.25 cents per dollar.

a. At this rate, how much tax would you pay on an $8 purchase? **b.** At this rate, how much tax would you pay on a $60 purchase?

15. Find the sum: $7.08 + 8.092$

16. Round 124.098 to the nearest tenth.

17. Find the quotient: $0.22\overline{)6710}$

18. Maggie spends 6 days working to every 2 days she spends traveling. In 360 days, how much time does she spend traveling?

19. Add: $6\frac{3}{8} + 4\frac{4}{5}$

20. Find the product: 3.45×2.1

21. Subtract: $4\frac{1}{5} - 2\frac{7}{9}$

22. Write the numeral expressed by

$$40{,}000 + 300 + \frac{8}{100} + \frac{4}{100{,}000}$$

8 Percent

When you take out a loan to buy a condominium or house or when you refinance a mortgage, banks charge certain points or origination fees in addition to the cost of borrowing the money for fifteen or thirty years. Each point is equal to 1% of the total amount borrowed and is generally required in advance, unlike the mortgage payments. Depending on your financial situation, selecting the lowest interest rate advertised might not be your most economical choice.

- *Research some newspapers. Which bank really offers the best alternative?*

Skills Check

Take this short quiz to see how well prepared you are for Chapter 8. The answers are given at the bottom of the page. So are the sections to review if you get any answers wrong.

1. Multiply: 18.461×100

2. Divide: $83.92 \div 1000$

3. Multiply: 152.9×10.75

4. Divide: $26\overline{)26.78}$

5. Find the product: $\frac{2}{100} \times \frac{52}{65}$

6. Change $\frac{5}{9}$ to a decimal rounded to the nearest hundredth.

7. Change 1.086 to a mixed number.

8. Reduce $\frac{36}{144}$ to lowest terms.

ANSWERS: 1. 1846.1 [Section 6.5] 2. 0.08392 [Section 6.6] 3. 1643.675 [Section 6.5] 4. 1.03 [Section 6.6] 5. $\frac{2}{125}$ [Section 5.1] 6. 0.56 [Section 6.6] 7. $1\frac{43}{500}$ [Section 6.1] 8. $\frac{1}{4}$ [Section 4.3]

8.1 The Relationship Between Percents and Decimal Numbers

The word "percent" literally means "per hundred." Its symbol is %, as in 52%. We write a percent to specify a ratio—a number of things compared to 100 of those things. That is,

$$n\% = \frac{n}{100}$$

The entire amount (of anything) is expressed as 100% (of that thing) because

$$100\% = \frac{100}{100} = 1$$

$$100\% = \frac{100 \text{ elements}}{100 \text{ elements}} = 1$$

Thus if a class of 40 students has 100% attendance, then all 40 students are in class. And if a group of 7 fish is 100% goldfish, then all 7 fish are goldfish.

Suppose that 25% of the students in a class are girls. Then what percent of the students are boys? We know that all the students are 100%, and the girls are 25%. So the boys must be the difference:

$$100\% - 25\% = 75\%$$

EXAMPLE 1

In a certain poll, a video by U2 received 85% of the votes as top choice. What percent of the top-choice votes was received by other videos?

SOLUTION

The total percent of top-choice votes was 100%. The percent received by U2 was 85%. Then the remaining percent must be 100% – 85%, or 15%.

So the other videos received 15% of the top-choice votes.

■ *Now do margin exercises 1a and 1b. The answers are given at the back of the book.*

SECTION GOALS

- To understand the meaning of percent
- To change percents to decimal numbers, and vice versa

1a. In a second poll, people were to choose between three music videos. A video by the B-52's received 36.3% of the vote. An M. C. Hammer video received 34.78%. What percent was received by the third video?

1b. In a certain survey, people were to choose between political parties. Democrats and Republicans tied with 43% of the vote each. How much of the vote did all other parties receive?

2a. Write 253%, 0.0087%, and 72% as decimal numbers.

2b. Change 365%, 8%, and 0.4% to decimal numbers.

3a. Rewrite 0.63 and 5.834 as percents.

3b. Thirty-two hundredths of a class are female. What are the percents of females and of males in the class?

Sometimes when we are working with percents, it is necessary to rewrite them as decimals. To do so, remember that the percent sign means "per hundred."

To write a percent as a decimal number

1. *Divide* the percent by 100 by moving the decimal point two places to the left.
2. Drop the percent sign.

EXAMPLE 2

Write 517% and 0.9% and $6\frac{1}{2}\%$ as decimal numbers.

SOLUTION

We need to move the decimal point two places to the left "in exchange" for the %.

$$517\% = 5.17. = 5.17$$
$$0.9\% = 0.00.9 = 0.009$$
$$6\frac{1}{2}\% = 6.5\% = 0.06.5 = 0.065$$

■ *Now do margin exercises 2a and 2b.*

We change from a whole number or a decimal number to a percent in the opposite way.

To write a decimal number as a percent

1. *Multiply* by 100 by moving the decimal point two places to the right.
2. Add a percent sign.

EXAMPLE 3

Write 0.46 and 17 and 3.829 as percents.

SOLUTION

We need to move the decimal point two places to the right and add a percent sign.

$$0.46 = .46\% = 46\%$$
$$17 = 17. = 17.00\% = 1700\%$$
$$3.829 = 3.82.9\% = 382.9\%$$

■ *Now do margin exercises 3a and 3b.*

Work the exercises that follow.

8.1 EXERCISES

In Exercises 1 through 20, write these decimal numbers as percents.

1. 1.85 **2.** 46.85 **3.** 2.46 **4.** 94.27

5. 0.348 **6.** $0.0\overline{5}$ **7.** $19.6\overline{6}$ **8.** 0.2674

9. 0.9 **10.** 0.0001 **11.** 0.1 **12.** 0.008

13. 5.1593 **14.** 1.1 **15.** 6.278 **16.** 200.1

17. 2 **18.** 4 **19.** 40,257 **20.** 100

In Exercises 21 through 36, write the percents as decimal numbers.

21. 150% **22.** 250% **23.** 135% **24.** 60%

25. 3% **26.** 8% **27.** 2% **28.** 7.2%

29. 0.5% **30.** 0.98% **31.** 1.096% **32.** 0.7%

33. $63\frac{1}{2}\%$

34. $18\frac{1}{5}\%$

35. $2\frac{1}{4}\%$

36. $5\frac{3}{4}\%$

In Exercises 37 through 41, solve the word problems.

37. The Earth's surface is 70% water. The rest is land.

a. Write the percent of the Earth's surface that is water as a decimal.

b. What percent of the Earth's surface is land?

c. What percent represents the entire Earth's surface, including both land and water?

d. What is the ratio of land to water on the Earth's surface?

38. The East Sahara has weather that includes 97% sunshine during daylight hours.

a. Write this percent as a decimal.

b. For what percent of the daylight hours is there not sunshine?

39. Efficiency is the ratio of output (work) to input (food or fuel). The efficiencies of several machines follow. Write each percent as a decimal.

Electric motor	80.3%
Steam turbine	39.87%
Gasoline engine	25.2%
Steam engine	13.44%
Human body	34.8%

40. Louis XIV was the king of France for 95% of his lifetime. Write this percent as a decimal.

41. The population of New York City is approximately 41% of the population of New York State, and it is 3% of the population of the entire United States.

a. What percent of the population of the United States does not live in New York City?

b. What percent of the population of New York State does not live in New York City?

c. Write the percent of the population of New York State that lives in New York City as a decimal.

8.2 The Relationship Between Percents and Fractions

Writing percents as fractions is very simple because the percent sign really means "divided by 100." We already saw that

$$n\% = \frac{n}{100}$$

To write a percent as a fraction

1. Write a fraction with the percent as numerator and with 100 as denominator.
2. Drop the percent sign and simplify.

EXAMPLE 1

Write 230% and 17% as fractions.

SOLUTION

We "exchange" the % sign for a denominator of 100 and simplify.

$$230\% = \frac{230}{100} = 2\frac{30}{100} = 2\frac{3}{10}$$

Thus 230% is $2\frac{3}{10}$ in fractional form.

$$17\% = \frac{17}{100}$$

This can't be simplified, so 17% in fractional form is $\frac{17}{100}$.

■ *Now do margin exercises 1a and 1b.*

An extra step is needed when the percent has a decimal part.

EXAMPLE 2

Write 0.4% as a fraction.

SOLUTION

Using the given procedure yields

$$0.4\% = \frac{0.4}{100}$$

and we get a decimal number in the numerator. To eliminate it, we multiply numerator and denominator by 10.

$$\frac{0.4}{100} = \frac{0.4 \times 10}{100 \times 10} = \frac{4}{1000} = \frac{1}{250}$$

So 0.4% in fraction form is $\frac{1}{250}$.

■ *Now do margin exercises 2a and 2b.*

SECTION GOAL

■ To write a percent as a fraction, and vice versa

1a. Write 25% and 368% as fractions.

1b. Write 1045% and 6% as fractions.

2a. Write 0.5% and 0.005% as fractions.

2b. Write 3.4% and 0.34% as fractions.

3a. $\frac{3}{30}$ of the days in April were rainy and $\frac{13}{30}$ were sunny. Write these fractions as percents.

APRIL						
S	M	T	W	T	F	S
			1 S	2 S	3 S	4 S
5 S	6 S	7 S	8 S	9 S	10 S	11 S
12 S	13 R	14 S	15 R	16 R	17	18
19	20	21	22	23	24	25
26	27	28	29	30		

3b. Rewrite $5\frac{5}{9}$ and $\frac{11}{17}$ as percents. Round to the nearest hundredth of a percent.

4a. Change 4.63% to a decimal and to a fraction.

4b. Change $56\frac{1}{7}\%$ to a fraction and to a decimal rounded to the nearest hundredth.

To write a fraction or mixed number as a percent

1. Write the fraction or mixed number as a decimal number.
2. Write the decimal number as a percent.

EXAMPLE 3

Write $\frac{5}{3}$ and $2\frac{1}{2}$ as percents.

SOLUTION

To write $\frac{5}{3}$ as a decimal number, we divide 5 by 3.

$$\frac{5}{3} = 3\overline{)5.00000} \quad 1.66666\ldots$$

We can round 1.66666 . . . to 1.6667.

Next we write the decimal number as a percent.

$$1.6667 = 166.67\%$$

(Note that the improper fraction we started with gave us a percent greater than 100%.)

$$2\frac{1}{2} = 2.5 = 250\%$$

So $\frac{5}{3}$ is 166.67% and $2\frac{1}{2}$ is 250%.

■ *Now do margin exercises 3a and 3b.*

EXAMPLE 4

Change $87\frac{1}{2}\%$ to a fraction and to a decimal.

SOLUTION

First we rewrite $87\frac{1}{2}\%$ as 87.5%. Then,

$$87.5\% = 0.875$$

and

$$0.875 = \frac{875}{1000} = \frac{7}{8}$$

So $87\frac{1}{2}\%$ is 0.875 as a decimal as $\frac{7}{8}$ as a fraction.

■ *Now do margin exercises 4a and 4b.*

Work the exercises that follow.

8.2 EXERCISES

In Exercises 1 through 24, write the percents as fractions in simplest form.

1. 5%

2. 1%

3. 7%

4. 6%

5. 15%

6. 30%

7. 45%

8. 198%

9. $\frac{2}{5}\%$

10. $\frac{1}{8}\%$

11. $\frac{3}{5}\%$

12. $\frac{8}{5}\%$

13. 214%

14. 825%

15. 190%

16. 32%

17. $20\frac{1}{2}\%$

18. $40\frac{1}{4}\%$

19. $32\frac{1}{5}\%$

20. $\frac{1}{5}\%$

21. 120.75%

22. 113.8%

23. 23.19%

24. 65.17%

In Exercises 25 through 42, write the fractions and mixed numbers as percents. Round quotients where necessary to the nearest hundredth of a percent.

25. $\frac{1}{20}$ **26.** $\frac{1}{4}$ **27.** $\frac{9}{5}$ **28.** $\frac{1}{15}$

29. $\frac{3}{16}$ **30.** $\frac{1}{25}$ **31.** $\frac{1}{7}$ **32.** $\frac{1}{9}$

33. $\frac{42}{17}$ **34.** $5\frac{3}{7}$ **35.** $6\frac{7}{8}$ **36.** $\frac{1}{100}$

37. $3\frac{3}{4}$ **38.** $4\frac{5}{9}$ **39.** $\frac{1}{200}$ **40.** $\frac{8}{11}$

41. $\frac{6}{13}$ **42.** $\frac{201}{37}$

43. About one-fifth of the hot deserts are covered by sand, and approximately nine hundred seventy-two thousandths of the world's water supply is found in the ocean. Rewrite these amounts as percents.

44. The weight of a tractor is about $1\frac{1}{5}$ times the weight of an elephant, and the weight of a greyhound bus is about 1.9 times the weight of an elephant. Write these amounts as percents; that is, write each weight as a percent of the weight of an elephant.

45. The Himalayan Karakoram range is the world's biggest group of mountains. It contains 96 of the world's highest 109 peaks. Express this ratio as a percent.

In Exercises 46 through 52, use the following table to write the amount of time requested as a fraction.

Stages of the Average Life Span

Stage	Percent
Pre-birth development and growth	1%
Infancy	3%
Childhood	14%
Adolescence	9%
Prime of life	31%
Middle age	29%
Old age	13%

46. The total percent of the life span that is not called "old age"

47. The time spent in the prime of life

48. The time spent during childhood and adolescence

49. The time spent in prenatal development and infancy

50. The time spent in middle and old age

51. The time spent in the years prior to adolescence

52. The time spent in the years after adolescence

8.2 MIXED PRACTICE

By doing these exercises, you will practice all the topics in this chapter up to this point.

53. Which of the fractions $\frac{11}{4}$, $1\frac{2}{3}$, and $\frac{14}{15}$ represent(s) a percent less than 100%?

54. Write 88% as a fraction reduced to lowest terms.

55. Write 8.5% as a decimal.

56. Write 26% as a decimal.

57. Write 14% as a fraction.

58. Write 36.8% as a fraction.

59. Which of these represent(s) approximately 70%?

$0.65 \quad \frac{2}{3} \quad 4.2$

60. Write $3\frac{4}{5}$ as a percent.

61. Write $\frac{18}{12}$ as a percent.

62. Write $\frac{3}{5}$ as a percent.

8.3 Solving Percent Problems Using the Percent Equation

The Percent Equation

Three types of questions identify percent problems. Examples of the three are

a. What is 8% of 200?
b. 2 is what percent of 135?
c. 16 is 5% of what amount?

All three types can be answered by using this general form of the **percent equation:**

Percent Equation

$$\text{Percent (as a decimal)} \times \text{base} = \text{amount}$$

The equation says that if we take a certain *percent* of a number called the *base*, the result is a number called the *amount*. We "take a percent" of a number by multiplying the percent (in decimal form) by the number.

Before we can do that, however, we need to identify the three parts of the percent equation in the percent problem. Here's how:

The *base* follows the word "of" in the percent problem. The *percent* is indicated by a % sign or the word "percent." The *amount* is the part that is related to the base by the percent; it is the last part we identify.

EXAMPLE 1

Identify the parts in the following percent problem: What is 8% of 200?

SOLUTION

The % sign tells us that 8% is the *percent*. The word "of" tells us that 200 is the *base*. Then the missing part is the *amount*, indicated by the word "What" in the problem.

So we have

$$\text{Percent} \times \text{base} = \text{amount}$$
$$8\% \times 200 = \text{what}$$

We generally use a letter (say, n) for the missing value. Then our equation here would be

$$8\% \times 200 = n$$

■ *Now do margin exercises 1a and 1b.*

To use the percent equation, we substitute the percent, base, and amount from the problem into the equation; then we solve for the missing numerical value. Here are all the steps:

SECTION GOALS

- To identify the parts of the percent equation
- To use the percent equation to find a missing amount, percent, or base

1a. Identify the parts of the percent problem "2 is what percent of 135?"

1b. Identify the parts of the percent problem "16 is 5% of what amount?"

2a. What is 18% of 315?

2b. What is 530% of 32?

To solve any percent problem

1. Identify the percent, base, and amount.
2. Substitute the known values into the general form of the percent equation.
3. Solve the equation for the missing value.

When the problem is solved, the answer should be checked to ensure that it is reasonable.

The Percent Equation: Finding the amount when the percent and the base are known

This is the simplest kind of problem to "translate" from a question to the percent equation. The percent is given, and the base follows "of".

EXAMPLE 2

What is 310% of 16?

SOLUTION

We first identify the percent, base, and amount.

The *percent* is 310%, or 3.10 in decimal form.
The *base* is 16.
The *amount* is the missing value.

Now we substitute into the general form of the percent equation.

$$\text{Percent (as a decimal)} \times \text{base} = \text{amount}$$
$$3.10 \times \quad 16 = n$$

We solve this equation by multiplying as indicated.

$$3.10 \times 16 = n$$
$$49.6 = n$$

Because 3×16 is 48, our answer of 49.6 is reasonable.

So 310% of 16 is 49.6.

■ *Now do margin exercises 2a and 2b.*

Note: You can check an answer by substituting the result into the percent equation. If that gives you a true statement, the answer is correct (provided you correctly identified the percent, base, and amount).

Our next example involves a percent that is less than 1%.

EXAMPLE 3

What is 0.52% of 1000?

SOLUTION

We identify the parts of the equation and substitute.

The *percent* is 0.52%, or 0.0052 in decimal form.
The *base* is 1000.
The *amount* is the missing value.

$$\text{Percent (as a decimal)} \times \text{base} = \text{amount}$$
$$0.0052 \times 1000 = n$$
$$5.2 = n$$

So 0.52% of 1000 is 5.2.

■ *Now do margin exercises 3a and 3b.*

3a. What is 0.23% of 17?

3b. 0.95% of 60 is what?

If your calculator has a [%] key, you can key in

1000 [×] .52 [%] [=]

The calculator will automatically move the decimal point two places and do the computation. The display should read 5.2, giving the same answer we got in Example 3.

But what if you do not have a [%] key on your calculator? Then you must enter the percent in decimal form. In Example 3 you would enter

1000 [×] 0.0052 [=]

Now work most or all of the *additional exercises* below before you go on. They are all problems of the *missing amount* type, and doing them will help you learn that type. Answers are at the bottom of page 512.

1. What is 430% of 4?
2. 0.61% of 0.7 is what?
3. What is 50% of 16?
4. What is 0.21% of 0.8?
5. 5.6% of 1000 is what?
6. What is 8.1% of 1000?
7. 150% of 34 is what?
8. Find 2% of 202.
9. $68\frac{3}{4}\%$ of 19 is what?

The Percent Equation: Finding the percent when the base and the amount are known

This type of problem is "translated" from question to equation. We use just the same steps as before to solve the problem. But we solve the *equation* a little differently.

EXAMPLE 4

90 is what percent of 1500?

SOLUTION

First we identify the percent, base, and amount: There is no percent sign or

4a. 300 is what percent of 1500?

4b. What percent of 180 is 63?

5a. What percent of 20 is 240?

5b. 0.3 is what percent of $\frac{50}{3}$?

word "percent" in the question, so the *percent* is the missing value. The *base* is 1500 because it is the value that follows the word "of." The *amount* must be 90. Knowing the parts, we can substitute them into the percent equation.

$$\text{Percent (as a decimal)} \times \text{base} = \text{amount}$$

$$n \times 1500 = 90$$

To solve this, we use what we know about inverse operations. We have n times 1500 on the left, but we want n alone. We can get it alone by dividing both sides of the equation by 1500.

$$\frac{n \times 1500}{1500} = \frac{90}{1500}$$

Because $\frac{n \times 1500}{1500} = n$ $\quad n = 0.06 \quad$ Because $\frac{90}{1500} = 0.06$

We have found the value of n in decimal number form. But we are looking for a percent. We must change 0.06 to a percent.

$$0.06 = 6\%$$

Check: $0.06 \times 1500 = 90$

So 90 is 6% of 1500. Correct

■ *Now do margin exercises 4a and 4b.*

Let's try another example. This time, the base is smaller than the amount.

EXAMPLE 5

What percent of $12\frac{1}{2}$ is 275?

SOLUTION

As usual, we first identify the parts of the problem. There is no percent sign, so we have

The *percent* is the missing value.
The *base* is $12\frac{1}{2}$ because it is the value following the word "of."
The *amount* is then 275.

Our equation is

$$\text{Percent (as a decimal)} \times \text{base} = \text{amount}$$

$$n \times 12\frac{1}{2} = 275$$

We solve it by dividing both sides of the equation by $12\frac{1}{2}$ (which we write as 12.5).

$$\frac{n \times 12.5}{12.5} = \frac{275}{12.5}$$

Because $\frac{n \times 12.5}{12.5} = n$ $\quad n = 22 \quad$ Because $\frac{275}{12.5} = 22$

Because we are looking for a percent, we must rewrite 22 as a percent. Moving the decimal point two places to the right gives 2200% for an answer.

So 275 is 2200% of $12\frac{1}{2}$.

■ *Now do margin exercises 5a and 5b.*

Now work most or all of the *additional exercises* below. They are all problems of the *missing percent* type and doing them will help you learn that type. Answers are at the bottom of page 512.

1. What percent of 27 is 24.3?
2. 14 is what percent of 5.6?
3. 300 is what percent of 160?
4. What percent of 500 is 175?
5. What percent of 2000 is 56?
6. What percent of 1000 is 81?
7. What percent of 100 is 2.5?
8. 100 is what percent of 2.5?
9. $\frac{1}{2}$ is what percent of 8?

The Percent Equation: Finding the base when the percent and the amount are known

This is the third of the three types of questions that can be asked in percent problems.

Again, we first identify the percent, base, and amount and then substitute them into the percent equation:

$$\text{Percent (as a decimal)} \times \text{base} = \text{amount}$$

EXAMPLE 6

$20 is 30% of what?

SOLUTION

We use "%" and "of" to identify the percent and base, and the remaining value is the amount.

The *percent* is 30%, which is 0.30 expressed as a decimal.
The word "what" follows "of," so the *base* is the missing value.
The *amount* is then $20.

Now we substitute these values.

$$\text{Percent (as a decimal)} \times \text{base} = \text{amount}$$
$$0.30 \times n = \$20$$

On the left side of this equation we have $0.30 \times n$. We want n alone. To get it, we divide both sides of the equation by 0.30.

$$\frac{0.30 \times n}{0.30} = \frac{\$20}{0.30}$$

Because $\frac{0.30 \times n}{0.30} = n$ $\qquad n = \$66.66\overline{6}$ $\qquad$ Because $\frac{\$20}{0.30} = \$66.66\overline{6}$

Because we are working with money, it is reasonable to round the answer to the nearest cent. So $20 is 30% of $66.67.

■ *Now do margin exercises 6a and 6b.*

Our next example involves a percent that is less than 1%.

6a. 12.8 is 42% of what? Round your answer to the nearest hundredth.

6b. $\frac{3}{5}$ is 50% of what?

7a. $\frac{1}{2}$% of what is 1.6?

7b. 0.82 percent of what is 0.2? Round your answer to the nearest hundredth.

EXAMPLE 7

0.5% of what is 0.2?

SOLUTION

We identify the percent, base, and amount and substitute into the percent equation.

The *percent* is 0.5%, or 0.005 as a decimal.
The *base* (after the word "of") is the missing value. We write n in the equation.
The *amount* is 0.2.

Then,

$$\text{Percent (as a decimal)} \times \text{base} = \text{amount}$$
$$0.005 \times \quad n = 0.2$$

We solve the equation by dividing both sides by 0.005.

$$\frac{0.005 \times n}{0.005} = \frac{0.2}{0.005}$$
$$n = 40$$

Check: $0.005 \times 40 = 0.2$ Correct

So 0.5% of 40 is 0.2.

■ *Now do margin exercises 7a and 7b.*

Now work most or all of the *additional exercises* below. Doing them will help you learn *missing base problems.*

Work the exercises that follow.

1. 30% of what is 2.7?

2. 33.6 is 80% of what?

3. 156 is 300% of what?

4. 250% of what is 186?

5. 30 is 15% of what?

6. 25% of what is 100?

7. 3.2% of what is 0.288?

8. 0.096 is 1.6% of what?

9. $8\frac{1}{2}$% of what is 18.7?

Missing Amount Problems (page 509): 1. 17.2 2. 0.00427 3. 8 4. 0.00168 5. 56 6. 81 7. 51 8. 4.04 9. 13.0625
Missing Percent Problems (page 511): 1. 90% 2. 250% 3. 187.5% 4. 35% 5. 2.8% 6. 8.1% 7. 2.5% 8. 4000% 9. 6.25%
Missing Base Problems (page 512): 1. 9 2. 42 3. 52 4. 74.4 5. 200 6. 400 7. 9 8. 6 9. 220

8.3 EXERCISES

In Exercises 1 through 40, solve the following percent problems of all three types. Remember to check to be sure your result is reasonable.

1. What percent of 19 is 12.92?

2. What percent of 33 is 32.01?

3. What percent of 20 is $3\frac{1}{5}$?

4. What percent of 400 is $10\frac{1}{2}$?

5. 150% of what is 51?

6. 200% of what is 96?

7. What percent of 9 is 2.7?

8. What percent of 42 is 33.6?

9. 90% of 27 is what?

10. 70% of 14 is what?

11. 79% of what is 20.54?

12. 86% of what is 14.62?

13. What percent of 16 is 8?

14. What percent of 45 is 9?

15. 2.5% of 100 is what?

16. 3.1% of 100 is what?

17. 4.3% of what is 4.515?

18. 5.4% of what is 24.3?

19. 3.2% of 9 is what?

20. 1.6% of 6 is what?

21. What percent of 200 is 0.8?

22. What percent of 400 is 3.2?

23. 0.2% of what is 0.0016?

24. 0.6% of what is 0.0042?

25. 10% of what is 5.3?

26. 40% of what is 52?

27. 300% of 52 is what?

28. 250% of 62 is what?

29. 0.9% of 3000 is what?

30. 0.7% of 7000 is what?

31. 0.8% of 0.5 is what?

32. 0.4% of 0.6 is what?

33. 187% of 1.35 is what?

34. 382% of 3.82 is what?

35. 1.3% of what is 0.065?

36. 2.1% of what is 0.168?

37. 0.2% of 15 is what?

38. 0.5% of 30 is what?

39. 0.3% of 2002 is what?

40. 0.7% of 6500 is what?

41. Find the missing amounts in the following list of purchases. To do so, use the equation

Tax rate (as a decimal) × price of item = tax

Purchase	Price of Item	Tax rate	Tax
car	$17,100	$6\frac{1}{4}\%$	______
groceries	$40.50	$8\frac{1}{2}\%$	____
coffee	$ .60	____	$ .06
clothing	$150.25	5.1%	____
CDs and tapes	$80	______	$1.50
airline ticket	______	8%	$13.50
property	________	$7\frac{3}{4}\%$	$775,000

8.3 MIXED PRACTICE

By doing these exercises, you will practice all the topics in this chapter up to this point.

42. Change $\frac{3}{7}$ to a percent, rounded to the nearest hundredth.

43. Change 189% to a fraction.

44. 2 is 0.8% of what?

45. Write $\frac{5}{6}$ as a percent, rounded to the nearest hundredth.

46. Change 87.4% to a decimal number.

47. What is 204% of 40?

48. 16.8 is what percent of 84?

49. Write 3.6% as a fraction.

50. Write 0.8% as a fraction and as a decimal number.

51. If 58% of all roses are red and 11% are white, what percent are neither red nor white?

52. Find 0.8% of 15.

8.4 Solving Percent Problems Using Proportions

SECTION GOAL

- To use proportions to solve percent problems

There is another way to solve problems that involve percent—with proportions.

To solve a percent problem using a proportion

1. Identify the percent, base, and amount.
2. Express the percent as a ratio with a denominator of 100:

$$P\% = \frac{P}{100}$$

3. Write the ratio:

$$\frac{\text{amount}}{\text{base}}$$

4. Equate the two ratios in a proportion and solve for the missing value.

EXAMPLE 1

What is 56% of 180?

SOLUTION

Because 56% is close to 50%, the answer will be about 50% of 180, or 90.

The *percent* is 56%. As a ratio with a denominator of 100, it is $\frac{56}{100}$.

The *base* is 180 because it follows the word "of."

The *amount* is the missing value.

The ratio $\frac{\text{amount}}{\text{base}}$ is $\frac{n}{180}$.

1a. What is 27% of 55?

1b. What is 8.25% of 8?

The proportion is:

$$\frac{56}{100} = \frac{n}{180}$$

We multiply to find the cross products.

$$\text{Product of extremes}: 56 \times 180 = 10{,}080$$
$$\text{Product of means}: 100 \times n$$

So, in a true proportion, we have

$$100 \times n = 10{,}080$$

We divide both sides by 100 to find n.

$$\frac{100 \times \text{n}}{100} = \frac{10{,}080}{100}$$

$$\text{Because } \frac{100 \times n}{100} = n \qquad n = 100.8 \qquad \text{Because } \frac{10{,}080}{100} = 100.8$$

Our estimate was "around 90", so the answer is reasonable.

Thus 56% of 180 is 100.8.

■ *Now do margin exercises 1a and 1b.*

In the next example, we are asked to find the whole.

EXAMPLE 2

$\frac{1}{15}$ is $37\frac{1}{2}\%$ of what number?

SOLUTION

This is tough to estimate, but we can see that the number will be larger than $\frac{1}{15}$ (because $\frac{1}{15}$ is only $37\frac{1}{2}\%$ of the number).

The percent is $37\frac{1}{2}\%$, or $\frac{37\frac{1}{2}}{100}$.
The *base* (following "of") is the missing value.
The *amount* is $\frac{1}{15}$.

The ratio $\frac{\text{amount}}{\text{base}}$ is $\frac{\frac{1}{15}}{n}$.

The proportion is then

$$\frac{37\frac{1}{2}}{100} = \frac{\frac{1}{15}}{n}$$

We solve for n by first finding the cross products.

$$\text{Product of extremes}: 37\frac{1}{2} \times n$$

$$\text{Product of means}: 100 \times \frac{1}{15} = \frac{100}{15} = \frac{20}{3}$$

In our true proportion, these products are equal.

$$37\frac{1}{2} \times n = \frac{20}{3}$$

To solve, we divide both sides by $37\frac{1}{2}$.

$$\frac{37\frac{1}{2} \times n}{37\frac{1}{2}} = \frac{\frac{20}{3}}{37\frac{1}{2}} \quad \text{Dividing by } 37\tfrac{1}{2}$$

$$n = \frac{\frac{20}{3}}{\frac{75}{2}} \quad \text{Changing to improper fraction}$$

$$= \frac{20}{3} \times \frac{2}{75} \quad \text{Multiplying by the reciprocal of the denominator}$$

$$n = \frac{40}{225} = \frac{8}{45} \quad \text{Simplifying}$$

So $\frac{1}{15}$ is $37\frac{1}{2}\%$ of $\frac{8}{45}$.

■ *Now do margin exercises 2a and 2b.*

2a. $87\frac{1}{2}\%$ of what is 42?

2b. 7.3 is $36\frac{1}{2}\%$ of what?

3a. $32\frac{1}{2}$ is what percent of 65?

3b. 50.009 is what percent of 27.5? Round to the nearest hundredth.

In this final example, we are asked to find the percent.

EXAMPLE 3

0.0002 is what percent of 0.00015?

SOLUTION

Because 0.0002 is larger than 0.00015, the answer will be more than 100%.

The percent is missing, so we have

$$n\% = \frac{n}{100}$$

The *base* is 0.00015 because that is what follows "of."

The *amount* must be 0.0002.

The ratio $\frac{\text{amount}}{\text{base}}$ is $\frac{0.0002}{0.00015}$.

We write the proportion and solve for the missing value.

$\frac{n}{100} = \frac{0.0002}{0.00015}$	The proportion
$n \times 0.00015 = 0.02$	Cross-multiplying
$\frac{0.00015 \times n}{0.00015} = \frac{0.02}{0.00015}$	Dividing by 0.00015
$n = 133.3\overline{3}$	Simplifying

We know that $0.3\overline{3}$ is $\frac{1}{3}$, so 0.0002 is $133\frac{1}{3}\%$ of 0.00015.

■ *Now do margin exercises 3a and 3b.*

Work the exercises that follow.

8.4 EXERCISES

In Exercises 1 through 30, solve each percent problem using the proportion method.

1. 68% of 150 is what?

2. 75% of 36 is what?

3. 12 is what percent of 48?

4. What percent of 30 is 198?

5. 90 is 60% of what?

6. 165 is 75% of what?

7. 0.96 is what percent of 150?

8. 88.8 is 148% of what?

9. What percent of 0.4 is 0.0002?

10. What percent of 0.0003 is 0.0006?

11. 3.68 is what percent of 32?

12. 0.2128 is what percent of 30.4?

13. What is 0.4% of 14?

14. What is 0.07% of 10?

15. What is 2.92% of 30?

16. What is 1.35% of 26?

17. 200% of 18.6 is what?

18. 105% of 46 is what?

19. What is 87.5% of 32?

20. What is 1.7% of 150?

21. 0.75 is 2.5% of what?

22. 0.03 is 1.2% of what?

23. What percent of 0.0018 is 0.009?

24. What percent of 0.0005 is 0.0001?

25. 50 is $62\frac{1}{2}\%$ of what?

26. 49 is $87\frac{1}{2}\%$ of what?

27. 5 is $\frac{1}{4}\%$ of what?

28. What is $\frac{1}{4}\%$ of 5?

29. What is 6% of $4.99?

30. What is 106% of $4.99?

8.4 MIXED PRACTICE

By doing these exercises, you will practice all the topics in this chapter up to this point.

31. 63 is 0.9% of what? (Use a percent equation here.)

32. What is 86% of 0.005? (Use a percent equation here.)

33. Use a percent equation to find 0.2% of 5.

34. Use a proportion to find 56% of 108.

35. Write 18.93% as a fraction.

36. Write $\frac{6}{9}$ as a percent, rounded to the nearest hundredth of a percent.

37. Write $17\frac{1}{5}$ as a percent.

38. Use a proportion to solve this problem. 7.2 is 25% of what number?

39. Write $86\frac{5}{8}\%$ as a decimal number.

40. Use a percent equation to find 0.4% of 11.

8.5 Applying Percents: Word Problems

We deal with percents in almost all aspects of everyday life. Sales taxes are computed as percents. Paycheck deductions are taken as percents. Percents are used to compare a wide range of measurements, from productivity and profit to free-throw scoring and drug effectiveness. It is almost impossible to avoid percents. In this section, we use them in some everyday problems.

The first example is a markup problem. The **markup rate** is the percent by which a dealer increases the cost to obtain the selling price.

EXAMPLE 1

A coat costs retailers \$180. It is sold for \$230. What is the markup rate?

Pricing Guide Worksheet	
Cost to Retailer	= \$180
Sale Price	= \$230
% Markup	=

SOLUTION

We are being asked, "What percent of the cost is the given markup amount?" Because the amount the coat has been marked up is \$230 – \$180, or \$50, we must find

"What percent of \$180 is \$50?"

Here, the *percent* is the missing value, the *base* is \$180, and the *amount* is \$50.

We substitute the known values into the percent equation.

$$\text{Percent (as a decimal)} \times \text{base} = \text{amount}$$
$$n \times 180 = 50$$

A good estimate of the answer is 50 ÷ 200, or 0.25. So 25% is the estimated markup rate.

We solve the equation by dividing both sides by 180.

$$\frac{n \times 180}{180} = \frac{50}{180}$$
$$n = 0.277\overline{7} \quad \text{Dividing}$$

Rounded to the ten-thousandths place, n is 0.2778.

The solution thus far is a decimal, but we are asked for a *percent*. So we must move the decimal point two places to the right and add a percent sign.

Thus the markup rate is 27.78%.

■ *Now do margin exercises 1a and 1b.*

SECTION GOAL

■ To apply percent in solving word problems

1a. A price ticket on a suit said "PRICE REDUCED \$75." If the original price was \$250, what is the discount rate (based on original price)?

1b. What is the new price of an item that was discounted 20% if the price was originally \$2500?

Remember that the **discount rate** is the percent by which the original cost is reduced.

In the next example, we have to find a missing piece of information first and then use it to solve the problem.

EXAMPLE 2

An inventory of spare parts showed that 168 had rusted. This represented 56% of the total number of parts. How many parts had not rusted?

SOLUTION

Solving this problem requires two steps. First, we must find the total number of parts. Then we can subtract the number of rusted parts from that, to find the number that did not rust. We know that 168 is 56% of the total number of parts.

The *percent* is 56%, which gives us the ratio $\frac{56}{100}$. The *base* is the missing value, and the *amount* is 168; they give us the ratio $\frac{168}{n}$.

We set up a proportion and solve.

$$\frac{56}{100} = \frac{168}{n} \quad \text{The proportion}$$

$$56 \times n = 100 \times 168 \quad \text{Cross-multiplying}$$

$$56 \times n = 16800 \quad \text{Multiplying out}$$

$$\frac{56 \times n}{56} = \frac{16800}{56} \quad \text{Dividing by 56}$$

$$n = \frac{16800}{56} = 300 \quad \text{Simplifying}$$

The total number of parts is 300.

Now we must subtract the number of rusted parts from the total.

$$300 - 168 = 132$$

So 132 parts were not rusted.

■ *Now do margin exercises 2a and 2b.*

Work the exercises that follow. Remember to check to be sure your answers are reasonable.

2a. On a certain day, 22% of the workers called in sick. If the company employs 150 people, how many people did not call in sick?

2b. Dresses on 24 mannequins have buttons. If 40% of the dresses on mannequins do not have buttons, how many mannequins are there?

8.5 EXERCISES

1. Meadows and pastures cover approximately 11,686,600 square miles. This is approximately 20% of the estimated land area of the Earth. Find the estimated land area of the Earth.

2. Twenty-six carburetors failed an emissions test. If this was 40% of the carburetors tested, how many were tested?

3. Find the sales tax on a dress that costs $124.80 if the tax rate is 5.5%.

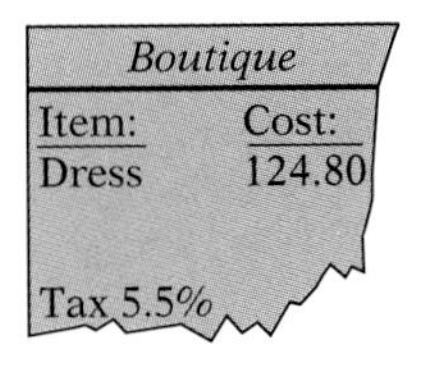

4. Find the total cost of a purchase of items totaling $149.98 in a state with a 3% sales tax.

5. A coat is marked "20% off." How much money will you save if the coat was originally priced at $175?

6. A bicycle originally priced at $220 was reduced by $52.80. What was the percent reduction?

7. A food processor is on sale for $160. This is 20% off the original price. What was the original price?

8. Your rent increases 15% after the first year and 10% more the second year. If your rent before any increases was $200, what is your rent after the two increases?

Use the following information for Exercises 9 and 10: Your paycheck shows that your gross pay was $1286.

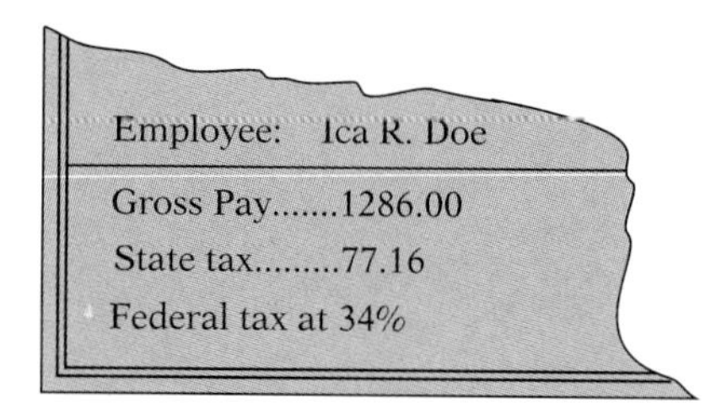

9. If you pay 34% in federal tax, what is the amount of your federal tax?

10. If you pay $77.16 in state tax, what is the state tax rate?

11. A TV set costs $800. How much tax will I have to pay if the tax rate in the state is $8\frac{1}{4}$%?

12. A stereo sold for $1500 last year. This year it sells for $1800. What is the rate at which it has been marked up?

13. An unofficial survey reported that Republican registration had increased 953% over the last 20 years. If there were 127,050 Republicans registered 20 years ago, how many are registered now?

14. A bill for a meal in a restaurant is $56. If you decide to leave a tip of 18%, what is the total cost of the meal?

15. It is estimated that 90% of the inhabitants of Australia live in urban areas. If Australia has 16,200,000 people, how many live in urban areas?

16. It is estimated that 80% of the population of China live in rural areas. If the population of China is estimated to be 1,062,000,000 people, approximately how many people do not live in rural areas?

17. Ice covers 10% of the land area of the Earth. If the land area of the Earth is approximately 58,433,000 square miles, how much is covered by ice?

Use the following information for Exercises 18 and 19: A couple plans to buy a house for $100,000. To do this, they must have between 15% and 25% for a down payment.

18. What is the minimum down payment they can have in dollars?

19. If they can afford to put $20,000 down, what percent of the cost of the house will they have made as a down payment?

20. The surface area of the Earth is approximately 196,949,970 square miles. Approximately 147,712,470 square miles is covered by oceans. Round these numbers to the nearest million, and determine what percent of the Earth is covered by oceans. Round to the nearest whole percent.

21. There were 116 medals awarded in the 1984 Winter Olympics. What percent were won by the former Soviet Union if it won 25 medals? Round your answer to the nearest hundredth of a percent.

22. In the 1988 New York Marathon, 6866 runners were between the ages of 40 and 49. If this represented approximately 29% of the runners, how many runners were there in all?

23. In the 1988 New York Marathon, 698 women said that they had been running over 20 years. This was approximately 9.3% of the number of both men and women who had been running over 20 years. How many people had been running over 20 years?

24. In the 1988 New York Marathon, 58 runners were 70 or over. Approximately 6 of these were over 80. What percent of those over 70 were over 80? Round to the nearest whole percent.

25. Anthracite is 95% carbon by weight. If a specimen weighs 24 ounces, how much of it is carbon?

26. 95% of the 75 million tons of fish harvested each year come from the oceans. How many millions of tons of fish come from other water sources?

8.5 MIXED PRACTICE

By doing these exercises, you will practice all the topics in this chapter up to this point.

27. Use a percent equation to find 560% of 85.

28. Use a proportion to find 140% of 112.

29. Use a percent equation to solve this problem:

15 is what percent of 90?

Round to the nearest hundredth of a percent.

30. A suit is reduced 35%. If the suit originally sold for $158, what is the discount?

31. Write $\frac{5}{8}$ as a percent.

32. Use a proportion to solve this problem:

0.65 is 0.5% of what?

33. How much of a commission will a realtor in a large city receive on a sale of a $246,000 house if the rate of commission is 12%?

34. Use a proportion to solve this problem:

0.45% of 18 is what?

35. Use a percent equation to solve this problem:

2.5 is what percent of 200?

36. Shopping in New Jersey for clothes is cheaper than shopping in New York, because New Jersey does not charge sales tax on clothes. How much will I save by buying an $85 suit in New Jersey if New York has an 8.25% sales tax?

37. During a recent flu epidemic, 14 children out of the 35 in the kindergarten were sick. What percent of the children were not sick?

Simple and Compound Interest

Every adult has had some experience with interest—either paying it or collecting it. **Interest** is payment for the temporary use of someone else's money. It is involved in all bank loans, and most savings and checking accounts are paid some interest.

Two types of interest are important in everyday activities. **Simple interest** is interest that is paid only on the borrowed amount (called the **principal**). It is calculated by using the formula

$$\text{Interest} = \text{principal} \times \text{interest rate} \times \text{time}$$

Suppose we want to find the amount of simple interest paid over a 3-year period on a principal of \$1000 at an interest rate of 10% per year. We substitute into the formula and get the following result:

$$\begin{aligned}\text{Interest} &= \text{principal} \times \text{rate} \times \text{time}\\ &= \$1000 \times 0.10 \times 3 = \$300\end{aligned}$$

Compound interest is interest that is paid on both the principal *and* any previous interest that has been added to the principal. This means that, over time, an amount of money increases faster with compound interest than with simple interest.

To show this, we can calculate the compound interest for the same situation we just examined: a principal of \$1000, a time of 3 years, and an interest rate of 10% per year. Now, however, the interest will be **compounded** (or computed and added to the principal) each year. We use the same simple-interest formula as before, but we must use it for one year at a time.

First year
Interest = \$1000 × 0.10 × 1 = \$100 New amount = \$1000 + \$100 = \$1100

Second year
Interest = \$1100 × 0.10 × 1 = \$110 New amount = \$1100 + \$110 = \$1210

Third year
Interest = \$1210 × 0.10 × 1 = \$121 New amount = \$1210 + \$121 = \$1331
Total interest = \$331

With compounding, the interest is \$331, which is \$31 more than is earned with simple interest.

Using the formula to compute compound interest can be time-consuming and tedious. A quicker method is to use tables of compound amounts. A small part of such a table appears as Table 1.

TABLE 1 Compound Amounts for $1

	Interest Rate per Period				
Time periods	**5%**	**6%**	**8%**	**10%**	**12%**
1	1.05	1.06	1.08	1.1	1.12
2	1.1025	1.1236	1.1664	1.21	1.2544
3	1.1576	1.1910	1.2597	1.331	1.4049
4	1.2155	1.2625	1.3605	1.4641	1.5735
5	1.2763	1.3382	1.4693	1.6105	1.7623

This table gives the amount to which $1 would grow at various interest rates, compounded annually. Let's use it to find the compound interest on $1000 at 10% for 3 years, compounded annually. We look for the intersection of the 10% interest rate column and the row for 3 time periods; there we find the value 1.331. This is the amount that $1 would be worth after 3 years of collecting (or accruing) compound interest. We then multiply that by the principal:

Principal × compound amount for $1 = amount
$1000 × 1.331 = $1331

As we calculated the long way, after 3 years at 10%, the total amount will be $1331. Using the table saved three separate calculations. The interest alone is the difference between the amount and the principal.

Amount − principal = interest
$1331 − $1000 = $331

EXERCISES

Answer the following questions by using Table 1 or by calculating.

1. Find the simple interest you will earn on $250 deposited in an account for 5 years at an annual rate of 8%.

2. Using the results of Exercise 1, find out how much more you would have earned if the interest had been compounded annually for the same amount of time.

BOXSPRINGS SAVINGS
Annual Compound Interest = 8% for 5 years
Deposit $250.00

3. Find the simple interest on $500 deposited in an account for 6 months at an annual rate of 8%. (*Hint:* 6 months is $\frac{6}{12}$ or $\frac{1}{2}$ year. Use this as the time period.)

4. Calculate the compound interest on $430 for 1 year at 5% interest per quarter. (Note that Table 1 is given in time periods. There are four quarters in a year.)

5. Find the simple interest on $1200 deposited in an account for 2.5 years at an annual rate of 11.5%.

6. You deposited $700 in the bank for 4 months. Find the total amount of money in the account if you earn 12% simple annual interest.

7. Find the amount of interest you will earn on $45,000 if you deposit it in an account earning 6% interest compounded annually and leave it there for 5 years.

8. What is the difference in the interest earned by $1 million at simple interest and at compound interest for a period of 4 years at 12% annual interest?

9. How much interest is earned by $18,000 in 4 years at a rate of 16% compounded annually? [*Hint:* The compound factor is (1 + interest as a decimal)^t, where t is the number of time periods. The compound factor here is $(1.16)^4$].

10. You purchase an apartment for $100,900 with a $10,000 down payment. The remainder of the money is to be paid in 5 years with interest at 10% compounded annually. How much will you have to pay then?

11. Calculate the amount you would earn in 1 year on a $20,000 CD earning 5% interest compounded annually. Would you prefer this CD or one paying 6% simple interest over the same time period? What is the difference in interest earned?

12. You can borrow $12,500 at either 10% simple interest or 8% compounded annually. Which would you choose for a 1-year loan? For a 5-year loan? Show why in each case. After how many years would you want to change the way your interest is figured? (Recall the hint for Exercise 9.)

8.6 Geometry and Measurement: Applying Percents and Using Diagrams

In solving word problems, we must sometimes combine topics and ideas from geometry, fractions, percents, or other areas of mathematics. It is also an excellent idea to make a sketch for a problem when one is not given.

Let's look at some examples.

Example 1

Five circles, all of radius 1 inch, are cut out of a rectangular piece of metal that measures 2.3 by 9.8 inches. None of the circles overlap. Find the area of the metal that remains.

Solution

There is no figure, so we draw one. We begin with the rectangle, because circles will be *cut out* of it.

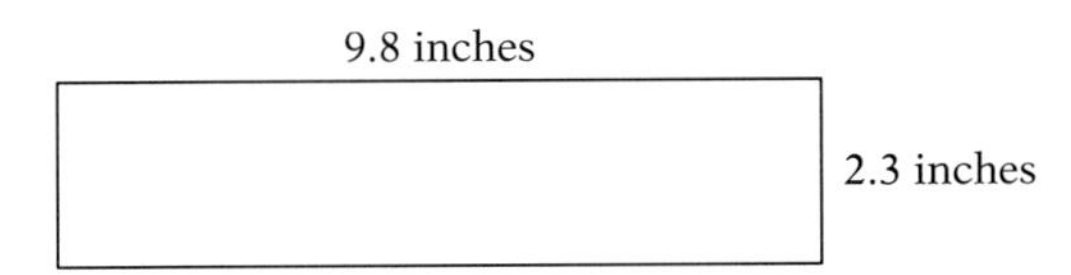

Next we draw in the five circles. We know that they are all circles and that they *don't overlap*. But we can place them wherever we like. Finally, we shade in the area we must find, as a reminder.

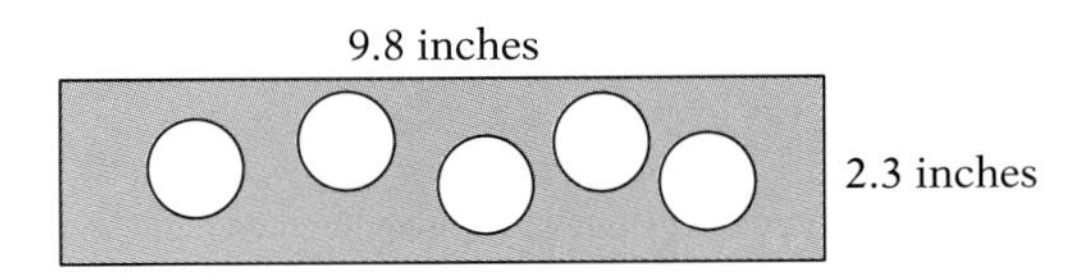

We can find the area of the metal that remains by subtracting the area of the five circles from the area of the rectangle.

$$\text{Area(shaded)} = \text{area(rectangle)} - \text{area(five circles)}$$

First we find the area of the rectangle.

$$A = \ell \times w = 2.3 \times 9.8 = 22.54 \text{ square inches}$$

Then we find the area of one circle.

$$A = \pi r^2$$

Because the radius of the circle is 1, the area is

$$A = 3.14 \times 1^2$$
$$= 3.14 \text{ square inches}$$

And the area of five circles is 5 times 3.14.

$$5 \times 3.14 = 15.70 \text{ square inches}$$

So the area of the remaining metal is $22.54 - 15.70 = 6.84$ square inches.

■ *Now do margin exercises 1a and 1b.*

Section Goals

- To solve word problems using diagrams
- To solve geometry problems involving percents

1a. Find the shaded area in this figure.

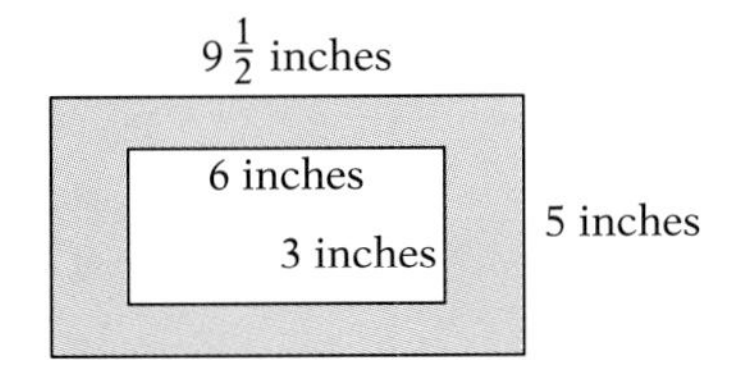

1b. In Example 1, what percent of the original rectangle remained after the circles were cut out? Round to the nearest whole percent.

2a. A red border is stitched around a towel that measures 9 inches by 13 inches. If the new border increases the area by 40%, what is the area of the new towel? How many square inches of it are red?

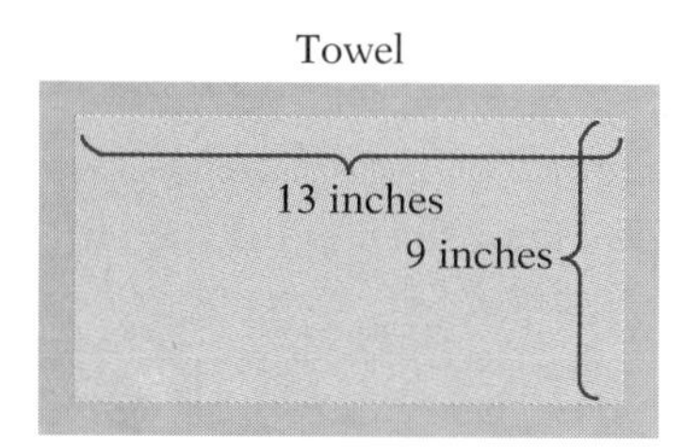

2b. A 14 foot by 20 foot rectangular mural for a school carnival is being colored by 20 third graders. If each of them has colored a square section that is $1\frac{1}{2}$ feet on a side, what percent of the mural is already colored? (Give your answer to the nearest whole percent.)

In the next example, we are given information about a geometric figure, along with a percent, and we are asked to find another piece of information.

EXAMPLE 2

A rectangular area rug is 4 feet wide by 6 feet long. A rectangular piece equivalent to 40% of the area of the original rug is cut off its length. If the width remains 4 feet, what is the length of the cut piece? What are the dimensions of the remaining portion of the original rug?

SOLUTION

This problem will definitely be easier to solve if we first make a sketch. We draw a rectangle, label one side 4 feet and one 6 feet, and draw a dashed line to show where the cut is made. The sketch also shows what we know about the percents.

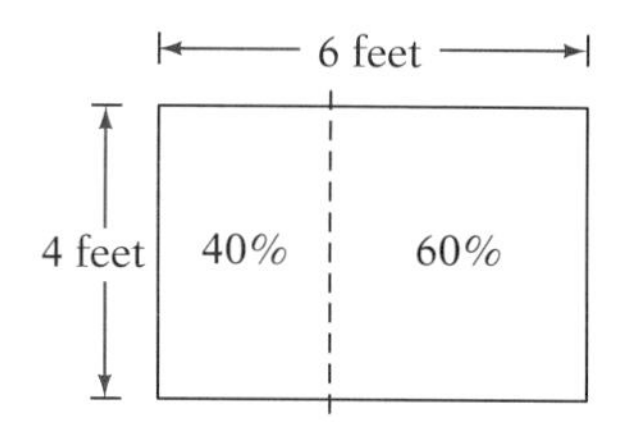

Let's first find the area of the cut-off piece. To do that, we must find the area of the original rug.

$$\text{Area} = \ell w$$
$$= 6 \times 4 = 24 \text{ square feet}$$

We know that 40% of the rug has been cut off, so we ask (and then must answer) the question "40% of 24 square feet is what?"

The percent is 40, the base is 24, and the amount is the unknown value.

$$\text{Percent (as a decimal)} \times \text{base} = \text{amount}$$
$$0.40 \times 24 = n$$
$$9.60 = n \qquad \text{Multiplying out}$$

So 9.60 square feet of the rug has been cut off. We can use this area to find the length that was cut off, knowing its width is 4 feet. We have

$$A = \ell \times w$$
$$9.6 = \ell \times 4 \qquad \text{Substituting for } A \text{ and } w$$
$$\frac{9.6}{4} = \frac{\ell \times 4}{4} \qquad \text{Dividing both sides by 4}$$
$$2.4 = \ell \qquad \text{Dividing out}$$

So the length of the cut-off piece is 2.4 feet.

To find the length of the remaining rug, we subtract.

$$6 - 2.4 = 3.6 \text{ feet}$$

Its "width" is still 4 feet, so the dimensions of the remainder of the original rug are 3.6 feet by 4 feet.

■ *Now do margin exercises 2a and 2b.*

Work the exercises that follow. Remember to check to be sure your answers are reasonable.

8.6 EXERCISES

1. A circular pool with a diameter of 21.08 feet is built inside a square patio with a side of 25.6 feet. How much patio space is there around the pool?

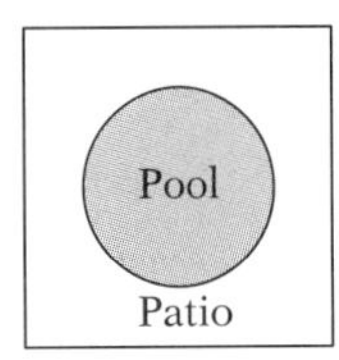

2. A square card table that is $3\frac{1}{4}$ feet on a side is serving as the workspace for a round jigsaw puzzle with a radius of $18\frac{1}{4}$ inches. What percent of the card table will be exposed when the puzzle is complete? Round to the nearest whole percent.

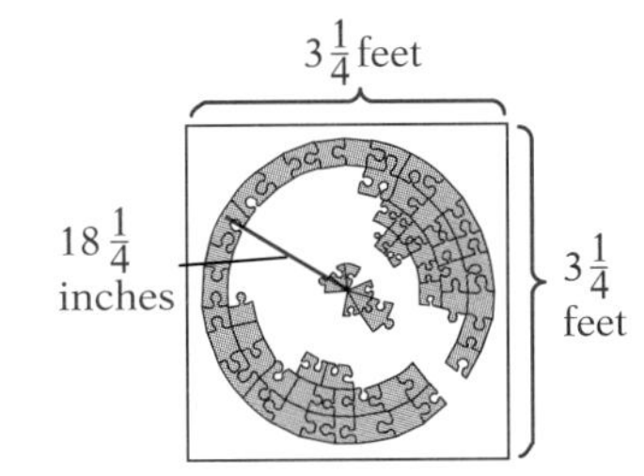

3. Flagpoles are being arranged in a square. If the square is 14 poles on a side, how many poles will be needed?

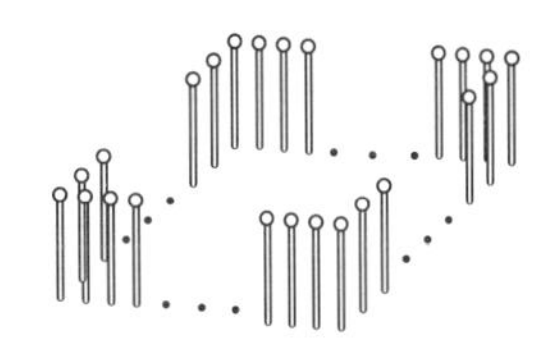

4. The diameter of a round tablecloth is 20.6 cm. If $15\frac{1}{2}\%$ of the tablecloth is hanging off the table, what is the actual area of the top of the table in square centimeters? Round to the nearest hundredth.

5. A wall that measures 22 feet by $15\frac{1}{2}$ feet is being painted. If 65.3% of the wall has already been painted, how many square feet are left to be painted?

6. If you want to arrange 68 stones in a square around a vegetable plot, how many stones will there be on a side?

7. A bathroom floor has tiles that are 1 foot square. Aubrey has put 14 tiles in place, covering 14% of the bathroom floor area. How large is the bathroom floor?

8. Sidney, DeNean, and Terri are each renting 13 videos out of a library of 125 videos. What percent of the total number of videos are they renting?

9. Stones are to be placed around a rectangular swimming pool. There will be 19 stones on each side along the longer sides of the pool, and 14 stones on each side along the shorter ends. How many stones will there be in all?

10. A city law states that classrooms must contain at least 20.5 square feet for every student in the class. If a room is designed to hold 45 students, how big must its area be?

11. A cylindrical container is 12 inches tall, and its round base has a diameter of 7.35 inches. If it is filled to 80% of its capacity, how much liquid is it holding? Round to the nearest inch.

12. Books are being placed on a shelf. Each book is $2\frac{1}{2}$ inches thick. Each cover is $\frac{1}{4}$ inch thick. How much shelf space will 2 dozen books take up on the shelf?

13. A cube with an edge of 3 inches has a spherical ball placed inside of it. The ball has a diameter of 2 inches. The cube is then filled with sand right up to and level with the top, covering the ball. What volume of sand is in the cube? Use 3.14 for π and round to the nearest hundredth.

14. Chicken wire is to be placed along a straight fence. There are 21 fence poles that are 27 inches apart. The poles are 3 inches in diameter. How much chicken wire is needed to reach exactly from the first pole to the last?

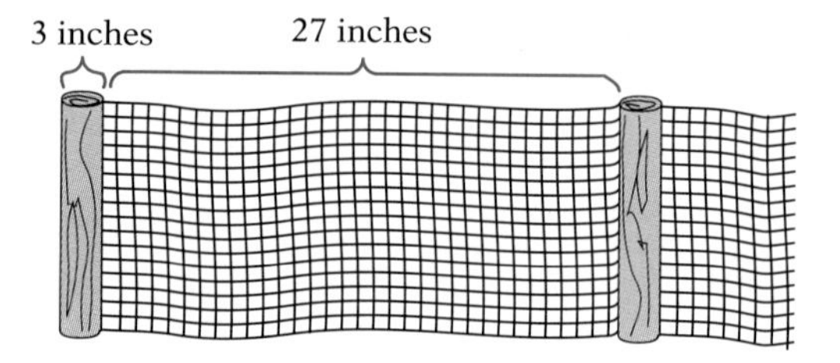

15. A quilt has panels shaped as shown in this figure. The inner panel is a square $9\frac{1}{2}$ square inches in area. The perimeter of the large square is 21 inches. What is the total area of the trapezoids?

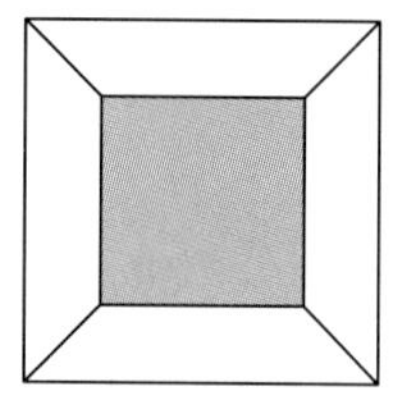

16. Colored and white square tiles are laid in the pattern shown here. The perimeter of the section shown is 106.8 inches. What is the area of one tile?

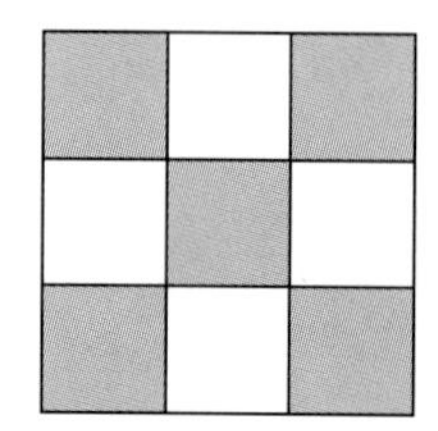

17. A rectangular room that has dimensions of $10\frac{1}{2}$ feet by $12\frac{3}{4}$ feet contains 100 tiles. What is the area of each tile?

18. A knitted afghan has 24 panels, each measuring $8\frac{1}{2}$ inches by $4\frac{1}{4}$ inches. If each square foot of knitted panel takes 1 skein of yarn, how many skeins will you need for this project?

8.6 MIXED PRACTICE

By doing these exercises, you will practice all the topics in this chapter up to this point.

19. What is 45% of the area of a circle of radius 1?

20. Use any method to find 65% of 18.

21. Between midnight Saturday and midnight Sunday, it rained 40% of the time. How many hours did it rain?

22. An inventory showed that 280 parts were defective. This was 56% of the total inventory. How many parts were in the inventory?

23. Use a proportion to find 120% of 36.

24. A dress is selling for $30. This price represents a 25% markup. What was the original price?

25. Use any method to solve this problem:

0.0012 is what percent of 0.6?

26. How much will I have to pay back at the end of the year if I borrow $1200 at the beginning of the year and I must pay a finance charge of 5% annually?

27. A towel is 35% flowered. Its dimensions are 18 inches by 24 inches. How much of the area is flowered?

28. Complete line 45 of the form that includes the following lines:

43. Annual salary, wages, tips: $50,007
44. 8% of the amount on line 43: $4000.56
45. 0.054 times the amount on line 44: ______

29. Write $\frac{3}{7}$ as a percent, rounded to the nearest hundredth.

30. Use a percent equation to solve this problem: 70 is 3.5% of what number?

31. Use any method to solve this problem: 64.8 is 7.2% of what number?

32. Find the area of the shaded region of this figure. The radius of the large circle is 3. The radius of the small circle is $1\frac{1}{2}$. Use 3.14 for π and round to the nearest tenth.

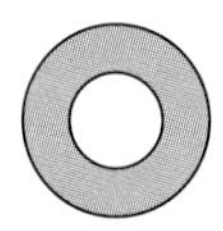

33. A rectangular room measures 14 feet by 18 feet. You want to put carpeting in 35% of the room. How many square feet of carpeting will you need?

An Application to Statistics: Reading Circle Charts and Histograms

Reading a Circle Chart

A common type of data display in business and in newspapers and magazines is a **circle chart** or **pie chart.** In a circle chart, the total is represented as a circle, and the parts that make up the total are shown as "slices." The numbers are usually given in the form of percents; the total is 100%.

EXAMPLE 1

The accompanying chart shows how a family spends its yearly income of $31,000.

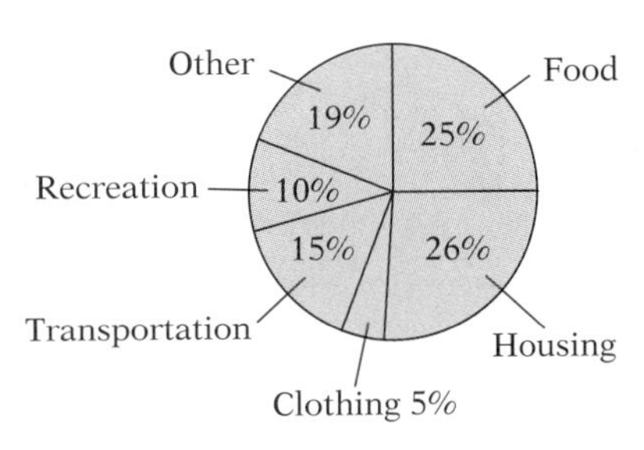

How much money does this family spend on transportation?

SOLUTION

The chart indicates that 15% of the income is spent on transportation. We must answer the question: 15% of $31,000 is what?

Writing as an equation and solving, we get

$$n = 0.15 \times 31{,}000 = 4650$$

So the family spends $4650 on transportation yearly.

■ *Now do margin exercises 1a and 1b.*

1a. Use the figure in Example 1 to find how much the family would spend on food per year if its yearly income were $20,000.

1b. How much would this same family spend on clothing and recreation together with a $20,000 income?

Reading a Histogram

Some kinds of graphs use the widths and heights of vertical bars to represent information. The graph here is such a **histogram.** The width of each bar indicates a height *interval* of about 12 inches. Specifically, the intervals are 0–11.9, 12.0–23.9, and so on. The height of each bar indicates how many bushes have heights that fall in that interval. The scale at the left indicates the number of bushes in the height interval.

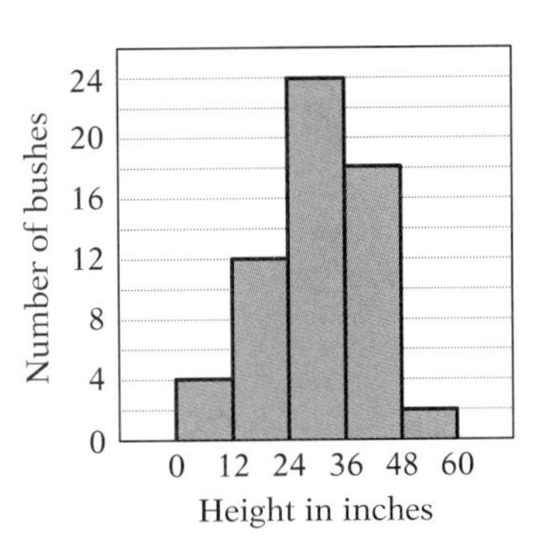

EXAMPLE 2

In the preceding histogram, what height interval contains the most bushes? How many bushes are in that height interval?

SOLUTION

To find the interval with the most bushes, we look for the tallest bar. That bar is the one representing the 24-to-35.9 inch interval. Next we look across the top of this tallest bar to the left-hand scale, and we find that its height represents 24 bushes on that scale. Therefore, 24 bushes are 24.0 to 35.9 inches tall.

■ *Now do margin exercises 2a and 2b.*

2a. Using the graph that precedes Example 2, find the total number of bushes that are less than 24 inches high.

2b. Using the graph that precedes Example 2, find the percent of bushes that are 36 inches high or taller.

Work the exercises that follow. Remember to check to be sure your answers are reasonable.

EXERCISES

The following pie chart represents the way a Third-World government intends to spend money received on the sale of oil to the United States. Use this information to answer Exercises 1 and 2.

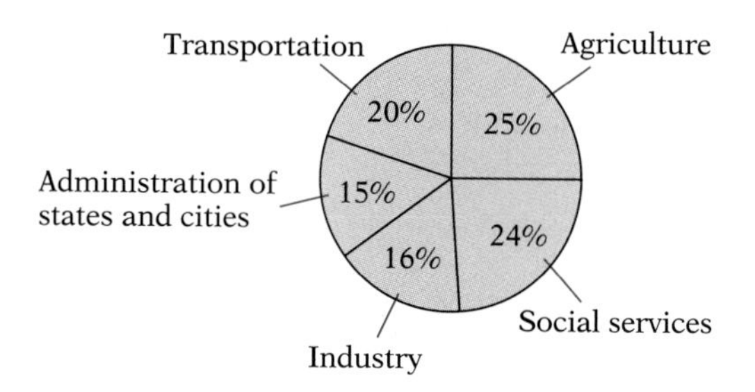

The total amount of money received is $2 billion.

1. How much will the country spend on social services and agriculture together?

2. How much more will be spent on industry than on the administration of states and cities?

The following circle chart shows how individuals spent their money for recreation in 1986. Use this information to answer Exercises 3 and 4 for a total annual recreation budget of $2000.

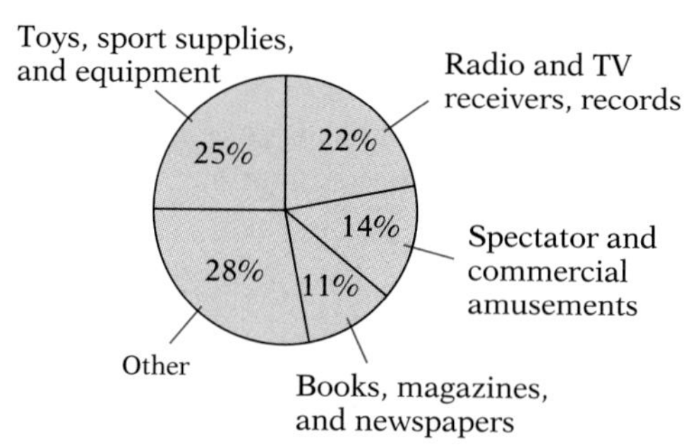

U.S. Census Bureau

3. How much money is left each month after the money for spectator and commercial amusements is spent?

4. How much is spent per week on radio and TV receivers and records if the same amount is spent each week?

The following figure shows how a peasant in a certain Third-World country spends his time. Assume that the entire circle represents a 16-hour day (he has to sleep), and use this information to answer Exercises 5 and 6.

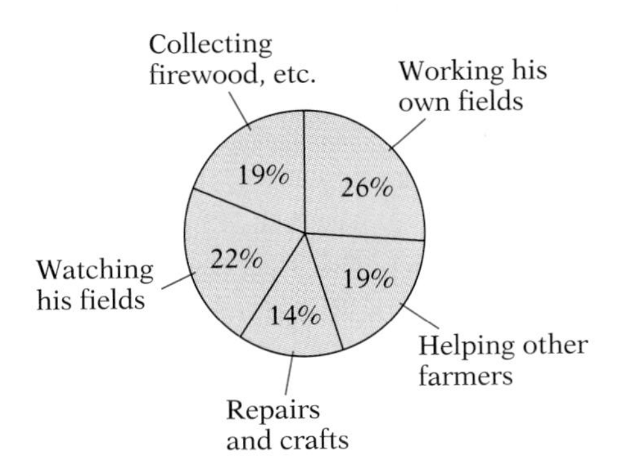

5. How many hours does he spend working his own fields and helping other farmers?

6. How many minutes does he spend collecting firewood, etc.?

Use the histogram on page 539 to answer Exercises 7 and 8.

7. How many bushes are shorter than 24 inches or taller than or equal to 48 inches?

8. What percent of the bushes are 24 to 47.9 inches high? Round to the nearest ten percent.

In the following histogram, the number of tenants who live in a small apartment building are graphed by age intervals, 0–14, 15–29, and so on. (A person who is exactly 15, 30, etc. is included in the older group.) Use this information to answer Exercises 9 through 12.

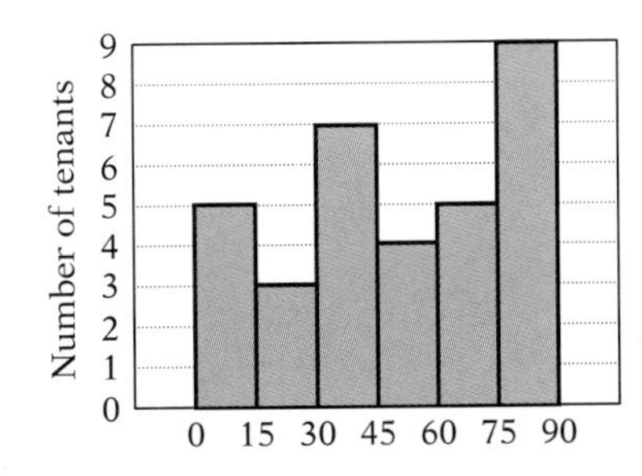

9. What is the width of each age interval in this graph? How many tenants are there altogether?

10. What percent of the tenants in this building are less than 45 years of age? Round to the nearest whole percent.

11. What percent of the tenants are at least 30 but less than 60 years of age? Round to the nearest whole percent.

12. What percent of the tenants in this building are less than 90 years of age but at least 60 years old? Round to the nearest whole percent.

13. See the circle chart preceding Exercise 1 on page 540. What is the ratio of the amount of money spent on social services to that spent on transportation?

14. See the circle chart in Example 1 on page 539. What is the ratio of the amount of money spent on transportation to that spent on food?

For Exercise 15 and 16, see the pie chart on page 541.

15. What is the ratio of the amount of time spent on repairs and crafts to the time spent working the peasant's own fields and helping other farmers?

16. What is the ratio of the amount of time a peasant spends sleeping to the amount of time shown in the chart?

17. See the histogram on page 541. What is the ratio of the number of tenants under 30 to the number of tenants under 75 years of age?

Chapter 8 Review

ERROR ANALYSIS

These problems have been worked incorrectly. Tell where each is wrong, and correct the mistake.

1. Change 0.82% to a decimal.

Incorrect Solution

$0.82\% = 0.82$

Error ______________________

____

Correct Solution

2. Find 62% of 50.

Incorrect Solution

$62 \times 50 = 3100$

Error ______________________

____

Correct Solution

3. 0.05% of 0.75 is what?

Incorrect Solution

$5(0.75) = n$
$3.75 = n$

Error ______________________

____

Correct Solution

4. 423% of what number is 120? Round to the nearest hundredth.

Incorrect Solution

$$\frac{423}{100} = \frac{120}{n}$$
$100 \times n = 50760$
$n = 507.60$

Error ______________________

____

Correct Solution

5. 30 is what percent of 600?

Incorrect Solution

$n \times 600 = 30$

$$600\overline{)30.00}\quad 0.05$$

30 is 0.05% of 600.

Error ______________________

____

Correct Solution

INTERPRETING MATHEMATICS

By working these exercises, you will test and strengthen your mathematics vocabulary.

1. Explain what is meant by the phrase "percent means per hundred." That is, what does percent mean to you?

2. Write a percent equation that is true. Then identify the base, the amount, and the percent.

3. There are three different forms of the percent equation. State each of them in your own words.

4. In geometry, we often speak of irregular polygons. Can you draw one? What makes it irregular?

REVIEW PROBLEMS

The exercises that follow will give you a good review of the material presented in this chapter. Work through them and check your answers at the back of the book.

Section 8.1

1. Write 3.4% as a decimal.

2. Write 0.576 as a percent.

3. Write 0.009% as a decimal.

4. Write 2.104% as a decimal.

5. Write 1.002 as a percent.

6. Write 0.0003% as a decimal.

Section 8.2

7. Write $\frac{1}{4}$ as a percent.

8. Arrange in order from greatest to least: 7.5%, $\frac{1}{2}$%, 0.75, 0.50

9. Write $\frac{3}{8}$ as a percent.

10. Write $3\frac{3}{4}$% as a fraction.

11. Change 5632% to a mixed number.

12. Which is larger as a percent, 2.5% or $\frac{1}{4}$?

Section 8.3

Use a percent equation to solve each of the following.

13. 32 is 6.4% of what?

14. What is 25% of 200?

15. 9.8 is what percent of 49?

16. 1800 is what percent of 90?

17. 0.018% of what number is 0.054?

18. Find 0.05% of 150.

Section 8.4

Use a proportion to solve each of the following.

19. 54 is 30% of what?

20. 60 is 30% of what?

21. 52.5% of 78 is what?

22. What is 0.03% of 52?

23. 0.004% of 16 is what?

24. 36 is what percent of 18?

Section 8.5

25. An inventory of cassettes shows that 150 are defective. This represents 25% of the total number of cassettes. How many cassettes are not defective?

26. In a class of 50 students, 60% were girls, 4 had blonde hair, and 10% were born in California. What percent of the students had blonde hair?

27. On a certain block, 240 flowers were planted of which 80 did not sprout. What percent of the flowers did not sprout?

28. In a bag of 40 marbles, 15 were blue. What percent were not blue?

29. A necklace originally priced at $180 is marked down by 15%. What is the new price?

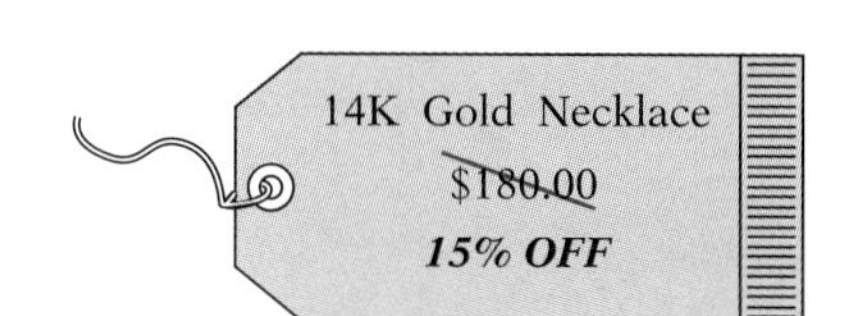

30. The number of women members in a certain club increased from 40,500 in 1940 to 222,500 in 1980. Give the approximate percent of increase.

Section 8.6

31. A lawn that measures 120 feet by 95 feet has been fertilized. If 85% of the lawn grows well, how many square feet do not grow well?

32. Flowers are to be planted around a square swimming pool. There will be 20 flowers on a side. How many flowers will be needed altogether?

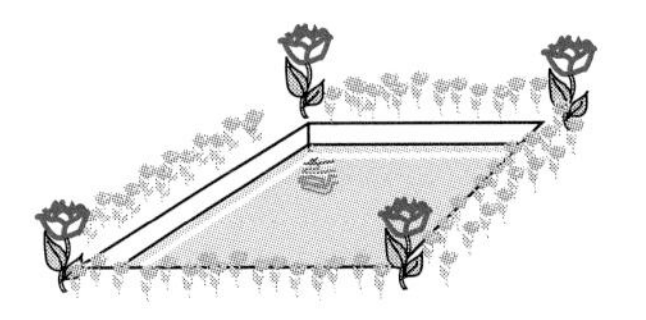

33. A sheet has a design that is 25% flowers. If the sheet measures 78 cm by 56 cm, what is the area that does not contain flowers? Give your answer in square centimeters.

34. A circle with a diameter of 10 inches is drawn inside a square with an area of 100 square inches. How much larger is the area of the square than that of the circle?

35. A rug has 45% of its area covered by furniture. The rug measures 5 yards 2 feet by 7 yards 2 feet. How many square feet are not covered by furniture?

36. A rectangle is cut into three equal-sized squares with a small piece left over. Each square has an area of 624 square inches. What is the area of the small piece that is left over if each square is 30% of the area of the original rectangle?

Mixed Review

37. Write $\frac{2}{5}$ as a percent.

38. 1025 is what percent of 80?

39. 140% of 45 is what?

40. 2.5 is 30% of what? Express as a mixed number.

41. 36 is what percent of 4?

42. 900 is what percent of 0.9?

43. 5 is what percent of 0.0025?

44. What is 0.002% of 193?

45. Find the tax rate if the sales tax is $1.60 on a $20 purchase.

Use the following information to answer Exercises 46 through 49.

The monthly finance charge on a credit card is calculated by multiplying the unpaid balance on the card by the interest rate (written as a decimal).

46. Find the monthly finance charge if the unpaid balance on a credit card in a particular month is $1000 and the rate is 17.8%.

47. Find the finance charge if the unpaid balance on a credit card is $1200 and the rate is 15.2%.

48. Find the finance charge if the unpaid balance is \$764 and the rate is 13.2%.

49. Find the total amount to be repaid this month if the unpaid balance is \$580 and the monthly finance charge is 19%.

50. In a group of 20 employees, 12 called in sick. What percent of the employees called in sick?

51. How much money will you have to borrow for a mortgage loan if you buy a house for \$235,000 and you can afford to put 15% down as a down payment?

52. A bag of candy contained 30 pieces of candy. Only 12 were lollipops. What percent of the candies were lollipops?

53. A store bought a refrigerator at a cost of \$850. The store then marked up the price 10%. What was the selling price after the markup?

54. A typewriter sold for \$125 last year. There was a 30% increase in the price this year. What is the new price?

55. A summer house that was bought last year had the price increased \$15,000 and was sold again. If the price increase was 15%, what was the original price?

56. A realtor generally receives 6% commission for every piece of property she sells.

a. If a piece of property sells for $175,000, how much of a commission does the realtor get?

b. If someone offers her $24,000 as a commission, what was the price of the property?

57. A ball team won 30 games out of the 40 that it played. What percent of the games did it lose?

58. A teacher graded a certain exam and found that out of 60 students, 24 failed. What percent of the students passed?

59. On a test, you left 6 questions blank. This accounted for 15% of the questions on the test. How many questions were on the test?

60. You have gone on a diet. You originally weighed 160 pounds and you lost 20 pounds. What percent of your weight did you lose?

61. Betting on a horse race can be a negative experience. If you started out with $500 in your pocket and lost 25% of your money, how much did you lose?

62. You received a 5.5% pay increase. If your old salary was $248 per week, what is your new salary?

CHAPTER 8 TEST

This exam tests your knowledge of the topics in Chapter 8. Express answers in simplest form. Use 3.14 as an approximation for π.

1. a. Write 62 as a percent.

b. Write 0.01% as a decimal.

c. Write 8.627% as a decimal.

2. a. Write $\frac{1}{4}\%$ as a fraction.

b Write $1\frac{3}{8}$ as a percent.

c. Write $\frac{2}{7}$ as a percent rounded to the nearest hundredth.

3. Use a percent equation to solve:

a. 64.75 is 35% of what?

b. What is 180% of 17?

c. 29 is what percent of 145?

4. Use the proportion method to solve:

a. What is 300% of 2.4?

b. 35 is what percent of 700?

c. 0.008 is 3.2% of what?

5. a. How much will my bill be if I purchase an item that costs $45 and a sales tax of $4\frac{1}{4}$% is charged on the item?

b. A coat selling for $180 is reduced 40%. What is the new selling price?

c. Twenty cupcakes out of a batch will have vanilla frosting. If this represents 40% of the total number of cupcakes, how many cupcakes will be in the batch?

6. a. Find the circumference of a circle with a diameter of 17.

b. Find the area of the largest circle that can be contained in a rectangle that measures 3 inches by 4 inches.

c. A circular flower bed is 6 feet in diameter. If you are are going to put peat moss on 75% of this flower bed, how many square feet of peat moss will you need?

CUMULATIVE REVIEW

CHAPTERS 1–8

The questions that follow will help you maintain the skills you have developed. Write all fractions in simplest form.

1. In a certain school 25 out of every 35 students are females. If there are 700 students, how many are males?

2. Factor 7200 into prime factors.

3. Find the square root of 9409.

4. Subtract: 200 – 42.003

5. What is the area in square inches of a trapezoid with bases of 2.3 and 4.5 inches and a height of 1 foot?

6. Factor 810 into the product of primes and write this using exponential notation.

7. Write two billion, three hundred thousand, thirty and two hundred-thousandths as a number.

8. In a certain hotel, there are 32 floors. Each floor has 26 rooms, one for each letter of the alphabet. How many rooms are there in the hotel?

9. Divide 3103 by 17,000 and round your answer to the nearest thousandth.

10. For every 3 turns of one gear, a smaller gear turns 7 times. At this rate, how many times will the larger gear turn if the smaller gear turns 84 times?

11. Add $92{,}000 + \frac{3}{100{,}000} + \frac{2}{10}$ + one million and write your answer as a decimal.

12. Write 0.052 as a percent.

13. What percent of 84 is 36? Round to the nearest whole percent.

14. Find the product: 40.008×300

15. Find the difference: $45600 - 7356.981$

16. 27 is what percent of 5.4?

17. A school athletic coach is making teams for a relay race. If there are 40 students on a team and there are 1717 students altogether, how many additional students will the coach need to form the last team?

18. Round 46124.09038 to the nearest thousandth and to the nearest ten thousand.

19. Simplify: $[(1 \div 4^2 + \frac{1}{3} - 0.2 \times 1)^2 + 6] \div 2$ Round to the nearest hundredth.

20. $6\frac{3}{8} - 4\frac{4}{5} \times \frac{1}{2} + 3 \div \frac{1}{8}$

21. 10 is 0.5% of what?

22. Subtract: $34\frac{1}{15} - 32\frac{17}{25}$

23. Write in words: 2,000,070,002.00300601

9 Operations with Rational Numbers

One factor that determines the color of the glaze on a piece of pottery is the temperature that the piece is fired to. The temperature may vary depending on how many pieces are on each shelf of the kiln, as well as the heat applied. To determine the temperature, potters often use cones of clay designed to melt at certain temperatures. A potter says that he fires to "cone 8" or "cone 10" when he fires his pots instead of indicating an actual temperature.

- *What other ways are temperatures measured without a thermometer?*

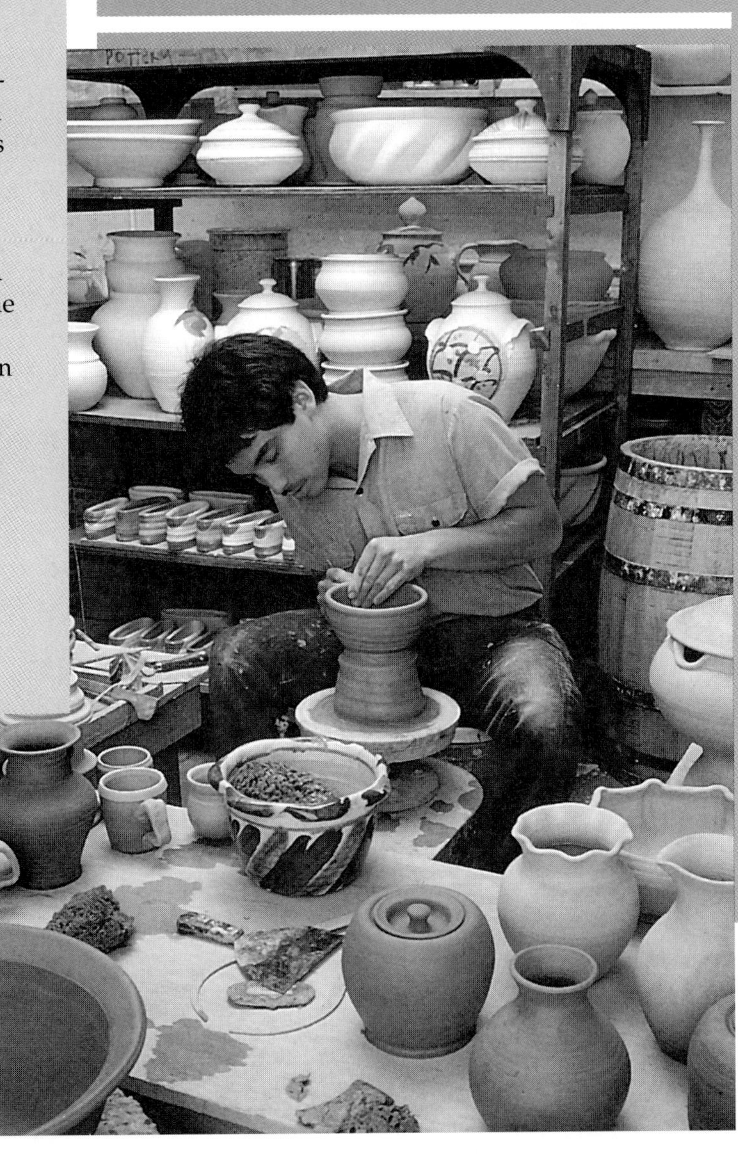

Skills Check

*T*ake this short quiz to see how well prepared you are for Chapter 9. The answers are given at the bottom of the page. So are the sections to review if you get any answers wrong.

1. Add: $2\frac{1}{3} + 4\frac{1}{6}$

2. Subtract: $3\frac{3}{4} - 1\frac{1}{8}$

3. Multiply: $8\frac{2}{5} \times 7\frac{1}{3}$

4. Divide: $9 \div 1\frac{1}{4}$

5. Rewrite $\left(\frac{1}{3}\right)^3 \left(\frac{2}{3}\right)^0$ without exponents.

6. Add: $6.28 + 7 + 0.48$

7. Subtract: $11 - 7.049$

8. Multiply: 76.45×2.03

9. Divide: $0.7\overline{)42.14}$

10. Simplify: $3 + 6 \times 2^3$

ANSWERS: 1. $6\frac{1}{2}$ [Section 4.6] 2. $2\frac{5}{8}$ [Section 4.8] 3. $61\frac{3}{5}$ [Section 5.2] 4. $7\frac{1}{5}$ [Section 5.4] 5. $\frac{1}{27}$ [Section 5.6] 6. 13.76 [Section 6.3] 7. 3.951 [Section 6.4] 8. 155.1935 [Section 6.5] 9. 60.2 [Section 6.6] 10. 51 [Section 3.4]

9.1 Rational Numbers

SECTION GOALS

- To graph rational numbers on a number line
- To use $<$ and $>$ in order relationships for rational numbers
- To give the absolute value of a rational number

In preceding chapters we worked only with zero and the positive numbers—that is, with numbers to the right of zero on the number line. However, there are also negative numbers—numbers that we can use to represent losses, to suggest decreases in temperature or speed, or to indicate the "opposite" direction. In this section, we include the negative numbers with zero and the positive numbers to obtain the **rational numbers.**

First, a definition:

> The **opposite** of any non-zero number a is the number $-a$. If a is a positive number, then its opposite, $-a$, is a negative number.

For example, the opposite of 3 is –3, and the opposite of –3 is 3. Just as the number 3 can be located on a number line, its opposite, the negative number –3, can also be located on the number line. As you may have guessed by now, the number –3 lies to the *left* of zero.

Plotting Numbers and Opposites

To plot the positive number 3, we count three units to the *right* of zero as shown on this number line.

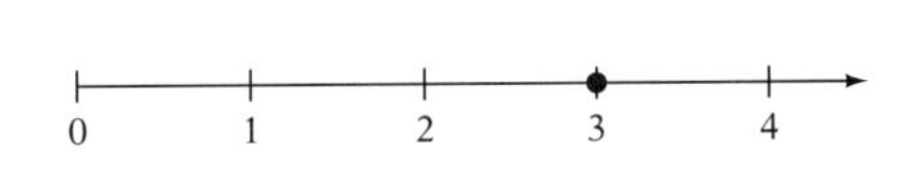

To accommodate the negative numbers, we must extend the number line to the left of zero, marking units as shown here.

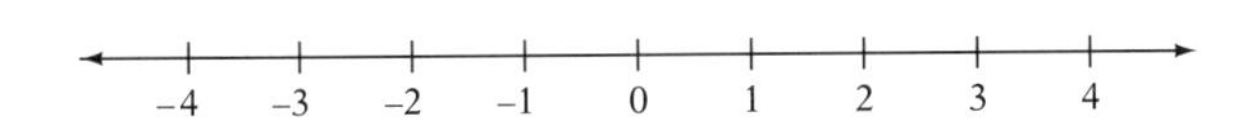

Then, to plot the negative number –3, we count three units to the *left* of zero.

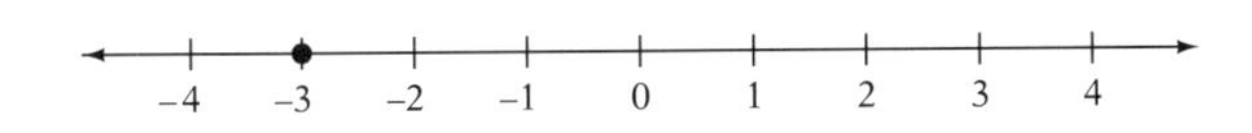

The number itself tells us how many units to count from zero. A minus sign tells us to count to the left of the zero.

> Numbers to the right of zero on the number line are **positive numbers**.
> Numbers to the left of zero on the number line are **negative** numbers; they have minus signs.
> Numbers with signs are called **signed numbers**.

1a. Locate $-5\frac{1}{3}$ and its opposite on the number line.

1b. What is $-[-(-2.5)]$? Locate it and its opposite on the number line.

When we read a number with the minus sign in front of it, we read the sign as "negative." So -4 is read "negative four." When we read a number with the plus sign in front of it, we simply say the number. So $+4$ is read "four." Where there is no sign, the number is considered positive. For example, 5 is considered to be $+5$.

Fractions and decimal numbers also have opposites. For example, the opposite of $\frac{2}{5}$ is $-\frac{2}{5}$, and the opposite of -1.1313 is 1.1313. These numbers can also be plotted on a number line.

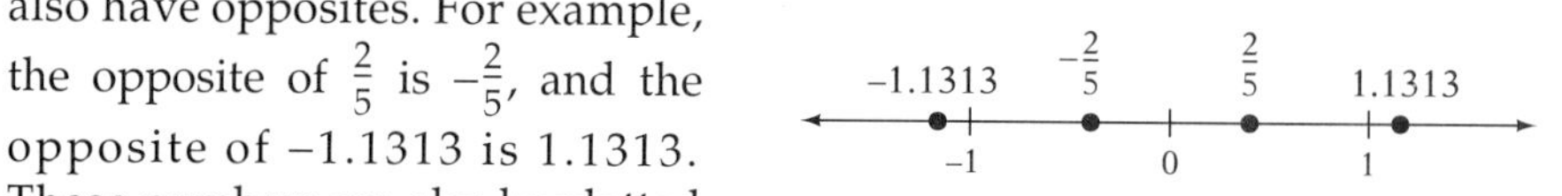

We use the minus sign to designate a negative number, as in -3. We also use it to mean the opposite of a number, as in $-(-3)$. But we know that the opposite of -3 is 3, so now we see that $-(-3) = 3$. More generally, for any number a,

$$-(-a) = a$$

EXAMPLE 1

Locate -2.5 and its opposite on the number line.

SOLUTION

The opposite of -2.5 is 2.5. These numbers are both located $2\frac{1}{2}$ units from zero. 2.5 is to the right of zero, and -2.5 is to the left as we can see on the following number line.

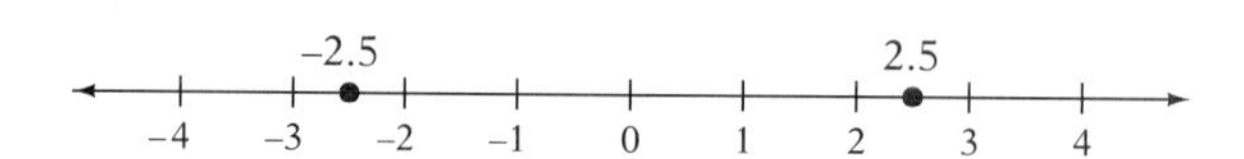

■ *Now do margin exercises 1a and 1b. The answers are given at the back of the book.*

The Rational Numbers

When we combine the positive whole numbers, and their negatives, and zero, we get the set of integers.

> **Integers**
>
> The **integers** are the set of numbers that contain the positive whole numbers, their opposites (the negative numbers), and zero.

Integers are part of a larger set of numbers called rational numbers. Examples of rational numbers include $\frac{1}{2}$, -79, 37.007, and -0.65. We define the rational numbers as follows:

Rational Numbers

The **rational numbers** are the set of all numbers that can be represented as fractions, $\frac{a}{b}$, where a and b are integers and b is not equal to zero.

All integers are rational numbers because any integer a can be represented as the fraction $\frac{a}{1}$.

Absolute Value

Every number and its opposite are the same distance from zero on the number line, but in opposite directions. We say, then, that they have the same absolute value.

Absolute Value

The **absolute value** of a number a, symbolized $|a|$, is the distance of a from zero on the number line. Because absolute value represents distance, the absolute value of a number can never be negative.

Note that the numbers –2.5 and 2.5 in Example 1 both have an absolute value of 2.5, because they are both 2.5 units from 0. That is, $|2.5| = |-2.5| = 2.5$.

In the next example, we will find the absolute value of a negative rational number.

EXAMPLE 2

Find $\left|-5\frac{1}{4}\right|$.

SOLUTION

Because $-5\frac{1}{4}$ is $5\frac{1}{4}$ units from zero (disregard the direction), its absolute value is $5\frac{1}{4}$.

Thus $\left|-5\frac{1}{4}\right| = 5\frac{1}{4}$.

■ *Now do margin exercises 2a and 2b.*

What happens if there is a minus sign in front of an absolute value sign? For example, what is $-|6|$? This is read "the opposite of the absolute value of 6." To find its value, we first find the absolute value of 6 and then find the opposite of that number.

EXAMPLE 3

Evaluate: $-|-7.6|$

2a. Find the absolute value of $4\frac{1}{5}$.

2b. Find the the absolute value of $\frac{-127}{16}$.

SOLUTION

The distance of –7.6 from zero is 7.6, so its absolute value is 7.6. So

$$|-7.6| = 7.6$$

Then the opposite of 7.6 is –7.6.

$$\text{So } -|-7.6| = -7.6$$

■ *Now do margin exercises 3a and 3b.*

3a. Evaluate: $-|-5.1|$

3b. Evaluate: $|-|-29.14||$

Ordering Rational Numbers

To finish this section, we will compare and order integers and rational numbers. We know from our work with whole numbers that a is greater than b if it is farther to the right on the number line. This is also true if a and b are rational numbers.

EXAMPLE 4

Which is greater, –8 or $-6\frac{1}{2}$?

SOLUTION

We first locate the numbers on the number line.

–8 $-6\frac{1}{2}$

–8 –7 –6 –5 –4 –3 –2 –1 0 1 2 3

Because $-6\frac{1}{2}$ is to the right of –8, $-6\frac{1}{2}$ is greater.

So $-6\frac{1}{2} > -8$.

■ *Now do margin exercises 4a and 4b.*

4a. Which is greater, –7.4 or –5.3?

4b. Which is less, –2.3 or 0?

EXAMPLE 5

Put the following rational numbers in order from least to greatest:

$$-37\tfrac{2}{3}, \quad 22\tfrac{1}{5}, \quad -12, \quad 25.3$$

SOLUTION

This time, we will not use a number line. Instead, we will "picture" the line.

The smallest number is $-37\frac{2}{3}$, because it would be farthest to the left on a number line. The greatest number is 25.3. (Why?) Arranging the other numbers in order, we have $-37\frac{2}{3}$, –12, $22\frac{1}{5}$, and 25.3.

So $-37\frac{2}{3} < -12 < 22\frac{1}{5} < 25.3$.

■ *Now do margin exercises 5a and 5b.*

Work the exercises that follow.

5a. Arrange in order from smallest to largest: $-12\frac{5}{7}$, $-12\frac{5}{12}$, –11.3, $-11\frac{2}{9}$

5b. Insert < or > to make the following statement true. $-35\frac{1}{8}$? $-29\frac{1}{14}$

9.1 EXERCISES

In Exercises 1 through 8, evaluate each expression.

1. The opposite of $-|-6|$
2. The opposite of $|14|$
3. The opposite of $|-8|$
4. The opposite of $|-5|$
5. $-\left|-23\frac{6}{11}\right|$
6. $-|16.07|$
7. $|-11.5|$
8. $\left|-9\frac{3}{7}\right|$

In Exercises 9 through 16, graph each number and its opposite.

9. $-\left|-5\frac{1}{2}\right|$
10. $-\left|-6\frac{3}{5}\right|$
11. $-\left|1\frac{3}{8}\right|$
12. $-|2.2|$
13. $|-|-3.2||$
14. $-\left|1\frac{3}{4}\right|$
15. $-|-5.7|$
16. $|-|-1.6||$

In Exercises 17 through 24, graph each number on a number line by showing the two integers the number lies between.

17. $-332\frac{2}{3}$
18. $-400\frac{2}{5}$
19. $-76\frac{1}{5}$
20. 133.5
21. $-|98.7|$
22. -67.98
23. $-|-254|$
24. $-103\frac{3}{4}$

In Exercises 25 through 32, insert the proper sign (< or >) in each blank space.

25. $-10 \underline{\qquad} 5$

26. $|-6775| \underline{\qquad} |-3|$

27. $35 \underline{\qquad} -24$

28. $47.6 \underline{\qquad} -17$

29. $-6 \underline{\qquad} 4\frac{1}{5}$

30. $-10\frac{2}{3} \underline{\qquad} 19$

31. $-18 \underline{\qquad} 26\frac{2}{5}$

32. $-17 \underline{\qquad} 32$

In Exercises 33 through 40, rewrite each set of rational numbers in order from smallest to largest.

33. $15, -11, -26$

34. $-32, -10, 18$

35. $27, -45, -13$

36. $135, -63, -27$

37. $-23\frac{1}{3}, |41|, 19\frac{2}{5}$

38. $16, -86\frac{2}{3}, -\left|36\frac{2}{5}\right|$

39. $-42\frac{1}{2}, -|57.4|, 45$

40. $-\left|38\frac{2}{5}\right|, 29\frac{3}{8}, -52.7$

In Exercises 41 through 48, simplify by removing parentheses and absolute value marks.

41. $-[-(-1465)]$

42. $-(-|-|-|5776|||)$

43. $-\{-[-(-|-|26||)]\}$

44. $-[-(-|137|)]$

45. $-\left(-\left|-\left|-78\frac{11}{12}\right|\right|\right)$

46. $-|-(|-|45||)|$

47. $-\{-[-(-|365|)]\}$

48. $-[-|-|-(-|273.5|)||]$

In Exercises 49 through 56, simplify and then use > or < to indicate the appropriate order.

49. $-(-|9484|) \underline{\qquad} -(-|-96|)$

50. $-\left[-\left(-6\frac{3}{4}\right)\right] \underline{\qquad} -|-(-300)|$

51. $-[-(-|-|280||)] \underline{\qquad} -(|-49|)$

52. $-[-|-|-(-85)||] \underline{\qquad} |-|-|-(-59|||$

53. $-[-(-|-844|)] \underline{\qquad} |-|-163||$

54. $[-|-|-(-29)||] \underline{\qquad} -[-(-|-925|)]$

55. $-\left[-\left(-3\frac{1}{3}\right)\right] \underline{\qquad} -\left|-\left[-\left(\left|-3\frac{1}{2}\right|\right)\right]\right|$

56. $-\left[-\left(\left|-\left(-9\frac{2}{3}\right)\right|\right)\right] \underline{\qquad} -[-(-|9.6|)]$

9.2 Adding and Subtracting Rational Numbers

Addition

Recall how we added two positive whole numbers, like 3 and 2, as shown on the following number line. We started at one of the numbers (say, 3) and then moved the other number of units (here, 2) to the right. The number we ended at was the sum.

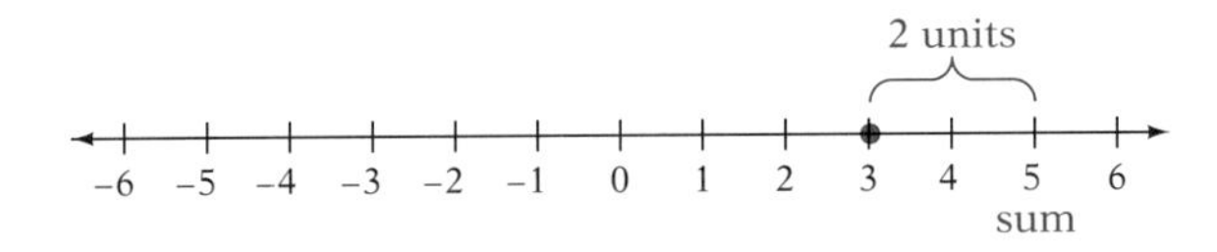

Let's extend that idea to the addition of any two integers, using the following number line. To add −3 and −2, we should begin at −3. Then, because the second number (−2) is negative, we should move *left* (in the negative direction) 2 units. We end up at −5.

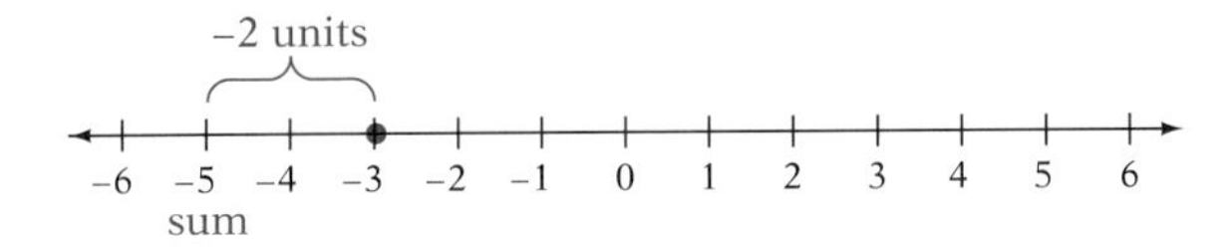

We could also have started at −2 and moved left 3 units. The answer would have been the same: $(-2) + (-3) = (-3) + (-2) = -5$.

Let's use the number line to add a negative number to a positive number: $4 + (-3)$. We begin at 4 and, because the second number (−3) is negative, we move 3 units to the left.

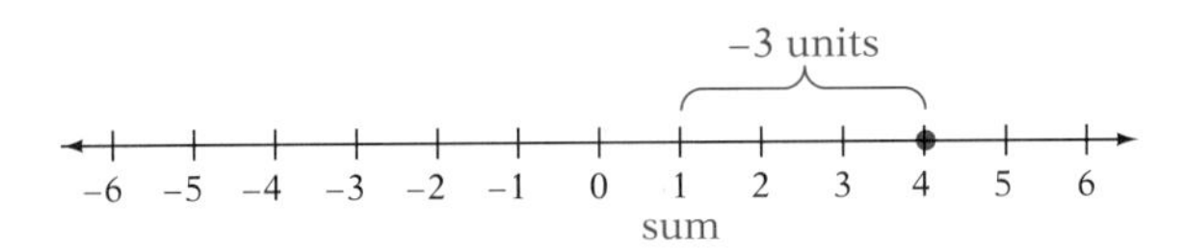

The result is positive 1. So $4 + (-3) = 1$.

To add two rational numbers ($a + b$)

1. Note the signs of the two numbers.
2. If the signs of the numbers are the same, add their absolute values.
3. If the signs of the numbers are different, subtract the smaller absolute value from the larger.
4. Give the result the same sign as the number with the greater absolute value.

SECTION GOALS

- To find the sum of two or more rational numbers
- To find the difference of two rational numbers

1a. Add: $-81 + 176$

1b. Add: $62 + (-18)$

2a. Add: $\left(-6\frac{3}{5}\right) + \left(-1\frac{1}{4}\right)$

2b. Add: $(-3.62) + (-4.6)$

In the first example, we add a negative integer to a positive integer.

EXAMPLE 1

Add: $26 + (-49)$

SOLUTION

Because the signs are different, we subtract absolute values.

$$|-49| - |26| = 49 - 26 = 23$$

The number with the greater absolute value (–49) is negative, so the result is negative: –23.

Thus $26 + (-49) = -23$.

■ *Now do margin exercises 1a and 1b.*

In the next example, we add a negative rational number to another negative rational number.

EXAMPLE 2

Add: $\left(-7\frac{1}{8}\right) + \left(-6\frac{3}{8}\right)$

SOLUTION

Because the signs are the same, we add the absolute values.

$$\left|-7\frac{1}{8}\right| + \left|-6\frac{3}{8}\right| = 7\frac{1}{8} + 6\frac{3}{8} = 13\frac{4}{8} = 13\frac{1}{2}$$

The number with the greater absolute value $\left(-7\frac{1}{8}\right)$ is negative, so the result is negative: $-13\frac{1}{2}$.

Thus $\left(-7\frac{1}{8}\right) + \left(-6\frac{3}{8}\right) = -13\frac{1}{2}$.

■ *Now do margin exercises 2a and 2b.*

The commutative and associative properties of addition of whole numbers also hold for addition of rational numbers (including, of course, the integers). Similarly, 0 is the identity element for addition.

In the next example, these properties enable us to add more than two numbers by assuring us that we can change the order of addition without altering the result.

Study Hint

When adding more than two integers or rational numbers, add the positive numbers together, add the negative numbers together, and combine the results.

EXAMPLE 3

Add: $(-4) + 6 + (-18)$

SOLUTION

Because there are two negative numbers, we begin by adding them:

$$(-4) + (-18)$$

The signs are the same, so we add the absolute values.

$$|-4| + |-18| = 4 + 18 = 22$$

The number with the greater absolute value is negative, so the result is negative: -22.

There is only one positive number, so we now combine it with the sum of the negative numbers. We have $(-22) + (6)$. The signs are different, so we subtract absolute values:

$$|-22| - |6| = 22 - 6 = 16$$

The number with the greater absolute value (-22) is negative, so our result is negative: -16.

Therefore, $(-4) + 6 + (-18) = -16$.

■ *Now do margin exercises 3a and 3b.*

3a. Add: $-14 + 9 + (-89)$

3b. Add: $-86 + \left(-23\frac{1}{3}\right) + 18\frac{2}{5}$

Subtraction

Before we can develop a rule for subtracting two integers, we need the concept of additive inverse.

Additive Inverses

Two numbers are **additive inverses** of each other if their sum is zero. The additive inverse of a is $-a$ because $a + (-a) = 0$. Similarly, the additive inverse of $-a$ is a.

Thus 3 and -3 are additive inverses because $3 + (-3) = 0$. The additive inverse of $-5\frac{2}{7}$ is $5\frac{2}{7}$ because $-5\frac{2}{7} + 5\frac{2}{7} = 0$. As we can see, the additive inverse of a number is also its opposite.

We can now discuss the subtraction of integers and rational numbers. In fact, we can now define subtraction in terms of addition.

Definition of Subtraction

For any integers a and b,

$$a - b = a + (-b)$$

In other words, we subtract a rational number b from another rational number a by adding the inverse of b to a. To put it more simply,

To Subtract Two Rational Numbers ($a - b$)

1. Replace the subtrahend b with its additive inverse $-b$.
2. Rewrite the operation as addition by changing the operation sign.
3. Add $a + (-b)$.

Let's try some examples. In the first example, we will subtract a negative rational number from a positive rational number.

EXAMPLE 4

Subtract: $18 - \left(-10\frac{1}{2}\right)$

SOLUTION

To subtract, we first replace $-10\frac{1}{2}$ with its additive inverse, $10\frac{1}{2}$, and rewrite the operation as addition.

$$18 - \left(-10\frac{1}{2}\right) = 18 + \left(+10\frac{1}{2}\right)$$

Then we add. Because the signs are the same, we add absolute values.

$$|18| + \left|10\frac{1}{2}\right| = 18 + 10\frac{1}{2} = 28\frac{1}{2}$$

The number with the greater absolute value (18) is positive, so we make the result positive: $28\frac{1}{2}$

Thus $18 - \left(-10\frac{1}{2}\right) = 28\frac{1}{2}$

■ *Now do margin exercises 4a and 4b.*

EXAMPLE 5

Subtract: $-48 - 26$

SOLUTION

To subtract, we first replace +26 with its additive inverse –26 and rewrite as addition: $-48 + (-26)$. Then we add. Because the signs are the same, we add the absolute values and make the result negative.

$$-48 - 26 = -48 + (-26) = |-48| + |-26| = 74$$

So $-48 - 26 = -74$.

■ *Now do margin exercises 5a and 5b.*

The **range** of a set of data is the difference between the greatest and least numbers in the set. It is found by subtracting the least number from the greatest number. Our final example is a range problem.

4a. Subtract: $1\frac{3}{8} - \left(-2\frac{1}{8}\right)$

4b. Subtract: $5.62 - (-3.7)$

5a. Subtract: $-68 - 34\frac{1}{2}$

5b. Subtract 378.32 from –1276.8.

6a. The markings on a safe go from 31 to 158. What is their range?

6b. The following numbers were recorded: –3.4, 8.9, 32.18, 45, –17.3, and –0.002. What is the range of these numbers?

EXAMPLE 6

Hourly readings on a pressure gauge were –1.80, –2.75, –2.25, and –1.25. What is the range of these readings?

SOLUTION

Our definition says that we must subtract the least value from the greatest. Because the least value is –2.75 and the greatest is –1.25, our problem becomes

$$-1.25 - (-2.75)$$

First, we replace –2.75 with its additive inverse, +2.75, and rewrite the operation as addition.

$$-1.25 - (-2.75) = -1.25 + 2.75$$

Now we add. Because the signs are different, we must subtract absolute values.

$$|2.75| - |-1.25| = 2.75 - 1.25 = 1.5$$

The result gets the same sign as 2.75, so it is positive: 1.5 is the range.

Therefore, readings from –2.75 to –1.25 have a range of 1.5 units. You will find that the range is always positive or zero.

■ *Now do margin exercises 6a and 6b.*

Work the exercises that follow.

9.2 EXERCISES

In Exercises 1 through 68, add or subtract as indicated.

1. $2.94 + (-1.8)$

2. $3.86 + (-2.6)$

3. $(-13.7) + 7.9$

4. $(-27) - 63$

5. $(-90) + 47 + (-38)$

6. $(-232) + 76 + (-86)$

7. $(-56) + (+87) + (-54)$

8. $(-4.2) + (-9.8)$

9. $\left(-5\frac{6}{7}\right) + \left(-4\frac{6}{11}\right)$

10. $\left(-12\frac{2}{3}\right) + \left(-6\frac{7}{9}\right)$

11. $\left(-18\frac{5}{6}\right) + \left(-5\frac{3}{8}\right)$

12. $23 - (-37)$

13. $(-18) + (-24) + (-67)$

14. $(-54) + (-78) + (-58)$

15. $(-24) + (-56) + (-89)$

16. $\left(-5\frac{2}{9}\right) + 1\frac{4}{9} + \left(-7\frac{5}{9}\right)$

17. $\left(-9\frac{5}{6}\right) + 3\frac{5}{8}$

18. $\left(-6\frac{2}{11}\right) + 5\frac{4}{7}$

19. $3\frac{4}{7} + \left(-15\frac{6}{7}\right)$

20. $\left(-4\frac{3}{4}\right) - \left(-1\frac{5}{8}\right)$

21. $97 + (-45) + 23$

22. $43 + (-28) + 47$

23. $63 + (-241) + (+75)$

24. $102 + (-98) + 77$

25. $(-2.4) + (-7.36)$

26. $(-11.3) + (-8.45)$

27. $(-6.8) + (-9.46)$

28. $(-89) + 90 + (-126)$

29. $\left(-2\frac{1}{3}\right) + \left(-3\frac{2}{3}\right) + 5\frac{1}{3}$

30. $8.1 + (-4.6) + (-2.3)$

31. $7.7 + (-6.3) + 9.2$

32. $\left(-9\frac{1}{2}\right) + \left(-4\frac{7}{12}\right)$

33. $+127 - (-94)$

34. $+136 - (-27)$

35. $67 - (-89)$

36. $(-32) + (-67) + (-41)$

37. $\left(-6\frac{2}{3}\right) - 4\frac{1}{2}$

38. $\left(-8\frac{1}{3}\right) - 3\frac{2}{7}$

39. $\left(-14\frac{5}{7}\right) - 29\frac{6}{11}$

40. $2.9 - 56 - 92.5$

41. $(-37) - 28$

42. $7\frac{5}{9} + \left(-27\frac{7}{9}\right)$

43. $(-18) - 75$

44. $(-15.2) + 14.7$

45. $6.3 - (-14.7)$

46. $2.4 - (-6.2)$

47. $-12 - 4.7$

48. $(-2.47) + (-3.21)$

49. $6.6 - 87 - 16.9$

50. $3.5 - 12.6 - 2.5$

51. $4.1 - 94 - 35.5$

52. $\left(-13\frac{1}{2}\right) - \left(-26\frac{3}{4}\right)$

53. $(-19.7) - (-8.2)$

54. $(-18.5) - (-9.4)$

55. $(-14.8) - (-12.4)$

56. $(-26.3) - (-6.3)$

57. $\left(-7\frac{5}{6}\right) - \left(-4\frac{2}{3}\right)$

58. $\left(-8\frac{2}{3}\right) - \left(-2\frac{1}{8}\right)$

59. $\left(-8\frac{2}{3}\right) - \left(-17\frac{5}{7}\right)$

60. $-\left(-13\frac{1}{2}\right) + \left(-26\frac{3}{4}\right)$

61. $-9.7 - 8.92$

62. $-8.41 - 5.76$

63. $-2.4 - 15.75$

64. $-8.3 - 18.42$

65. $9.7 - (-11.3) - 15.9$

66. $15\frac{8}{9} - \left(-5\frac{2}{9}\right) + 46\frac{1}{3}$

67. $9\frac{4}{5} - \left(-20\frac{3}{5}\right) - 2\frac{3}{5}$

68. $11.6 - (-7.9) - (-4.7)$

69. Markings from -17 to 101 are indicated along a number line. What is the range of these markings?

70. What is the range of these test scores: 43, 25, 100, 18, 42, and 32?

71. What is the range of these fractional measures: $2\frac{1}{3}$, $5\frac{4}{7}$, $8\frac{3}{8}$, and $1\frac{1}{5}$?

72. A program randomly lists the numbers -102, 47, $188\frac{1}{2}$, 4.28, and -3.18. What is the range of these numbers?

9.2 MIXED PRACTICE

By doing these exercises, you will practice all the topics in this chapter up to this point.

73. Subtract: $(-65) - \left(-94\frac{1}{2}\right)$

74. Add: $(85.8) + (-33.95)$

75. Graph $-|4|$ on a number line.

76. Find the sum: $(-123) + \left(46\frac{1}{6}\right)$

77. Arrange $-|-|-85||$, $-|37|$, and -42 in order from smallest to largest.

78. Subtract: $\left(11\frac{1}{4}\right) - \left(-52\frac{1}{2}\right)$

79. Insert < or > between the numbers $-|26|$ and $-|-63|$.

80. Graph $-|-5|$ on a number line.

81. Find the sum: $(-56.85) + 79 + (-129.6)$

82. Arrange $|-27|$, $|-|32||$, and $-|-14|$ in order from largest to smallest.

83. Add: $35\frac{2}{3} + \left(-42\frac{1}{2}\right) + \left(-88\frac{1}{7}\right)$

84. Subtract 23.85 from (-27.5).

9.3 Multiplying and Dividing Rational Numbers

Multiplication

As you saw in Chapter 2, multiplication is a way of performing repeated addition. For example, 4×3 is the same as $3 + 3 + 3 + 3$, or 12. Similarly, $4 \times (-3)$ is the same as $(-3) + (-3) + (-3) + (-3)$, or -12. Thus multiplication of signed numbers is just like that of positive numbers, except for the sign of the result.

To multiply two rational numbers ($a \times b$)

1. Find the product of the absolute values of the factors.
2. If the factors have the same sign, make the product positive.
3. If the factors have different signs, make the product negative.

In our first example, we multiply two negative rational numbers.

EXAMPLE 1

Multiply: $\left(-2\frac{1}{3}\right)\left(-7\frac{2}{5}\right)$

SOLUTION

We first must change the mixed numbers to improper fractions.

$$\left(-2\frac{1}{3}\right)\left(-7\frac{2}{5}\right) = \left(-\frac{7}{3}\right)\left(-\frac{37}{5}\right)$$

Then we multiply the absolute values.

$$\left|-\frac{7}{3}\right|\left|-\frac{37}{5}\right| = \left(\frac{7}{3}\right)\left(\frac{37}{5}\right)$$

$$= \frac{7 \times 37}{3 \times 5} = \frac{259}{15}$$

We next simplify the result by changing it to a mixed number and reducing.

$$\frac{259}{15} = 17\frac{4}{15}$$

Because the signs of the original factors are the same, we make the result positive: $17\frac{4}{15}$.

So $\left(-2\frac{1}{3}\right)\left(-7\frac{2}{5}\right) = 17\frac{4}{15}$

■ *Now do margin exercises 1a and 1b.*

The properties of multiplication of whole numbers also hold for rational numbers. They enable us, for example, to multiply rational numbers in any order. Thus, to multiply more than two numbers, we simply multiply two at a time.

SECTION GOALS

- To find the product of two or more rational numbers
- To find the quotient of two rational numbers

1a. Multiply: $\left(-3\frac{1}{4}\right)\left(-6\frac{1}{3}\right)$

1b. Multiply: $(-5.4)\,(9.05)$

2a. Multiply: (–0.3) (–5) (–4.5)

2b. Multiply: $\left(-4\frac{1}{2}\right)(20)\left(-3\frac{5}{9}\right)$

3a. Divide: $\left(-2\frac{1}{7}\right) \div \left(-\frac{1}{5}\right)$

3b. Divide: (–0.5) ÷ 12.5

EXAMPLE 2

Multiply: (–3.1) (–4.2) (–0.2)

SOLUTION

We begin by multiplying any two of these numbers together; we will pick the first two negative numbers: (–3.1) (–4.2). We first multiply their absolute values. Because both factors have the same sign, we make the result positive:

$$|-3.1||-4.2| = (3.1)(4.2) = 13.02$$

We then multiply this result by the remaining number: (–0.2). Multiplying the absolute values, we get

$$|13.02||-0.2| = (13.02)(0.2) = 2.604$$

Because these two factors have different signs, we make this result negative: –2.604.

So (–3.1) (–4.2) (–0.2) = –2.604

■ *Now do margin exercises 2a and 2b.*

Division

When we divide rational numbers, we use essentially the same procedure for finding the sign of the answer as for multiplication.

To divide two rational numbers ($a \div b$)

1. Find the quotient of the absolute values of the numbers.
2. Make the quotient positive if the original numbers have the same sign.
3. Make the quotient negative if the original numbers have different signs.

In our final example, we divide a negative rational number by a positive one.

EXAMPLE 3

Divide: (–3.6) ÷ (1.5)

SOLUTION

We divide the absolute values of the numbers.

$$|-3.6| \div |1.5| = 3.6 \div 1.5 = 2.4$$

Because the signs of the original numbers are different, we make the result negative: –2.4

Thus, (–3.6) ÷ (1.5) = –2.4.

■ *Now do margin exercises 3a and 3b.*

Work the exercises that follow.

9.3 EXERCISES

In Exercises 1 through 64, multiply or divide as indicated.

1. (6) (–73)

2. (5) (–74)

3. (–25) (–2) (6)

4. (–38) (–2) (–15)

5. $\left(-8\frac{2}{5}\right)\left(-1\frac{6}{7}\right)$

6. $\left(-2\frac{11}{12}\right)\left(5\frac{1}{5}\right)$

7. (–112) ÷ (7)

8. (–224) ÷ (8)

9. (+230) ÷ (–10)

10. (+84) ÷ (7)

11. (–74) ÷ (0.08)

12. (–98) ÷ (0.05)

13. (–190) ÷ (–38)

14. –(–315) ÷ (–9)

15. –(0.7) (–5) (0.8)

16. –(0.3) (–9) (0.2)

17. $-\left(-6\frac{1}{3}\right)\left(-1\frac{4}{9}\right)$

18. $\left(-6\frac{1}{2}\right)\left(-5\frac{1}{5}\right)$

19. $\left(-\frac{5}{11}\right)\left(\frac{2}{3}\right)\left(\frac{22}{25}\right)$

20. $-\left(-\frac{1}{2}\right)\left(\frac{8}{11}\right)\left(\frac{3}{4}\right)$

21. $(7) \div \left(-4\frac{1}{3}\right)$

22. $(-6) \div \left(-3\frac{2}{5}\right)$

23. (23) (–8) (–10)

24. (38) (6) (–4)

25. $-(-0.16) \div (-0.2)$

26. $(-0.38) \div (-0.19)$

27. $(-22)\ (2)\ (-8)$

28. $-(-32)\ (5)\ (-3)$

29. $-(-520) \div (40)$

30. $(-204) \div (12)$

31. $-(-1)\left(4\frac{3}{10}\right)$

32. $(-9)\left(8\frac{4}{15}\right)$

33. $3\frac{3}{5} \div \left(-2\frac{7}{10}\right)$

34. $\left(-1\frac{1}{2}\right) \div \left(-3\frac{1}{4}\right)$

35. $2\frac{2}{5} \div \left(-6\frac{1}{5}\right)$

36. $7\frac{5}{7} \div \left(7\frac{1}{2}\right)$

37. $-(-82) \div (0.04)$

38. $(-63) \div (0.08)$

39. $-(384) \div (16)$

40. $(+99) \div (-9)$

41. $(-2)\ (-14)\ (19)$

42. $-(-17)\ (-6)\ (8)$

43. $-(4)\ (-71)$

44. $(+3)\ (-68)$

45. $-(-9)\ (3)\ (-4)$

46. $(-15)\ (8)\ (-9)$

47. $(-0.42) \div (-0.02)$

48. $-(-0.72) \div (-0.03)$

49. $(7)\ (-5)\ (-13)$

50. $(13)\ (-9)\ (12)$

51. $(-3) \div \left(-1\frac{4}{7}\right)$

52. $(8) \div \left(-2\frac{5}{6}\right)$

53. $(0.12)(-23)(0.37)$

54. $-(0.56)(-69)(0.43)$

55. $(-16{,}740) \div (-180)$

56. $(-100.75) \div (-25)$

57. $\left(-\frac{2}{5}\right)\left(\frac{5}{9}\right)\left(\frac{1}{12}\right)$

58. $-\left(-\frac{4}{7}\right)\left(\frac{1}{2}\right)\left(\frac{3}{8}\right)$

59. $-(-6)\left(3\frac{5}{12}\right)(-1.2)$

60. $(-7)\left(2\frac{9}{14}\right)(2.2)$

61. $(-10.5)(-387)(-0.08)$

62. $(-1.1)(-598)(+0.06)$

63. $(-0.24)(+4844)(-0.01)$

64. $(-2.8)(-183)(-0.02)$

9.3 MIXED PRACTICE

By doing these exercises, you will practice all the topics in this chapter up to this point.

65. Graph $-9\frac{2}{5}$ on the number line.

66. Arrange $-28\frac{1}{3}$, $\left|-26\frac{1}{2}\right|$, and $-\left|-\left|-26\frac{1}{3}\right|\right|$ in order from smallest to largest.

67. Arrange $|-47|$, $\left|-\frac{80}{2}\right|$, and $-(-|36|)$ in order from largest to smallest.

68. Find the quotient of $(-1197.2) \div 19.5$, rounded to the nearest tenth.

69. Find the sum: $2.84 + (-1.96) + (-35)$

70. Find the difference: $\left(-384\frac{1}{2}\right) - 58\frac{2}{9}$

71. Find the sum: $\left(-213\frac{3}{8}\right) + \left(-187\frac{2}{7}\right)$

72. Find the quotient: $(-484) \div \left(-11\frac{2}{3}\right)$

73. Find the product: $(-37.4)\,(-41.3)$

74. Find the quotient of $(-1776) \div (-37.4)$, rounded to the nearest hundredth.

75. Find the difference: $(-127.7) - (-69.9)$

76. Find the product: $(-89)(8)(-17.5)$

77. Find the difference: $215\frac{1}{4} - \left(-36\frac{3}{8}\right)$

9.4 Applying Operations with Rational Numbers: Word Problems

SECTION GOAL

- To solve applications of addition, subtraction, multiplication, and division of integers and rational numbers

We solve word problems with rational numbers in the same way as we solve any word problem. However, we must be careful of the signs of the numbers.

EXAMPLE 1

You have a meeting in 10 minutes. Your boss calls, and you spend $4\frac{1}{2}$ minutes on the phone. Then your meeting is delayed 15 minutes. You work $13\frac{1}{3}$ minutes on the computer, spend $5\frac{1}{2}$ minutes getting coffee, and then sit down to read until your meeting starts. How long do you have for reading?

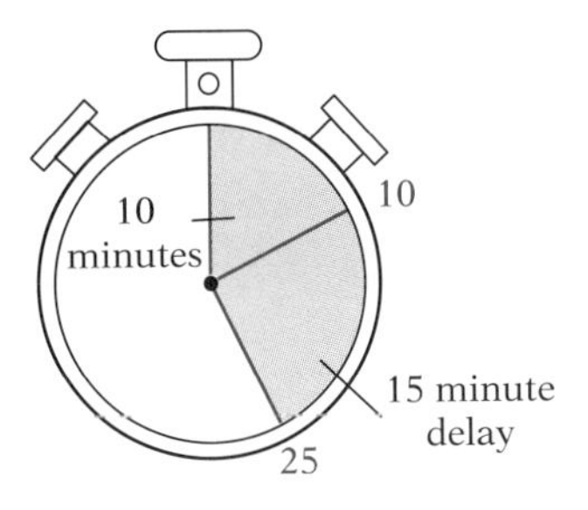

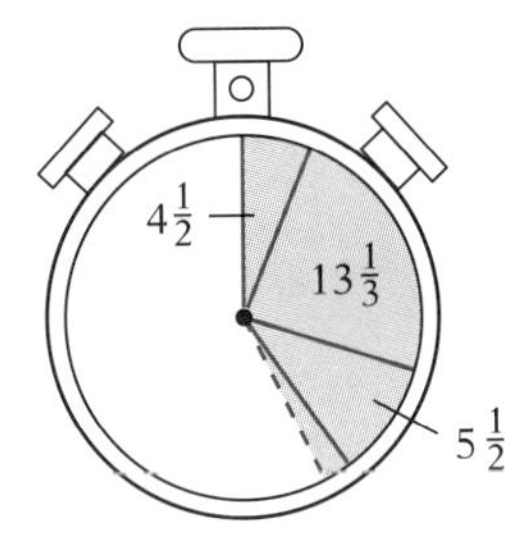

SOLUTION

We can think of these times as additions to, and subtractions from, the present time. If we do so, we can break down the information as follows:

meeting in 10 min	$+10$
$4\frac{1}{2}$ min on phone	$-4\frac{1}{2}$
meeting delayed 15 min	$+15$
$13\frac{1}{3}$ min on computer	$-13\frac{1}{3}$
$5\frac{1}{2}$ min for coffee	$-5\frac{1}{2}$

If we now add these signed numbers, the sum will give us the time left for reading.

$$10 + \left(-4\frac{1}{2}\right) + 15 + \left(-13\frac{1}{3}\right) + \left(-5\frac{1}{2}\right)$$

We add the positive numbers first:

$$10 + 15 = 25$$

Then we add the negative numbers:

$$\left(-4\frac{1}{2}\right) + \left(-13\frac{1}{3}\right) + \left(-5\frac{1}{2}\right) = \left(-4\frac{3}{6}\right) + \left(-13\frac{2}{6}\right) + \left(-5\frac{3}{6}\right)$$

$$= -22\frac{8}{6} = -23\frac{2}{6} = -23\frac{1}{3}$$

Then we add 25 and $-23\frac{1}{3}$.

$$25 + \left(-23\frac{1}{3}\right) = 24\frac{3}{3} - 23\frac{1}{3}$$

$$= 1\frac{2}{3}$$

So you have only $1\frac{2}{3}$ minutes left for reading.

■ *Now do margin exercises 1a and 1b.*

1a. Blaine was graphing information on a number line. She started at 5 and graphed a point, moved right 16 and graphed a point, and then moved left 36 and graphed a third point. Where is her third point located?

1b. In a certain history book, information was given about the world from 283 B.C. to 1991 A.D. What is the range of years covered in the book?

EXAMPLE 2

The temperature in a laboratory flask was measured as 100.4 degrees. It increased by 2.4 degrees and then decreased by twice the amount of the increase. What was the final temperature?

SOLUTION

We begin by rewriting the numbers with signs showing whether they are increases (positive numbers) or decreases (negative numbers).

The beginning temperature was 100.4 (positive), the increase was 2.4 (positive), and the decrease was twice the increase, or $2 \times 2.4 = 4.8$. But because this was a decrease, we write it with a minus sign: -4.8.

To find the final temperature, we add $100.4 + 2.4 + (-4.8)$.

We begin by adding the positive numbers.

$$100.4 + 2.4 = 102.8$$

We then add -4.8 to the result.

$$\begin{aligned} 102.8 + (-4.8) &= |102.8| - |-4.8| \\ &= 102.8 - 4.8 = 98.0 \end{aligned}$$

Because the sign of the number with the larger absolute value is positive, we make the result positive: 98.0.

So the current temperature is 98 degrees.

■ *Now do margin exercises 2a and 2b.*

2a. A newspaper reported the following action on a stock today.

Closing price:

$39\frac{7}{8}$

Change:

down $\frac{5}{8}$

What was the opening price?

2b. Your pulse rate was 60 beats per minute before you exercised, then it increased 70 beats during exercise, and then it dropped 35 beats after some rest. What was your pulse rate after the rest?

3a. On your mathematics final, your teacher marked the following numbers of points: +10, –3, +24, –12, –2, –1, +16, –2, +23, and +30. What was your score?

3b. Before your tonsils were removed, you habitually ran high fevers. Suppose that your temperature, starting at 98.6 degrees, went up 4.2 degrees, then dropped 2.5 degrees, and then rose 1.5 degrees again in the evening. What was your temperature then?

EXAMPLE 3

According to the U.S. Bureau of the Census, the number of insured commercial banks in the United States in 1978 was 14,391. The changes in this number over the next 7 years were as follows:

1979	–27	1980	+70	1981	–20
1982	+38	1983	+13	1984	+16
1985	–76				

How many insured commercial banks were there in 1985?

SOLUTION

We shall first find the sum of the positive numbers (the gains) and the sum of the negative numbers (the losses) and add those two sums. Then we shall add the result to the "starting" number.

First we add the gains.

$$70 + 38 + 13 + 16 = +137$$

And then we add the losses.

$$(-27) + (-20) + (-76) = -123$$

The sum of the gains and the losses is

$$137 + (-123) = 14$$

We add this result to the "starting" number.

$$14{,}391 + 14 = 14{,}405$$

■ *Now do margin exercises 3a and 3b.*

Work the exercises that follow.

9.4 EXERCISES

Work each of the following problems. Be careful to use the proper signs.

1. You are walking to get exercise. You walk $1\frac{7}{8}$ miles east and then three times as many miles west. You then turn around and walk back $1\frac{1}{2}$ miles. How far, and in what direction, from your starting point are you? (*Hint*: Think of your starting point as zero on a number line.)

2. Alan is 5 feet 7 inches tall. His brother Buzzie is $2\frac{1}{2}$ inches taller. Their brother, Bobby is 1 inch shorter than Buzzie, and their sister Susan is $7\frac{1}{2}$ inches shorter than Alan. What is the range of heights in this family?

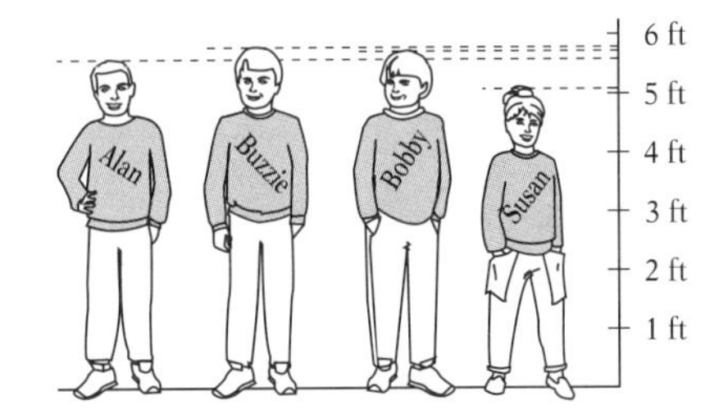

3. The flu "bug" has been playing havoc with your temperature. Your temperature went up to 102.6, then dropped 3 degrees and then rose 2.4 degrees again in the evening. How far above or below normal (98.6°) is your temperature?

4. You have two flagpoles that you want to erect. One is $18\frac{1}{2}$ feet long and the other is $8\frac{1}{3}$ feet long. Each must stick into the ground $4\frac{1}{4}$ feet. What is the total number of feet that will be showing above the ground on the two flagpoles?

5. A plane is flying at an altitude of 35,000 feet above sea level. It flies over Denver, the "mile-high" city. At that time, a $12\frac{1}{2}$ foot antenna in Denver is how far below the plane?

6. Your sister weighed 4.5 pounds at birth. Your birth weight was 1.5 times hers. Your brother weighed $1\frac{1}{2}$ pounds more than you. What is the range of birth weights in your family?

7. A bakery often receives shipments of flour. It started the week with 150 pounds of flour. It then used $65\frac{1}{2}$ pounds of flour on Monday, received $38\frac{1}{2}$ pounds of flour on Tuesday, used $60\frac{1}{8}$ pounds on Tuesday, and received 112 pounds on Wednesday. How much flour is left in the inventory?

8. You enter an elevator on the 14th floor and the elevator goes up 6 floors, down 5 floors, and up 7 floors again. On what floor are you now?

9. In 1961, people flew as high as 203.2 miles and dived as deep as 728 feet below sea level, using a gas mixture for breathing. What is the range of these record achievements?

10. Mt. McKinley in North America has an elevation of 20,320 feet above sea level. This is 20,602.5 feet higher than the lowest point in North America, Death Valley. Find the elevation of Death Valley.

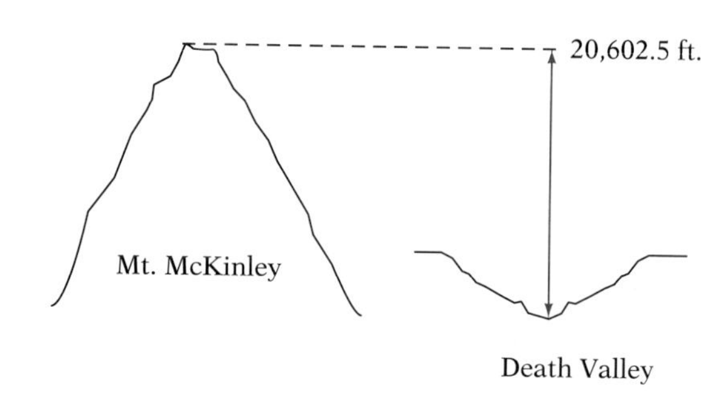

11. The speed limit is 55 miles per hour. A man drives 3 hours at this speed, $2\frac{1}{2}$ hours at 10 miles an hour below the speed limit, and then 20 more hours at $1\frac{1}{2}$ miles an hour faster than he was driving but still under the speed limit. On the average, how far below the speed limit was he traveling per hour?

12. You are driving from your home in North Bergen. If you go east 15.2 miles and then 6 miles more and turn around and then go west 12.5 miles, stop, and drive 22.7 miles further west. How many miles, in what direction, from your home are you?

13. The following heights above and below sea level were recorded by a scientific team: 35 feet below, 46.3 feet above, 19 feet below, 25.17 feet above, 0.3 feet above, 135.4 feet below, 35 feet below, and 6.24 feet above. What is the range of these readings? What was the mean height recorded?

14. In 1954 a man flew 93,000 feet high in a rocket plane. Another dived 13,287 feet below sea level in a Bathyscaphe. What is the range of these records?

15. You enter a staircase on the middle step. You walk down 12 steps, up 15 steps, and down 22 steps to the bottom. How many steps are there?

16. Asia has a high point of 29,028 feet and a low point of 1302 feet below sea level; North America has a high point of 20,320 feet and a low point of 282 feet below sea level. Which continent has the greater range in elevation? What is its range?

17. A basement in a house is $7\frac{1}{2}$ feet below the ground. The structure above the ground is $3\frac{1}{2}$ times as tall as the basement is below ground. How many feet are there between the floor of the basement and the top of the house?

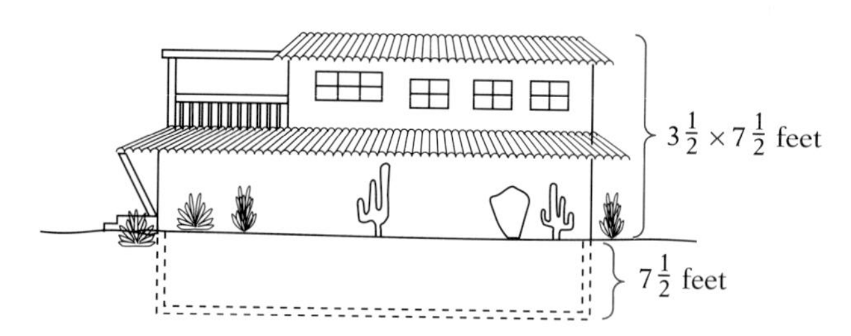

18. Organize the following data in a table, and determine the number of people who immigrated to the United States from France in the period 1931–1980. The number of people who immigrated from France to the United States during the period 1931–1940 was 12,623. During 1941–1950, the number rose 26,186; during 1951–1960, the number increased 12,312. From 1961 to 1970, the number dropped by 5,884, and during the period 1971 to 1980, the number dropped 20,168.

19. You need to write checks for $54.34, $78.65, and $124.56 and to deposit $160. After these transactions, you will be overdrawn by $23.89. (The balance in your account will be –$23.89.) How much money did you have in your account to start with?

Check #	Item	Withdrawal	Deposit	Balance
673	*Zach's Records*	–54.34		
674	*Auto Service Co.*	–78.65		
675	*Jamal*	–124.56		
	Paycheck (Deposit)		+160	–23.89

20. Howe Caverns, in New York, has depths reaching approximately $198\frac{1}{2}$ feet below the ground. If a person is riding in an aerial balloon 1000 feet above the caverns, what is the difference in altitude between the person in the balloon and a person walking through the caverns?

For Exercises 21 through 23, use the temperatures in the following table, which were reported for the first three weeks of December in New York state.

	Sun	Mon	Tues	Wed	Thurs	Fri	Sat
Wk 1	41.2	45	42	49	53.7	60.1	59.1
Wk 2	50	56.1	58	55.8	35	24.1	13.1
Wk 3	–13.3	–10	–7.9	–5.6	4.5	14.3	24.1

21. What was the range of temperatures on Sunday for the three weeks shown?

22. What was the range of temperatures for the entire period shown?

23. Find the mean and median of the temperatures during week 3. What is the mode?

9.4 MIXED PRACTICE

By doing these exercises, you will practice all the topics in this chapter up to this point.

24. Add: $\left(-5\frac{5}{6}\right) + \left(-43\frac{3}{4}\right)$

25. Subtract: $\left(6\frac{7}{9}\right) - \left(-9\frac{5}{8}\right)$

26. Insert the proper sign, < or >, between $|-12|$ and $|-6|$.

27. A T-shirt firm started the week with 300 T-shirts. It took orders for 150 and 134 T-shirts on the telephone and sold 200 T-shirts in the showroom. How many more shirts does the company need to fill its phone orders?

28. Multiply: $\left(\frac{3}{4}\right)\left(-\frac{8}{9}\right)\left(\left|-3\frac{5}{6}\right|\right)$

29. Add: $(0.36) + (-1.27) + (4.3)$

30. Find the quotient: $(256) \div \left(-41\frac{3}{5}\right)$

31. I owed $325.46 to a company. Then I wrote the company two checks: one for $123.12 and another for $185.14. How much do I still owe?

32. Multiply: $(-0.24)(5.1)(0.004)$

33. Find the sum: $(62) + (|-15|) + (-|-42|)$

34. Subtract: $(-6.4) - (3.002)$

35. Divide: $\left(-6\frac{2}{15}\right) \div \left(1\frac{1}{8}\right)$

36. A miner travels 195 yards to the mine entrance and then another 92 yards into the mine. How far does he travel to work and back?

9.5 Integer Exponents and the Order of Operations

SECTION GOALS

- To simplify rational expressions containing exponents
- To simplify negative exponents
- To use the standard order of operations to evaluate an expression

Integer Exponents

The exponent 2 in the expression $(-3)^2$ tells us to use -3 as a factor twice. Thus

$$(-3)^2 = (-3)\,(-3), \text{ or } 9$$

What happens when there is a minus sign outside the parentheses, as, for example, in $-(4)^2$? We read this as "the opposite of 4^2." We square 4 and find the opposite. This gives us -16.

In general, if a is any rational number and n is any positive integer, then

$$(-a)^n = \underbrace{(-a)\,(-a)\cdots(-a)}_{n \text{ factors}}$$

$$-(a^n) = -\underbrace{(a \times a \times \cdots \times a)}_{n \text{ factors}}$$

Note that $-n^a$ means the same thing as $-(n)^a$ and $-(n^a)$.

Also, by definition,

If a is any non-zero integer, $a^0 = 1$.

We use these definitions in our first example.

EXAMPLE 1

Simplify:

a. $(-3)^3$ **b.** -2^4 **c.** $|-(-9)^2|$ **d.** 6^0

SOLUTION

a. $(-3)^3$ means -3 used as a factor 3 times.

$$(-3)\,(-3)\,(-3) = (9)\,(-3) = -27$$

So $(-3)^3$ is -27.

b. -2^4 means raise 2 to the fourth power and *then* find the opposite of the result.

$$(2)\,(2)\,(2)\,(2) = (4)\,(2)\,(2) = (8)\,(2) = (16)$$

The opposite of 16 is -16.

c. $|-(-9)^2|$ means square -9, then find the opposite of that number, and then take its absolute value.

1a. Simplify: -3^5

1b. Simplify: $-|-6|^2$

$(-9)(-9) = 81$ Squaring
$-(81) = -81$ Finding the opposite
$|-81| = 81$ Taking absolute value

d. $6^0 = 1$ by definition.

■ *Now do margin exercises 1a and 1b.*

In all our examples, we have encountered only positive exponents. Can we raise a number to a negative exponent? To see the answer, look at the pattern formed as we decrease an exponent by 1, in stages:

$$3^4 = (3)(3)(3)(3) = 81$$
$$3^3 = (3)(3)(3) = 27$$
$$3^2 = (3)(3) = 9$$
$$3^1 = 3$$
$$3^0 = 1$$

As the exponents get smaller, the numerical value of the expression also decreases. In each case, we can find the required value by dividing the previous value by the base. For example,

3^4 is 81 and 3^3 is $81 \div 3 = 27$
3^2 is $27 \div 3 = 9$
3^1 is $9 \div 3 = 3$

Suppose we continue with the same pattern right into the negative exponents. We find that

$$3^1 = 3$$
$$3^0 = 3 \div 3 = 1$$
$$3^{-1} = 1 \div 3 = \frac{1}{3}$$
$$3^{-2} = \frac{1}{3} \div 3 = \frac{1}{3} \times \frac{1}{3} = \frac{1}{9}$$
$$3^{-3} = \frac{1}{9} \div 3 = \frac{1}{9} \times \frac{1}{3} = \frac{1}{27}$$
$$3^{-4} = \frac{1}{27} \div 3 = \frac{1}{27} \times \frac{1}{3} = \frac{1}{81}$$

If you look back now and compare 3^4 and 3^{-4}, 3^3 and 3^{-3}, and 3^2 and 3^{-2}, you will see a pattern: The number raised to the negative exponent is the reciprocal of the same number raised to the positive exponent. That's the key:

For a, not equal to zero, and n any integer,

$$a^{-n} = \frac{1}{a^n}$$

Therefore, we can find 3^{-4} by computing

$$3^{-4} = \frac{1}{3^4} = \frac{1}{3 \times 3 \times 3 \times 3} = \frac{1}{81}$$

EXAMPLE 2

Simplify:

a. 4^{-3} **b.** -8^{-2} **c.** -5^{-2}

SOLUTION

a. $4^{-3} = \frac{1}{(4)^3} = \frac{1}{(4)\,(4)\,(4)} = \frac{1}{64}$

b. $-8^{-2} = -\left[\frac{1}{(8)^2}\right] = -\left[\frac{1}{(8)(8)}\right]$

$= -\frac{1}{64}$

c. $-5^{-2} = \frac{1}{(-5)^2} = \frac{1}{(-5)\,(-5)} = \frac{1}{25}$

■ *Now do margin exercises 2a and 2b.*

When we raise a negative rational number to a power, we follow this procedure:

If $\frac{-a}{b}$ is a rational number, then

$$\left(\frac{-a}{b}\right)^n = \underbrace{\left(\frac{-a}{b}\right)\left(\frac{-a}{b}\right)\cdots\left(\frac{-a}{b}\right)}_{n \text{ factors}} = \frac{(-a)^n}{b^n}$$

$$-\left(\frac{a}{b}\right)^n = -\underbrace{\left(\frac{a}{b}\right)\left(\frac{a}{b}\right)\cdots\left(\frac{a}{b}\right)}_{n \text{ factors}} = -\frac{a^n}{b^n}$$

Accordingly,

$$\left(\frac{-2}{3}\right)^3 = \left(\frac{-2}{3}\right)\left(\frac{-2}{3}\right)\left(\frac{-2}{3}\right)$$

$$= \frac{(-2) \times (-2) \times (-2)}{3 \times 3 \times 3} = \frac{-8}{27}$$

We raise a rational number to a negative exponent just as we raise integers to a negative exponent:

$$\left(\frac{a}{b}\right)^{-n} = \frac{1}{\left(\frac{a}{b}\right)^n} = \frac{b^n}{a^n}$$

2a. Simplify: $-\left|-(-8)^{-2}\right|$

2b. Simplify: $-[-(9)^{-3}]$

3a. Simplify:
$3(2-4)^2 - 8(9-16)^3$

3b. Simplify: $9\,|11-7|^2 - 2|6-9^2|$

Accordingly,

$$\left(\frac{3}{4}\right)^{-1} = \left(\frac{4}{3}\right)^{1} = \frac{4}{3}$$

and

$$\left(\frac{3}{5}\right)^{-3} = \left(\frac{5}{3}\right)^{3} = \frac{5\times5\times5}{3\times3\times3} = \frac{125}{27}$$

You should also realize that

$$\frac{-a}{b} = \frac{a}{-b} = -\frac{a}{b}$$

The definition of zero as an exponent also holds for rational numbers:

If $\frac{a}{b}$ is a non-zero rational number, then

$$\left(\frac{a}{b}\right)^{0} = 1$$

Simplifying Expressions with Rational Numbers

The standard order of operations covers rational numbers as well as the other numbers we have worked with.

Standard Order of Operations

1. Do all operations inside parentheses, working outward.
2. Simplify any expressions containing exponents or absolute value signs.
3. Do all multiplications and divisions, working from left to right.
4. Do all additions and subtractions, working from left to right.

EXAMPLE 3

Simplify: $2(5-3^2) + 2(4-6)^2$

SOLUTION

Because there are parentheses and exponents, we simplify these first.

$2(5-3^2) + 2(4-6)^2$	
$2(5-9) + 2(-2)^2$	Working within parentheses
$2(-4) + 2(4)$	Simplifying

Continuing, we do the multiplications and then the addition.

$$2(-4) + 2(4) = -8 + 8 = 0$$

So $2(5-3^2) + 2(4-6)^2 = 0$.

■ *Now do margin exercises 3a and 3b.*

Our next example includes negative and zero exponents.

EXAMPLE 4

Simplify: $(3-4)^2 + 2\left(3 - \frac{4}{5}\right)^2 - 5^{-2}(-1)^0$

SOLUTION

We simplify within the parentheses first.

$$(3-4)^2 + 2\left(3 - \frac{4}{5}\right)^2 - 5^{-2}(-1)^0$$

$$(-1)^2 + 2\left(2\frac{1}{5}\right)^2 - 5^{-2}(-1)^0 \quad \text{Clearing parentheses}$$

Then we simplify the exponents.

$$1 + 2\left(\frac{11}{5}\right)^2 - \frac{1}{25}(1)$$

$$1 + 2\left(\frac{121}{25}\right) - \frac{1}{25}$$

Next we do multiplications and then additions and subtractions from left to right.

$$1 + 2\left(\frac{121}{25}\right) - \frac{1}{25}$$

$$1 + \frac{242}{25} - \frac{1}{25} \quad \text{Multiplying}$$

$$1 + 9\frac{17}{25} - \frac{1}{25} \quad \text{Simplifying}$$

$$10\frac{17}{25} - \frac{1}{25} = 10\frac{16}{25} \quad \text{Simplifying}$$

So $(3-4)^2 + 2\left(3 - \frac{4}{5}\right)^2 - 5^{-2}(-1)^0 = 10\frac{16}{25}$.

■ *Now do margin exercises 4a and 4b.*

4a. $9|-2| + 2 - 5(-2)^{-2}$

4b. $-6|2| - 5^{-3} + 7$

EXAMPLE 5

Simplify, using a calculator: $\{(-1.67)^2 - [7(2.7)^4 - (0.5)^3]\} + 19.21^2$

SOLUTION

We will first use the calculator to clear as many exponents as possible. We shall start in the square brackets by finding $(2.7)^4$; we key in

2.7 [×] 2.7 [×] 2.7 [×] 2.7 [=]

The display reads

53.1441

This number must be multiplied by 7, so we key in

[×] 7 [=]

5a. Simplify: $[|-3|^{-2} + 6(3 - 8)]^2$

5b. Simplify: $5|3 - 6| - 2(-4)^{-3}$

obtaining, as our display,

372.0087

Next we want to subtract, from this, the result of $(0.5)^3$, so we key in

[−] .5 [×] .5 [×] .5 [=]

and the display reads

371.8837

This number, which is the result of $[7(2.7)^4 - (0.5)^3]$, is to be subtracted from the result of $(-1.67)^2$. (Note that this square is a positive number.) To indicate subtraction, we next key in

[±]

and the display is

−371.8837

Then we key in

[+] 1.67 [×] 1.67 [=]

The display now reads

−369.0948

This entire amount is to be added to the result of 19.21^2, so we key in

[+] 19.21 [×] 19.21 [=]

Our final display, the result, is

−0.0707

So $\{(-1.67)^2 - [7(2.7)^4 - (0.5)^3]\} + 19.21^2$ is equal to −0.0707.

■ *Now do margin exercises 5a and 5b.*

NOTE: If you did not obtain each of the displays shown here, your calculator works differently from the one we are using. You should write down the results of each separate step, and re-enter your numbers to obtain the final answer.

Work the exercises that follow.

9.5 EXERCISES

In Exercises 1 through 4, write each expression without an exponent.

1. a. $(-3)^{-3}$ b. $\left(-\frac{1}{3}\right)^{-1}$ c. -2^{-4} d. $\left(-\frac{1}{2}\right)^{-1}$

2. a. -5^2 b. $-(-5)^2$ c. $-(-5)^{-3}$ d. $\left(-\frac{3}{5}\right)^{-1}$

3. a. $-\left(\frac{2}{5}\right)^2$ b. $-\left(-\frac{2}{5}\right)^2$ c. $-\left(-\frac{2}{5}\right)^{-3}$ d. $\left(-\frac{2}{5}\right)^{-1}$

4. a. $|-6^3|$ b. $-|-6|^2$ c. $|-6|^{-3}$ d. $\left|-\frac{1}{6}\right|^0$

In Exercises 5 through 50, use the standard order of operations to simplify each expression.

5. $14 - 7(5)^3 + 5^5$
6. $(-13) + |-9| - 4^3$
7. $2(6 - 4)^1 - 5^1$
8. $12 - |6^{-1}| + 27$

9. $3(9 - 5)^3 - 3^4$
10. $7\,(3 + 6) - 7^2$
11. $50 - 9(2^{-2}) + 3^{-2}$
12. $28 - |4^{-3}| + 9$

13. $16 - |-8^{-2}| + 11^{-1}$

14. $4(7 + 3) - 6^3$

15. $-13 + 37 \times 8^{-2}$

16. $17 - |40|(-1)$

17. $-33 + 25 \times 6^{-3}$

18. $-21 + 14 \times 43^0$

19. $44 - (39)(-2)^{-5}$

20. $20 - |15||-3^3|$

21. $38 - |22|(-5)$

22. $-29 + 31 \times 2^{-4}$

23. $49^0 - (2 - 3)^2$

24. $18^2 - (4 - 6)^3$

25. $7^3 - (8 - 9)^{-9}$

26. $26 + |-18| \times 1^{-3}$

27. $-30 \div [-5 + (-6) \times (-2)]$

28. $-18 \div 3 + (-9) \times 3$

29. $-28 \div 4 + [-(5 \times (-1))]$

30. $-44 \div 2 + 3^2 \times (-7)$

31. $48 + (-10) \times 2^{-2}$

32. $(-19) + 47 \times \left(\frac{1}{3}\right)^{-3}$

33. $(-32) + |5^{-1}| - 46$

34. $5^2 - (5-7)^0$

35. $[-34 - |12 - 10|^{-2}]^2$

36. $-(-7 - |6-4|^{-5})$

37. $[-45 - |(5-2)|^{-3}]^2$

38. $-(-24) - \left|8 - \frac{1}{3}\right|^{-2}$

39. $25\left(\frac{1}{5^2}\right) + 4 - \frac{1}{2}$

40. $\left|-\left|9\left(-\frac{1}{3}\right)^3\right| + 3 - 1\right|$

41. $-\left|10\left|\left(-\frac{1}{2}\right)^1\right| + 8 - 7\right|$

42. $9\left|\left(-\frac{1}{3}\right)^4\right| + 7^3 - \frac{1}{2}$

43. $-2\left\{\left[8^{-2} - \left(\frac{1}{3}\right)\right] \div |-4|\right\}$

44. $-[5^{-3} - |-12|^2 \div |-5|]$

45. $[|-7|^2 + |-6| - 3^2]^2$

46. $-6(11 - 9(2)) + 11^2$

47. $[(-18)^3 \div (-12)]^2 - 12^2$

48. $[|-11|^2 + (-5) - 8^2]$

49. $[6 - 2|3|^2 - 5 \times 2] + 7^4$

50. $\{12^3 \times [4(8) \div 8^3]\}^{-2}$

9.5 MIXED PRACTICE

By doing these exercises, you will practice all the topics in this chapter up to this point.

51. Subtract: $\left(7\frac{3}{4}\right) - \left(-14\frac{1}{4}\right)$

52. Multiply: $(-5)\left(2\frac{13}{25}\right)(-2)$

53. Arrange $-(-|-4|)$, $-|-|4.1||$, and $-|-4|$ in order from smallest to largest.

54. Use your calculator to evaluate $[6(-3)^{-3} \div 3^{-4}]^2 - 4^0$.

55. Man first walked on the moon in 1969. If we identify this as the year 0 in the development of space travel, how would we express the year 1953 in terms of space travel?

56. Divide: $\left(-\frac{6}{7}\right) \div \left(-\frac{5}{13}\right)$

57. Subtract: $\left(11\frac{5}{12}\right) - 6\frac{5}{12}$

58. Evaluate: $[9(-1)^{-1} - (-7)^2]$

59. Multiply: $(-11.5)(-1.2)$

60. Find the quotient: $(234) \div (-9)$

61. Add: $\left(7\frac{2}{3}\right) + \left(-3\frac{1}{8}\right)$

9.6 Solving Equations with One Operation

SECTION GOALS

- To solve equations by using the addition principle
- To solve equations by using the multiplication principle

We have solved many equations in this text, but only informally. Here we discuss and use an algebraic procedure for solving more difficult equations.

The Addition Principle

Consider the equation

$$x + 3 = 5$$

This equation asks "what number plus three will give a result of five?" The letter, x, called the **variable,** stands for the unknown number. "Solving the equation for x" means determining what number used in place of x—that is, what value of x—will make the equation a true statement. That number is called the **solution** of the equation. You can probably see by inspection that here the unknown number is 2. However, most equations don't have such an obvious solution. In such cases, we can solve an equation of the type by using the **addition principle of equality.**

$$x + a = b$$

Addition Principle of Equality

For rational numbers a, b, and c,

If $a = b$, then $a + c = b + c$.

We use this principle to get our equation in the form

$$x = \text{a number}$$

That is, we add the same number to both sides of the equation to get the variable alone on one side of the equation and everything else on the other side.

Study Hint

The sum of a number and its additive inverse is zero:
$a + (-a) = 0$ and $(-a) + a = 0$.

Let's apply the addition principle to an equation.

EXAMPLE 1

Solve for x: $x + 3 = 5$

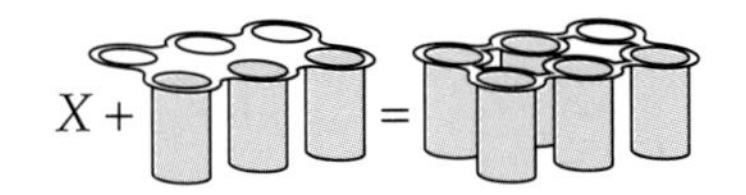

1a. Solve for n: $n + 9 = -4$

1b. Solve for x: $x + 7 = 5$

2a. Solve for m: $m - 27 = 16$

2b. Solve for y: $y - 18 = -32$

SOLUTION

To isolate x, we need $x + 0$ on the left. To get it, we add -3 (the additive inverse of 3) to both sides of the equation.

$$\begin{aligned} x + 3 &= 5 \\ -3 &= -3 \quad \text{Adding } -3 \\ x + 0 &= 5 + (-3) \\ x &= 2 \end{aligned}$$

To check the solution, substitute the value 2 for x in the original equation.

If the resulting statement is true, then 2 is a solution to that equation.

Check:

$$\begin{aligned} x + 3 &= 5 \\ 2 + 3 &= 5 \\ 5 &= 5 \quad \text{Correct} \end{aligned}$$

So $x = 2$ is the solution of $x + 3 = 5$.

■ *Now do margin exercises 1a and 1b.*

The next equation involves a subtraction. We handle it in exactly the same way.

EXAMPLE 2

Solve for y: $y - 4 = 10$

SOLUTION

By the definition of subtraction of integers, $y - 4$ can be written as $y + (-4)$. Therefore, we can get y alone by adding $+4$, the additive inverse of -4, to both sides of the equation.

$$\begin{aligned} y - 4 &= 10 \\ 4 &= 4 \quad \text{Adding 4} \\ y + 0 &= 10 + 4 \\ y &= 14 \end{aligned}$$

Check:

$$\begin{aligned} y - 4 &= 10 \\ 14 - 4 &= 10 \\ 10 &= 10 \quad \text{Correct} \end{aligned}$$

Thus, $y = 14$ is the solution to $y - 4 = 10$.

■ *Now do margin exercises 2a and 2b.*

Next we will solve an equation with decimal numbers.

EXAMPLE 3

Solve for r: $0.6 + r = -2.1$

SOLUTION

Here we must add the additive inverse of 0.6 to both sides of the equation.

$$\begin{aligned} 0.6 + r &= -2.1 \\ -0.6 \quad &= -0.6 \qquad \text{Adding } -0.6 \\ \hline 0 + r &= -2.1 + (-0.6) \\ r &= -2.7 \end{aligned}$$

Check:

$$\begin{aligned} 0.6 + r &= -2.1 \\ 0.6 + (-2.7) &= -2.1 \\ -2.1 &= -2.1 \qquad \text{Correct} \end{aligned}$$

So $r = -2.7$ is the solution to $0.6 + r = -2.1$.

■ *Now do margin exercises 3a and 3b.*

3a. Solve for n: $4.9 + n = 3$

3b. Solve for n: $n - 3.6 = 8$

The Multiplication Principle

In an expression of the general form ax, a is called the **coefficient** of x. Thus 4 is the coefficient of x in $4x$, and 17 is the coefficient of y in $17y$. To solve an equation like $4x = 8$, we have to give x the coefficient 1. To do so, we use the **multiplication principle of equality**.

Multiplication Principle of Equality

For rational numbers a, b and c,

If $a = b$, then $c \times a = c \times b$.

To use this principle, we multiply both sides of the equation by the multiplicative inverse of the coefficient of x. That gives us an equation in the form

$$x = \text{a number}$$

4a. Solve for x: $12x = 144$

4b. Solve for n: $32n = -288$

Study Hint

The multiplicative inverse of a is $\frac{1}{a}$.
The multiplicative inverse of $\frac{1}{a}$ is a.
The product of a number and its multiplicative inverse is 1.

$$(a)\left(\frac{1}{a}\right) = \left(\frac{1}{a}\right)(a) = 1$$

EXAMPLE 4

Solve for x: $3x = 6$

SOLUTION

First we multiply both sides of the equation by $\frac{1}{3}$ (the multiplicative inverse of 3).

$$3x = 6$$
$$\left(\frac{1}{3}\right)3x = 6\left(\frac{1}{3}\right) \qquad \text{Multiplying by } \frac{1}{3}$$

Then we simplify.

$$x = \frac{6}{3} \qquad \text{Multiplying}$$
$$x = 2 \qquad \text{Simplifying}$$

To check the solution, we substitute for x in the original equation. If we get a true statement, the solution is correct.

$$3x = 6$$
$$3(2) = 6$$
$$6 = 6 \qquad \text{Correct}$$

So, $x = 2$ is the solution to $3x = 6$.

■ *Now do margin exercises 4a and 4b.*

In the next example, the coefficient is both negative and fractional. Its reciprocal is also negative and fractional.

EXAMPLE 5

Solve for t: $-\frac{3t}{14} = 28$

SOLUTION

First we rewrite the equation as

$$\left(-\frac{3}{14}\right)t = 28$$

Next we multiply both sides of the equation by $-\frac{14}{3}$ (the multiplicative inverse of $-\frac{3}{14}$).

$$\left(-\frac{14}{3}\right)\left(-\frac{3}{14}\right)t = 28\left(-\frac{14}{3}\right) \qquad \text{Multiplying by } -\frac{14}{3}$$

When we simplify, we get

$$t = -\frac{392}{3}, \text{ or } -130\frac{2}{3}$$

Check:

$$\frac{-3t}{14} = 28$$

$$\frac{-3\left(-130\frac{2}{3}\right)}{14} = 28$$

$$\frac{(-3)\left(\frac{-392}{3}\right)}{14} = 28$$

$$\frac{392}{14} = 28$$

$$28 = 28 \qquad \text{Correct}$$

So $t = 130\frac{2}{3}$ is the solution to the equation $-\frac{3t}{14} = 28$

■ *Now do margin exercises 5a and 5b.*

Our final example involves a decimal-number coefficient. The multiplicative inverse of a decimal number is simply 1 divided by that decimal number. Also, the variable is on the right side of the equation. We use exactly the same procedure.

5a. Solve for r: $-\frac{4}{5}r = 482$

5b. Solve for n: $-\frac{8}{9}n = 408$

6a. Solve for n: $-36 = -0.04n$

6b. Solve for y: $3.2y = 76$

EXAMPLE 6

Solve for n: $40 = -0.02n$

SOLUTION

We multiply both sides of the equation by $-\frac{1}{0.02}$ (the multiplicative inverse of -0.02) and simplify.

$$40 = -0.02n$$

$$\left(\frac{-1}{0.02}\right)(40) = (-0.02)n\left(\frac{-1}{0.02}\right) \quad \text{Multiplying by } \frac{-1}{0.02}$$

$$-\frac{40}{0.02} = n \quad \text{Simplifying}$$

$$-2000 = n \quad \text{Dividing}$$

Check

$$40 = 0.02n$$

$$40 = 0.02\,(-2000)$$

$$40 = 40 \quad \text{Correct}$$

So $n = -2000$ is the solution to the equation $40 = -0.02n$.

■ *Now do margin exercises 6a and 6b.*

Work the exercises that follow.

9.6 EXERCISES

In Exercises 1 through 12, use the addition principle to solve these equations.

1. $y + 9 = 15$
2. $15 = x + 4$
3. $1.7 = y - 92$
4. $x - 87 = 1.2$
5. $y + 4 = -2.2$
6. $-1.7 = x + 9$
7. $-36 = y - 0.38$
8. $x - 28 = 1.8$
9. $348 + y = -132$
10. $-\frac{3}{8} = \frac{1}{8} + y$
11. $\frac{1}{7} + x = -\frac{5}{7}$
12. $-86 = x + (-117)$

In Exercises 13 through 24, use the multiplication principle to solve these equations.

13. $19y = 57$
14. $75 = 1.5x$
15. $1.4y = 56$
16. $-9y = -12$
17. $0.2y = 14$
18. $27 = 0.03x$
19. $\frac{3}{4}y = 24$
20. $\frac{2}{7}x = 36$
21. $560 = -\frac{x}{14}$
22. $-\frac{x}{8} = -176$
23. $\frac{2}{5}y = 40$
24. $-0.08x = 160$

In Exercises 25 through 40, use either the addition principle or the multiplication principle to solve these equations.

25. $4y = 0$
26. $0 = x + 3$
27. $y - 2.7 = -3.4$
28. $x - 1.6 = -3.5$

29. $-\frac{y}{2} = 25$

30. $-\frac{x}{5} = 15$

31. $\frac{3}{4} + y = \frac{1}{8}$

32. $\frac{3}{4} + x = 0$

33. $33y = -264$

34. $x + 289 = -293$

35. $y - 1.8 = -2.7$

36. $x - 1.9 = -2.6$

37. $34.5 + y = 9.8$

38. $-23.7 = -\frac{x}{3}$

39. $96.4 = -0.2y$

40. $-60.40 = x + 23.4$

9.6 MIXED PRACTICE

By doing these exercises, you will practice all the topics in this chapter up to this point.

41. Subtract: $\left(-2\frac{1}{4}\right) - \left(-4\frac{3}{5}\right)$

42. Multiply: $(-2.6)(-10.2)(-|-5.6|)$

43. Find the quotient: $(-357) \div (-17)$

44. Insert the proper sign, $<$ or $>$, between $-|-|-19||$ and $-|-17|$.

45. Add: $\left(-19\frac{2}{5}\right) + 35\frac{4}{9}$

46. Find the difference: $37\frac{1}{14} - \left(-83\frac{2}{3}\right)$

47. Solve for x: $-14 = x + 5$

48. Multiply: $\left(-\frac{1}{9}\right)\left(\frac{40}{41}\right)\left(-\frac{9}{10}\right)$

49. Evaluate: $|-(-3)^{-2} - 2(4-8)^3| - 9^2$

50. Solve for x: $-\frac{x}{11} = 55$

51. Add: $(-8.9) + (-7.2) + (-2.6)$

9.7 Solving Equations with Two Operations

Section Goal

- To solve equations by using both the addition and multiplication principles

We solve equations of the type $x + a = b$ with the addition principle and those of the type $ax = b$ with the multiplication principle. However, to solve an equation of the form

$$ax + b = c$$

we must use *both* the addition principle and the multiplication principle.

We always use the addition principle first, to isolate the term that contains the variable. Then we use the multiplication principle to give the variable the coefficient 1.

Example 1

Solve for x: $2x + 4 = 8$

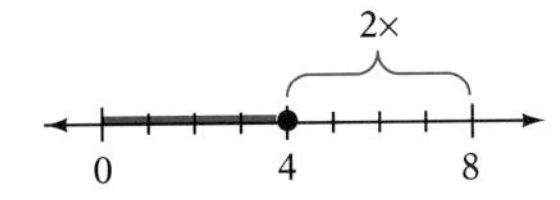

SOLUTION

To isolate the variable $2x$, we add -4 to both sides of the equation.

$$\begin{array}{r} 2x + 4 = \ \ 8 \\ \underline{-4 = -4} \\ 2x + 0 = \ \ 4 \end{array} \qquad \text{Adding } -4$$

Then, to obtain a coefficient of 1 for x, we multiply both sides of the equation by $\frac{1}{2}$ (the multiplicative inverse of 2).

$$2x = 4$$

$$\left(\frac{1}{2}\right)2x = 4\left(\frac{1}{2}\right) \qquad \text{Multiplying by } \frac{1}{2}$$

$$x = \frac{4}{2}$$

Then we simplify, getting

$$x = 2$$

Check:

$$2x + 4 = 8$$

$$2(2) + 4 = 8$$

$$8 = 8 \qquad \text{Correct}$$

So $x = 2$ is the solution to $2x + 4 = 8$.

■ *Now do margin exercises 1a and 1b.*

In the next example the variable has a fractional coefficient.

1a. Solve for x: $3x + 5 = 9$

1b. Solve for y: $17 = 8 + 5y$

2a. Solve for t: $-16 + \frac{4t}{5} = 7$

2b. Solve for x: $4 = -8 - \frac{3x}{2}$

3a. Solve for x: $-250 = 1.5x + 0.05$
Round to the nearest tenth.

3b. Solve for x:
$0 = -3.6x + 1.696 - 4$

EXAMPLE 2

Solve for n: $-\frac{2n}{3} + 2 = 9$

SOLUTION

To isolate the term containing the variable, we add −2 (the additive inverse of 2) to both sides of the equation.

$$\begin{aligned} -\frac{2n}{3} + 2 &= 9 \\ -2 &= -2 \quad \text{Adding } -2 \\ \hline -\frac{2n}{3} + 0 &= 7 \end{aligned}$$

Next we rewrite $-\frac{2n}{3}$ as $\left(-\frac{2}{3}\right)n$. Then we multiply both sides of the equation by $-\frac{3}{2}$, the multiplicative inverse of the coefficient $-\frac{2}{3}$. This results in a coefficient of 1 for n.

$$\begin{aligned} -\frac{2n}{3} &= 7 \\ \left(-\frac{2}{3}\right)n &= 7 && \text{Rewriting} \\ \left(-\frac{3}{2}\right)\left(-\frac{2}{3}\right)n &= 7\left(-\frac{3}{2}\right) && \text{Multiplying by } -\frac{3}{2} \\ n = -\frac{21}{2} &= -10\frac{1}{2} && \text{Simplifying} \end{aligned}$$

We can check that this is correct by substituting in the original equation.

So $n = -10\frac{1}{2}$ in the equation $-\frac{2n}{3} + 2 = 9$.

■ *Now do margin exercises 2a and 2b.*

In the next example we show only the work.

EXAMPLE 3

Solve for x: $-1.6 = 2.3x + 0.7$

SOLUTION

Isolating the variable on one side of the equation gives us

$$\begin{aligned} -1.6 &= 2.3x + 0.7 && \text{Original equation} \\ -0.7 &= -0.7 && \text{Adding } -0.7 \text{ to both sides} \\ \hline -2.3 &= 2.3x && \text{Simplifying} \\ \left(\frac{1}{2.3}\right)(-2.3) &= 2.3x\left(\frac{1}{2.3}\right) && \text{Multiplying by } \frac{1}{2.3} \\ -1 &= x && \text{Simplifying} \end{aligned}$$

You should check that $x = -1$ in the solution of the equation $-1.6 = 2.3x + 0.7$.

■ *Now do margin exercises 3a and 3b.*

Work the exercises that follow.

9.7 EXERCISES

In exercises 1 through 52, solve each of the following equations.

1. $18 = 3x + 6$
2. $4r + 12 = 16$
3. $-4n + \frac{12}{13} = 32$
4. $-9m + \frac{6}{7} = 21$
5. $5x + 1.9 = 1.6$
6. $2.4r + 1.2 = 1.8$
7. $\frac{n}{6} + 7 = -8$
8. $-12 = \frac{m}{4} + 3$
9. $0.6x - 1.8 = 1.2$
10. $7r - 1.4 = 3.5$
11. $-\frac{8}{9} + 11n = 26$
12. $\frac{2}{3} + 8m = 17$
13. $4 + 2.4r = 12.4$
14. $\frac{3}{8}m + 9 = 5$
15. $-27 + \frac{5}{9}n = -34$
16. $-35 + \frac{3}{5}m = -62$
17. $6 - 3.2x = 18$
18. $6 = 8 - 6r$
19. $\frac{5}{8}n - 6 = 29$
20. $\frac{4}{9}m - 12 = 32$
21. $\frac{x}{3} + 8 = 11$
22. $\frac{r}{4} + 7 = 18$
23. $\frac{2}{7}n + 15 = 29$
24. $\frac{5}{6}m + 21 = 36$
25. $-29 = \frac{x}{7} + 11$
26. $\frac{r}{3} + 9 = -16$
27. $6n + 2.1 = 1.2$
28. $5.1m + 1.3 = 6.4$
29. $\frac{x}{2} - 15 = -23$
30. $\frac{r}{9} - 26 = -44$
31. $\frac{n}{3} - 32 = -41$
32. $-57 = \frac{m}{8} - 41$
33. $-\frac{3}{5} + \frac{1}{14}x = 28$
34. $\frac{7}{8} + \frac{r}{12} = 15$
35. $2n - 1.2 = 2.8$
36. $5m - 0.9 = 1.6$
37. $-3x + \frac{9}{10} = 15$
38. $-\frac{r}{6} + \frac{11}{16} = 16$
39. $57 = 5n + 42$
40. $0.6m + 18 = 30$
41. $\left(\frac{1}{4}\right)x + 8 = 16$
42. $\left(\frac{3}{4}\right)r + 7 = 22$
43. $\frac{n}{6} + 9 = 26$
44. $\frac{m}{5} + 8 = 32$

45. $\left(\frac{3}{4}\right)x + 1.7 = 1.4$

46. $-\left(\frac{2}{9}\right)r + 13 = 12$

47 $-6\frac{2}{3} + 8.1n = 3.4$

48. $-8.3 + 3.1m = 1$

49. $-18 = -8 + \left(\frac{2}{3}\right)x$

50. $-1.9 + \left(\frac{4}{7}\right)r = -2.1$

51. $-1.4 - 9.2n = -5.08$

52. $-8.6 - 6.2m = -5.5$

9.7 MIXED PRACTICE

By doing these exercises, you will practice all the topics in this chapter up to this point.

53. Add: $(-45.67) + (-394.78) + 23.48$

54. Solve for x: $463 = 94 + x$

55. Evaluate: $-9 + (-4 + 2.4)^2 - 2^{-2}$

56. Solve for x: $2x + 9\frac{1}{2} = -8\frac{1}{4}$

57. Subtract: $\left(-34\frac{2}{7}\right) - 18\frac{3}{4}$

58. Solve for y: $168 = -7y$

59. Evaluate: $\{[-7 - (8^2)]^2 + 11^2\}\left(-9\frac{1}{2}\right)$

60. Multiply: $(-34)\left(-5\frac{1}{3}\right)(18)$

61. Solve for y: $263 = \frac{y}{3} - 84$

62. Evaluate: $(4)\,[(-3) + (-4)^2 - 0.02]^2$

63. Multiply: $(-34.5)(-2.1)(0.8)$

64. Divide: $-19.86 \div (-0.002)$

9.8 Geometry and Measurement: Working with Temperature

SECTION GOALS

- To solve word problems involving temperature change
- To use formulas to convert among the Fahrenheit, Celsius, and Kelvin temperature scales

In this section, we discuss the measurement of temperatures. We will convert temperature measurements among the three most common scales.

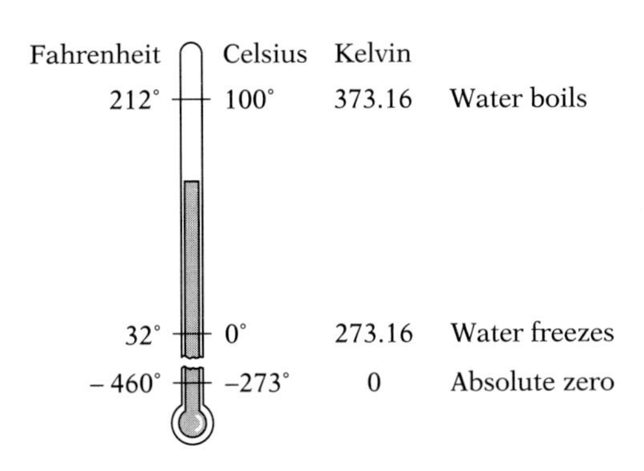

In the United States, we measure temperature on the Fahrenheit scale, developed in 1714 by a German physicist named Gabriel Fahrenheit. He chose thirty-two degrees (32°F) as the freezing point of water, and 212 degrees (212°F) as the boiling point of water.

In other countries, temperature is measured on the Celsius scale. This scale was developed by a Swedish astronomer, Andre Celsius, in 1742. On this scale, zero degrees (0°C) is the freezing point of water, and 100 degrees (100°C) is the boiling point of water.

Another scale, used mainly in the Sciences, is the Kelvin scale, developed in the late 1800s. Absolute zero (0 K) on the Kelvin scale is theoretically the lowest possible temperature—the one at which all molecular movement ceases. The freezing point of water is 273.16 K, and its boiling point is 373.16 K. Temperatures on the Fahrenheit and Celsius scales can be negative numbers (below zero). *Note:* The degree sign is *not* used with Kelvin temperatures.

To change a temperature from degrees Celsius (°C) to degrees Fahrenheit (°F)

Use either of the following formulas:

$$°F = \left(\frac{9}{5}\right)°C + 32$$

or

$$°F = (°C \times 1.8) + 32$$

1a. Polar climates have average warmest temperatures of less than 10.5 degrees Celsius. Change this to degrees Fahrenheit.

1b. The outside temperature on Tuesday was –5 degrees Celsius. What was the temperature in degrees Fahrenheit?

EXAMPLE 1

A thermometer calibrated in degrees Celsius says that your temperature is 40°C. Express this in degrees Fahrenheit.

SOLUTION

First we substitute the Celsius temperature into the formula.

$$°F = \left(\frac{9}{5}\right)°C + 32$$
$$= \left(\frac{9}{5}\right)40 + 32$$

Then we solve for °F, using the standard order of operations.

$$°F = \left(\frac{9}{5}\right) 40 + 32$$
$$= \frac{360}{5} + 32 \qquad \text{Multiplying}$$
$$= 72 + 32 \qquad \text{Simplifying}$$
$$= 104 \qquad \text{Adding}$$

Your temperature is 104 degrees Fahrenheit.

■ *Now do margin exercises 1a and 1b.*

To change a temperature from degrees Fahrenheit to degrees Celsius

Use either of the following formulas

$$°C = \frac{5}{9}(°F - 32)$$

or

$$°C = (°F - 32) \div 1.8$$

We use one of these in the next example.

EXAMPLE 2

The lowest temperature ever recorded in New Hampshire was –46 degrees Fahrenheit in 1925. Express this temperature to the nearest degree Celsius.

SOLUTION

We substitute the Fahrenheit temperature into the formula and simplify.

$$
\begin{aligned}
°C &= (°F - 32) \div 1.8 & \\
&= (-46 - 32) \div 1.8 & \text{Substituting} \\
&= (-78) \div 1.8 & \text{Simplifying} \\
&= -43.3\overline{3} & \text{Dividing}
\end{aligned}
$$

The temperature was approximately –43°C.

■ *Now do margin exercises 2a and 2b.*

In our last example, we will change a Celsius temperature to kelvins.

2a. Dalol, Ethiopia, the hottest spot in the world, has an average yearly temperature of 94 degrees Fahrenheit. A place that is 110 degrees colder is what temperature in degrees Celsius?

2b. During a run on a very hot day, your temperature can rise to 106.2 degrees Fahrenheit. What is this temperature in degrees Celsius?

3a. A temperature of –85 degrees Celsius is what in kelvins?

3b. A temperature of 100.8 K is what in degrees Celsius?

To change a temperature from degrees Celsius to kelvins

Use the formula

$$K = °C + 273.16$$

EXAMPLE 3

A temperature of –100°C is equivalent to what temperature on the Kelvin scale?

SOLUTION

We substitute the Celsius temperature into the proper formula and simplify.

$$\begin{aligned} K &= °C + 273.16 \\ &= -100 + 273.16 \\ &= 173.16 \end{aligned}$$

So –100°C is equivalent to 173.16 K. (Note, no degree sign is used with the K.)

■ *Now do margin exercises 3a and 3b.*

Work the exercises that follow.

9.8 EXERCISES

Work the following word problems. When necessary to round, round answers to the nearest hundredth of a degree.

1. The coldest temperature ever recorded was –128.5°F in Vostok, Antarctica. Change this temperature to degrees Celsius.

2. Because of dehydration, Alberto Salazar's body temperature dropped to 88°F after he won the 1982 Boston Marathon on a very hot, sunny day. What is this temperature in degrees Celsius?

3. The hottest temperature on record in the inhabitable world was 136°F, recorded in El Azizia, Libya, in 1892. Change this temperature to degrees Celsius.

4. Mid-latitude areas are described as "warm climate" areas if they have an average cold temperature of not less than –3°C and not more than 18°C. Find the difference between these two extremes in degrees Fahrenheit.

5. The average temperature recorded in New Hampshire in the winter varies from –5°C to 21°C. What is the range of average temperatures there in kelvins?

21°
15°
10°
5°
0°
–5°

6. Cold, snowy climates have, in the warmest months, temperatures no higher than 10°C and, in the coldest months, temperatures no higher than –3°C. Find the range of these two extremes.

7. The boiling point of ether is 35.6°C, or 96.1°F. Express this temperature in kelvins.

Table 1 shows how cold it feels when the wind is blowing at a given speed. For example, at 10°F, with the wind blowing 10 miles per hour, the temperature feels (and chills) as though it were –9°F. Use this wind chill chart to answer Exercises 8 through 11.

TABLE 1 Wind Chills

	Wind Speed (mph)								
Thermometer Reading (°F)	**5**	**10**	**15**	**20**	**25**	**30**	**35**	**40**	**45**
35	33	22	16	12	8	6	4	3	2
30	27	16	9	4	1	–2	–4	–5	–6
25	21	10	2	–3	–7	–10	–12	–13	–14
20	19	3	–5	–10	–15	–18	–20	–21	–22
15	12	–3	–11	–17	–22	–25	–27	–29	–30
10	7	–9	–18	–24	–29	–33	–35	–37	–38
5	0	–15	–25	–31	–36	–41	–43	–45	–46
0	–5	–22	–31	–39	–44	–49	–52	–53	–54
–5	–10	–27	–38	–46	–51	–56	–58	–60	–62
–10	–15	–34	–45	–53	–59	–64	–67	–69	–70
–15	–21	–40	–51	–60	–66	–71	–74	–76	–78
–20	–26	–46	–58	–67	–74	–79	–82	–84	–85
–25	–31	–52	–65	–74	–81	–86	–89	–92	–93
–30	–36	–58	–72	–81	–88	–93	–97	–100	–102
–35	–42	–64	–78	–88	–96	–101	–105	–107	–109
–40	–47	–71	–85	–95	–103	–109	–113	–115	–117
–45	–52	–77	–92	–103	–110	–116	–120	–123	–125

8. How cold will it feel with a temperature of 20 degrees below zero Fahrenheit and a 5-mph wind?

9. How cold will it feel with a temperature of 15°F and a 45-mph wind?

10. A temperature of –5°F and a 5-mph wind will feel how cold?

11. The temperature is 10 below zero Fahrenheit, with a wind of 15 mph. What difference in temperature will be felt if the wind increases to 40 mph?

12. The tropical rainy climate has average monthly temperatures that never fall below 18°C. Express this temperature in degrees Fahrenheit.

13. Wood's metal has a melting point of 150°F. This is such a low melting point that a spoon made of it will melt in a cup of hot tea. Express this melting point in kelvins.

14. The average temperatures recorded in northern sections of Maine are –10°C in the winter and 18°C in the summer. What is the mean of these two temperatures in degrees Fahrenheit?

15. The ideal temperature of a fluid for quick absorption is 40°F. If a fluid has a temperature of 300 K, is it ideal, too hot, or too cool?

16. The coldest inhabited spot, Oymyakon, Siberia, had a temperature of –96°F in 1964. The highest temperature recorded in the United States was 134°F in Death Valley on July 10, 1913. Find the range of these temperatures in degrees Celsius.

17. a. The element calcium melts at 810°C. What is this temperature in degrees Fahrenheit?
b. The melting point of aluminum is 660.2°C. What is this temperature in kelvins?

18. Corals live at 23°C. Express this temperature in degrees Fahrenheit.

19. Order these temperatures from coldest to hottest: –5° Celsius, –10° Fahrenheit, 200 kelvins, 350° Fahrenheit, 0° Fahrenheit. Then find their median in kelvins.

20. The following temperatures were recorded using a Fahrenheit thermometer: 20°, 1°, 17°, –25°, and –58°. What are the mean, median, and range of these temperatures in degrees Celsius?

21. The hottest city in the United States is Key Word, Florida. It has an annual average temperature of 77.4°F. Write this temperature in degrees Celsius.

22. The longest hot spell on record was 162 consecutive days of 100°F in Marble Bar, Australia. Change this temperature to degrees Celsius.

23. The lowest temperature ever recorded in the United States was –79.8°F at Prospect Creek Camp, Alaska, on January 13, 1971. Change this temperature to degrees Celsius.

24. On a certain day, the outside temperature was 25°F colder in Chicago than in New York. What was the temperature in Chicago if it was 15°F in New York?

25. The coldest spot in the world is Plateau Station, Antarctica, which has an average yearly temperature of –70 degrees Fahrenheit. Express this temperature in degrees Celsius.

9.8 MIXED PRACTICE

By doing these exercises, you will practice all the topics in this chapter up to this point.

26. Find the Celsius temperature that is equivalent to 42°F.

27. Solve for x: $3x = 0$

28. Find the product: $(-14)(13.5)$

29. Solve for x: $\frac{x}{15} + 6 = 149$

30. Evaluate: $\left[(-8)(2) + (-4) \times \left(2\frac{2}{3}\right)\right]^0 + \left(\frac{1}{3}\right)^{-2}$

31. Which is coldest, 16°C, 65°F, or 200 K?

32. Add: $\left(74\frac{2}{3}\right) + \left(-36\frac{1}{8}\right)$

33. Subtract: $(-7.3) - (5.8)$

34. Solve for x: $-7 + 3x = 9$

35. Divide: $\left(-6\frac{3}{5}\right) \div \left(-3\frac{1}{10}\right)$

36. Solve for y: $46.88 = 46y - \frac{1}{2}$

37. Use your calculator to evaluate: $[-2^3 + 0.06(-3)]^2 + 5^{-2}$

38. Find the product: $\left(-16\right)\left(-27\right)\left(+9\frac{1}{3}\right)$

39. Find a Kelvin temperature between 45°C and 19°F.

??? Problem-Solving Preparation: Mixed Applications

In this problem-solving section, you will be solving problems based on any and all of the previous sections. Because we cannot give you an example of each problem, we will just offer you another interesting problem-solving procedure. This is one that you might need on a trip.

Foreign Exchange Rates

Country	Foreign Exchange for $1	Country	Foreign Exchange for $1	Country	Foreign Exchange for $1
Argentina (austral)	550.96	Holland (guilder)	2.00	Singapore (dollar)	1.82
Australia (dollar)	1.20	Hong Kong (dollar)	7.37	South Africa (rand)	2.53
Austria (schilling)	12.50	India (rupee)	14.97	South Korea (won)	602.05
Belgium (franc)	37.17	Indonesia (rupiah)	1,515	Spain (peseta)	109.91
Brazil (cruzado)	4.05	Ireland (pound)	0.67	Sweden (kroner)	6.07
Britain (pound)	0.59	Israel (shekel)	1.65	Switzerland (franc)	1.55
Canada (dollar)	1.13	Italy (lira)	1,287	Tahiti (franc)	111.37
Chile (peso)	259.81	Japan (yen)	133.98	Taiwan (dollar)	23.65
Colombia (peso)	359.32	Jordan (dinar)	0.60	Thailand (baht)	23.65
Denmark (krone)	6.89	Mexico (peso)	2,387	Turkey (lira)	1,634
Ecuador (sucre)	328.08	New Zealand (dollar)	1.58	Venezuela (bolivar)	33.06
Egypt (pound)	2.14	Norway (kroner)	6.52	W. Germany (mark)	1.78
Finland (mark)	4.03	Philippines (peso)	18.74	Yugoslavia (dinar)	18519
France (franc)	6.09	Portugal (escudo)	145.77		
Greece (drachma)	145.65	Saudi Arabi (riyal)	2.39		

Travelers from the United States who visit other countries may need to determine the money exchange rates to budget for their trips. The accompanying table shows various exchange rates—that is, how much $1 was worth in foreign currencies—for late September 1989.

EXAMPLE 1

A Mexican hotel advertises rooms at 225,000 pesos per night. How much is this in U.S. dollars?

SOLUTION

Using the table, we first find the exchange rate for Mexico: 2,387 pesos per $1.

We then set up a proportion with the information we have and solve for the unknown value.

$$\frac{2{,}387 \text{ pesos}}{1 \text{ dollar}} = \frac{225{,}000 \text{ pesos}}{n \text{ dollar}}$$

$$\frac{2{,}387}{1} = \frac{225{,}000}{n} \quad \text{Dropping units}$$

$$2{,}387 \times n = 225{,}000 \times 1 \quad \text{Cross-multiplying}$$

$$\frac{2{,}387 \times n}{2{,}387} = \frac{225{,}000}{2{,}387} \quad \text{Dividing by 2,387}$$

$$n = 94.260 \quad \text{Simplifying}$$

So 225,000 pesos is $94.26.

■ *Now do margin exercises 1a and 1b.*

1a. Find the cost of a hotel room in Tahitian francs if the room costs $150 in the United States.

1b. Ten pounds in Ireland is equivalent to how many dollars in the United States?

EXERCISES

1. One job offer for a beautician includes a salary of $16,000 plus benefits amounting to $5667. A second offer is for a salary of $21,000. How much more money does the first job offer altogether?

Use the preceding table, if necessary, to solve Exercises 2 through 6.

2. In which countries was the dollar worth less than one of the foreign currency units?

3. A tourist brought back 9000 Japanese yen from a recent trip. How much is this in U.S. currency?

4. A visitor to London, England, bought jewelry for 250 pounds. How much did she pay in U.S. currency?

5. A trip to Norway resulted in the purchase of an embroidered blouse for 260.8 kroner. That same blouse was available in New York for $92. Was it more or less expensive in New York?

6. Money that we had invested in Canada was exchanged for 118 U.S. dollars. About how many Canadian dollars is this?

7. A farmer is planting his vegetables for the spring. He has planted 45 rows of beans with 25 plants in each. He has also planted 23 rows of corn with 20 plants in each. Will there be more corn or more beans if all his plants grow?

8. A pair of running shoes costs $79 and will be good for 200 miles. A second pair costs $99 and will be good for 250 miles. Which will give the best performance in miles per dollar?

9. Packages are being stacked with 16 in the first row, 15 in the second row, and so on until the last row, which has 1 package. How many packages will there be altogether in the stack?

10. A man wants to arrange 36 stones into the shape of a square with the same number on each side to surround a flower bed. How many stones will there be on each side?

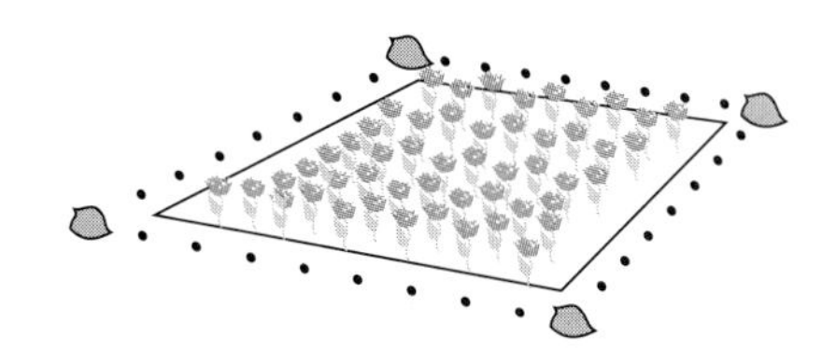

11. The chances of winning a certain lottery are 1 in 10 million. What are the chances of not winning?

LOTTERY	
GAME 1:	**2, 10, 12, 28, 34, 45**
GAME 2:	**9, 14, 15, 23, 31, 54**

12. Some rabbits are multiplying (on a calculator, of course). They have found a pattern that gives them answers of 10, 0.1, 0.001, etc. What are they multiplying by?

13. Each year, on my cat Rail's birthday, I feed her 10 treats for each year old she is. When she is 10 years old, how many birthday treats will she have eaten altogether?

Chapter 9 Review

ERROR ANALYSIS

These problems have been worked incorrectly. Tell where each is wrong, and correct the mistake.

1. Subtract: $18 - 26$

Incorrect Solution

$18 - 26 = 18 + 26 = 44$

Error ____________________

____

Correct Solution

2. Add: $(-36) + (+24)$

Incorrect Solution

$(-36) + (+24) = -60$

Error ____________________

____

Correct Solution

3. Subtract: $-27 - (+13)$

Incorrect Solution

$-27 - (+13) = +27 + (-13) = -14$

Error ____________________

____

Correct Solution

4. Simplify: $-|-6|^2$

Incorrect Solution

$-|-6|^2 = (-6)^2 = 36$

Error ____________________

____

Correct Solution

5. Multiply: $(-18)\,(-2)$

Incorrect Solution

$(-18)\,(-2) = -36$

Error ____________________

____

Correct Solution

6. Rewrite 3^{-2} without exponents.

Incorrect Solution

$3^{-2} = -(3)^2 = -9$

Error ____________________

____

Correct Solution

INTERPRETING MATHEMATICS

By working these exercises, you will test and strengthen your mathematics vocabulary.

1. How are positive numbers and negative numbers alike? In what ways could they be considered opposites?

2. What is meant by a signed number? What is the additive inverse of a signed number?

3. How are integers and rational numbers alike? How are they different? Name a rational number that is not an integer. Can you name a number that is an integer but is not a rational number?

4. Use the word distance to explain absolute value.

5. What is an integer exponent? Write a number with an integer exponent.

6. Explain the addition principle of equality in your own words.

7. How does the multiplication principle of equality help in solving equations?

8. How are the Celsius, Fahrenheit, and Kelvin scales alike?

REVIEW PROBLEMS

The exercises that follow will give you a good review of the material presented in this chapter. Work through them and check your answers at the back of the book.

Section 9.1

1. Arrange $\left|4\frac{1}{2}\right|$, $|-4|$, and $-|-4|$ in order so that the symbol $>$ can be used.

2. Graph -9 and its inverse on the number line.

3. Simplify: $-(-|-8|)$

4. Arrange $|15|$, $-|16|$, and $-|13|$ in order from smallest to largest.

5. Graph -18 and its inverse on the number line.

6. Find $-|-|-11||$.

Section 9.2

In Exercises 7 through 12, add or subtract as indicated.

7. $-3\frac{1}{2} - \left(-3\frac{1}{2}\right)$

8. $(-7.3) + (5.8)$

9. $\left(7\frac{3}{4}\right) - \left(-14\frac{1}{4}\right)$

10. $(-26) + 18 + (-34)$

11. $(-48) + 79$

12. $(36) - (-18)$

Section 9.3

In Exercises 13 through 18, multiply or divide as indicated.

13. $(0.1)(-2)(0.34)$

14. $\left(-\frac{8}{9}\right) \div \left(-\frac{11}{13}\right)$

15. $(2)(-15)(-9)$

16. $(234) \div (-9)$

17. $(-18)(-26)$

18. $(-3)(10)$

19. A mortgage company advertises that its current home mortgage rate is 11.75%, which represents a drop of 0.6% from the previous rate. What was the old rate?

20. A certain stock has fluctuated as follows: up $\frac{1}{2}$, down $\frac{3}{8}$, up $\frac{1}{4}$, up $\frac{1}{8}$. If the opening price was $3\frac{5}{8}$, what is the current price?

21. Your heart rate changes with the amount of energy that you exert. If your heart rate was 70 beats per minute before you exercised, and increased 40 beats per minute during exercise, and then dropped 10 beats per minute after exercise, what was your heart rate after exercise?

22. An airplane is at an altitude of 29,000 feet above sea level. It flies over a valley that has a low point 25 feet below sea level. How far above the valley "floor" is the airplane?

23. The temperatures in a certain part of the world have been known to range from –23°F to 102°F. What is the range of these temperatures?

24. A certain plant can grow anywhere from 35 feet below sea level to 72 feet above sea level. What is the range of its "growth zone"?

Section 9.5

In Exercises 25 through 30, evaluate each expression. Round answers where appropriate.

25. $(-3)^{-3} + (-2)(5) - (2)(1.9)$

26. $\left(-\frac{2}{3}\right)^2 + \left(\frac{3}{2}\right)^{-1} - (2 \times 3)^{-1}$

27. $-8\frac{2}{5} \times 4(-9 + 6) + 2^{-2}$

28. $(-7)^2 + 2(3 - 4)^2 + 9 - 6$

29. $1^4 + (-1)^3 - 1^{-3} + (-1)^{-5}$

30. $1^8 + (-1)^{11} - 1^{-13} + (-1)^6$

Section 9.6

In Exercises 31 through 36, solve for x.

31. $3x + 6 = 0$

32. $3 - x = 12$

33. $19x - 57 = 0$

34. $x - 25 = 5$

35. $3 - x - 18 = 0$

36. $\frac{1}{2}x = 4$

Section 9.7

In Exercises 37 through 42, solve for x.

37. $5x - 12 = 32$

38. $2.5 - 6x = 7.5$

39. $-3x - 12 + 15 = 0$

40. $\frac{x}{2} = \frac{4}{5} + 2$

41. $18x - 27 + 12 = 0$

42. $\frac{2x}{12} + 45 = 7$

Section 9.8

In Exercises 43 through 45, round answers to the nearest tenth of a degree.

43. Find the Celsius equivalent of 78°F.

44. Find the Fahrenheit equivalent of 45 K.

45. Find the Fahrenheit equivalent of 23°C.

In Exercises 46 through 48, arrange the temperatures listed in order from coldest to hottest.

46. 23°C, 45°F, 290 K, 185 K, 35°F, 10°C

47. 53°C, 590 K, 49°F, 285 K, 3°F, 12°C

48. 24°F, 123°C, 20 K, 5°F, 15 K, 1°C

Mixed Review

49. Add: $(-38) + (-46) + (+109)$

50. Subtract: $(126) - 385$

51. Multiply: $(-12)(-11)$

52. Divide: $(-2400) \div (-12)$

53. The temperature in a hot house on a certain day is 95°F. By night, the temperature was 74°F. What was the change in temperature?

54. Arrange $-|-|-82||$, $-|97|$, and -43 in order from smallest to largest.

55. Subtract 39.86 from (-37.3).

56. Multiply: $\left(\frac{9}{4}\right)\left(-\frac{8}{9}\right)\left(\left|-9\frac{1}{6}\right|\right)$

57. Divide $37\frac{5}{6}$ by $\left(-4\frac{9}{10}\right)$.

58. Evaluate: $63 + |-17| + (-|-43|)$

59. Evaluate: $[6(-9)^{-1} + 3(-4)]^3 - 4^0$

60. Ed's freshman GPA was 3.15. It went up by 0.34, down by 0.15, and up by 0.08 in his next three years. What was his final GPA?

61. Evaluate: $\left[9(-1)^{-1} - (-7)^3\left(-3\frac{3}{7}\right)\right](|(-6)^{-1}|)$

62. A meter is checked and is found to read over or under the actual level by these amounts: +0.004, −0.36, −0.004, +0.5, and −0.003. After all these readings, what is the amount that this scale is off?

63. Which is coldest, 0°C, 0°F or 0 kelvins?

64. Solve for y: $46.93 = 46y - \frac{1}{3}$.

65. In Africa, the highest point of elevation is Mt. Kilimanjaro at 19,340 feet above sea level. The lowest point is a spot in Egypt that is 436 feet below sea level. A man is standing at the lowest spot. He is 6 feet $3\frac{1}{3}$ inches tall. Another man, 5 feet $8\frac{1}{4}$ inches tall, is standing on top of Mt. Kilimanjaro. Find the difference in the elevations of the tops of these men's heads.

Chapter 9 Test

This exam tests your knowledge of the topics in Chapter 9. Write answers in simplest form.

1a. Graph -5.6 and its opposite on the number line.

b. Arrange $-|-16|$, $|-13|$, and $-|-27|$ in order from largest to smallest.

c. Arrange -153, $-152\frac{1}{2}$, and $-|-165|$ in order from smallest to largest.

2a. Add: $29 + (-18)$

b. Add: $(-3.6) + (+1.85)$

c. Subtract: $\left(-10\frac{1}{3}\right) - \left(-8\frac{2}{5}\right)$

3a. Multiply: $(-2)(-149)$

b. Multiply: $\left(-\frac{8}{19}\right)(+76)$

c. Multiply: $\left(10\frac{1}{2}\right)\left(-6\frac{1}{5}\right)$

4a. Is 10°C warmer or cooler than 45°F?

b. The stock for a certain company closed yesterday at $31\frac{1}{8}$. Today it opened $3\frac{3}{8}$ points lower than yesterday's closing, and then it went to a level $6\frac{2}{3}$ points higher than yesterday's closing price. What is the range of these values?

c. The changes in a reading on a scale were -2, $3\frac{1}{2}$, 10.5, -16, -15.5, and -4.5. What does the scale read now in relation to its starting point?

5a. Evaluate: $[2(-3) + 13] \times [5 + (7)^{-2}]$

b. Evaluate: $(-3)^{-2} + 4(-5)(-8)^0$

c. Evaluate: $(-4)^3 - 2(-7 - 5)^2$

6a. Solve for m: $3\frac{1}{2}m = 7$

b. Solve for n: $5.2n + 16 = -10$

c. Solve for x: $\frac{x}{2.5} = 93.4$

ANSWERS to Margin Exercises

Section 1.1 *pages 3–6*

1a. 76: 6 ones; 163: 6 tens; 6,894: 6 thousands **1b.** 3275: 2; 653227: 2; 5,841: 464, 4
2a. 3,000,000 + 80,000 + 9,000 + 400 + 6 **2b.** 206,300 **3a.** Forty-seven thousand, six hundred four **3b.** 305,006

Section 1.2 *pages 11–12*

1a. 773,612,998 **1b.** 317,762,998 **2a.** 2,839,485 > 2,039,438 > 204,958 **2b.** 2,928,374 **3a.** 2,371,146,985 < twenty-eight billion three (28,000,000,003) **3b.** two billion, nine hundred thousand, forty

Section 1.3 *pages 17–20*

1a. 74,000 **1b.** Both would round to 10,000 **2a.** 3,470,000 **2b.** 3,645,934 **3a.** 9,000,000 **3b.** 956,607
4a. 1,000; 0 **4b.** 0

Section 1.4 *pages 25–30*

1a. 121 **1b.** 382 **1c.** 86 **1d.** 383 **2a.** 981 **2b.** 10,593 **3a.** 264 **3b.** 86,178 **4a.** 2,000,000
4b. 563,213 and 964,596

Section 1.5 *pages 35–38*

1a. 33 **1b.** 31 **2a.** 25 **2b.** 59 **3a.** 89 **3b.** 1246 **4a.** 17,377 **4b.** 391,818 **5a.** 3,993,479
5b. 7,506,219

Section 1.6 *pages 41–43*

1a. 1470 people **1b.** 172 feet **2a.** 2,858,236 square miles **2b.** $74,100,000 *Saturday Night Fever;* $149,882,000 together **3a.** 1972 **3b.** 42 years old

Problem-Solving Preparation *pages 51–52*

1a. 1,690,000 **1b.** $1764

Section 1.7 *pages 53–56*

1a. irregular pentagon **1b.** equilateral triangle **2a.** area **2b.** perimeter

Section 2.1 *pages 71–76*

1a. 630 **1b.** 30,000 **2a.** 42,000 **2b.** five zeros **3a.** 48 **3b.** 3 **4a.** 483 **4b.** 756,054

Section 2.2 *pages 81–84*

1a. 2077 **1b.** 69 or 39 depending on how you set up your problem **2a.** 33,366 **2b.** 7 **3a.** 13,773,632
3b. 820,690,140 **4a.** 3,171,176 **4b.** 253,368

Section 2.3 *pages 89–94*

1a. 71 **1b.** 91 **2a.** 509 r 2 **2b.** $21\overline{)84021}$

Section 2.4 *pages 97–100*

1a. 45 r 368 **1b.** 388 r 245 **2a.** 4080 **2b.** 80 **3a.** 1876 r 13 **3b.** 1658

Section 2.5 *pages 103–104*

1a. $45 **1b.** $12,000 **1c.** $132 **1d.** 65 outfits **2a.** 12 packages **2b.** 101 boxes **2c.** 560 gallons **2d.** 12 pens

Section 2.6 *pages 111–114*

1a. 7 pounds 11 ounces **1b.** 9 pounds 4 ounces **2a.** 1 day 3 hours 28 minutes **2b.** 2 quarts 1 pint 1 cup **3a.** 2 weeks 6 days **3b.** 23 portions

Section 2.7 *pages 121–124*

1a. 60 feet **1b.** 74 yards **2a.** 44 inches **2b.** 29 feet **3a.** 64,480 square yards **3b.** 529 square inches **4a.** 72 square feet **4b.** 23 feet

An Application to Statistics *pages 129–132*

1a. 93 kilowatt hours **1b.** 360 kilowatt hours **2a.** 1986 **2b.** 1970

Section 3.1 *pages 145–148*

1a. 225; 39,005 **1b.** 220; 3900; 4040 **2a.** 111; 222,222 **2b.** 2,310,000,000; 3752; 992,200 **3a.** a. no; b. yes; c. no; d. yes **3b.** a. no; b. no; c. no; d. yes

Section 3.2 *pages 151–154*

1a. 1, 2, 3, 4, 6, 9, 12, 18, 36 **1b.** 1, 2, 4, 37, 74, 148 **2a.** composite **2b.** composite **3a.** composite **3b.** 113 **4a.** $2 \times 2 \times 5 \times 5$ **4b.** $3 \times 5 \times 167$ **5a.** $2 \times 2 \times 2 \times 2 \times 2 \times 2 \times 3$ **5b.** Three: 2, 3, and 457

Section 3.3 *pages 161–162*

1a. $2^3 \times 3^3$ **1b.** yes **2a.** 2025 **2b.** 1008 **3a.** 4096 **3b.** 531,441

Section 3.4 *pages 167–170*

1a. 5 **1b.** $(5 + 3) \times 11 + 2 - 3$ **2a.** 5 **2b.** 37 **3a.** 38 **3b.** 588

Problem-Solving Preparation *pages 175–178*

1a. $P = 1200$ **1b.** $S = 23$

Section 3.5 *pages 179–180*

1a. 24 **1b.** 31 **2a.** 86 **2b.** 77 **3a.** no **3b.** no

Section 3.6 *pages 185–188*

1a. 84 inches **1b.** 76 inches **2a.** 676 square feet **2b.** 49 square inches **3a.** 15 in. **3b.** 35 in. **4a.** 63 units **4b.** 20 units

Section 3.7 *pages 193–198*

1a. 4 **1b.** 6 **2a.** 16 **2b.** 5 **3a.** 40 **3b.** 72 **4a.** 8 **4b.** 40 **5a.** 63 **5b.** 35 **6a.** 105 **6b.** 1260 **7a.** 108 **7b.** 132

An Application to Statistics *pages 203–206*

1a. 1984 **1b.** 95¢ **2a.** 1984 **2b.** coffee

Section 4.1 *pages 221–226*

1a. $\frac{21500}{22000}$ **1b.** $\frac{200}{300}$ **2a.** $\frac{93}{293}$ **2b.** $\frac{2}{19}$ **3a.** $\frac{9}{60}$ **3b.** $\frac{496}{500}$ **4a.** $1\frac{1}{2}; \frac{3}{2}$ **4b.** $2\frac{3}{4}; \frac{11}{4}$ **5a.** $23\frac{1}{4}$

5b. $97\frac{4}{7}$ **6a.** $\frac{69}{5}$ **6b.** $\frac{71}{4}$

Section 4.2 *pages 231–234*

1a. $\frac{6}{16}, \frac{12}{32}, \frac{30}{80}, \frac{60}{160}$; answers may vary **1b.** $\frac{15}{18}, \frac{35}{42}, \frac{500}{600}, \frac{60}{72}$; answers may vary **2a.** $\frac{5}{35}$ **2b.** $\frac{46}{22}$

3a. $\frac{5}{6}$ and $\frac{4}{6}$ **3b.** $\frac{24}{42}$ and $\frac{14}{42}$ **4a.** $\frac{420}{540}$ and $\frac{450}{540}$ **4b.** $\frac{30}{70}$ and $\frac{63}{70}$

Section 4.3 *pages 239–240*

1a. $\frac{1}{3}$ **1b.** $\frac{9}{140}$ Divisibility rule for 3 shows 27 and 420 are both divisible by 3. **2a.** $\frac{1}{25}$ **2b.** No. Numerator and denominator are divisible by 3. **2c.** $\frac{3}{8}$ **2d.** No. $\frac{150 \div 10}{270 \div 10} = \frac{15}{27}$ $\frac{15 \div 3}{27 \div 3} = \frac{5}{9}$ So $\frac{150}{270} = \frac{5}{9}$

Section 4.4 *pages 243–244*

1a. $\frac{1}{8}$ **1b.** $\frac{1}{3} > \frac{1}{4} > \frac{1}{6}$ **2a.** $15\frac{9}{10} > 15\frac{7}{8}$ **2b.** $2\frac{3}{105} < 2\frac{5}{140}$ **3a.** 3 **3b.** 0

Section 4.5 *pages 249–252*

1a. 1 **1b.** $\frac{1}{2}$ **2a.** $1\frac{1}{2}$ **2b.** $1\frac{1}{4}$ **3a.** $1\frac{11}{18}$ **3b.** $\frac{25}{28}$ **4a.** $\frac{17}{19}$ **4b.** $\frac{37}{60}$

Section 4.6 *pages 255–256*

1a. $42\frac{16}{21}$ **1b.** $19\frac{44}{45}$ **2a.** $20\frac{27}{44}$ **2b.** $28\frac{19}{24}$

Section 4.7 *pages 261–262*

1a. $\frac{1}{3}$ **1b.** $\frac{5}{17}$ **2a.** $\frac{1}{25}$ **2b.** $\frac{4}{15}$ **3a.** $\frac{17}{24}$ **3b.** $\frac{1}{42}$

Section 4.8 *pages 265–268*

1a. $41\frac{1}{12}$ **1b.** $28\frac{17}{56}$ **2a.** $3\frac{8}{11}$ **2b.** $11\frac{3}{5}$ **3a.** $18\frac{11}{14}$ **3b.** $74\frac{15}{16}$

Section 4.9 *pages 271–272*

1a. $23\frac{23}{100}$ mph **1b.** $\frac{4}{5}$ minute **2a.** $181\frac{2}{5}$ miles **2b.** $35\frac{2}{5}$ inches **3a.** $4\frac{1}{6}$ cups

3b. $167\frac{1}{2}$ pounds

An Application to Statistics *pages 277–278*

1a. $\frac{7}{10}$ **1b.** $\frac{997}{1000}$

Section 5.1 *pages 291–292*

1a. 2 **1b.** 3 **2a.** $\frac{1}{6}$ **2b.** $\frac{1}{8}$ **3a.** $\frac{4}{105}$ **3b.** $\frac{7}{45}$

Section 5.2 *pages 295–298*

1a. $104\frac{19}{30}$ **1b.** $89\frac{1}{4}$ **2a.** $2\frac{3}{5}$ **2b.** $29\frac{17}{18}$ **3a.** 1 **3b.** 1 **4a.** $\frac{100}{39,601}$ **4b.** $973\frac{9}{11} = \frac{10712}{11}$

$\frac{10712}{11} \times \frac{11}{10712} = 1$

Section 5.3 *pages 301–302*

1a. $3\frac{1}{8}$ **1b.** $\frac{3}{4}$ **2a.** $\frac{1}{10}$ **2b.** $\frac{2}{35}$ **3a.** 85 **3b.** 21 **4a.** $3\frac{1}{3}$ **4b.** $1\frac{1}{27}$

Section 5.4 *pages 305–306*

1a. $6\frac{2}{5}$ **1b.** $17\frac{1}{2}$ **2a.** $\frac{1}{18}$ **2b.** $\frac{4}{31}$ **3a.** $\frac{43}{50}$ **3b.** $\frac{11}{12}$

Section 5.5 *pages 309–312*

1a. 50,150 ounces **1b.** 825 pounds **2a.** 13 lengths **2b.** 52 feet **3a.** 9 mph **3b.** $76\frac{1}{4}$ mph

4a. $\frac{2}{3}$ pound **4b.** 150 pieces; none

Section 5.6 *pages 319–322*

1a. $2\frac{1255}{7776}$ **1b.** $3\frac{526}{625}$ **2a.** $6\frac{1}{6}$ **2b.** 3 **3a.** 520 **3b.** $2\frac{48}{49}$

Section 5.7A *pages 327–328*

1a. $26\frac{1}{4}$ square yards **1b.** $22\frac{1}{2}$ square meters **2a.** 140 square inches **2b.** $3\frac{3}{16}$ square units

Section 5.7B *pages 333–336*

1a. $34\frac{21}{64}$ cubic feet **1b.** $\frac{343}{4096}$ cubic foot **2a.** $166\frac{2}{3}$ cubic feet **2b.** $59\frac{1}{2}$ cubic inches **3a.** $32\frac{2}{3}$ square inches

3b. $433\frac{1}{2}$ square inches **4a.** 1186 square inches **4b.** $7\frac{37}{75}$ square inches

Section 6.1 *pages 353–356*

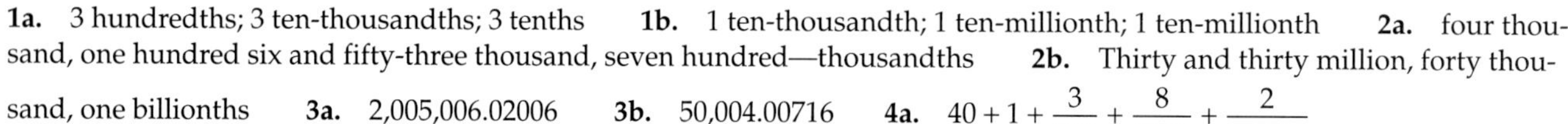

1a. 3 hundredths; 3 ten-thousandths; 3 tenths **1b.** 1 ten-thousandth; 1 ten-millionth; 1 ten-millionth **2a.** four thousand, one hundred six and fifty-three thousand, seven hundred—thousandths **2b.** Thirty and thirty million, forty thousand, one billionths **3a.** 2,005,006.02006 **3b.** 50,004.00716 **4a.** $40 + 1 + \frac{3}{100} + \frac{8}{1000} + \frac{2}{10,000}$

4b. $600 + 3 + \frac{8}{100} + \frac{5}{1,000,000}$ **5a.** $\frac{103}{10,000}$ **5b.** $1\frac{20,307}{100,000}$

Section 6.2 *pages 361–364*

1a. $98.003 < 98.04 < 98.404$ **1b.** $75.214 > 75.204 > 75.024$ **2a.** 3000 **2b.** 15,700 **3a.** 0.74679 **3b.** 8.1417 **4a.** 12.539995 **4b.** 132.599000 **5a.** \$4.6 thousand; \$6.5 thousand; \$8.2 thousand; \$8.6 thousand **5b.** 1.5 million; 2.5 million; 5.7 million; 3.6 million

Section 6.3 *pages 371–372*

1a. 12.1 **1b.** 6.73 **2a:** 7805.7 **2b.** 8184.5 **3a.** 1837.03 **3b.** 39,177.17 **4a.** 1160.6228 **4b.** 9313.2242

Section 6.4 *pages 377–380*

1a. 17.5 **1b.** 7.7 **2a.** 165.614 **2b.** 3.889 **3a.** 910.015 **3b.** 760.0064 **4a.** 37.18 **4b.** 24.9791

Section 6.5 *pages 383–386*

1a. 36.72 **1b.** 391.2 **2a.** 0.1007 **2b.** 0.547808 **3a.** 0.0257427 **3b.** 37.7157238 **4a.** 85.1674846228 **4b.** 453.3110918 **5a.** 357.796422 **5b.** 2.5991616

Section 6.6 *pages 391–396*

1a. 9.3 **1b.** 7.1 **2a.** 6 **2b.** 3 **3a.** 73.5 **3b.** 33 **4a.** 3000 **4b.** 50,000 **5a.** 134.25 **5b.** 4275444.4

Section 6.7 *pages 399–400*

1a. 874.63 degrees **1b.** 2,795,460,000 (or 2795.46 million) miles **2a.** 800,300,000 people **2b.** 315 inches **3a.** \$5983.15 **3b.** \$0.21, or 21 cents

Section 6.8 *pages 409–412*

1a. 1.1025 **1b.** 4.124961 **2a.** 149.54 **2b.** 605.16 **3a.** 0.375 **3b.** 8.25 **4a.** 0.833 **4b.** 0.8182

Section 6.9 *pages 415–418*

1a. 36.11 inches **1b.** 5.65828 meters **2a.** 15.1976 square inches **2b.** 33.16625 square feet **3a.** The area becomes 9 times greater. **3b.** No. When the radius is 3.4 feet, the area is 36.2984 feet.

An Application to Statistics *pages 423–426*

1a. 80.4 inches **1b.** 4.09 inches **2a.** 32 **2b.** 1933 **3a.** 79.3 inches **3b.** 5.025 inches **4a.** 2 **4b.** 1 and 2

Section 7.1 *pages 445–448*

1a. $\frac{1}{8}$, 1:8, 1 to 8 **1b.** $\frac{3}{4}$, 3:4, 3 to 4 **2a.** $\frac{1}{50}$; 1:50, 1 to 50 **2b.** $\frac{9}{2}$; 9:2; 9 to 2 **3a.** 60 miles per hour **3b.** $1\frac{11}{30}$ teaspoons per mug **4a.** 50¢ per pint **4b.** \$3 per pound

Section 7.2 *pages 453–456*

1a. no **1b.** no **2a.** no **2b.** no **3a.** $t = 162$ **3b.** $r = 47\frac{1}{2}$ **4a.** $t = 75$ **4b.** $r = 1.2$ **5a.** $\frac{9}{11.5}$ **5b.** 26:4

Section 7.3 *pages 459–464*

1a. 180,000 mm **1b.** 120,000,000 m **2a.** 250 pints **2b.** 5.15625 gallons **3a.** 168 g **3b.** 65cm^2

Section 7.4 *pages 469–470*

1a. 216 people **1b.** \$180 **2a.** 130 prescriptions **2b.** 107,143 males

Section 7.5 *pages 477–480*

1a. $\frac{11}{1344}$ cubic inches **1b.** $523{,}333\frac{1}{3}$ cubic miles **2a.** 3768 in^3 **2b.** 865.5 in^3 **3a.** 314 cm^2 **3b.** \$193
4a. 1004.8 in^2 **4b.** 314 cm^2

Section 8.1 *pages 497–498*

1a. 28.92% **1b.** 14% **2a.** 2.53; 0.000087; 0.72 **2b.** 3.65; 0.08; 0.004 **3a.** 63%; 583.4% **3b.** Females, 32%; males 68%

Section 8.2 *pages 501–502*

1a. $\frac{1}{4}; 3\frac{17}{25}$ **1b.** $10\frac{9}{20}; \frac{3}{50}$ **2a.** $\frac{1}{200}; \frac{1}{20{,}000}$ **2b.** $\frac{17}{500}; \frac{17}{5000}$ **3a.** 10%; $43.3\overline{3}$% **3b.** 555.56%; 64.71%
4a. 0.0463; $\frac{493}{10{,}000}$ **4b.** $\frac{393}{700}$; 0.56

Section 8.3 *pages 507–512*

1a. percent; missing part; base: 135; amount: 2 **1b.** percent: 5%; base: missing part; amount: 16 **2a.** 56.7
2b. 169.6 **3a.** 0.039 **3b.** 0.57 **4a.** 20% **4b.** 35% **5a.** 1200% **5b.** 1.8% **6a.** 30.48 **6b.** $1\frac{1}{5}$
7a. 320 **7b.** 24.39

Section 8.4 *pages 517–520*

1a. 14.85 **1b.** 0.66 **2a.** 48 **2b.** 20 **3a.** 50% **3b.** 181.85%

Section 8.5 *pages 523–532*

1a. 30% **1b.** \$2000 **2a.** 117 people **2b.** 40 mannequins

Section 8.6 *pages 533–534*

1a. $29\frac{1}{2}$ square inches **1b.** 30% **2a.** 163.8 square inches; 46.8 square inches **2b.** 16%

An Application to Statistics *pages 539–542*

1a. \$5000 **1b.** \$3000 **2a.** 16 bushes **2b.** $33\frac{1}{3}$%

Section 9.1 *pages 557–560*

1a. 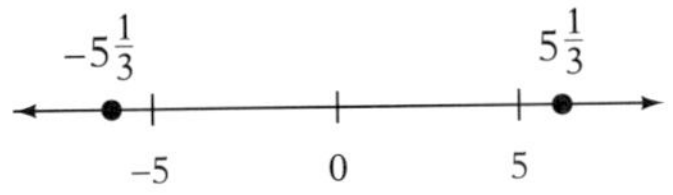

1b. 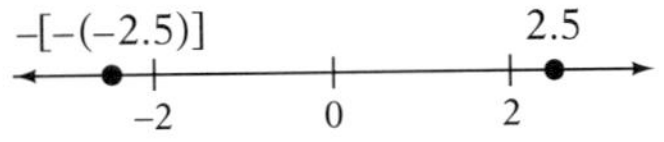

2a. $4\frac{1}{5}$ **2b.** $\frac{127}{16}$ **3a.** –5.1

3b. 29.14 **4a.** –5.3 **4b.** –2.3 **5a.** $-12\frac{5}{7}, -12\frac{5}{12}, -11.3, -11\frac{2}{9}$ **5b.** <

Section 9.2 *pages 563–568*

1a. 95 **1b.** 44 **2a.** $-7\frac{17}{20}$ **2b.** –8.22 **3a.** –94 **3b.** $-90\frac{14}{15}$ **4a.** $3\frac{1}{2}$ **4b.** 9.32 **5a.** $-102\frac{1}{2}$

5b. –1655.12 **6a.** 127 **6b.** 62.3

Section 9.3 *pages 573–574*

1a. $20\frac{7}{12}$ **1b.** –48.87 **2a.** –6.75 **2b.** 320 **3a.** $10\frac{5}{7}$ **3b.** –0.04

Section 9.4 *pages 579–582*

1a. –15 **1b.** 2274 years **2a.** $40\frac{1}{2}$ **2b.** 95 beats **3a.** 83 points **3b.** 101.8 degrees

Section 9.5 *pages 587–592*

1a. –243 **1b.** –36 **2a.** $-\frac{1}{64}$ **2b.** $\frac{1}{729}$ **3a.** 2756 **3b.** –6 **4a.** $18\frac{3}{4}$ **4b.** $-5\frac{1}{125}$ **5a.** 893.34568

5b. 15.15625

Section 9.6 *pages 597–602*

1a. $n = -13$ **1b.** $x = -2$ **2a.** $m = 43$ **2b.** $y = -14$ **3a.** $n = -1.9$ **3b.** $n = 11.6$ **4a.** $x = 12$ **4b.** $n = -9$

5a. $r = -602\frac{1}{2}$ **5b.** $n = -459$ **6a.** $n = 900$ **6b.** $y = 23.75$

Section 9.7 *pages 605–606*

1a. $x = 1\frac{1}{3}$ **1b.** $y = 1\frac{4}{5}$ **2a.** $t = 28\frac{3}{9}$ **2b.** $x = -8$ **3a.** $x = -166.7$ **3b.** x = –0.64

Section 9.8 *pages 609–612*

1a. 50.9°F **1b.** 23°F **2a.** –26.67°C **2b.** 41.22°C **3a.** 188.16K **3b.** –172.36°C

Problem-Solving Preparation *pages 617–620*

1a. 16,705.50 Tahitian francs **1b.** $14.93

ANSWERS to Chapter 1 Odd-Numbered Exercises

Section 1.1 *pages 7–10*

1. 2: millions; 1: hundred thousands; 3: thousands; 7: tens **3.** 1: hundred thousands; 8: ten thousands; 2: hundreds; 7: tens **5.** 1: hundred millions; 2: hundred thousands; 3: thousands; 6: tens **7.** 1: billions; 9: millions; 4: thousands; 6: ones **9.** 36,198; 30,000 + 6,000 + 100 + 90 + 8 **11.** one thousand, seven hundred thirty; 1,000 + 700 + 30 **13.** twenty-four billion, seven hundred seventeen million; 20,000,000,000 + 4,000,000,000 + 700,000,000 + 10,000,000 + 7,000,000 **15.** 63,000,000; 60,000,000 + 3,000,000 **17.** nine million, nine hundred seventy-six thousand, one hundred thirty-nine; 9,000,000 + 900,000 + 70,000 + 6,000 + 100 + 30 + 9 **19.** 1,083,208,840,000; 1,000,000,000,000 + 80,000,000,000 + 3,000,000,000 + 200,000,000 + 8,000,000 + 800,000 + 40,000 **21.** three million, six hundred sixty-six thousand, five hundred; 3,000,000 + 600,000 + 60,000 + 6,000 + 500 **23.** two thousand, two hundred twenty-two; 2,000 + 200 + 20 + 2 **25.** incorrect **27.** incorrect **29.** correct

Section 1.2 *pages 13–16*

1. moderate gale **3.** fresh breeze **5.** whole gale **7.** strong gale **9.** 200,000 < 292,000,000,000 **11.** 298,624 > 27,472 **13.** 124,830,000 < 209,051,000 **15.** 1,526 > 1,223 > 1,173; 1,173 < 1,223 < 1,526 **17.** 27,365 > 19,287 > 12,726; 12,726 < 19,287 < 27,365 **19.** 152,859,384 > 19,928,704; 19,928,704 < 152,859,384 **21.** 19,283,746 > 4,227,368; 4,227,368 < 19,283,746 **23.** 19,987,667 > 19,987,632 > 19,908,767; 19,908,767 < 19,987,632 < 19,987,667 **25.** 6,368,429 > 6,364,289 > 6,362,984; 6,362,984 < 6,364,289 < 6,368,429 **27.** violet **29.** red **31.** blue **33.** yellow **35.** no ten thousands **37.** 1,908 > 1,809 > 1,098 **39.** 29,028 **41.** 11,006,050

Section 1.3 *pages 21–24*

1. 19,800; 20,000; 20,000 **3.** 39,800; 40,000; 40,000 **5.** 96,700; 97,000; 100,000 **7.** 100,000; 100,000; 100,000 **9.** 1870 **11.** 1880 **13.** 5,698,000,000 **15.** 1,400,000 **17.** 62,100,000,000 **19.** 815,000,000,000 **21.** 34,900,000 **23.** 2,286,000 **25.** 13,000 **27.** 1,000,000,000 **29.** greatest: 53,499,999 compact discs; least; 52,500,000 compact discs **31.** tallest: 14,999 feet; shortest: 5,000 feet **33.** 430 **35.** six hundred thirty-two thousand, ninety **37.** 4,370,000 **39.** 3,000,000 **41.** eight ten thousands, or 80,000 **43.** 463,875 < 4,006,387 < 4,063,875

Section 1.4 *pages 31–34*

1. 88 **3.** 99 **5.** 151 **7.** 101 **9.** 997 **11.** 1165 **13.** 1240 **15.** 4473 **17.** 16770 **19.** 12122 **21.** 1572 **23.** 1809 **25.** 2122 **27.** 1830 **29.** 104; 1417 **31.** 939; 1721 **33.** 29 **35.** 12; 76 **37.** 1919 **39.** 332 **41.** 1120 **43.** 1061 **45.** 100 **47.** 110 **49.** a. **51.** b. **53.** 237,417 **55.** 4,992,262 **57.** 1,598,762 + 883,260 **59.** 312,161 + 1,217,788 **61.** 370 **63.** 12,894 **65.** 237,000 **67.** 10,000 **69.** 2 million **71.** 17,000

Section 1.5 *pages 39–40*

1. 121 **3.** 613 **5.** 478 **7.** 85 **9.** 89 **11.** 491 **13.** 244 **15.** 416 **17.** 3873 **19.** 1828 **21.** 5973 **23.** 1104 **25.** 11688 **27.** 8147 **29.** 86251; 9635; 76616 **31.** 54213; 1862; 52351 **33.** 1,355,638 **35.** correct **37.** 50,643 **39.** 56,000 **41.** 702,631 **43.** 35,700

Section 1.6 *pages 45–50*

1. 1912 **3.** 1904 **5.** 2464 feet **7.** 1897 kilometers **9.** 513,800,000 records **11.** 16,406 kilometers **13.** 330,000 square miles **15.** 9,269,000 members **17.** 85 feet **19.** 523,384 miles **21.** 105 feet **23.** 19,311 feet **25.** 566 feet **27.** 43,503 units **29.** 11,621,000 **31.** 177 people **33.** 1540 **35.** 170 vehicles **37.** 578 feet **39.** 90,410 square miles **41.** $20,000,000 **43.** 10,000 feet **45.** greatest: 45,000,000; least: 14,000,000 **47.** 4680 **49.** 6461 miles **51.** 290 calories **53.** 105,004 **55.** 496,000 **57.** 5631 > 5621 > 5611 **59.** 51,500 acres

Problem-Solving Preparation *page 52*

1. $327,000,000 **3.** $400,000,000 **5.** $280

Section 1.7 *pages 57–58*

1. rectangle (quadrilateral); irregular **3.** parallelogram (quadrilateral); irregular **5.** trapezoid (quadrilateral), irregular **7.** hexagon; irregular **9.** perimeter **11.** volume **13.** area **15.** perimeter **17.** 488,077 **19.** 53 **21.** thirty-one million, two hundred thirty-four thousand, four hundred ninety-six **23.** 45556 < 45656 < 56576 **25a.** scalene **25b.** isosceles **25c.** equilateral **27.** 44,313 **29.** 14,000,000,084

Chapter Review *pages 59–66*

Error Analysis *pages 59–60*

1. *Error:* incorrect grouping and borrowing

Correct Solution:

```
  2,000,000
-       123
  1,999,877
```

3. *Error:* incorrect alignment

Correction Solution:

```
   47
  123
+ 276
  446
```

5. *Error:* incorrect subtraction algorithm

Correct Solution:

```
  426
- 293
  133
```

Interpreting Mathematics *pages 60–61*

Responses to Interpreting Mathematics items will vary.

Review Problems *pages 61–66*

1. two hundred six thousand, one hundred twenty-two **3.** one million, seven hundred sixty-five **5.** 8,605 **7.** 29,002 is larger **9.** 182,760,000 < 192,730,000 < 274,880,000 < 1,772,660,000 **11.** 199,273 > 198,742 > 1,938 > 927 **13.** 1260; 1300; 1000; 0; 0 **15.** 129,840; 129,800; 130,000; 130,000; 100,000 **17.** 75,230; 75,200; 75,000; 80,000; 100,000 **19.** 277 **21.** 368 **23.** 12,726 **25.** 85,019 **27.** 4392 **29.** 1,222,967 **31.** $1018 **33.** 16,915 square feet **35.** 1,750,000 more families **37a.** c-hexagon; **37b.** a-octagon; **37c.** d-heptagon; **37d.** b-rhombus **39.** perimeter **41.** perimeter **43.** volume **45.** 295 square feet **47.** 6547 ccf/kwh **49.** 621 million metric tons

Chapter 1 Test *pages 67–68*

1a. 2,500,006 **1b.** 700,011 **1c.** sixty-two million, thirty thousand, seven **2a.** 1002 < 1234 < 1422 **2b.** 1,000,030 < 1,000,031 **2c.** 1,029, 399 > 277,738 **3a.** 850 **3b.** 10,000 **3c.** 345,440,000 **4a.** 1315 **4b.** 240,194 **4c.** 19,239,556 **5a.** 177 **5b.** 11,269 **5c.** 91,999,822 **6a.** $3,857 **6b.** 187,262,752 **6c.** 19 years **7a.** perimeter **7b.** area **7c.** volume

ANSWERS to Chapter 2 Odd-Numbered Exercises

Section 2.1 *pages 77–79*

1. 95 **3.** 112 **5.** 270,000,000 **7.** 40,000,000,000 **9.** 426 **11.** 189 **13.** 879 **15.** 968 **17.** 18,000 **19.** 8,000,000 **21.** commutative property of multiplication **23.** identity element for multiplication **25.** 8525 pounds **27.** 2919 buttons **29.** 18,000,108 **31.** 450,099 **33.** 85,280 **35.** 131,264 **37.** 59,388 **39.** 172,719 **41.** 34,784 **43.** 315,014 **45.** 8,999,991 **47.** 131,572

Section 2.2 *pages 85–88*

1. 7015 **3.** 35,206 **5.** 21,303 **7.** 67,326 **9.** 17,119 **11.** 45,560 **13.** 756,678 **15.** 83,434 **17.** 48,692 **19.** 64,032 **21.** 575,260 **23.** 382,080 **25.** 32,824,077 **27.** 90,185,748 **29.** 68×79 **31.** 76×89 **33.** 137; 7124 **35.** 780; 50; 39000 **37.** 625; 37; 4375; 1875; 23125 **39.** answers will vary **41.** 603,603,603 **43.** 4,830,563,500 **45.** 8910 **47.** 212,040 **49.** 53,740,908 **51.** 41,676,336 **53.** 1035 **55.** 7395 **57.** 11,400 **59.** 11,110,000 **61a.** 1001; **61b.** 2002; **61c.** 3003; **61d.** 4004; **61e.** 7007; **61f.** 9009 **63.** 661,696,729 **65.** 176,828,519,127 **67.** 1630 **69.** 121,176,000 **71.** 145,122 **73.** 55,000; answers may vary **75.** 116,808

Section 2.3 *pages 95–96*

1. 19 **3.** 13 **5.** 37 r 2 **7.** 153 r 3 **9.** 39 r 1 **11.** 57 **13.** 30 r 6 **15.** 53 r 3 **17.** 789 **19.** 687 **21.** 608 **23.** 306 **25.** 326,001 **27.** 180,012 **29.** 1403 r 3 **31.** 828 r 50 **33.** 69,875 **35.** 150,004 **37.** 96,667 **39.** 10,101 **41.** 30,602 **43.** 92,373 **45.** 688,500 **47.** 2000 **49.** 11,248,000 **51.** 450

Section 2.4 *pages 101–102*

1. 102 **3.** 209 **5.** 907 **7.** 503 **9.** 4000 **11.** 3 **13.** 58 r 36 **15.** 20 r 100 **17.** 190 **19.** 607 **21.** 50 **23.** 2000 **25.** 404 **27.** 111 r 9435 **29.** 37; 7400; 74; 00; 200 **31.** 79; 8295; 79; 395; 395; 105 **33.** 107 r 7254 **35.** 6865 **37.** 100 **39.** 723,260 **41.** 78900; answers will vary **43.** 138,152 **45.** 1,960,900,000

Section 2.5 *pages 105–110*

1. 250 people **3.** 40 chips **5.** 30 apes **7.** 620,500 corporations **9.** 823,924 more people **11.** 83 yards **13.** 37 touchdowns **15.** Oakland had 2,627,513 fewer people than L.A. **17.** 288,805 **19.** $278,775 **21.** 85,100,120 people **23.** beef, rice, apple pie, either vegetable **25.** $126,735 **27.** 8,620,200 kilometers per family **29.** about 5 times **31.** 51,255,000 square miles **33.** 37; 34; 34; 43 **35.** 53 pounds **37.** 343 kilometers **39.** 110 people **41.** 99 hours a year **43.** 105 r 3 **45.** 451,060 **47.** $240 **49.** 107 **51.** 283,524,184 **53.** 4 feet

Section 2.6 *pages 115–120*

1. 9 yards 11 inches **3.** 8 yards 2 feet **5.** 13 gallons 1 quart **7.** 11 pints **9.** 8 weeks 4 days **11.** 31 hours 8 minutes **13.** 12 days 12 hours **15.** 8 minutes 27 seconds **17.** 1 foot 11 inches **19.** 6 miles 4776 feet **21.** 4 pints 1 cup **23.** 5 gallons 1 quart **25.** 142,857 weeks and 1 day left over **27.** 21 ounces **29.** 8 feet 10 inches **31.** 180 miles 2360 feet **33.** 93 pints **35.** 48 pounds 9 ounces **37.** 103 weeks 5 days **39.** 131 hours 28 minutes **41.** 12,800,000 ounces **43.** 28 gallons 2 quarts **45.** 3 feet 1 inch **47.** 15 miles 40 feet **49.** 2 quarts 1 pint **51.** 2 gallons 1 quart **53.** 3 minutes 40 seconds **55.** 2 weeks 4 days **57.** 10 tons **59.** 1 pound 2 ounces **61.** 1 year 4 months **63.** 2 days 4 hours **65.** 71 hours 48 minutes 12 seconds **67.** 22 weeks 1 day **69.** 1003 **71.** 3 hot dogs **73.** 6 quarts

Section 2.7 *pages 125–128*

1. 12 inches **3.** 414 square yards **5.** 60 feet **7.** 20 square feet **9.** 23 inches **11.** 273 square units **13.** 1260 yards **15.** 2580 yards **17.** 1,739,040 **19.** 29 pounds 4 ounces **21.** 504 square feet **23.** 180 units **25.** 59,190,865 **27.** 3 years 30 weeks

An Application to Statistics *pages 129–132*

1. 5 kilowatt hours **3.** less **5.** $5000 **7.** 1972 **9.** aerobic exercising

Chapter 2 Review *pages 133–138*

Error Analysis *pages 133–134*

1. *Error:* incorrect multiplication by 80

Correction Solution:

```
  236
 × 84
  944
 1888
19824
```

3. *Error:* error in placement of zero in quotient

Correct Solution:

```
    103
16)1648
   16
     4
     0
    48
    48
     0
```

5. *Error:* incorrect placement of partial divisors

Correct Solution:

```
  102
3)306
  3
   0
   0
    6
    6
    0
```

Interpreting Mathematics *pages 134–135*

Responses to Interpreting Mathematics items will vary.

Review Problems *pages 136–138*

1. 219,023 **3.** 899,991 **5.** 139,132 **7.** 2544 **9.** 188,481,384 **11.** 185,976 **13.** 903 **15.** 22 **17.** The remainder is 5 **19.** 2,046 **21.** 1045 r 5 **23.** The remainder is 4 **25.** $2,000 **27.** 1450 feet **29.** A family of 3 is allowed $351 more per person **31.** 121 miles 1560 feet **33.** 9 pounds 10 ounces **35.** 5 ounces per hamburger **37.** 144 square inches **39.** 8 feet 2 inches **41.** 2 yards

Chapter 2 Test *pages 139–140*

1a. 894,000; **1b.** 98,506,000; **1c.** 59,376,000,000 **2a.** 73,980 **2b.** 10,603,661 **2c.** 985,188,000 **3a.** 2517; **3b.** 82; **3c.** 30 **4a.** 305 **4b.** 2067 **4c.** The remainder is 92. **5a.** 1250 pieces; **5b.** 224 minutes; **5c.** $5 each **6a.** 2 quarts **6b.** 139 pounds **6c.** 1 yard 2 feet **7a.** 414 square inches; **7b.** 104 inches; **7c.** 720 square inches

Cumulative Review *pages 141–142*

1. 1,049,500 **3.** 1002 **5.** 1 day 6 hours **7.** 295,743,000 tons **9.** 216 square inches **11.** 89,316 **13.** 1793 **15.** 7 hours 43 minutes **17.** 134

ANSWERS to Chapter 3 Odd-Numbered Exercises

Section 3.1 *pages 149–150*

1a. true; **1b.** false **3a.** true; **3b.** false **5.** yes **7.** no **9.** no **11.** no **13.** no **15.** yes **17.** yes **19.** no **21.** no **23.** yes **25.** yes **27.** no **29.** no **31.** no **33.** 3, 9 **35.** 2, 3, 4, 5, 6, 8, 9, 10, 12 **37.** true **39.** false **41.** true **43.** true **45.** true **47.** false **49.** true **51.** true **53.** 2644; 700

Section 3.2 *pages 155–159*

1. 1, 2, 3, 4, 6, 9, 12, 18, 36 **3.** 1, 2, 3, 4, 6, 11, 12, 22, 33, 44, 66, 132 **5.** 1, 2, 4, 7, 11, 14, 22, 28, 44, 77, 154, 308 **7.** 1, 2, 4, 5, 10, 11, 20, 22, 44, 55, 110, 220 **9.** 1, 2, 3, 4, 6, 8, 9, 12, 18, 24, 36, 72 **11.** composite **13.** composite **15.** prime **17.** prime **19.** $180 = 2 \times 2 \times 3 \times 3 \times 5$ **21.** $108 = 2 \times 2 \times 3 \times 3 \times 3$ **23.** $396 = 2 \times 2 \times 3 \times 3 \times 11$ **25.** $504 = 2 \times 2 \times 2 \times 3 \times 3 \times 7$ **27.** $324 = 2 \times 2 \times 3 \times 3 \times 3 \times 3$ **29.** $135 = 3 \times 3 \times 3 \times 5$ **31.** $405 = 3 \times 3 \times 3 \times 3 \times 5$ **33.** $175 = 5 \times 5 \times 7$ **35.** $675 = 3 \times 3 \times 3 \times 5 \times 5$ **37.** $275 = 5 \times 5 \times 11$ **39.** $1000 = 2 \times 2 \times 2 \times 5 \times 5 \times 5$ **41.** $1500 = 2 \times 2 \times 3 \times 5 \times 5 \times 5$ **43.** $3575 = 5 \times 5 \times 11 \times 13$ **45.** $1288 = 2 \times 2 \times 2 \times 7 \times 23$ **47.** $1764 = 2 \times 2 \times 3 \times 3 \times 7 \times 7$ **49.** $10{,}000 = 2 \times 2 \times 2 \times 2 \times 5 \times 5 \times 5 \times 5$ **51.** $15{,}000 = 2 \times 2 \times 2 \times 3 \times 5 \times 5 \times 5 \times 5$ **53.** $1001 = 7 \times 11 \times 13$ **55.** $5050 = 2 \times 5 \times 5 \times 101$ **57.** $100{,}000 = 2 \times 2 \times 2 \times 2 \times 2 \times 5 \times 5 \times 5 \times 5 \times 5$ **59.** $900{,}000 = 2 \times 2 \times 2 \times 2 \times 2 \times 3 \times 3 \times 5 \times 5 \times 5 \times 5 \times 5$ **61.** yes **63.** $2 \times 2 \times 3 \times 5 \times 41$ **65.** no **67.** yes **69.** $3 \times 5 \times 5 \times 5$

Section 3.3 *pages 163–166*

1. $2^5 \cdot 3 \cdot 5$ **3.** $2^2 \cdot 7^2$ **5.** $2^4 \cdot 5^2$ **7.** $2 \cdot 5^3$ **9.** $2^3 \cdot 5^2$ **11.** $3^4 \cdot 5$ **13.** $2 \cdot 5^2$ **15.** $2^2 \cdot 13$ **17.** 5^3 **19.** $3^3 \cdot 7$ **21.** $2^2 \cdot 23$ **23.** 3^3 **25.** $3 \cdot 5^2$ **27.** $2 \cdot 3^2 \cdot 5$ **29.** 41 **31.** 256 **33.** 4096 **35.** 343 **37.** 3375 **39.** 729 **41.** 1 **43.** 531,441 **45.** 32,768 **47.** 32,768 **49.** 648 **51.** 1600 **53.** 4096 **55.** 2,299,968 **57.** 38,723,328 **59.** 36,905,625 **61.** 16 **63.** $2^2 \times 5 \times 11$ **65.** 144 **67.** 64 **69.** composite **71.** $2^4 \times 3^2 \times 5$

Section 3.4 *pages 171–174*

1. 9 **3.** 11 **5.** 34 **7.** 21 **9.** 20 **11.** 70 **13.** 62 **15.** 22 **17.** 17 **19.** 111 **21.** 47 **23.** 57 **25.** 0 **27.** 8 **29.** 6 **31.** 28 **33.** 42 **35.** 81 **37.** 240 **39.** 251 **41.** 0 **43.** 2025 **45.** 1089 **47.** 377 **49.** 900 **51.** 89 **53.** 4553 **55.** 208 **57.** 82 **59.** 6744 **61.** 48,400 **63.** 961 **65.** procedures may vary **67.** procedures may vary **69.** procedures may vary **71.** 7 **73.** 2048 **75.** 1, 2, 4, 13, 26, 52 **77.** 4×9; 2×18; 1×36; 3×12; 6×6

Problem-Solving Preparation *pages 176–178*

1. 4 grams per cubic centimeter **3.** $2370 **5.** 31,640 grams per unit **7.** 41 gallons **9.** 765 miles **11.** 86

Section 3.5 *pages 181–183*

1. 6 **3.** 13 **5.** 22 **7.** 28 **9.** 18 **11.** 23 **13.** 29 **15.** 32 **17.** 46 **19.** 41 **21.** 67 **23.** 88 **25.** 51 **27.** 54 **29.** 65 **31.** 75 **33.** 66 **35.** 71 **37.** 52 **39.** 79 **41.** 84 **43.** 89 **45.** 94 **47.** 92 **49.** 34 **51.** 98 **53.** no **55.** yes **57.** no **59.** yes **61.** 22 **63.** $2 \times 3 \times 3 \times 7$ **65.** 39 **67.** 405 **69.** 38,416 **71.** $2^4 \times 5^2$ **73.** 2,654,208

Section 3.6 *pages 189–192*

1. 86 yards 1 foot 6 inches **3.** 105 feet **5.** 380 yards **7.** 2401 square inches **9.** 9 units **11.** 100 yards **13.** 14 feet **15.** 5 **17.** 25 **19.** 63 **21.** 35 **23.** 30 units **25.** 121 square feet **27.** 26 units **29.** 23

Section 3.7 *pages 199–202*

1. 11 **3.** 2 **5.** 34 **7.** 19 **9.** 17 **11.** 2 **13.** 4 **15.** 5 **17.** 6 **19.** 1 **21.** 5 **23.** 1 **25.** no **27.** no **29.** yes **31.** no **33.** 456 **35.** 100 **37.** 42 **39.** 168 **41.** 160 **43.** 52 **45.** 82 **47.** 34 **49.** 130 **51.** 396 **53.** 114 **55.** 240 **57.** 1540 **59.** 870 **61.** 4 **63.** 12 **65.** 252 **67.** 4,550,013 **69.** 84 **71.** yes **73.** 1180 **75.** 280

An Application to Statistics *pages 205–206*

1. 85¢ **3.** 40¢ **5.** $3.40 **7.** 1924 and 1928 1948 and 1952

Chapter 3 Review *pages 207–212*

Error Analysis *pages 207–208*

1. *Error:* misunderstanding definition of exponent

Correct Solution:

$3^3 = 3 \times 3 \times 3 \times 3 \times 3 = 243$

3. *Error:* incorrect multiplication by 2

Correct Solution:

$2^3 = 2 \times 2 \times 2 = 8$

5. *Error:* misinterpretation of exponent outside parentheses

Correct Solution:

$$85 - (4-2)^3 + 8$$
$$85 - 2^3 + 8$$
$$85 - 8 + 8$$
$$77 + 8$$
$$85$$

Interpreting Mathematics *pages 208–209*

Responses to Interpreting Mathematics items will vary.

Review Problems *pages 210–212*

1. divisible by 2 only **3.** divisible by 3 only **5.** divisible by 2, 4, 5, 8, 10 **7.** prime **9.** $2 \times 2 \times 107$
11. $2 \times 2 \times 2 \times 2 \times 2 \times 3 \times 5 \times 5 \times 5$ **13.** 8 **15.** 324 **17.** 5×89 **19.** 28 **21.** 1 **23.** 17 **25.** 58
27. 41 **29.** This is not a perfect square, so no whole-number square root exists. **31.** 196 square feet **33.** 62 feet
35. 256 square feet **37.** 336 **39.** 125 **41.** LCM = 300; GCF = 5 **43.** $2^2 \times 3^3 \times 5$ **45.** $2^3 \times 5^3$ **47.** 18
49. 61 feet **51.** 22,801 square feet **53.** LCM = 300; GCF = 5 **55.** Yes, this is a right triangle. $25^2 = 24^2 + 7^2$

Chapter 3 Test *pages 215–216*

1a. 20,000 and 1,888,872; **1b.** no, **1c.** true **2a.** 1, 2, 4, 7, 8, 14, 28, 56 **2b.** composite **2c.** $3 \times 3 \times 5 \times 5$
3a. 5×7^2; **3b.** 1080; **3c.** 144 **4a.** 17 **4b.** 226 **4c.** 59 **5a.** 200; **5b.** 12; **5c.** 27
6a. 264 feet **6b.** 1369 square inches **6c.** 17 inches **7a.** 6; **7b.** 4; **7c.** 25 **8a.** 60 **8b.** 51
8c. 144

Cumulative Review *pages 217–218*

1. 5174 **3.** 70,040 **5.** 3 days 11 hours **7.** twenty-seven thousand, three hundred sixty-five; one million, fifty
9. 29 inches **11.** 1, 2, 4, 5, 8, 10, 20, 25, 40, 50, 100, 200 **13.** 2052 **15.** $25,063,248 **17.** 121 **19.** 403

ANSWERS to Chapter 4 Odd-Numbered Exercises

Section 4.1 *pages 227–230*

1. $\frac{2}{2}; \frac{0}{2}$ **3.** $\frac{2}{4}; \frac{2}{4}$ **5a.** $\frac{3}{4}$ **5b.** $\frac{1}{4}$ **7a.** $\frac{2}{7}$ **7b.** $\frac{5}{7}$ **9.** $\frac{9}{19}$ **11.** $\frac{32}{43}$ **13.** $\frac{11}{17}$ **15a.** $\frac{5}{40}$

15b. $\frac{35}{40}$ **17.** $\frac{30}{365}$ **19.** $2\frac{1}{2}; \frac{5}{2}$ **21.** $4\frac{6}{7}$ **23.** $15\frac{3}{8}$ **25.** $\frac{93}{4}$ **27.** $\frac{93}{5}$ **29.** $\frac{24}{5}$ **31.** $\frac{107}{10}$

33. $\frac{29}{3}$ **35.** $\frac{116}{5}$ **37.** Answers will vary **39.** Answers will vary **41.** $7\frac{2}{13}$ **43.** $7\frac{10}{17}$

Section 4.2 *pages 235–238*

1. 0 **3.** 27 **5.** $\frac{60}{90}$ **7.** $\frac{45}{81}$ **9.** 88 **11.** 77 **13.** 45 **15.** 48 **17.** 6 **19.** 12 **21.** $\frac{5}{20}; \frac{12}{20}$

23. $\frac{18}{45}; \frac{20}{145}$ **25.** $\frac{7}{12}; \frac{10}{12}$ **27.** $\frac{5}{8}; \frac{6}{8}$ **29.** $\frac{8}{30}; \frac{21}{30}$ **31.** $\frac{25}{40}; \frac{36}{40}$ **33.** $\frac{12}{300}; \frac{210}{300}; \frac{255}{300}$ **35.** $\frac{220}{300}; \frac{200}{300}; \frac{27}{300}$

37. $\frac{5}{120}; \frac{8}{120}; \frac{90}{120}$ **39.** $\frac{70}{900}; \frac{45}{900}; \frac{54}{900}$ **41.** 6 **43.** 12 **45.** 32 **47.** 48 **49.** 120 **51.** 315

53. 9 **55.** 18 **57.** $\frac{11}{12}; \frac{3}{12}$ **59.** $\frac{184}{9}$ **61.** $\frac{186}{6}$ **63.** $\frac{49}{100}; \frac{37}{100}; \frac{14}{100}$

Section 4.3 *pages 241–242*

1. $\frac{3}{4}$ **3.** $\frac{2}{3}$ **5.** $\frac{21}{41}$ **7.** $\frac{14}{27}$ **9.** $\frac{1}{3}$ **11.** $\frac{1}{7}$ **13.** $\frac{21}{50}$ **15.** $\frac{13}{27}$ **17.** $\frac{113}{174}$ **19.** $\frac{50}{83}$

21. $\frac{1}{5}\ \frac{1}{3}\ \frac{1}{4}\ \frac{3}{5}\ \frac{2}{3}\ \frac{3}{4}$ **23.** $\frac{3}{5}\ \frac{2}{3}\ \frac{1}{5}\ \frac{1}{3}\ \frac{1}{4}\ \frac{3}{4}$ **25.** $\frac{1}{5}\ \frac{2}{3}\ \frac{1}{3}\ \frac{1}{4}\ \frac{3}{5}\ \frac{3}{4}$ **27.** $\frac{1}{5}\ \frac{3}{5}\ \frac{2}{3}\ \frac{3}{4}\ \frac{1}{3}\ \frac{1}{4}$ **29.** $78\frac{1}{3}$ **31.** $\frac{334}{23}$

33. $\frac{152}{7}$ **35.** $\frac{2}{3}$ **37.** $\frac{409}{100{,}000}$

Section 4.4 *pages 245–247*

1 > **3.** < **5.** < **7.** > **9.** < **11.** < **13.** $\frac{3}{4}$ **15.** $12\frac{11}{12}$ **17.** $\frac{7}{15}$ **19.** $\frac{3}{14}$ **21.** $\frac{11}{36}$

23. $\frac{35}{64}$ **25.** $42\frac{5}{44} < 42\frac{3}{11}$ **27.** $23\frac{17}{26} < 23\frac{9}{13}$ **29.** $33\frac{5}{8} > 33\frac{1}{3} > 33\frac{7}{24}$ **31.** $11\frac{2}{5} > 11\frac{1}{4} > 11\frac{3}{20}$

33. $\frac{2}{9} < \frac{1}{2} < \frac{3}{5}$ **35.** $\frac{5}{14} < \frac{8}{20} < \frac{7}{10}$ **37.** Moon, Mercury, Venus, Earth **39.** Yes. The door will fit. **41.** Mars

43. Yes **45.** $\frac{5}{8}$ **47.** greater than $\frac{1}{2}$ **49.** less than $\frac{1}{2}$ **51.** 15 **53.** 32 **55.** 43 **57.** $25\frac{4}{5}$

59. $\frac{52}{56}; \frac{35}{56}; \frac{13}{14} > \frac{5}{8}$ **61.** $\frac{31}{12}$ **63.** $\frac{4}{5}$ **65.** $7\frac{1}{3}$

Section 4.5 *pages 253–254*

1. $1\frac{5}{16}$ **3.** $1\frac{7}{9}$ **5.** $1\frac{2}{5}$ **7.** $1\frac{1}{3}$ **9.** $1\frac{6}{35}$ **11.** $1\frac{13}{24}$ **13.** $1\frac{5}{18}$ **15.** $1\frac{1}{20}$ **17.** $\frac{61}{80}$ **19.** $1\frac{1}{12}$

21. $1\frac{13}{16}$ **23.** $2\frac{3}{20}$ **25.** $2\frac{1}{4}$ **27.** $1\frac{19}{24}$ **29.** $1\frac{61}{99}$ **31.** $\frac{149}{170}$ **33.** $\frac{107}{210}$ **35.** $1\frac{109}{132}$ **37.** $1\frac{1}{6}$

39. $\frac{5}{7}$ **41.** $\frac{1}{4}$ **43.** $\frac{2}{9} < \frac{1}{4} < \frac{2}{3}$

Section 4.6 *pages 257–260*

1. $3\frac{7}{8}$ **3.** $4\frac{5}{14}$ **5.** $9\frac{1}{2}$ **7.** $8\frac{19}{22}$ **9.** $11\frac{3}{4}$ **11.** $17\frac{11}{20}$ **13.** $16\frac{16}{35}$ **15.** $22\frac{81}{88}$ **17.** $28\frac{49}{90}$

19. $23\frac{55}{84}$ **21.** $31\frac{23}{24}$ **23.** $21\frac{9}{20}$ **25.** $30\frac{5}{12}$ **27.** $24\frac{1}{4}$ **29.** $17\frac{1}{20}$ **31.** $22\frac{1}{24}$ **33.** $12\frac{23}{30}$

35. $8\frac{13}{120}$ **37.** $26\frac{17}{24}$ **39.** $21\frac{3}{10}$ **41.** $13\frac{61}{120}$ **43.** $34\frac{23}{24}$ **45.** $46\frac{25}{36}$ **47.** $33\frac{3}{4}$ **49.** $12\frac{1}{8}$

51. $2\frac{1}{4}$ **53.** $\frac{3}{7}$ **55.** $\frac{1}{3}$ **57.** $17\frac{11}{15}$ **59.** $\frac{4}{5} > \frac{3}{4} > \frac{2}{7}$

Section 4.7 *pages 263–264*

1. $\frac{2}{3}$ 3. $\frac{1}{3}$ 5. $\frac{2}{5}$ 7. $\frac{7}{10}$ 9. $\frac{38}{63}$ 11. $\frac{7}{36}$ 13. $\frac{3}{20}$ 15. $\frac{1}{25}$ 17. $\frac{2}{21}$ 19. $\frac{11}{36}$

21. $\frac{1}{10}$ 23. $\frac{13}{36}$ 25. $\frac{11}{90}$ 27. $\frac{5}{99}$ 29. $\frac{1}{123}$ 31. $\frac{72}{221}$ 33. $\frac{119}{250}$ 35. $\frac{41}{84}$ 37. $55\frac{17}{24}$

39. $\frac{28}{63}$; $\frac{27}{63}$ 41. $\frac{37}{40}$

Section 4.8 *pages 269–270*

1. $1\frac{2}{3}$ 3. $2\frac{4}{7}$ 5. $43\frac{5}{8}$ 7. $35\frac{4}{15}$ 9. $8\frac{5}{8}$ 11. $9\frac{11}{27}$ 13. $14\frac{1}{14}$ 15. $4\frac{29}{48}$ 17. $18\frac{1}{12}$

19. $22\frac{17}{24}$ 21. $26\frac{4}{5}$ 23. $7\frac{9}{10}$ 25. $32\frac{42}{55}$ 27. $23\frac{11}{30}$ 29. $33\frac{17}{26}$ 31. $20\frac{7}{9}$ 33. $15\frac{1}{9}$ 35. $31\frac{7}{11}$

37. $18\frac{4}{27}$ 39. $11\frac{1}{15}$ 41. $16\frac{13}{16}$ 43. $44\frac{2}{9}$ 45. $23\frac{23}{24}$ 47. $28\frac{5}{36}$ 49. $22\frac{1}{48}$ 51. $9\frac{14}{75}$ 53. $1\frac{1}{5}$

55. $7\frac{4}{9}$ 57. $52\frac{1}{60}$ 59. $36\frac{29}{84}$

Section 4.9 *pages 273–276*

1. 3 feet $10\frac{1}{2}$ inches 3. 59 days 5. $3\frac{1}{4}$ inches 7. $152\frac{3}{10}$ yards 9. $151\frac{7}{8}$ pounds 11. $3\frac{5}{8}$ ounces

13. $9\frac{3}{4}$ inches 15. $\frac{2}{3}$ hour 17. $8\frac{1}{6}$ ounces 19. $59\frac{5}{12}$ feet 21. $41\frac{7}{12}$ miles 23. $20\frac{5}{12}$ years old; in $3\frac{1}{4}$ years, Shepherd will be $19\frac{3}{4}$ and Stephanie will be $23\frac{2}{3}$. 25. $\frac{7}{30}$ 27. $658\frac{3}{10}$ miles 29. $66\frac{5}{12}$ feet 31. $67\frac{49}{80}$

33. $17\frac{3}{10}$ miles 35. $23\frac{7}{8}$ 37. $11\frac{73}{84}$

An Application to Statistics *pages 277–278*

1. $\frac{3}{4}$ 3. $\frac{5}{6}$ 5. $\frac{999{,}999}{1{,}000{,}000}$ 7. $\frac{11}{12}$ 9. 20 11. $\frac{1}{2}$

Chapter 4 Review *pages 279–284*

Error Analysis *page 279–280*

1. ***Error:*** misunderstanding concept of reducing

Correct Solution:

$$\frac{18 \div 18}{54 \div 18} = \frac{1}{3}$$

3. ***Error:*** mistakenly adding numerators and denominators

Correct Solution:

$$\frac{1}{3} = \frac{5}{15}$$
$$+\frac{2}{5} = \frac{6}{15}$$
$$\frac{11}{15}$$

5. ***Error:*** error in re-grouping

Correct Solution:

$$7 = 6\frac{4}{4}$$
$$-\frac{3}{4} = \frac{3}{4}$$
$$6\frac{1}{4}$$

Interpreting Mathematics *pages 280–281*

Responses to Interpreting Mathematics items will vary.

Review Problems *pages 281–284*

1. $\frac{1}{6}$ **3.** $18\frac{2}{3}$ **5.** $\frac{521}{7}$ **7.** $\frac{4}{60}, \frac{45}{60}$ **9.** $\frac{24}{54}$ and $\frac{60}{135}$ **11.** $\frac{37}{63}$ **13.** $\frac{3}{10}$ **15.** $17\frac{1}{3} < 17\frac{3}{7} < 17\frac{4}{9}$

17. $1\frac{2}{45} < 1\frac{3}{48} < 1\frac{20}{120}$ **19.** 32 **21.** $1\frac{4}{9}$ **23.** $1\frac{5}{12}$ **25.** $20\frac{5}{9}$ **27.** $24\frac{13}{36}$ **29.** $27\frac{37}{99}$ **31.** $\frac{11}{112}$

33. $\frac{1}{20}$ **35.** $6\frac{1}{6}$ **37.** $2\frac{8}{9}$ **39.** $2\frac{499}{500}$ **41.** $30\frac{1}{12}$ inches **43.** $78\frac{13}{24}$ degrees **45.** $\frac{1}{12}$

47. $\frac{6810}{690}$ **49.** $\frac{68}{3230}, \frac{57}{3230}$ **51.** $\frac{2}{3}; \frac{17}{24}; \frac{17}{24}$ is larger **53.** $197\frac{5}{6}$

Chapter 4 Test *pages 285–286*

1a. $5\frac{2}{7}$ **1b.** $\frac{61}{9}$ **1c.** $\frac{44}{110}, \frac{90}{110}, \frac{85}{110}$ **2a.** $\frac{2}{5}$ **2b.** $45\frac{7}{18}$ **2c.** 20 **3a.** $1\frac{1}{49}$ **3b.** $\frac{23}{30}$ **3c.** $\frac{33}{34}$

4a. $22\frac{5}{9}$ **4b.** $8\frac{29}{70}$ **4c.** $28\frac{16}{45}$ **5a.** $\frac{28}{75}$ **5b.** $\frac{1}{14}$ **5c.** $\frac{9}{1000}$ **6a.** $1\frac{39}{40}$ **6b.** $12\frac{37}{42}$

6c. $1\frac{7}{45}$ **7a.** 6 inches **7b.** $7\frac{7}{8}$ quarts **7c.** $9\frac{1}{14}$

Cumulative Review *pages 287–288*

1. 1362 **3.** 334,480 **5.** 2 feet 7 inches **7.** $23\frac{1}{3}$ **9.** 46 inches **11.** $97\frac{1}{3}$ **13.** 1,006,009 **15.** $11\frac{15}{28}$

17. 99,840 **19.** 16 pounds 11 ounces

ANSWERS to Chapter 5 Odd-Numbered Exercises

Section 5.1 *pages 293–294*

1. 9 **3.** 4 **5.** 6 **7.** 6 **9.** $\frac{8}{81}$ **11.** $\frac{1}{9}$ **13.** $\frac{5}{7}$ **15.** $\frac{5}{16}$ **17.** 1 **19.** 1 **21.** $\frac{3}{7}$

23. $\frac{1}{7}$ **25.** $\frac{3}{16}$ **27.** $\frac{2}{5}$ **29.** $\frac{2}{5}$ **31.** $\frac{160}{441}$ **33.** $\frac{6}{385}$ **35.** $\frac{1}{140}$

Section 5.2 *pages 299–300*

1. $\frac{53}{8}$ **3.** $\frac{64}{9}$ **5.** $78\frac{2}{3}$ **7.** $5\frac{1}{4}$ **9.** $14\frac{2}{7}$ **11.** $8\frac{3}{10}$ **13.** $\frac{2}{5}$ **15.** $\frac{19}{90}$ **17.** $\frac{2}{3}$ **19.** $\frac{1}{5}$

21. $\frac{1}{4}$ **23.** $17\frac{3}{35}$ **25.** $2\frac{1}{7}$ **27.** $\frac{1}{5}$ **29.** $\frac{3}{19}$ **31.** $\frac{2}{5}$ **33.** 50 **35.** $53\frac{9}{10}$ **37.** $5\frac{5}{7}$ **39.** 33

41. 63 **43.** $12\frac{2}{9}$ **45.** $9\frac{1}{11}$ **47.** $177\frac{1}{3}$ **49.** $3\frac{1}{2}$ **51.** $33\frac{1}{4}$ **53.** $9\frac{1}{3}$ **55.** $5\frac{4}{7}$ **57.** $\frac{a}{b}$

59. $\frac{2}{15}$ **61.** $2\frac{19}{40}$ **63.** 1

Section 5.3 *pages 303–304*

1. $\frac{27}{17}=1\frac{10}{17}$ **3.** $\frac{49}{37}=1\frac{12}{37}$ **5.** 16 **7.** 49 **9.** $\frac{1}{3}$ **11.** $\frac{1}{7}$ **13.** 5 **15.** 7 **17.** 2 **19.** $\frac{7}{10}$

21. $\frac{5}{7}$ **23.** $3\frac{6}{7}$ **25.** $1\frac{2}{21}$ **27.** $\frac{9}{10}$ **29.** $\frac{7}{8}$ **31.** $\frac{14}{17}$ **33.** $9\frac{33}{53}$ **35.** $\frac{9}{44}$ **37.** $1\frac{1}{11}$ **39.** $\frac{71}{81}$

41. $\frac{19}{55}$ **43.** $\frac{3}{59}$ **45.** $\frac{20}{571}$ **47.** $\frac{25}{1596}$ **49.** $\frac{5}{162}$ **51.** $\frac{6}{35}$ **53.** $2\frac{1}{6}$ **55.** $1\frac{2}{7}$ **57.** $27\frac{1}{5}$

Section 5.4 *pages 307–308*

1. 9 **3.** 41 **5.** $1\frac{3}{16}$ **7.** $7\frac{1}{3}$ **9.** $4\frac{6}{7}$ **11.** $\frac{16}{31}$ **13.** $\frac{28}{55}$ **15.** $2\frac{10}{19}$ **17.** $\frac{14}{57}$ **19.** $\frac{1}{8}$

21. $\frac{9}{59}$ **23.** $\frac{1}{2}$ **25.** $1\frac{4}{5}$ **27.** $\frac{35}{113}$ **29.** $\frac{3}{28}$ **31.** $14\frac{1}{4}$ **33.** $\frac{5}{6}$ **35.** $3\frac{1}{2}$ **37.** $\frac{3}{8}$

39. $\frac{2}{3}$ **41.** $1\frac{1}{10}$ **43.** 3 **45.** $2\frac{79}{122}$ **47.** $\frac{1}{95}$ **49.** $36\frac{3}{4}$ **51.** $\frac{11}{48}$ **53.** $\frac{8}{35}$ **55.** $\frac{76}{119}$

57. $2\frac{6}{11}$ **59.** $\frac{48}{55}$ **61.** $2\frac{1}{32}$ **63.** $\frac{1}{21}$

Section 5.5 *pages 313–318*

1. 90 million **3.** 6; none **5.** 6 **7.** 231 square feet **9.** 13 feet $2\frac{1}{16}$ inches **11.** 3051 ounces

13. $33\frac{53}{59}$ Saturn years **15.** 5600 people **17.** $258\frac{33}{200}$ ounces **19.** 25 steps **21.** $136{,}723\frac{1}{2}$ miles **23.** $19

25. 2100 sheets **27.** 85 letters **29.** $\$13{,}506\frac{2}{3}$ **31.** 12,000,000 square miles **33.** $\frac{2}{33}$ **35.** 20 **37.** 3 feet

39. $\frac{32}{45}$ **41.** $16\frac{1}{3}$ **43.** $\frac{16}{35}$

Section 5.6 *pages 323–325*

1. $\frac{1}{4}$ **3.** $\frac{1}{27}$ **5.** $54\frac{19}{25}$ **7.** $29\frac{23}{49}$ **9.** $\frac{2197}{5184}$ **11.** $\frac{64}{5929}$ **13.** $\frac{5184}{60{,}025}$ **15.** $\frac{1331}{4608}$ **17.** $12\frac{1}{9}$

19. 12 **21.** $8\frac{3}{4}$ **23.** $35\frac{5}{7}$ **25.** 2 **27.** $49\frac{2}{3}$ **29.** 23 **31.** 183 **33.** $34\frac{1}{9}$ **35.** $53\frac{1}{16}$

37. $50\frac{1}{2}$ **39.** $59\frac{3}{4}$ **41.** $5\frac{1}{2}$ **43.** $1\frac{4}{33}$ **45.** $38\frac{1}{3}$ **47.** $417\frac{9}{10}$ **49.** $3\frac{1}{2}$ **51.** 122,500 **53.** 1174

55. 144 **57.** $154\frac{1}{6}$ feet **59.** $\frac{9}{16}$ **61.** $15\frac{1}{8}$ **63.** 72 slices **65.** 2 cups **67.** $20\frac{1}{3}$

Section 5.7A *pages 329–332*

1. $145\frac{5}{6}$ square inches **3.** $2\frac{26}{35}$ inches **5.** The trapezoid ($36\frac{1}{4}$ square inches) has a greater area than the triangle ($32\frac{3}{8}$ square inches). **7.** They have equal bases and equal heights, so they have the same area. **9.** $5^2+12^2=13^2$; 30 square yards **11.** $29\frac{19}{24}$ square feet **13.** $9\frac{1}{32}$ square inches **15a.** 256 cm by 192 cm **15b.** 300 squares

15c. bases, 5 cm and 8 cm; height, $1\frac{1}{2}$ cm; area, $9\frac{3}{4}$ square centimeters

Section 5.7B *pages 337–340*

1. volume, 49 cubic feet; surface area, 91 square feet **3.** 5382 BTU **5.** volume, $3\frac{3}{8}$ cubic inches; surface area, $13\frac{1}{2}$ square inches **7.** $2\frac{2}{3}$ feet wide; surface area, $72\frac{2}{3}$ square feet **9.** the cube **11.** 51,840 cubic inches **13.** 75 cubic feet **15.** $\frac{256}{625}$ **17.** $33\frac{1}{8}$ inches **19.** $12\frac{7}{50}$ **21.** 612 characters **23.** $382\frac{1}{2}$ cubic inches **25.** $2\frac{6}{17}$ **27.** $7\frac{7}{16}$ square feet

Chapter 5 Review *pages 341–346*

Error Analysis *pages 341–342*

1. *Error:* multiplied whole numbers and fractions separately

Correct Solution:

$$2\frac{2}{3} \times 3\frac{1}{7} = \frac{8}{3} \times \frac{22}{7} = \frac{176}{21} = 8\frac{8}{21}$$

3. *Error:* cross-multiplied

Correct Solution:

$$\frac{2}{3} \times \frac{3}{4} = \frac{6}{12} = \frac{1}{2}$$

5. *Error:* did not invert divisor

Correct Solution:

$$3\frac{1}{4} \div 2\frac{2}{3} = \frac{13}{4} \div \frac{8}{3} = \frac{13}{4} \times \frac{3}{8}$$

$$= \frac{39}{32} = 1\frac{7}{32}$$

Interpreting Mathematics *page 342*

Responses to Interpreting Mathematics items will vary.

Review Problems *pages 343–346*

1. $\frac{3}{20}$ **3.** $\frac{1}{18}$ **5.** $\frac{160}{441}$ **7.** $7\frac{7}{8}$ **9.** $2\frac{1}{4}$ **11.** 152 **13.** $\frac{2}{3}$ **15.** $1\frac{1}{3}$ **17.** 42 **19.** $22\frac{2}{3}$ **21.** $2\frac{5}{8}$ **23.** $2\frac{1}{11}$ **25.** 54 feet **27.** 65 flowers **29.** 67 stringed instruments **31.** $\frac{64}{729}$ **33.** $\frac{125}{216}$ **35.** $105\frac{1}{12}$ **37.** 66 square units **39.** volume, 93 cubic inches; surface area, $164\frac{5}{12}$ square inches **41.** $26\frac{2}{3}$ square yards **43.** $5\frac{1}{10}$ **45.** 30¢ **47.** $24\frac{1}{2}$ **49.** 3 ounces **51.** 6 **53.** $5\frac{1}{5}$

Chapter 5 Test *pages 347–348*

1a. $\frac{4}{7}$ **1b.** $\frac{1}{30}$ **1c.** $19\frac{1}{5}$ **2a.** $2\frac{4}{9}$ **2b.** $1\frac{5}{16}$ **2c.** 6 **3a.** $18\frac{3}{4}$ inches **3b.** $4\frac{11}{24}$ ounces **3c.** 13 countries **4a.** $\frac{27}{64}$ **4b.** $3\frac{13}{27}$ **4c.** $\frac{7}{25}$ **5a.** 90 square inches **5b.** 192 cubic feet **5c.** 600 square inches

Cumulative Review *pages 349–350*

1. 8 hr. 7 min **3.** 2226 **5.** 1 gallon 1 quart **7.** $220\frac{5}{6}$ square feet **9.** $2\frac{7}{16}$ glasses **11.** 8 **13.** $8\frac{1}{2}$ **15.** 45 cm **17.** 864 **19.** 78

ANSWERS to Chapter 6 Odd-Numbered Exercises

Section 6.1 *pages 357–360*

1a. 3 is in the tens place and means thirty (30); **1b.** 3 is in the ones place and means three (3); **1c.** 2 is in the hundredths place and means two hundredths (0.02); **1d.** 7 is in the hundred-thousandths place and means seven hundred-thousandths (0.00007) **3a.** 9 is in the ones place and means nine (9); **3b.** 8 is in the tenths place and means eight tenths (0.8); **3c.** 3 is in the thousandths place and means three thousandths (0.003); **3d.** 7 is in the ten-thousandths place and means seven ten-thousandths (0.0007) **5a.** 9 is in the hundreds place and means nine hundred (900); **5b.** 3 is in the tenths place and means three tenths (0.3); **5c.** 2 is in the thousandths place and means two thousandths (0.002); **5d.** 6 is in the hundred-thousandths place and means six hundred-thousandths (0.00006) **7a.** 9 is in the thousands place and means nine thousand (9,000); **7b.** 1 is in the ones place and means one (1); **7c.** 4 is in the thousandths place and means four thousandths (0.004). **9.** $100 + 7 + \frac{5}{10} + \frac{8}{100} + \frac{9}{1000} + \frac{4}{10{,}000}$ **11.** $200 + 40 + 4 + \frac{9}{10} + \frac{9}{100} + \frac{8}{1000}$ **13.** $60{,}000 + 9000 + 100 + 4 + \frac{5}{10} + \frac{3}{100} + \frac{6}{1000}$ **15.** $60 + 1 + \frac{7}{10} + \frac{3}{100} + \frac{5}{10{,}000} + \frac{4}{100{,}000} + \frac{6}{1{,}000{,}000}$ **17.** two millionths **19.** one and two hundred three thousandths **21.** 0.83 **23.** 0.0084 **25.** forty-four millionths **27.** one hundred eleven hundred-millionths **29.** 0.0013 **31.** 3.153 **33.** ten thousand, six hundred eighty-seven hundred-millionths **35.** ten and thirty-eight thousand, seven hundred seventy-seven millionths **37.** 0.015 **39.** 0.033 **41.** $18\frac{63}{1{,}000}$ **43.** $16\frac{60{,}407}{100{,}000}$ **45.** $\frac{9}{10{,}000}$ **47.** $\frac{7{,}032{,}102}{100{,}000{,}000}$ **49.** $28\frac{875}{1000}$; $34\frac{677}{1000}$ British measure is greater. **51.** 231; $277\frac{42}{100}$; British measure is greater. **53.** $3150\frac{4}{10}$; $2219\frac{4}{10}$; U.S. measure is greater.

Section 6.2 *pages 365–370*

1. 17.83 < 27.41 < 36.52 **3.** 19.74 < 35.18 < 44.13 **5.** > **7.** < **9.** 0.08643 < 0.8643 **11.** 0.042017 < 0.04217 **13.** 0.0007 > 0.0004 **15.** 0.009 > 0.001 **17.** < **19.** > **21.** The male in Canada **23.** rubber, petroleum, alcohol, beechwood **25.** Mercury **27.** anomalistic year, sidereal year, trophical year **29.** gold **31.** 1985, 1987, 1983, 1986 **33.** roaster **35.** fryer **37.** $16.0 thousand **39.** $14,500 **41.** $13.8 thousand **43.** $15.9 thousand **45.** $20,000 **47.** 6.85; 6.94999; 6.9049 **49.** 1.99; 1.957,867 **51.** 2.1949; 2.18997; 2.1854 **53.** 68.0 **55.** 1.88 > 1.08 > 1.008 **57.** 42.901 **59.** 96.563 > 96.536 > 96.057 > 96.0563 **61.** $9\frac{67}{1000}$

Section 6.3 *pages 373–376*

1. 40.7799 **3.** 107.76 **5.** 47.35 **7.** 43.83 **9.** 28.006 **11.** 84.009 **13.** 25.67 **15.** 102.575 **17.** 6.7509 **19.** 0.7976 **21.** 98.65711 **23.** 513.238 **25.** 165.03 **27.** 5238.22211 **29.** 18,865.48 **31.** 7250.1968 **33.** 4241.99 **35.** 36,810.568 **37.** 105,133.393 **39.** 100,821.3448 **41.** 178,756.9548 **43.** 767,787.83683 **45.** 28.80; 3.67 **47.** 14.08; 4.76; 18.84 **49.** 28.07; 108.540; 0.008 **51.** 529.04; 318.630; 0.719 **53.** forty-one and three hundred ninety-six thousandths **55.** 18.024, 18.202, 18.204 **57.** 0.063 **59.** $\frac{12418}{1000}$, or $12\frac{418}{1000}$ **61.** 0.8684

Section 6.4 *pages 381–382*

1. 10.32 **3.** 5.12 **5.** 0.064 **7.** 627.28 **9.** 508.48 **11.** 8241.15 **13.** 386.46 **15.** 208.379 **17.** 56.2829 **19.** 9.876 **21.** 310.029 **23.** 52.439577 **25.** 920.29 **27.** 850.36 **29.** 92,455.718 **31.** 631.565 **33.** 5424.177 **35.** 10.95633 **37.** 1645.43757 **39.** 3074.9846 **41.** thirty-seven and twenty-six thousandths **43.** 14.443 **45.** 38.485 **47.** 990.612

Section 6.5 *pages 387–390*

1. 12.6 **3.** 25.5 **5.** 0.165 **7.** 0.840 **9.** 0.584 **11.** 6.00 **13.** 20.40 **15.** 198.94 **17.** 0.19095 **19.** 34.276 **21.** 22.248 **23.** 580.405 **25.** 48.076 **27.** 36.450 **29.** 5.355 **31.** 692.529 **33.** 4,960,645.382 **35.** 4,535,002.210 **37.** 389.6410 **39.** 18.99044 **41.** 4549.3331 **43.** 62,929.2015 **45.** 2.047140 **47.** 6.18658315 **49.** 1.24312 **51.** 2612.842323 **53.** 6426.3468 **55.** 516.0918 **57.** 4933.00808 **59.** 263.27275 **61.** 11.11; 112.11; 1122.11; 11222.11; 112222.11; 1122222.11. Explanations will vary. **63.** 682.65; 6832.65; 68332.65; 683332.65; 6833332.65; 68333332.65; 683333332.65. Explanations will vary. **65.** 445.824

67. $80 + 4 + \frac{5}{1000}$ **69.** 12.1 **71.** 94.07 **73.** 78.783 **75.** 310.74

Section 6.6 *pages 397–398*

1. 23.1 **3.** 21.6 **5.** 0.04 **7.** 0.32 **9.** 521 **11.** 127 **13.** 3 **15.** 14 **17.** 209.5 **19.** 12.25 **21.** 75.6 **23.** 1.4 **25.** 3.5 **27.** 0.81 **29.** 0.92 **31.** 40 **33.** 19.28 **35.** 14,701.67 **37.** 0.91 **39.** 0.0003 **41.** 630 **43.** 33.22 **45.** 174.871 **47.** 0.034892 **49.** $106.0305 < 106.035 < 106.35$

Section 6.7 *pages 401–408*

1. 7784.1 miles **3.** 101.25 times **5.** 286.44 miles **7.** 18,060,000 (or 18.06 million) **9.** 2.7 points **11.** 49.45 years **13.** 20 times **15.** 4000 times **17.** $335.84 **19.** $1215.99 **21.** 0.045mm **23.** 108.007 **25.** $1274.25 **27.** 49.6 grams **29.** 4762.8 grams **31.** 666,000 bills **33.** 4.52 kilograms **35.** 4.92 pounds **37.** 1.82 degrees **39.** 14.2 miles **41.** 1.17 miles per hour **43.** 1.42 seconds **45.** 1098.122 square miles **47.** 512.945 **49.** 0.17380 **51.** 154.063 **53.** 898.656

Section 6.8 *pages 413–414*

1. 1.44 **3.** 1 **5.** 1 **7.** 0.000002744 **9.** 0.0529 **11.** 0.000000000125 **13.** 148.84 **15.** 5.165 **17.** 354.658 **19.** 132.44 **21.** 1 **23.** 17,490 **25.** 20.72 **27.** 32.8329 **29.** 0.5 **31.** 0.1111 **33.** 0.1667 **35.** 0.6 **37.** 3.6 **39.** 16.3333 **41.** 7.8182 **43.** 102.75 **45.** $17.07 **47.** $3.71 **49.** 0.0081 **51.** 1065 **53.** 20.305 **55.** 50.2952

Section 6.9 *pages 419–422*

1. 46.4092 inches **3.** $\frac{11}{24}$ inches **5.** 11.15 inches **7.** 6782.4 miles **9.** 62,137 miles **11.** 69.34 yards

13. 0.553896 square inch **15.** 0.38465 square inch **17.** 0.43921064 square inch **19.** 27.5625 square inches **21.** 7926.98 miles **23.** 683.14625 square feet **25.** 6.3375 inches **27.** 15.1 cm **29.** The square of the radius doubles when the area doubles. **31.** When the radius doubles, the circumference doubles. **33.** 15.3 cm **35.** 63.585 square centimeters **37.** 40.144 **39.** 0.945 inch **41.** 11.3354 square inches **43.** 27 years old **45.** 11.52 cm

An Application to Statistics *pages 423–428*

1. mean, 1485.898 feet; median 1625.4 feet **3.** median, 20.5 electoral votes; mode, 11 **5.** median, 15, 216.5; mean, 16,576 finishers **7.** median: 133 crew members **9.** $\frac{13}{125}$ **11.** $\frac{11}{200}$ **13.** 0.215 **15.** 0.2315

Chapter 6 Review *pages 429–438*

Error Analysis pages 429–430

1. *Error:* treated decimal point as a barrier

   Correct Solution:
   $$\begin{array}{r} 0.7 \\ +0.7 \\ \hline 1.4 \end{array}$$

3. *Error:* incorrect number of decimal places

   Correct Solution:
   $$\begin{array}{r} 4.35 \\ \times 2.3 \\ \hline 1305 \\ 870 \\ \hline 10.005 \end{array}$$

5. *Error:* disregarded the understood zeros after decimal point

   Correct Solution
   $$\begin{array}{r} 100.00 \\ -\ 6.84 \\ \hline 93.16 \end{array}$$

Interpreting Mathematics pages 430–431

Responses to Interpreting Mathematics items will vary.

Review Problems pages 432–438

1. $6 + \frac{3}{1000}$ **3.** 18.0725 **5.** 2 tens; 1 tenth; 3 hundredths; 5 ten-thousandths **7.** 25 **9.** 18.50; 18.500; 18.5000; 18.49997 **11.** 17.60; 17.605; 17.6049; 17.60495; 17.604949 **13.** 37.99 **15.** 4241.99 **17.** 93.568 **19.** 58.05 **21.** 44.58 **23.** 468,7791 **25.** 0.05508 **27.** 3.2 **29.** 12148.4 **31.** 0.93 **33.** 280 **35.** 16327.2 **37.** 5.538 times **39.** 78 liters **41.** 29.5 feet per second **43.** 45.93 **45.** 238.328 **47.** 0.6 **49.** area, 17.7 square feet; circumference, 14.9 feet **51.** 92.63 feet **53.** 15,700 feet (about 3 miles) **55.** 104.7316 **57.** $3000 + 60 + 2 + \frac{4}{100}$ **59.** $36.49 **61.** 16 times **63.** 9.5 feet **65a.** telephone workers **65b.** hotel employees, $232; radio employees, $404; telephone employees, $488 **67a.** 317.3 inches **67b.** 6.5 inches **69.** 20.5 hour **71.** 29.9°F

Chapter 6 Test pages 439–440

1a. $200 + 50 + 4 + \frac{9}{100} + \frac{8}{1000}$ **1b.** 304.3008 **1c.** $3\frac{78}{1000}$ **2a.** 3.014 is the smaller **2b.** $8.05001 > 8.04 > 8.006$ **3a.** 11.391 **3b.** 12.9083 **3c.** 92.42711 **4a.** 0.0014 **4b.** 694.98 **4c.** 22.734 **5a.** 32,000 square miles **5b.** 3,250,000 people **5c.** 35 years old **6a.** 63.696 **6b.** 13.69 **6c.** 44.84 **7a.** 38.465 square inches **7b.** 26.376 cm **7c.** 10.15 cm

Cumulative Review pages 441–442

1. 1400 **3.** 125,870 **5.** $11\frac{13}{15}$ **7.** 15 **9.** 27.25 **11.** 0.18006 **13.** 380 **15.** $\frac{125}{1000} = \frac{1}{8}$ **17.** $13.4 > 13.04 > 13.004$ **19.** $3\frac{3}{8}$ square inches

ANSWERS to Chapter 7 Odd-Numbered Exercises

Section 7.1 pages 449–452

1. $\frac{1}{2}$; 1:2; 1 to 2 **3.** $\frac{12}{5}$; 12:5; 12 to 5 **5.** $\frac{94}{3125}$; 94:3125; 94 to 3125 **7.** $\frac{61}{84}$; 61:84; 61 to 84 **9.** $\frac{17}{4}$; 17:4; 17 to 4 **11.** $\frac{1497}{376}$; 1497:376; 1497 to 376 **13.** 50 miles per hour **15.** 0.02 hours per mile **17.** 5.24 miles per hour **19.** $0.19 per foot **21.** 11.42 feet per dollar **23.** $0.06 per tablet **25.** 0.12 miles per minute **27.** $833\frac{1}{3}$ miles per month **29.** 15¢ per ounce **31.** 23 miles per gallon **33.** 3.6 daisies per balloon **35.** 0.007 feet per pound

Section 7.2 *pages 457–458*

1. yes **3.** no **5.** no **7.** no **9.** no **11.** yes **13.** $t = 30.6$ **15.** $k = 145.6$ **17.** $t = 24\frac{2}{3}$

19. $F = 1\frac{1}{3}$ **21.** $k = \frac{3}{16}$ **23.** $f = 21.07$ **25.** $t = 1\frac{2}{7}$ **27.** $t = 3\frac{3}{4}$ **29.** $r = 0.56$ **31.** $t = 78.125$

33. > **35.** < **37.** < **39.** < **41.** $n = 3$ **43.** $\frac{\text{1 4x4}}{\text{11 vehicles}}$ **45.** yes **47.** $\frac{\text{1 own}}{\text{3 do not own}}$

49. $n = 2.2$

Section 7.3 *pages 465–468*

1. $10\frac{5}{22}$ miles **3.** 23,056 yards **5.** 6853.33 yards **7.** 3.66 miles **9.** 360 inches **11.** $132\frac{1}{2}$ inches

13. $19\frac{2}{3}$ feet **15.** 7.48 inches **17.** 1524 cm **19.** $10\frac{17}{18}$ yards **21.** 61.25 miles per hour **23.** 80 kilometers per hour **25.** $4\frac{15}{28}$ ounces **27.** 0.0875 pound **29.** $351\frac{53}{60}$ minutes **31.** 165.56 pounds **33.** 364 grams

35. yes **37.** $n = 900$ **39.** 12 oz for $1.29

Section 7.4 *pages 471–475*

1. 132 sets **3.** 3703 sets **5.** 1200 pints **7.** 8640 liters **9.** 218.9 pounds **11.** 25,200 people

13. 191.25 minutes **15.** the 3-lb container **17a.** 60 wins **17b.** 60 losses **19a.** 0.0000015 mile

19b. $666,666.67 **21.** $17\frac{1}{7}$ miles per hour **23.** 100 feet **25.** 7084 bream **27.** $0.09 per ounce **29.** $\frac{11}{41}$

31. $58.81 per hour **33.** $\frac{\text{4 pairs of high-heeled shoes}}{\text{7 pairs of shoes}}$ **35.** 90 laps **37.** $43.51

Section 7.5 *pages 481–484*

1. 78.5 square inches **3.** It becomes 4 times as large **5.** 490.625 square inches **7.** 67.1175 cubic inches **9.** 0.04 gallons **11.** 9.42 square units **13a.** 0.00001256 square unit **13b.** 0.00020096 square unit **13c.** (b) is 16 times (a) **15.** 2,354,861,840,000 mi^2 **17.** 50,240,000 mi^2 **19.** 200,960,000 mi^2 **21.** 3,017,540,000 mi^2 **23.** 17,662,500,000 mi^2 **25.** 48.984 pounds **27.** yes **29.** calculator π: 249.55 in.^2; π = 3.14: 249.68 in.^2; Answers may vary according to rounding. **31.** calculator π: 1428.89 cm^2; π 3.14: 1429.62 cm^2; Answers may vary according to rounding. **33.** yes **35.** 19¢ **37.** 20 people per hour **39.** 88 kilometers per hour **41.** $1.23 per gallon **43.** 7.85 mm^3

Chapter 7 Review *pages 485–490*

Error Analysis *page 485*

1. *Error:* computed incorrectly

Correct Solution

$$\frac{700{,}000}{3500} = \frac{n}{5000}$$

$$\frac{7000}{35} = \frac{n}{5000}$$

$35{,}000{,}000 = 35 \times n$

$\$1{,}000{,}000 = n$

3. *Error:* The numbers in the ratio do not correspond.

Correct Solution

$$\frac{9}{36} = \frac{n}{100}$$

$900 = 36 \times n$

$25 = n$

There are 25 girls.

Interpreting Mathematics *pages 486–487*

Responses to Interpreting Mathematics items will vary.

Review Problems *pages 487–490*

1. $\frac{\$296}{\text{month}}$ **3.** 125 to 114; $\frac{125}{114}$; 125:114 **5.** $\frac{80 \text{ pages}}{\text{minute}}$ **7.** $n = 40$ **9.** 1.05 to 3.00 **11.** $n = 56$ **13.** 4.8 meters **15.** 11.32 meters **17.** 221,760 inches **19.** 75 students **21.** 860 mufflers **23.** $375 per person **25.** volume, 1259.19 cubic inches; surface area, 563.82 square inches. **27.** 5024 square miles **29.** 10,173.6 inches **31.** $50\frac{1}{2}$ to 80; $50\frac{1}{2}$: 80; $\frac{50\frac{1}{2}}{80}$ **33.** $\frac{\$4 \text{ rent}}{\$17 \text{ earnings}}$ **35.** $1.3\overline{3}$ seconds/word **37.** 53 words per minute **39.** 7:8 **41.** no **43.** $\frac{48 \text{ beats}}{\text{minute}}$ **45.** at $1.70 per gallon **47.** $39.48

Chapter 7 Test *pages 491–492*

1a. $\frac{5 \text{ coffee}}{3 \text{ tea}}$ **1b.** $\frac{6 \text{ non-chocolate}}{7 \text{ chocolate}}$ **1c.** 5 miles per hour **2a.** 15:25 is larger **2b.** $n = 189$ **2c.** yes

3a. 15,000 grams **3b.** 20 quarts **3c.** 15 meters **4a.** 108 minutes **4b.** $1088.64 **4c.** 36.8 feet

5a. 317.925 cubic centimeters **5b.** 28.26 square centimeters **5c.** 659.4 square inches.

Cumulative Review *pages 493–494*

1. 1031 **3.** 158 **5.** 20,000 + 3,000 + 500 + 9 **7.** 650 r 3 **9.** $1\frac{1}{3}$ **11.** $2.7 > 2.07 > 2.007$ **13.** 5523.019 **15.** 15.172 **17.** 30,500 **19.** $11\frac{7}{40}$ **21.** $1\frac{19}{45}$

ANSWERS to Chapter 8 Odd-Numbered Exercises

Section 8.1 *pages 498–500*

1. 185% **3.** 246% **5.** 34.8% **7.** $1966.\overline{6}$% **9.** 90% **11.** 10% **13.** 515.93% **15.** 627.8% **17.** 200% **19.** 4,025,700% **21.** 1.5 **23.** 1.35 **25.** 0.03 **27.** 0.02 **29.** 0.005 **31.** 0.01096 **33.** 0.635 **35.** 0.0225 **37a.** 0.7 **37b.** 30% **37c.** 100% **37d.** $\frac{3}{7}$ **39.** 80.3% = 0.803; 39.87% = 0.3987; 25.2% = 0.252; 13.44% = 0.1344; 34.8% = 0.348 **41a.** 97% **41b.** 59% **41c.** 0.41

Section 8.2 *pages 503–506*

1. $\frac{1}{20}$ **3.** $\frac{7}{100}$ **5.** $\frac{3}{20}$ **7.** $\frac{9}{20}$ **9.** $\frac{1}{250}$ **11.** $\frac{3}{500}$ **13.** $2\frac{7}{50}$ **15.** $1\frac{9}{10}$ **17.** $\frac{41}{200}$ **19.** $\frac{161}{500}$ **21.** $1\frac{83}{400}$ **23.** $\frac{2319}{10,000}$ **25.** 5% **27.** 180% **29.** 18.75% **31.** 14.29% **33.** 247.06% **35.** 687.5% **37.** 375% **39.** 0.5% **41.** 46.15% **43.** 20%; 97.2% **45.** 88.07% **47.** $\frac{31}{100}$ **49.** $\frac{1}{25}$ **51.** $\frac{9}{50}$ **53.** $\frac{14}{15}$ **55.** 0.085 **57.** $\frac{7}{50}$ **59.** 0.65 and $\frac{2}{3}$ **61.** 150%

Section 8.3 *pages 513–516*

1. 68% **3.** 16% **5.** 34 **7.** 30% **9.** 24.3 **11.** 26 **13.** 50% **15.** 2.5 **17.** 105 **19.** 0.288 **21.** 0.4% **23.** 0.8 **25.** 53 **27.** 156 **29.** 27 **31.** 0.004 **33.** 2.5245 **35.** 5 **37.** 0.03 **39.** 6.006 **41.** car = \$1068.75; groceries = \$3.45; coffee = 10%; clothing = \$7.67; CD's and tapes = 1.875%; airline ticket = \$168.75; property = \$10,000,000 **43.** $1\frac{89}{100}$ **45.** 83.33% **47.** 81.6 **49.** $\frac{9}{250}$ **51.** 31%

Section 8.4 *pages 521–522*

1. 102 **3.** 25% **5.** 150 **7.** 0.64% **9.** 0.05% **11.** 11.5% **13.** 0.056 **15.** 0.876 **17.** 37.2 **19.** 28 **21.** 30 **23.** 500% **25.** 80 **27.** 2000 **29.** \$.30 **31.** 7000 **33.** 0.01 **35.** $\frac{1893}{10{,}000}$ **37.** 1720% **39.** 0.86625

Section 8.5 *pages 525–528*

1. 58,433,000 square miles **3.** \$6.87 **5.** \$35 **7.** \$200 **9.** \$437.24 **11.** \$66 **13.** 1,337,837 Republicans **15.** 14,580,000 people **17.** 5,843,300 square miles **19.** 20% **21.** 21.55% **23.** 7505 runners **25.** 22.8 ounces **27.** 476 **29.** 16.67% **31.** 62.5% **33.** \$29,520 **35.** 1.25% **37.** 60%

Section 8.6 *pages 535–538*

1. 306.53 square feet **3.** 52 poles **5.** 118.327 square feet **7.** 100 square feet **9.** 62 stones **11.** about 407 cubic inches **13.** 22.81 cubic inches **15.** 18.0625 square inches **17.** 1.33875 square feet **19.** 1.413 square units **21.** 9.6 hours **23.** 43.2 **25.** 0.2% **27.** 151.2 square inches **29.** 42.86% **31.** 900 **33.** 88.2 square feet

An Application to Statistics *pages 539–542*

1. \$980,000,000 **3.** \$143.33 **5.** 7.2 hours **7.** 18 bushes **9.** 15 years; 33 tenants **11.** 33% **13.** $\frac{6}{5}$ **15.** $\frac{14}{45}$ **17.** $\frac{1}{3}$

Chapter 8 Review *pages 543–550*

Error Analysis *page 543*

1. *Error:* confused decimal percents with decimals

Correct Solution

0.82% = 0.0082

3. *Error:* converted percent to decimal incorrectly

Correct Solution

0.05% of 0.75 is what?
$0.0005 \times 0.75 = n$
$0.000375 = n$

5. *Error:* converted decimal to percent incorrectly

Correct Solution

$30 = n\%$ of 600
$30 = n \times 600$
$\frac{30}{60} = \frac{n \times 600}{600}$
$0.05 = n$
$5\% = n$

Interpreting Mathematics *pages 544*

Responses to Interpreting Mathematics items will vary.

Review Problems *pages 544–550*

1. 0.034 **3.** 0.00009 **5.** 100.2% **7.** 25% **9.** 37.5% **11.** $56\frac{8}{25}$ **13.** 500 **15.** 20% **17.** 300

19. 180 **21.** 40.95 **23.** 0.00064 **25.** 450 cassettes **27.** $33\frac{1}{3}\%$ **29.** $153 **31.** 1,710 square feet

33. $3276 cm^2$ **35.** 215.05 square feet **37.** 40% **39.** 63 **41.** 900% **43.** 200,000% **45.** 8%
47. $15.20 **49.** $589.19 **51.** $199,750 **53.** $935 **55.** $100,000 **57.** 25% **59.** 40 questions
61. $125

Chapter 8 Test *pages 551–552*

1a. 6200% **1b.** 0.0001 **1c.** 0.08627 **2a.** $\frac{1}{400}$ **2b.** 137.5% **2c.** 28.57% **3a.** 185 **3b.** 30.6

3c. 20% **4a.** 7.2 **4b.** 5 **4c.** 0.25 **5a.** $46.92 **5b.** $108 **5c.** 50 cupcakes **6a.** 53.68 units
6b. 7.065 square units **6c.** 21.195 square feet

Cumulative Review *pages 553–554*

1. 200 males **3.** 97 **5.** 40.8 square inches **7.** 2,000,300,030.00002 **9.** 0.183 **11.** 1,092,000.20003
13. 43% **15.** 38,243.019 **17.** 3 students **19.** 3.02 **21.** 2000 **23.** two billion, seventy thousand, two and three hundred thousand, six hundred one hundred-millionths.

ANSWERS to Chapter 9 Odd-Numbered Exercises

Section 9.1 *pages 561–562*

1. 6 **3.** −8 **5.** $-23\frac{6}{11}$ **7.** 11.5 **9.** $-5\frac{1}{2}$; number line: $-5\frac{1}{2}$ 0 $5\frac{1}{2}$ **11.** $-1\frac{3}{8}$; number line: $-1\frac{3}{8}$ 0 $1\frac{3}{8}$

13. 3.2; number line: −3.2 0 3.2 **15.** −5.7; number line: −5.7 0 5.7 **17.** $-332\frac{2}{3}$; number line: −333 −332 0

19. $-76\frac{1}{5}$; number line: −77 −76 0 **21.** −|98.7|; number line: −99 −98 **23.** −|−254|; number line: −255 −254 −253 0 **25.** < **27.** >

29. < **31.** < **33.** $-26 < -11 < 15$ **35.** $-45 < -13 < 27$ **37.** $-23\frac{1}{3} < 19\frac{2}{5} < |41|$ **39.** $-|57.4| < -42\frac{1}{2} < 45$

41. −1465 **43.** 26 **45.** $78\frac{11}{12}$ **47.** 365 **49.** $9484 > 96$ **51.** $-280 < -49$ **53.** $-844 < 163$

55. $-3\frac{1}{3} > -3\frac{1}{2}$

Section 9.2 *pages 569–572*

1. 1.14 **3.** −5.8 **5.** −81 **7.** −23 **9.** $-10\frac{31}{77}$ **11.** $-24\frac{5}{24}$ **13.** −109 **15.** −169 **17.** $-6\frac{5}{24}$

19. $-12\frac{2}{7}$ **21.** 75 **23.** −103 **25.** −9.76 **27.** −16.26 **29.** $-\frac{2}{3}$ **31.** 10.6 **33.** 221 **35.** 156

37. $-11\frac{1}{6}$ **39.** $-44\frac{20}{77}$ **41.** −65 **43.** −93 **45.** 21 **47.** −16.7 **49.** −97.3 **51.** −125.4 **53.** −11.5

55. −2.4 **57.** $-3\frac{1}{6}$ **59.** $9\frac{1}{21}$ **61.** −18.62 **63.** −18.15 **65.** 5.1 **67.** $27\frac{4}{5}$ **69.** 118 **71.** $7\frac{7}{40}$

73. $29\frac{1}{2}$ **75.** $-|4|$ 0 **77.** $-|-|-85|| < -42 < -|37|$ **79.** $-|26| > -|-63|$ **81.** −107.45

83. $-94\frac{41}{42}$

Section 9.3 *pages 575–578*

1. −438 **3.** 300 **5.** $15\frac{3}{5}$ **7.** −16 **9.** −23 **11.** −925 **13.** 5 **15.** 2.8 **17.** $-9\frac{4}{27}$ **19.** $-\frac{4}{15}$

21. $-1\frac{8}{13}$ **23.** 1840 **25.** −0.8 **27.** 352 **29.** 13 **31.** $4\frac{3}{10}$ **33.** $-1\frac{1}{3}$ **35.** $-\frac{12}{31}$ **37.** 2050

39. −24 **41.** 532 **43.** 284 **45.** −108 **47.** 21 **49.** 455 **51.** $1\frac{10}{11}$ **53.** −1.0212 **55.** 93

57. $-\frac{1}{54}$ **59.** −24.6 **61.** −325.08 **63.** 11.6256 **65.** $-9\frac{2}{5}$ −10 −9 0

67. $|-47| > \left|\frac{-80}{2}\right| > -(-|-36|)$ **69.** −34.12 **71.** $-400\frac{37}{56}$ **73.** 1544.62 **75.** −57.8 **77.** $251\frac{5}{8}$

Section 9.4 *pages 583–586*

1. $2\frac{1}{4}$ miles west **3.** 3.4 degrees above **5.** $29{,}707\frac{1}{2}$ feet **7.** $174\frac{7}{8}$ pounds **9.** 931.2 miles **11.** $7\frac{3}{5}$ miles per hour **13.** range, 181.7 feet; mean, −18.30 feet **15.** 37 steps **17.** $33\frac{3}{4}$ feet **19.** $\$73.66$ **21.** 63.3 degrees

23. mean, 0.87 degrees; median, −5.6 degrees; mode, 24.1 **25.** $16\frac{29}{72}$ **27.** 184 shirts **29.** 3.39 **31.** $\$17.20$

33. 35 **35.** $-5\frac{61}{135}$

Section 9.5 *pages 593–596*

1a. $-\frac{1}{27}$ **1b.** −3 **1c.** $-\frac{1}{16}$ **1d.** −2 **3a.** $-\frac{4}{25}$ **3b.** $-\frac{4}{25}$ **3c.** $\frac{125}{8}$ **3d.** $-\frac{5}{2}$ **5.** 2264 **7.** −1

9. 111 **11.** $47\frac{31}{36}$ **13.** $16\frac{53}{704}$ **15.** $-12\frac{27}{64}$ **17.** $-32\frac{191}{216}$ **19.** $45\frac{7}{32}$ **21.** 148 **23.** 0 **25.** 344

27. $-4\frac{2}{7}$ **29.** −2 **31.** $45\frac{1}{2}$ **33.** $-77\frac{4}{5}$ **35.** $1173\frac{1}{16}$ **37.** $2028\frac{244}{729}$ **39.** $4\frac{1}{2}$ **41.** −6 **43.** $\frac{61}{384}$

45. 2116 **47.** 236,052 **49.** 2379 **51.** 22 **53.** $-|-|4.1|| < -|-4| < -(-|4|)$ **55.** −16 **57.** 5

59. 13.8 **61.** $4\frac{13}{24}$

Section 9.6 *pages 603–604*

1. $y = 6$ **3.** $y = 93.7$ **5.** $y = -6.2$ **7.** $y = -35.62$ **9.** $y = -480$ **11.** $x = -\frac{6}{7}$ **13.** $y = 3$ **15.** $y = 40$

17. $y = 70$ **19.** $y = 32$ **21.** $x = -7840$ **23.** $y = 100$ **25.** $y = 0$ **27.** $y = -0.7$ **29.** $y = -50$ **31.** $y = -\frac{5}{8}$

33. $y = -8$ **35.** $y = -0.9$ **37.** $y = -24.7$ **39.** $y = -482$ **41.** $2\frac{7}{20}$ **43.** 21 **45.** $16\frac{2}{45}$ **47.** $x = -19$

49. $46\frac{8}{9}$ **51.** -18.7

Section 9.7 *pages 607–608*

1. $x = 4$ **3.** $n = -7\frac{10}{13}$ **5.** $x = -0.06$ **7.** $n = -90$ **9.** $x = 5$ **11.** $n = 2\frac{4}{9}$ **13.** $r = 3.5$ **15.** $n = -12\frac{3}{5}$

17. $x = -3.75$ **19.** $n = 56$ **21.** $x = 9$ **23.** $n = 49$ **25.** $x = -280$ **27.** $n = -0.15$ **29.** $x = -16$

31. $n = -27$ **33.** $x = 400\frac{2}{5}$ **35.** $n = 2$ **37.** $x = -4\frac{7}{10}$ **39.** $n = 3$ **41.** $x = 32$ **43.** $n = 102$ **45.** $x = -0.4$

47. $n = 1\frac{59}{243}$ **49.** $x = -15$ **51.** $n = 0.4$ **53.** -416.97 **55.** -6.69 **57.** $-53\frac{1}{28}$ **59.** -49.039

61. $y = 1041$ **63.** 57.96

Section 9.8 *pages 613–616*

1. −89.17°C **3.** 57.78°C **5.** 26K **7.** 308.76K **9.** −30°F **11.** It will feel 24°F colder (−69°F)
13. 338.72K **15.** too hot **17a.** 1490°F; **17b.** 933.36K **19.** 200K, −10°F, 0°F, −5°C, 350°F; median, 255.38K
21. 25.22°C **23.** −62.11°C **25.** −56.67°C **27.** $x = 0$ **29.** $x = 2145$ **31.** 200K **33.** −13.1 **35.** $2\frac{4}{31}$
37. 66.9524 **39.** Any value from 265.94K to 318.16 K

Problem Solving Preparation *pages 618–620*

1. \$667 **3.** \$67.17 **5.** more expensive **7.** more beans **9.** 136 packages **11.** 9,999,999 in 10,000,000
13. 550

Chapter 9 Review *pages 621–628*

Error Analysis *page 621*

1. ***Error:*** incorrect operating sign.

Correct Solution:

$$\begin{aligned} 18 - 26 &= 18 - (+26) \\ &= 18 + (-26) \\ &= -8 \end{aligned}$$

3. ***Error:*** sign of the minuend is incorrect.

Correct Solution:

$$\begin{aligned} -27 - (+13) &= -27 + (-13) \\ &= -40 \end{aligned}$$

5. ***Error:*** incorrect application of rules for multiplication.

Correct Solution:

$$(-18)(-2) = 36$$

Interpreting Mathematics *page 622*

Responses to Interpreting Mathematics items will vary.

Review Problems *pages 623–628*

1. $\left|4\frac{1}{2}\right| > |-4| > -|-4|$ **3.** 8 **5.** 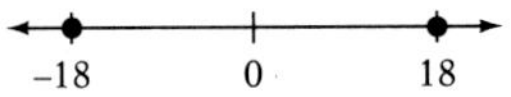

7. 0 **9.** 22 **11.** 31 **13.** –0.068 **15.** 270 **17.** 468 **19.** 12.35% **21.** 100 beats per minute **23.** 125°F **25.** $-13\frac{113}{135}$ **27.** $101\frac{1}{20}$

29. –2 **31.** $x = -2$ **33.** $x = 3$ **35.** $x = -15$ **37.** $x = 8\frac{4}{5}$ **39.** $x = 1$ **41.** $x = \frac{5}{6}$ **43.** 25.6°C

45. 73.4°F **47.** 3°F < 49°F < 285 K < 12°C < 53°C < 590 K **49.** 25 **51.** 132 **53.** 21°F **55.** –77.16

57. $-7\frac{106}{147}$ **59.** $-2033\frac{8}{27}$ **61.** $-197\frac{1}{2}$ **63.** 0 kelvins **65.** 19,755 feet, $4\frac{11}{12}$ inches

Chapter 9 Test *pages 629–630*

1a. 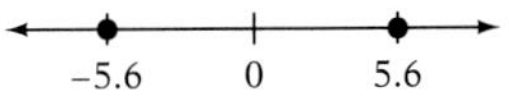

1b. $|-13| > -|-16| > -|-27|$ **1c.** $-|-165| < -153| < -152\frac{1}{2}$ **2a.** 11

2b. –1.75 **2c.** $-1\frac{14}{15}$ **3a.** 298 **3b.** –32 **3c.** $-65\frac{1}{10}$ **4a.** warmer **4b.** $10\frac{1}{24}$ points **4c.** 24 less

5a. $35\frac{1}{7}$ **5b.** $-19\frac{8}{9}$ **5c.** –352 **6a.** $m = 2$ **6b.** $n = -5$ **6c.** $x = 233.5$

Index